Inkjet Technology for Digital Fabrication

Inkjet Technology for Digital Fabrication

Edited by

IAN M. HUTCHINGS and GRAHAM D. MARTIN

Inkjet Research Centre, Institute for Manufacturing, University of Cambridge, United Kingdom

A John Wiley & Sons, Ltd., Publication

Registered office: John Wiley & Sons Ltd, The Atrium, Southern Gate, Chichester, West Sussex, PO19 8SQ, United Kingdom

For details of our global editorial offices, for customer services and for information about how to apply for permission to reuse the copyright material in this book please see our website at www.wiley.com.

Library of Congress Cataloging-in-Publication Data

Inkjet technology for digital fabrication / edited by Ian M Hutchings, Graham D. Martin.
pages cm
Includes index.
ISBN 978-0-470-68198-5 (cloth)
1. Microfluidics. 2. Microfabrication. 3. Ink-jet printing. 4. Three-dimensional printing. 5. Coating processes.
I. Hutchings, Ian M. II. Martin, Graham D. (Graham Dagnall)
TJ853.4.M53I545 2013
620.1'06 – dc23

2012029022

A catalogue record for this book is available from the British Library.

Cloth ISBN: 978-0-470-68198-5

Typeset in 10/12pt Times-Roman by Laserwords Private Limited, Chennai, India

Contents

About the Editors xiii
List of Contributors xv
Preface xvii

1. Introduction to Inkjet Printing for Manufacturing **1**
Ian M. Hutchings and Graham D. Martin

1.1 Introduction 1
1.2 Materials and Their Deposition by Inkjet Printing 3
1.2.1 General Remarks 3
1.2.2 Deposition of Metals 3
1.2.3 Deposition of Ceramics 6
1.2.4 Deposition of Polymers 7
1.3 Applications to Manufacturing 8
1.3.1 Direct Deposition 9
1.3.2 Inkjet Mask Printing 12
1.3.3 Inkjet Etching 13
1.3.4 Inverse Inkjet Printing 14
1.3.5 Printing onto a Powder Bed 15
1.4 Potential and Limitations 15
References 17

2. Fundamentals of Inkjet Technology **21**
Graham D. Martin and Ian M. Hutchings

2.1 Introduction 21
2.2 Surface Tension and Viscosity 23
2.3 Dimensionless Groups in Inkjet Printing 25
2.4 Methods of Drop Generation 27
2.4.1 Continuous Inkjet (CIJ) 27
2.4.2 Drop-on-Demand (DOD) 28
2.4.3 Electrospray 33
2.5 Resolution and Print Quality 34
2.6 Grey-Scale Printing 35
2.7 Reliability 36
2.8 Satellite Drops 38

2.9 Print-Head and Substrate Motion 39
2.10 Inkjet Complexity 42
References 42

3. Dynamics of Piezoelectric Print-Heads **45**
J. Frits Dijksman and Anke Pierik

3.1 Introduction 45
3.2 Basic Designs of Piezo-Driven Print-Heads 47
3.3 Basic Dynamics of a Piezo-Driven Inkjet Print-Head (Single-Degree-of-Freedom Analysis) 49
3.4 Design Considerations for Droplet Emission from Piezo-Driven Print-Heads 60
3.4.1 Droplet Formation 60
3.4.2 Damping 66
3.4.3 Refilling 67
3.4.4 Deceleration Due to Elongational Effects Prior to Pinching Off 70
3.4.5 Summary 71
3.5 Multi-Cavity Helmholtz Resonator Theory 71
3.6 Long Duct Theory 77
3.7 Concluding Remarks 83
References 84

4. Fluids for Inkjet Printing **87**
Stephen G. Yeates, Desheng Xu, Marie-Beatrice Madec, Dolores Caras-Quintero, Khalid A. Alamry, Andromachi Malandraki and Veronica Sanchez-Romaguera

4.1 Introduction 87
4.2 Print-Head Considerations 88
4.2.1 Continuous Inkjet (CIJ) 88
4.2.2 Thermal Inkjet (TIJ) 88
4.2.3 Piezoelectric Drop-on-Demand (Piezo-DOD) 89
4.3 Physical Considerations in DOD Droplet Formation 89
4.4 Ink Design Considerations 95
4.5 Ink Classification 95
4.5.1 Aqueous Ink Technology 96
4.5.2 Non-aqueous Ink Technologies 100
4.6 Applications in Electronic Devices 105
4.6.1 Organic Conducting Polymers 105
4.6.2 Conjugated Organic Semiconductors 106
4.6.3 Inorganic Particles 107
References 108

5. When the Drop Hits the Substrate **113**
Jonathan Stringer and Brian Derby

5.1 Introduction 113
5.2 Stable Droplet Deposition 114
5.2.1 Deposition Maps 114
5.2.2 Impact of Millimetre-Size Droplets 116
5.2.3 Impact of Inkjet-Sized Droplets 119
5.3 Unstable Droplet Deposition 120
5.4 Capillarity-Driven Spreading 122
5.4.1 Droplet–Substrate Equilibrium 122
5.4.2 Capillarity-Driven Contact Line Motion 124
5.4.3 Contact Angle Hysteresis 125
5.5 Coalescence 126
5.5.1 Stages of Coalescence 126
5.5.2 Coalescence and Pattern Formation 128
5.5.3 Stable Bead Formation 128
5.5.4 Unstable Bead Formation 130
5.6 Phase Change 131
5.6.1 Solidification 132
5.6.2 Evaporation 132
5.7 Summary 134
References 135

6. Manufacturing of Micro-Electro-Mechanical Systems (MEMS) **141**
David B. Wallace

6.1 Introduction 141
6.2 Limitations and Opportunities in MEMS Fabrication 142
6.3 Benefits of Inkjet in MEMS Fabrication 143
6.4 Chemical Sensors 144
6.5 Optical MEMS Devices 147
6.6 Bio-MEMS Devices 151
6.7 Assembly and Packaging 152
6.8 Conclusions 156
Acknowledgements 156
References 156

7. Conductive Tracks and Passive Electronics **159**
Jake Reder

7.1 Introduction 159
7.2 Vision 159
7.3 Drivers 160
7.3.1 Efficient Use of Raw Materials 160

7.3.2 Short-Run and Single-Example Production 161
7.3.3 Capital Equipment 162
7.4 Incumbent Technologies 162
7.5 Conductive Tracks and Contacts 162
7.5.1 What Is Conductivity? 162
7.5.2 Conductive Tracks in the Third Dimension 163
7.5.3 Contacts 163
7.6 Raw Materials: Ink 164
7.6.1 Particles 164
7.6.2 Dispersants 168
7.6.3 Carriers (Liquid Media) 170
7.6.4 Other Additives 170
7.7 Raw Materials: Conductive Polymers 172
7.8 Raw Materials: Substrates 172
7.9 Printing Processes 174
7.10 Post Deposition Processing 174
7.10.1 Sintering 174
7.10.2 Protective Layers 175
7.11 Resistors 175
7.12 Capacitors 176
7.13 Other Passive Electronic Devices 176
7.13.1 Fuses, Circuit Breakers, and Switches 176
7.13.2 Inductors and Transformers 177
7.13.3 Batteries 177
7.13.4 Passive Filters 177
7.13.5 Electrostatic Discharge (ESD) 177
7.13.6 Thermal Management 178
7.14 Outlook 178
References 178

8. Printed Circuit Board Fabrication **183**
Neil Chilton

8.1 Introduction 183
8.2 What Is a PCB? 183
8.3 How Is a PCB Manufactured Conventionally? 185
8.4 Imaging 185
8.4.1 Imaging Using Phototools 187
8.4.2 Laser Direct Imaging 188
8.5 PCB Design Formats 188
8.6 Inkjet Applications in PCB Manufacturing 189
8.6.1 Introduction 189
8.6.2 Legend Printing 190
8.6.3 Soldermask 194
8.6.4 Etch Resist 195
8.7 Future Possibilities 202
References 205

9. Active Electronics **207**
Madhusudan Singh, Hanna M. Haverinen, Yuka Yoshioka and Ghassan E. Jabbour

9.1 Introduction 207
9.2 Applications of Inkjet Printing to Active Devices 211
9.2.1 OLEDs 211
9.2.2 Other Displays 213
9.2.3 Energy Storage Using Batteries and Supercapacitors 214
9.2.4 Photovoltaics 215
9.2.5 Sensors 217
9.2.6 Transistors, Logic, and Memory 219
9.2.7 Contacts and Conductors 221
9.2.8 In Situ Synthesis and Patterning 223
9.2.9 Biological Applications 223
9.3 Future Outlook 224
References 225

10. Flat Panel Organic Light-Emitting Diode (OLED) Displays: A Case Study **237**
Julian Carter, Mark Crankshaw and Sungjune Jung

10.1 Introduction 237
10.2 Development of Inkjet Printing for OLED Displays 238
10.3 Inkjet Requirements for OLED Applications 241
10.3.1 Introduction 241
10.3.2 Display Geometry 241
10.3.3 Containment and Solid Content 241
10.4 Ink Formulation and Process Control 243
10.5 Print Defects and Control 246
10.6 Conclusions and Outlook 249
Acknowledgements 250
References 250

11. Radiofrequency Identification (RFID) Manufacturing: A Case Study **255**
Vivek Subramanian

11.1 Introduction 255
11.2 Conventional RFID Technology 256
11.2.1 Introduction 256
11.2.2 RFID Standards and Classifications 256
11.2.3 RFID Using Silicon 258
11.3 Applications of Printing to RFID 260
11.4 Printed Antenna Structures for RFID 260
11.4.1 The Case for Printed Antennae 260
11.4.2 Printed RFID Antenna Technology 261
11.4.3 Summary of Status and Outlook for Printed Antennae 262

11.5 Printed RFID Tags 263
11.5.1 Introduction 263
11.5.2 Topology and Architecture of Printed RFID 264
11.5.3 Devices for Printed RFID 267
11.6 Conclusions 273
References 273

12. Biopolymers and Cells **275**
Paul Calvert and Thomas Boland

12.1 Introduction 275
12.2 Printers for Biopolymers and Cells 277
12.2.1 Printer Types 277
12.2.2 Piezoelectric Print-Heads 277
12.2.3 Thermal Inkjet Print-Heads 279
12.2.4 Comparison of Thermal and Piezoelectric Inkjet for Biopolymer Printing 279
12.2.5 Other Droplet Printers 280
12.2.6 Rapid Prototyping and Inkjet Printing 281
12.3 Ink Formulation 282
12.3.1 Introduction 282
12.3.2 Printed Resolution 283
12.3.3 Major Parameters: Viscosity and Surface Tension 283
12.3.4 Drying 285
12.3.5 Corrosion 285
12.3.6 Nanoparticle Inks 285
12.3.7 Biopolymer Inks 285
12.4 Printing Cells 289
12.4.1 Cell-Directing Patterns 289
12.4.2 Cell-Containing Inks 289
12.4.3 Effects of Piezoelectric and Thermal Print-Heads on Cells 290
12.4.4 Cell Attachment and Growth 291
12.4.5 Biocompatibility in the Body 292
12.5 Reactive Inks 292
12.6 Substrates for Printing 296
12.7 Applications 297
12.7.1 Tissue Engineering 297
12.7.2 Bioreactors 298
12.7.3 Printed Tissues 298
12.8 Conclusions 299
References 299

13. Tissue Engineering: A Case Study **307**
Makoto Nakamura

13.1 Introduction 307
13.1.1 Tissue Engineering and Regenerative Medicine 307
13.1.2 The Third Dimension in Tissue Engineering and Regenerative Medicine 308

13.1.3 The Current Approach for Manufacturing 3D Tissues 309
13.1.4 A New Approach of Direct 3D Fabrication with Live Cell Printing 309
13.2 A Feasibility Study of Live Cell Printing by Inkjet 310
13.3 3D Biofabrication by Gelation of Inkjet Droplets 313
13.4 2D and 3D Biofabrication by a 3D Bioprinter 314
13.4.1 Micro-Gel Beads 314
13.4.2 Micro-Gel Fiber and Cell Printing 315
13.4.3 2D and 3D Fabrication of Gel Sheets and Gel Mesh 316
13.4.4 Fabrication of 3D Gel Tubes 316
13.4.5 Multicolor 3D Biofabrication 316
13.4.6 Viscosity in Inkjet 3D Biofabrication 318
13.5 Use of Inkjet Technology for 3D Tissue Manufacturing 319
13.5.1 Resolution and DOD Color Printing 319
13.5.2 Direct Printing of Live Cells 319
13.5.3 High-Speed Printing 319
13.5.4 3D Fabrication Using Hydrogels 320
13.5.5 Linkage to Digital Data Sources 321
13.5.6 Applicability to Various Materials including Humoral Factors and Nanomaterials 321
13.5.7 Use of Pluripotent Stem Cells in Bioprinting 322
13.6 Summary and Future Prospects 322
Acknowledgements 323
References 323

14. Three-Dimensional Digital Fabrication **325**
Bill O'Neill

14.1 Introduction 325
14.2 Background to Digital Fabrication 326
14.3 Digital Fabrication and Jetted Material Delivery 329
14.4 Liquid-Based Fabrication Techniques 330
14.4.1 PolyJet™: Objet Geometries 330
14.4.2 ProJet™: 3D Systems 333
14.4.3 Solidscape 3D Printers 333
14.5 Powder-Based Fabrication Techniques 335
14.5.1 ZPrinter™: Z Corporation 335
14.5.2 Other Powder-Based 3D Printers 338
14.6 Research Challenges 338
14.7 Future Trends 340
References 341

15. Current Inkjet Technology and Future Directions **343**
Mike Willis

15.1 The Inkjet Print-Head as a Delivery Device 343
15.2 Limitations of Inkjet Technology 344
15.2.1 Jetting Fluid Constraints 344

15.2.2 Control of Drop Volume 345
15.2.3 Variations in Drop Volume 345
15.2.4 Jet Directionality and Drop Placement Errors 345
15.2.5 Aerodynamic Effects 347
15.2.6 Impact and Surface Wetting Effects 348
15.3 Today's Dominant Technologies and Limitations 348
15.3.1 Thermal DOD Inkjet 348
15.3.2 Piezoelectric DOD Inkjet 350
15.4 Other Current Technologies 351
15.4.1 Continuous Inkjet 351
15.4.2 Electrostatic DOD 351
15.4.3 Acoustic Drop Ejection 352
15.5 Emerging Technologies 353
15.5.1 Stream 353
15.5.2 MEMS 354
15.5.3 Flextensional 356
15.5.4 Tonejet 356
15.6 Future Trends for Print-Head Manufacturing 357
15.7 Future Requirements and Directions 358
15.7.1 Customisation of Print-Heads for Digital Fabrication 358
15.7.2 Reduce Sensitivity of Jetting to Ink Characteristics 359
15.7.3 Higher Viscosities 359
15.7.4 Higher Stability and Reliability 360
15.7.5 Drop Volume Requirements 360
15.7.6 Lower Costs 361
15.8 Summary of Status of Inkjet Technology for Digital Fabrication 361
References 362

Index **363**

About the Editors

Ian M. Hutchings has been GKN Professor of Manufacturing Engineering at the University of Cambridge since 2001, working in the Institute for Manufacturing which is part of the Department of Engineering. Previously, he was Reader in Tribology at the Department of Materials Science and Metallurgy. His research interests are interdisciplinary, crossing the boundaries between engineering, materials science and applied physics. He established the Inkjet Research Centre at Cambridge in 2005 with support from the UK Engineering and Physical Sciences Research Council and several academic and industrial collaborators.

Graham D. Martin has been Director of the Inkjet Research Centre at the Institute for Manufacturing, University of Cambridge since 2005. He has a PhD in solid state physics and has worked in inkjet-related research, consultancy and product development for many years. The companies he has worked for include Cambridge Consultants Ltd (Non Impact Printing Systems group leader), Elmjet (Technical Director) and Videojet (Director of Technology).

List of Contributors

Khalid A. Alamry, Organic Materials Innovation Centre, School of Chemistry, University of Manchester, United Kingdom

Thomas Boland, Department of Biomedical Engineering, University of Texas at El Paso, USA

Paul Calvert, College of Engineering, University of Massachusetts Dartmouth, USA

Dolores Caras-Quintero, Organic Materials Innovation Centre, School of Chemistry, University of Manchester, United Kingdom

Julian Carter, Technology Consultant, United Kingdom

Neil Chilton, Printed Electronics Ltd., United Kingdom

Mark Crankshaw, R&D – Engineering, Xaar PLC, United Kingdom

Brian Derby, School of Materials, University of Manchester, United Kingdom

J. Frits Dijksman, Philips Research Europe, High Tech Campus 11, The Netherlands

Hanna M. Haverinen, Solar and Alternative Energy Engineering Research Center, King Abdullah University of Science and Technology (KAUST), Saudi Arabia

Ian M. Hutchings, Inkjet Research Centre, University of Cambridge, United Kingdom

Ghassan E. Jabbour, Department of Electrical Engineering, Department of Materials Science and Engineering, Solar and Alternative Energy Engineering Research Center, King Abdullah University of Science and Technology (KAUST), Saudi Arabia

Sungjune Jung, Department of Physics, University of Cambridge, United Kingdom

Marie-Beatrice Madec, Organic Materials Innovation Centre, School of Chemistry, University of Manchester, United Kingdom

Andromachi Malandraki, Organic Materials Innovation Centre, School of Chemistry, University of Manchester, United Kingdom

Graham D. Martin, Inkjet Research Centre, University of Cambridge, United Kingdom

Makoto Nakamura, Graduate School of Science and Engineering for Research, University of Toyama, Japan

Bill O'Neill, Department of Engineering, University of Cambridge, United Kingdom

Anke Pierik, University of Twente, Faculty of Science and Technology, Physics of Fluids, The Netherlands

Jake Reder, Celdara Medical, LLC, USA

Veronica Sanchez-Romaguera, Organic Materials Innovation Centre, School of Chemistry, University of Manchester, United Kingdom

Madhusudan Singh, Department of Materials Science and Engineering, University of Texas at Dallas, USA

Jonathan Stringer, Department of Mechanical Engineering, University of Sheffield, United Kingdom

Vivek Subramanian, Department of Electrical Engineering and Computer Sciences, University of California, USA

David B. Wallace, MicroFab Technologies, Inc., USA

Mike Willis, Pivotal Resources Ltd, United Kingdom

Desheng Xu, Organic Materials Innovation Centre, School of Chemistry, University of Manchester, United Kingdom

Stephen G. Yeates, Organic Materials Innovation Centre, School of Chemistry, University of Manchester, United Kingdom

Yuka Yoshioka, Heraeus Materials Technology, USA

Preface

From its initial use for product marking and date coding in the 1980s, and its development and widespread adoption for the desktop printing of text and images in the following two decades, inkjet technology is now having an increasing impact on commercial printing for many applications including labels, print-on-demand books and even newspapers. With great intrinsic flexibility and very short set-up times, inkjet printing is also challenging conventional methods for more specialised uses such as ceramic tile decoration and textile printing.

Exactly the same processes by which individual drops of liquid are produced and directed onto a substrate under digital control can be used to deposit materials other than the coloured 'inks' used for text and graphics. Metals, ceramics and polymers, with a wide range of functionality, can all be printed by inkjet methods, and exciting possibilities are also raised by the ability to print biological materials, including living cells. We are at the dawn of a digital age for printing, and it is the aim of this book to show how the changes which are happening in that world will lead to equally revolutionary changes in the ways in which we can manufacture products. Digital fabrication offers the possibilities of tailoring materials at a microscopic level, and positioning them exactly where they are required, with exactly the right properties. It has the potential to generate structures and functions which cannot be attained by other methods, and which are limited only by the creativity and ingenuity of the designer. It forms a new and powerful addition to the portfolio of methods available for manufacturing.

We are very grateful to the authors who have contributed to this volume, which we hope will help to define this rapidly moving field of research and provide a valuable resource for those who want to explore it further. It is impossible to forecast how it will develop, even over the next 10 years. What appears certain to us is that it will not stand still.

Ian M. Hutchings and Graham D. Martin
Cambridge
April 2012

1

Introduction to Inkjet Printing for Manufacturing

Ian M. Hutchings and Graham D. Martin
Inkjet Research Centre, University of Cambridge, United Kingdom

1.1 Introduction

The basic principles of conventional printing have remained the same for hundreds of years: the various different printing processes which we take for granted all involve the repeated reproduction of the same image or text many times. Usually, this is achieved by transferring a pattern of liquid or semi-liquid ink from some master pattern to the paper or other substrate through direct contact. Changes to the printed product can be achieved only by changing the master pattern, which involves making physical changes within the printing machine.

In contrast, the inkjet printer which is now ubiquitous in the modern home and office works on a fundamentally different principle. Each small droplet of ink, typically 10–100 μm in diameter, is created and deposited under digital control, so that each pattern printed in a sequence can just as readily be different from the others as it can be the same. The principles of inkjet printing were first developed commercially during the 1970s and 1980s, with the practical applications of marking products with dates and bar codes, and addressing bulk mail. As indicated in Figure 1.1, the technology used for these purposes, which demand high operating speeds but can tolerate quite low resolution in the printed text, is now fully mature: these printers, which use continuous inkjet (CIJ) technology, are widely used as standard equipment on production lines worldwide. The next development, from the mid-1980s onwards, involved drop-on-demand (DOD) printing which is capable of much higher resolution than the early coders and placed the

Inkjet Technology for Digital Fabrication, First Edition. Edited by Ian M. Hutchings and Graham D. Martin.

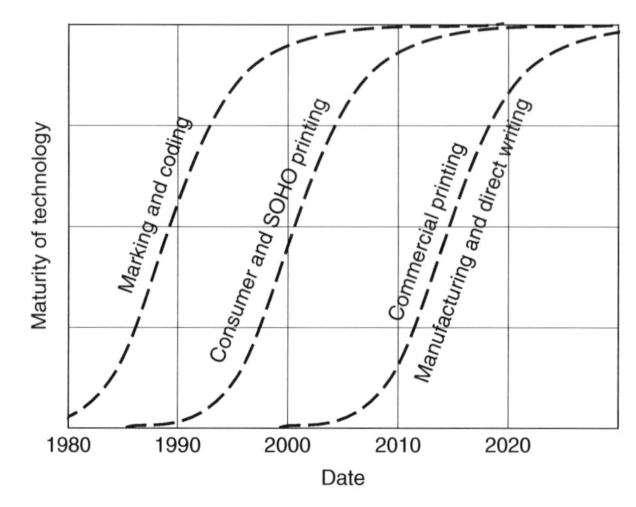

Figure 1.1 *The applications of inkjet technology have developed in three waves: initially for marking and coding, followed by desktop printing of text and graphics in the home and small office environment and, currently, increasing use in commercial printing and manufacturing.*

capability for digital reproduction of text and images, at low cost, into the domestic and small office environment. The principles of both the CIJ and DOD printing technologies are described in Chapter 2.

The subject of this book is the third wave of technology development shown in Figure 1.1; the use of inkjet printing as a manufacturing process. This advance, which is occurring in parallel with the use of inkjet for commercial printing in direct competition with such processes as offset lithography, employs the same basic principles of drop generation as the earlier applications, but with an emphasis on the features of reliability, accuracy, flexibility and robustness which are essential for successful industrial application. Many of the applications discussed in this volume are still under development, and there is undoubted scope for further innovation. Several features of inkjet printing make it particularly attractive for manufacturing.

Firstly, it is a digital process. The location of each droplet of 'ink' (i.e. the material being deposited) can be predetermined on a two-dimensional grid. If necessary the location can be changed in real time, for example to adjust for distortion or misalignment of the substrate, or to ensure that a certain height of final deposit is achieved. Because it is a digital process, each product in a sequence can easily be made different from every other, in small or even in major ways; bespoke products are generated just as readily as multiple replicas of the same design. Since the pattern to be printed is held in the form of digital data, there may be significant cost savings over processes which involve the use of a physical mask or template.

Secondly, it is a non-contact method; the only forces which are applied to the substrate result from the impact of very small liquid drops. Thus fragile substrates can be processed which would not survive more conventional printing methods. The substrate need not even be solid: we shall see examples in Chapter 13 where materials are printed into a liquid bath, and in Chapter 14 where the substrate is a bed of powder. Material can be deposited onto non-planar (rough or textured) substrates, since the process can be operated with a stand-off distance between the print-head and substrate of at least 1 mm. In conventional contact-based printing, the printed material may also be transferred by accidental contact, potentially causing poor quality or contamination; such problems are avoided in a non-contact process.

Thirdly, a wide range of materials can be deposited. By selection of an appropriate print-head, liquids with viscosities from 1 to 50 mPa s or higher can be printed. Several different methods can be used to generate printed structures. Multiple combinations of materials can be used, and inkjet printing can also be combined with other process steps, so that in principle complex heterogeneous and composite structures can be produced, with different materials distributed in all three dimensions.

A further benefit is that inkjet printing is modular and scalable. Multiple print-heads can be assembled to print in tandem, for example by placing them side-by-side to print a wider pattern, or one after the other to print different materials in sequence. These concepts are standard for graphical printing, where four or more colours are commonly used, but can also be readily extended to the manufacturing context.

In this introductory chapter, we shall briefly review the range of materials which can be deposited by inkjet printing and the various methods by which inkjet processes can be used for both additive and subtractive fabrication in manufacturing. The processes of jet and droplet formation, and the various types of print-heads, are introduced in Chapter 2. Later chapters describe the formulation of printable fluids and examine particular applications of inkjet printing in much more detail.

1.2 Materials and Their Deposition by Inkjet Printing

1.2.1 General Remarks

Inkjet technology has been used to deposit a very wide range of materials, including metals, ceramics and polymers, for many different applications. Biological materials, including living cells, have also been successfully printed; they are the subject of Chapters 12 and 13, and we shall not consider them further here. The most important restriction is that the substance being printed must be in liquid form (or contain small solid particles in a liquid medium) with appropriate rheological properties at the point of printing. As discussed in Section 1.3, the material which is printed need not be the same as the final material required: there are several process routes in which a precursor material is deposited, followed by other steps to achieve the final product.

1.2.2 Deposition of Metals

Figure 1.2 illustrates several routes by which inkjet printing can be used to form metallic deposits. These involve direct printing from a melt, printing a suspension of metallic particles which are then sintered to bond them together, printing a metal compound which is then chemically reduced to form the metal and printing a suitable catalyst followed by electroless plating to deposit the metal. Any of these processes can also be combined with one or more secondary electroless plating or electro-plating steps to produce a thicker metallic deposit, which can even be of a different metal.

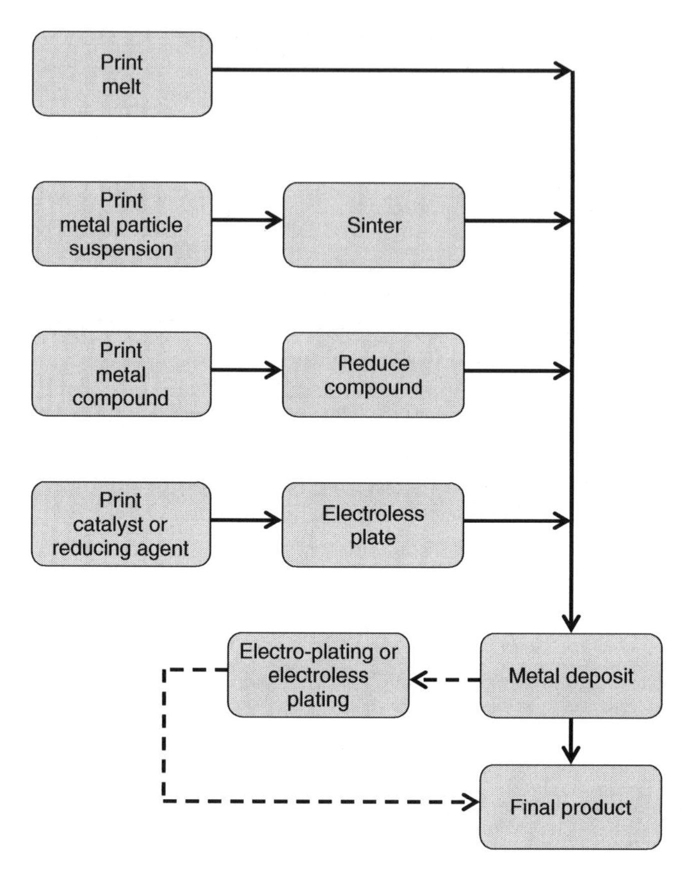

Figure 1.2 *Process routes by which inkjet printing can be used to deposit metals.*

In principle, the simplest method is to print droplets of molten metal directly onto the substrate. An early application of the direct printing of liquid metal was to form solder droplets for chip connection bumps, via filling, connector tracks and rework on electronic printed circuits, and several researchers have reported the printing of lead–tin solder alloys by both CIJ and DOD methods (Hayes, Wallace and Cox, 1999; Liu and Orme, 2001a; Lee *et al*., 2008). Other metals with relatively low melting points such as indium, tin, lead and zinc have also been printed by CIJ and DOD methods (Tseng, Lee and Zhao, 2001; Cheng, Li and Chandra, 2005). Alloys with higher melting points pose significant challenges for print-head design, and although piezoelectric drive (as described in Chapter 2) may still be useful, precautions must be taken to isolate the transducer from the high melt temperature. Other actuation methods have also been used, such as direct pneumatic ejection in DOD printing (Cheng, Li and Chandra, 2005). The deposition of aluminium, both pure and alloyed, has been demonstrated in a droplet-based net-form manufacturing process in which the drops are generated and deflected by piezoelectric CIJ technology, with drops 190 μm in diameter being generated at 17 kHz, corresponding to a mass throughput of 1.5 kg/h (Liu and Orme, 2001b). Deposition in an inert atmosphere is necessary to avoid oxidation of the molten droplets, and this requirement, together with the complications introduced by operating the print-head at high temperature, limits the attractiveness of the direct melt printing process for many metals.

Metallic particles suspended in a suitable fugitive liquid can be inkjet printed and used for both structural and electrical applications. Small particles are generally favoured as the suspensions are more stable, that is, the particles do not sediment, and nozzle clogging is avoided. Industry rules of thumb suggest that particles smaller than 1/10 (and preferably smaller than 1/50) of the nozzle diameter are required to avoid blockage. Another very important advantage of small particle size, as discussed further in Chapter 7, is that the high surface-to-volume ratio leads to a lowered sintering temperature. There is considerable interest in the development of nanoparticle inks with good electrical properties, oxidation resistance and low sintering temperatures for printable electrical conductors (Park *et al*., 2007). For example, inks based on silver nanoparticles, typically 5–50 nm in size, can be sintered to form deposits of high electrical conductivity at temperatures lower than 300 °C, and even as low as 150 °C, which allows them to be used with some polymer substrates (Fuller, Wilhelm and Jacobson, 2002; Sanchez-Romaguera, Madec and Yeates, 2008). Other nano-particulate metals which have been successfully printed include gold and copper (Park *et al*., 2007). Although prices are still high, several silver nano-particle inks are commercially available, and conductivities following sintering can be as high as 50% of that of bulk silver. Lowering the sintering temperature widens the range of possible substrates, and there is also interest in methods of sintering which do not involve bulk heating: examples include the use of a laser that focuses on the ink deposit or scans rapidly over the surface, and microwave heating (Perelaer *et al*., 2008). A novel process involving treatment with aqueous halide solution at room temperature has also been shown to produce conductivities similar to those of thermal sintering (Zapka *et al*., 2008). Conductive particulate inks in which the solvent does not evaporate but cures to form a binder will give lower conductivity, but this is sometimes desirable, an example being the use of carbon nanotubes in a polymer matrix for the *in situ* fabrication of electrical resistors.

The third method of achieving a metallic deposit by inkjet printing is to print a precursor: a solution of a compound of the metal, usually silver, which is then decomposed by heating. This is further discussed in Chapter 4. Inks based on silver nitrate and on organic silver compounds have been successfully printed and processed to yield conductive metallic deposits by CIJ (Mei, Lovell and Mickle, 2005) and DOD processes (Smith *et al*., 2006, 2008). Decomposition of the printed compound to form the silver deposit is usually achieved by heating, although photolytic processing by laser irradiation has also been reported (Stringer, Xu and Derby, 2007). It has been shown that the need for a separate decomposition step can be eliminated by printing an organometallic silver ink directly onto a substrate heated to 130 °C, a temperature compatible with the use of several common polymer substrates (Perelaer *et al*., 2009).

The final approach shown in Figure 1.2 is to print a non-conductive but chemically active deposit (e.g. a catalyst or reducing agent) which is then subjected to a secondary treatment in a low-temperature electroless plating bath, typically to deposit copper or nickel. The printing process produces a template for the subsequent plating and defines the area to be coated; the plating time can be varied to control the thickness of metal deposit. Excellent conductivity can be achieved in this way, and the low process temperature is a significant advantage for some applications. As an example, silver nanoparticle ink has been printed as a 'seed' layer, followed by electroless plating of nickel: final deposit heights up to 76 μm were reported (Lok *et al*., 2007). A modification of this approach is to use inkjet printing to deposit a reducing agent and then subsequently a metal salt: this method has been used to form silver conductive tracks by printing ascorbic acid followed by silver nitrate solution (Bidoki, Nouri and Heidari, 2010).

1.2.3 Deposition of Ceramics

There are several different routes by which ceramic materials can be deposited by inkjet printing, which are analogous to some of the methods used for metals and are illustrated in Figure 1.3.

Suspensions of fine ceramic particles can be directly jetted, provided that the viscosity and surface tension of the mixture lie within the correct range. Alumina particles with median sizes between 0.3 and 1.5 μm have been mixed with wax and kerosene to produce jettable suspensions containing up to 40% ceramic particles by volume, which were then sintered after deposition to produce a material with a final relative density of 80% (Ainsley, Reis and Derby, 2002). A similar approach has been successfully used to deposit lead zirconate titanate (PZT) which was then sintered to effectively full density (Wang and Derby, 2005). Thermal DOD (defined in Chapter 2) has been used to print suspensions of 0.5 μm silicon nitride powder in an aqueous medium, and good mechanical properties have been claimed in the sintered product (Cappi *et al*., 2008).

Chemical precursors can also be printed and then transformed to the final ceramic product. One example is barium strontium titanate, a dielectric ceramic which has been deposited by printing a metal–organic precursor onto a ceramic substrate, followed by pyrolysis and annealing, to form a capacitor structure (Kaydanova *et al*., 2007). Cerium oxide has been produced by inkjet printing of precursor solutions onto a heated substrate, giving the desired crystalline deposit without any further heat treatment (Gallage *et al*., 2008). Several investigators have used sol–gel precursors followed by thermal treatment,

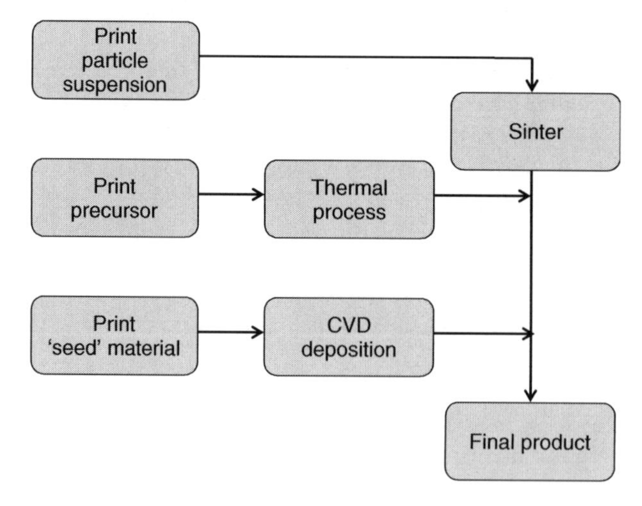

Figure 1.3 Process routes by which inkjet printing can be used to deposit ceramics.

for example to deposit PZT (Bathurst, Lee and Kim, 2008) and barium titanate films (Keat *et al.*, 2007).

A further method, which has been used to generate patterned deposits of nano-crystalline diamond, is to print a suspension of fine nano-diamond particles (4–5 nm in size) onto a silicon substrate and then to use these as 'seed' particles for the growth of a continuous diamond film by a conventional chemical vapour deposition (CVD) process. It has been suggested that this process may be developed to produce 3D diamond structures by sequential inkjet printing and CVD processing (Chen, Tzeng and Cheng, 2009).

1.2.4 Deposition of Polymers

Figure 1.4 summarises the methods by which polymeric materials can be deposited. Waxes, and other relatively short-chain polymers with molecular weights of a few hundred Daltons, can form readily jettable melts and can be used for some applications such as mask printing and rapid prototyping. Long-chain polymers, however, cannot be jetted directly since even as a melt their viscosity is usually too great, and alternative routes are needed to deposit these polymers by inkjet printing (Calvert, 2001). They can be dissolved, or colloidally dispersed to form a latex, in suitable solvents, although even in solution the presence of a small concentration of high molecular weight polymer may introduce sufficient viscoelasticity to inhibit good droplet formation (De Gans, Duineveld and Schubert, 2004). Electronically functional polymers, such as conductors (e.g. conjugated polymers such as poly(3,4-ethylenedioxythiophene) (PEDOT) poly(styrenesulphonate)(PSS), (PEDOT:PSS) and polyaniline), semiconductors and organic light-emitting diode (OLED) materials, can be printed in solution by

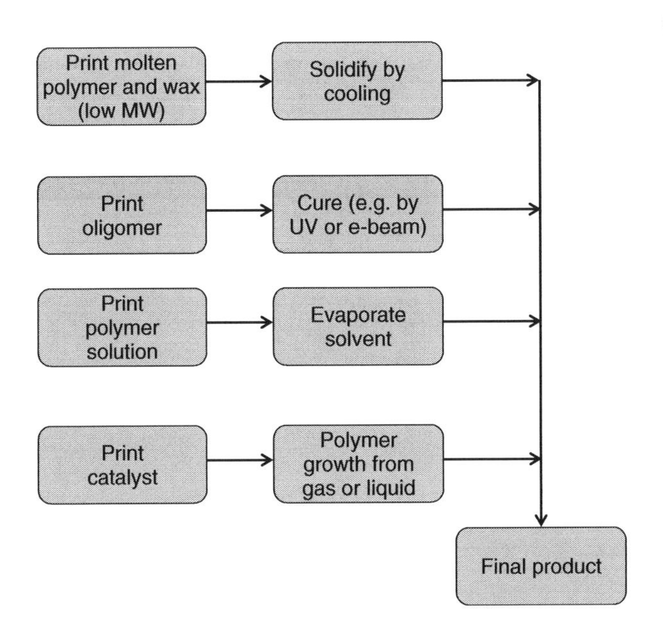

Figure 1.4 *Process routes by which inkjet printing can be used to deposit polymers.*

inkjet methods. There is major commercial interest in printing organic semiconductors for such applications as display backplanes, and also in fabricating large-area polymer light-emitting diode (PLED) displays, as discussed in Chapters 9 and 10.

For structural or optical applications or to achieve dielectric properties, thermoset polymers can be cross-linked *in situ* after printing, by thermal treatment, by electron beam treatment or by UV curing a formulated ink containing a photo initiator. UV curing is increasingly common in graphical printing applications, and can involve a brief 'pinning' exposure immediately after printing to arrest migration of the drop edge, followed by subsequent full curing, perhaps after further layers of material have been printed. Chapter 4 contains further details.

Finally, inkjet printing of a suitable catalyst has been used to initiate local formation of conductive films of polyacetylene in subsequent gas-phase treatment (Huber, Amgoune and Mecking, 2008).

1.3 Applications to Manufacturing

A summary of possible process routes which employ inkjet printing for manufacturing is shown schematically in Figure 1.5. These routes are applicable in principle to any materials, although some have been explored rather little.

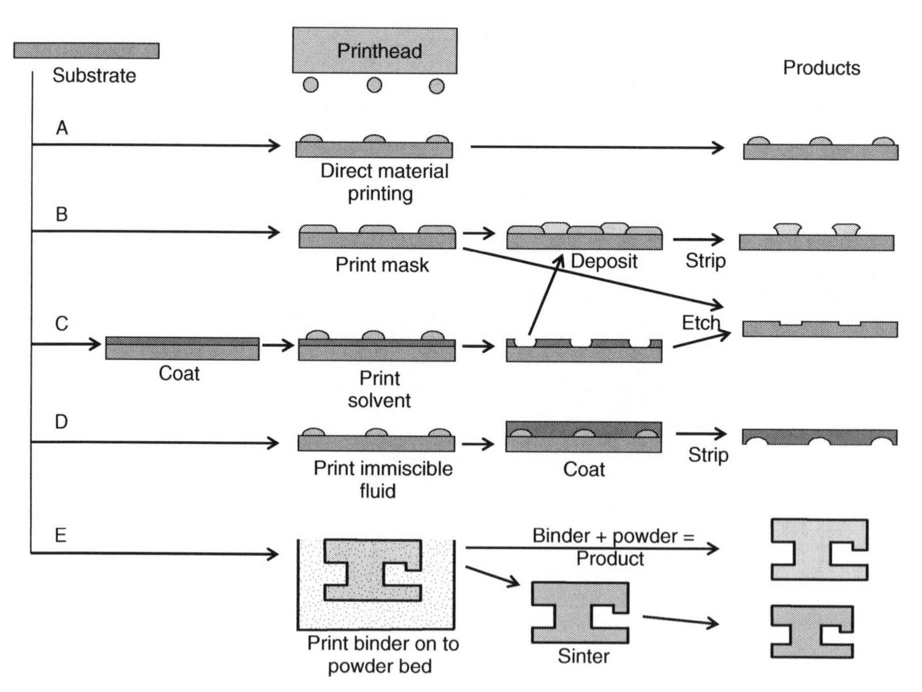

Figure 1.5 *Classification of process routes by which inkjet printing can be used to create structures: (A) direct material printing; (B) printing of a mask followed by material deposition or etching; (C) inkjet etching; (D) inverse inkjet printing and (E) printing onto a powder bed (See plate section for coloured version).*

1.3.1 Direct Deposition

The process shown as route A in Figure 1.5 involves the direct deposition of material onto a substrate in a digitally defined pattern (by any of the methods outlined in Section 1.2). This is the simplest process and has been widely exploited: it forms the basis of most of the examples discussed in other chapters of this book. Many manufacturing applications of inkjet printing are to be found in the electronics industry, and involve products in the form of thin films. The printing of conducting tracks has been widely explored, and is currently the most important application of the inkjet printing of metals. Both the printing of silver nanoparticle ink followed by sintering, as shown in Figure 1.6, and the printing of a catalyst followed by electroless plating can give good electrical conductivity; circuit elements produced in these ways have been demonstrated, including ultra-high frequency (UHF) transmission lines and antennae (Mäntysalo and Mansikkamäki, 2009), organic thin film transistors with silver electrodes (Kim *et al.*, 2007) and metal–insulator–metal crossovers (Sanchez-Romaguera *et al.*, 2008).

Little attention has been paid to the possibility of depositing mechanical components by the printing of metal particle ink, although some examples have been provided by Fuller, Wilhelm and Jacobson (2002), including thermal-expansion-driven actuators with features up to 1 mm tall, fabricated by multiple deposition (400 layers) of silver nanoparticle ink.

Direct printing of molten metal has been used to form solder bumps for electronics chip interconnection; the rapid heat transfer to the cooler surface leads to rapid solidification of the metal in contact with the substrate and thus to pinning of the spreading contact line, giving a well-defined smooth deposit with a high contact angle (Figure 1.7) (Liu and Orme, 2001a; Wallace *et al.*, 2006; Yokoyama *et al.*, 2009).

Repeated DOD deposition of liquid metal droplets has also been used to build vertical solder columns up to a few millimetres in height, corresponding to aspect ratios of about 20 (Lee *et al.*, 2008), while macroscopic artefacts tens of millimetres in size have been fabricated by CIJ printing of liquid aluminium alloys (Liu and Orme, 2001b).

Figure 1.6 *Conducting tracks, 0.5 mm wide, made by drop-on-demand printing of silver nanoparticle ink onto a ceramic substrate, followed by thermal sintering (Courtesy of Printed Electronics Ltd.).*

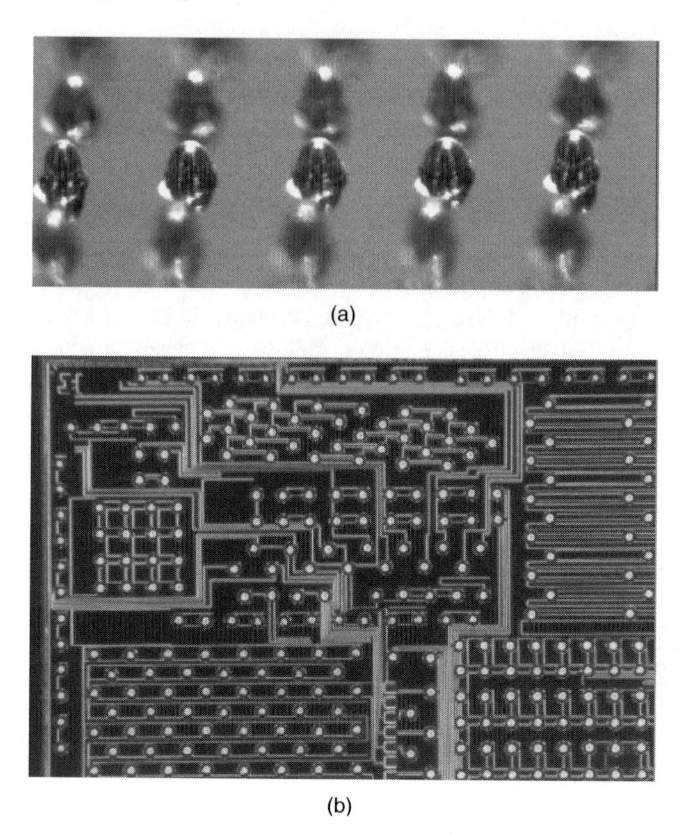

(a)

(b)

Figure 1.7 (a) Lead–tin solder bumps (100 μm diameter) printed by piezoelectric inkjet at 250 μm separation (centre-to-centre) and (b) multiple solder bumps (70 μm diameter) printed onto a test substrate (Reproduced with permission from Hayes et al. (1999) Copyright (1999) IMAPS Courtesy of MicroFab).

Oxide ceramics in thin film form have many potential applications, for example as dielectrics, piezoelectric materials and catalysts. Examples of their direct deposition by inkjet processes have been given in Section 1.2.3. Thick deposits of ceramics can be built up by repeated printing of particulate suspensions, and the fabrication of 3D artefacts by printing followed by sintering has been demonstrated for alumina (Ainsley, Reis and Derby, 2002), silicon nitride (as shown in Figure 1.8) (Cappi *et al*., 2008), PZT and titania (Figure 1.9) (Lejeune *et al*., 2009). Applications include mechanical components, piezoelectric transducers and photocatalysts.

Thin films of polymers have wide application in electronics as dielectrics and conductors, and also as functional materials: for example as semiconductors and light emitters. Polymers as well as ceramics provide materials for micro-scale sensors and actuators (Wilson *et al*., 2007). Processes discussed in Section 1.2.4 have all been used to deposit polymer films. Thicker deposits have potential for optical applications, and inkjet printing has been used to form optical waveguides and small plano-convex lenses, as single lenses or as multiple lens arrays, as discussed in Chapter 6 and illustrated in Figure 1.10 (Wallace *et al*., 2006; Chen *et al*., 2008).

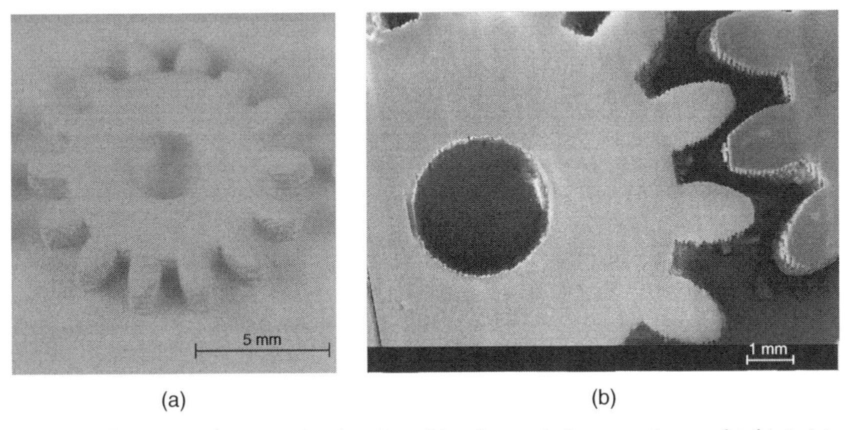

Figure 1.8 Silicon nitride gearwheels printed by thermal drop-on-demand inkjet: (a) green (as-printed, unsintered) and (b) sintered components (Reproduced with permission from Cappi et al. (2008) Copyright (2008) Elsevier Ltd.).

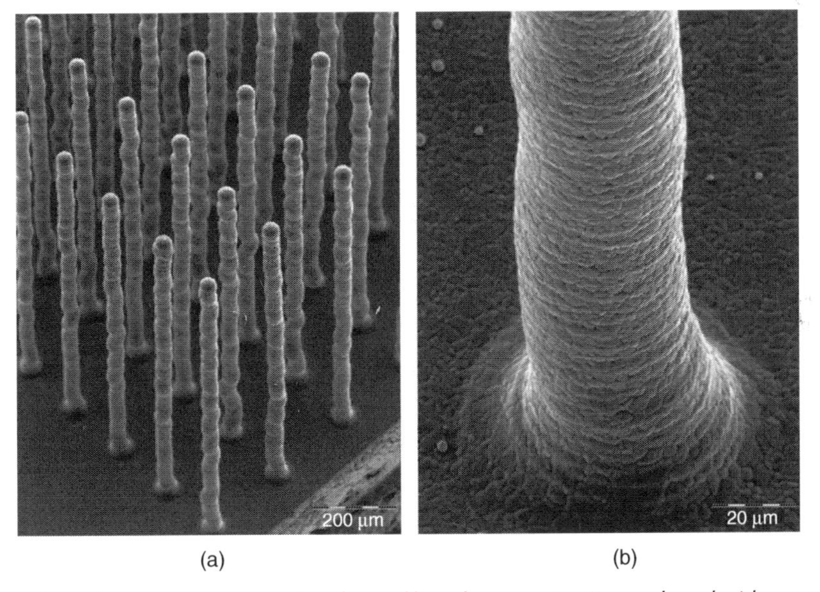

Figure 1.9 (a,b) Sintered titania pillars formed by inkjet printing (Reproduced with permission from Lejeune et al. (2009) Copyright (2009) Elsevier Ltd.).

An important challenge in many applications in which droplets of suspensions or solutions are deposited, followed by evaporation of the solvent, is to achieve solid deposits which are as flat and even as possible. There is a natural tendency for solutes or particles to deposit from evaporating drops towards the rim of the drop where the contact line becomes pinned: this is the 'coffee stain' effect discussed further in Chapter 5 (Tekin, Smith and Schubert, 2008). One explanation for this effect is that solvent close to the droplet edge evaporates more readily than from the centre, leading to transport

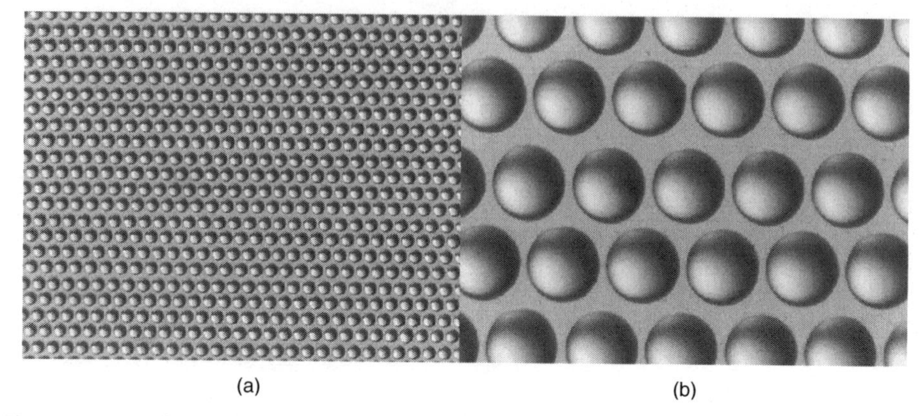

Figure 1.10 (a,b) Portion of array of 200 000 polymer lenses, each 104 μm diameter, formed by inkjet printing (Reproduced with permission from Wallace et al. (2006). Copyright (2006) SMTA Courtesy of MicroFab).

of suspended material towards the boundary. In some applications special measures, for example through the use of mixed solvents, must be taken to reduce this effect, while in others it can be actively exploited to give a desirable variation in film thickness (Lu, Chen and Lee, 2009).

A variant of the direct deposition process involves printing drops of liquid into another liquid: that is, the substrate is also a liquid. If the two liquids are immiscible and each printed drop adheres to the previously deposited material, then at least in principle this provides a potential method to manufacture fragile products, which are supported by the liquid bath as they grow. A method can also be used in which the printed liquid reacts with the liquid bath on contact to form a solid material. This type of approach, in which sodium alginate solution is printed into a bath of calcium chloride solution to form regions of hydrogel, has been used by Nakamura and colleagues to produce scaffolds and cell-containing structures for biological applications. It is discussed in detail in Chapter 13.

1.3.2 Inkjet Mask Printing

While direct deposition has been widely explored, it is by no means the only route by which inkjet printing can be used to create digitally defined structures. An alternative method (route B in Figure 1.5) is to print a mask, and then to use this to define areas to be etched, or onto which a further material can be deposited (e.g. by electroless plating or electro-plating). Direct inkjet printing of masks has been explored for electronic printed circuit board production (the application discussed in detail in Chapter 8), but little work has been done on the wider application of inkjet mask printing to the texturing of surfaces. CIJ has been used with a solvent-based polymer ink to deposit masks onto steel rolls for subsequent etch patterning; the deposited drop diameter was ~150 μm (Muhl and Alder, 1995).

More recently, several ink types were investigated for mask printing, and 120–150 μm features were printed onto metallic substrates with a UV-cured polymer

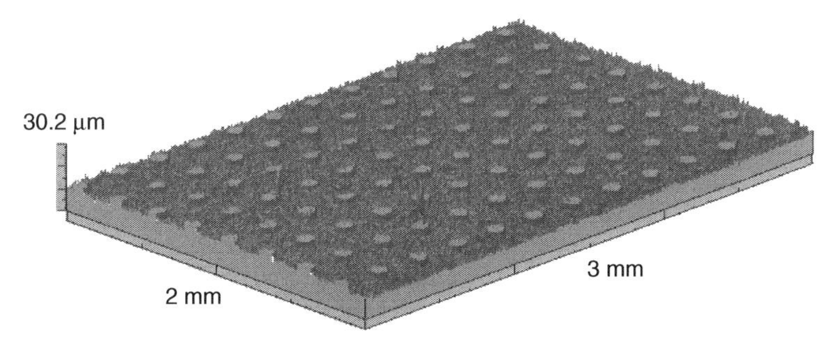

Figure 1.11 Steel surface patterned by inkjet masking followed by etching. The circular masked regions are 60 μm in diameter (Reproduced with permission from Costa and Hutchings (2008). Copyright (2008) Society for Imaging Science and Technology).

ink (James, 2004). An example of a steel substrate patterned by inkjet printing of a mask followed by acid etching is shown in Figure 1.11 (Costa and Hutchings, 2008). The circular masked regions, formed with a solvent-based polymer ink, were 60 μm in diameter; the thinnest parallel gap which could be formed between mask lines was ~20 μm, and the smallest square gap ~40 μm across. Solvent-based and UV-cured inks are not, however, generally designed for masking applications. Phase-change (e.g. wax-based) inks have been designed specifically as jettable etch resists; they are printed at high temperature onto a cold substrate and give good edge definition as drop spreading is arrested once the drop begins to solidify. One major application of inkjet mask printing is in photovoltaic solar cell fabrication.

1.3.3 Inkjet Etching

Route C in Figure 1.5 is known as inkjet etching and involves the printing of drops of solvent onto a suitable thin (usually polymeric) coating. Local dissolution of the coating, followed by evaporation of the solvent, leads to redistribution of the coating material at the edges of the crater and thus to the formation of a hole in the coating. While there may still be some residual coating material at the centre of the crater after the first evaporation event, this can usually be removed by repeated printing of solvent drops. First explored as a process for making holes in the fabrication of thin film transistors, inkjet etching has more recently been further investigated and shown to offer various possibilities of forming regular arrays of features, such as circular or rectangular holes and linear grooves, as illustrated in Figure 1.12 (De Gans, Hoeppner and Schubert, 2007). Patterning of polymer layers in this way has been proposed for the production of biochips and micro-patterned cell arrays, but it could also be used for mask fabrication, to be followed, as in direct mask printing, by subsequent etching of an underlying substrate or by deposition of a different material.

Variations of the inkjet etching process have also been developed. For example, if small drops of solvent are printed onto a thick polymer substrate, they will evaporate leaving an array of smooth-surfaced craters which can be used either directly as micro-lenses or as a template to form convex lenses by use of a suitable cast-replicating material; the focal length of the lenses can be altered by varying the drop deposition

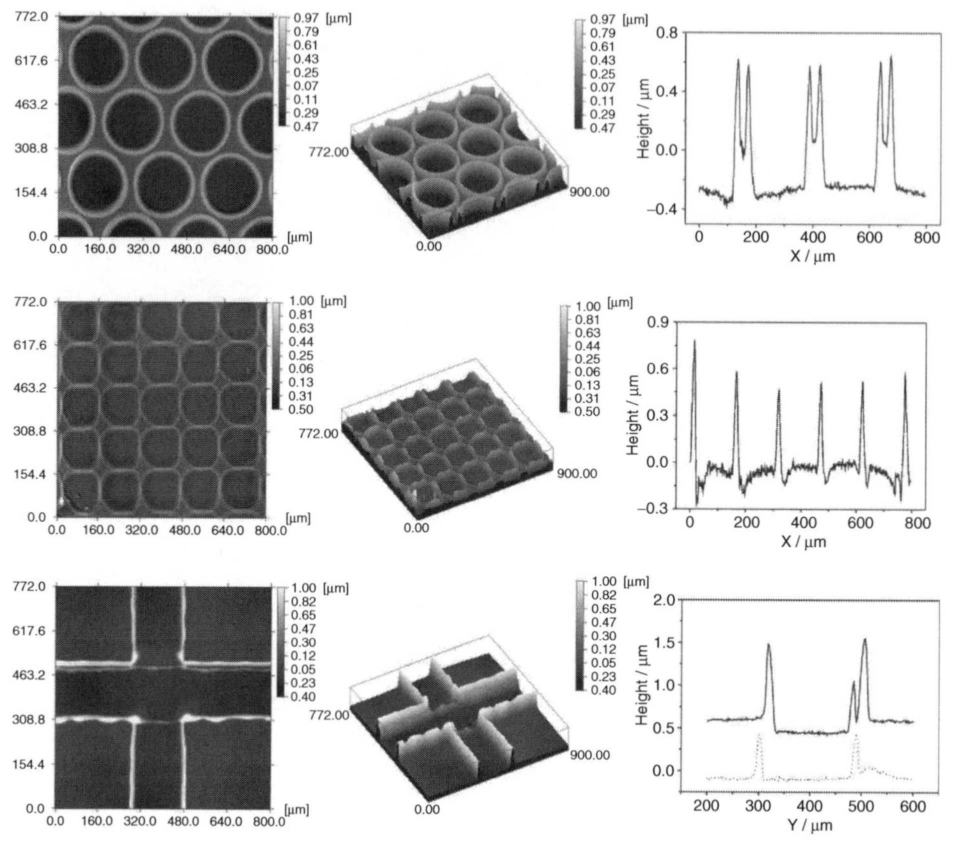

Figure 1.12 Examples of features etched into a polymer surface (polystyrene) by inkjet printing of solvent droplets (Reproduced with permission from De Gans et al. *(2007). Copyright (2007) Royal Society of Chemistry).*

sequence (Periccet-Camar *et al*., 2007). Inkjet printing can also be used, not to remove a polymeric coating material but to plasticise it so that an etchant can permeate it to attack the underlying substrate. The plasticisation can also be reversed, for example by heating the coating to drive off the plasticiser, creating a very versatile process for fabricating certain types of structure. This has been explored for forming openings in buried semiconductor layers in photovoltaic solar cells (Lennon *et al*., 2008).

1.3.4 Inverse Inkjet Printing

Route D in Figure 1.5 can be termed inverse inkjet printing since it forms holes or cavities in a solid material in the locations where the drops of ink are deposited. The process has been used to fabricate polymeric micro-sieves, by printing an array of sessile drops onto a substrate, applying a continuous film of a polymer which is immiscible with the drops, curing it to solidify it and then removing it from the substrate, as shown in Figure 1.13 (Jahn *et al*., 2009). In this application the pore size in the sieve is controlled

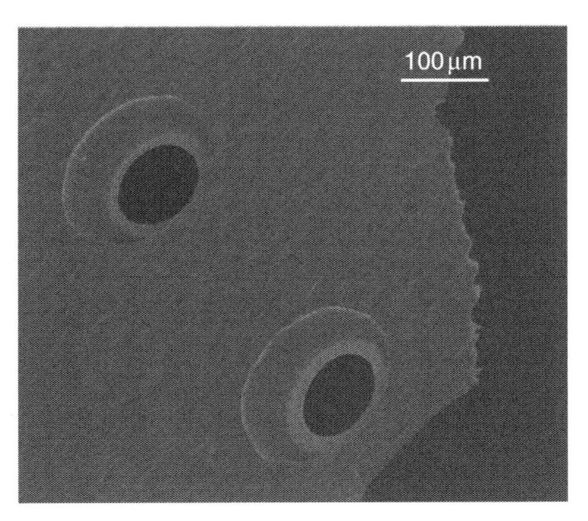

Figure 1.13 *Example of polymer micro-sieve made by inverse inkjet printing (Reproduced with permission from Jahn* et al. *(2009). Copyright (2009) American Chemical Society).*

by the height of the printed drop and the thickness of the polymer layer, but the process could also be adapted, by using a thicker layer, to produce an array of concavities which could then be used either as concave mirrors or, by replication, to generate an array of convex lenses.

1.3.5 Printing onto a Powder Bed

The final process sequence, shown as route E in Figure 1.5, involves using inkjet printing to deposit drops of a liquid binder material (e.g. a UV-curable polymer) in a pattern onto a flat bed of a solid powder. This forms the basis for an important group of 3D printing processes and is discussed in detail in Chapter 14. After the binder has cured or dried, a thin layer of fresh powder is spread evenly over the surface of the bed and the process is repeated. If a suitable series of 2D patterns is printed in this way, a 3D object is created within the powder bed, consisting of powder particles bonded by the printed ink. This can then be removed and either used as it is (although it may have rather low strength) or processed further (e.g. by sintering or infiltration) to produce the final product. There are several commercial processes based on this principle. Figure 1.14 shows two examples of objects made in this way: a fully coloured model made by printing with water-based inks onto a plaster powder in which hydration of the plaster causes the particles to bond, and a stainless steel mesh which was heated to sinter the particles and thus densify the product after printing. Infiltration of the bonded product with a second liquid, which is drawn into the spaces between the powder particles by capillary action, provides a further potential method of post-treatment.

1.4 Potential and Limitations

We have seen how inkjet printing can in principle be used to deposit a very wide range of materials, singly or in combination, either directly or as a step in a more complex

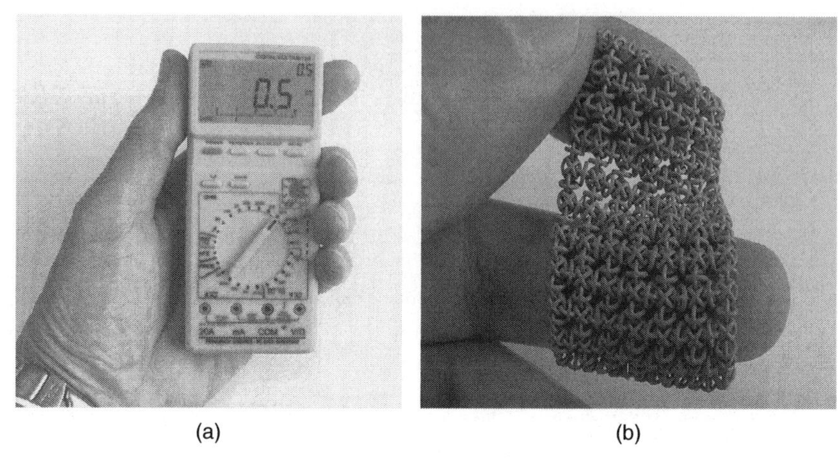

(a) (b)

Figure 1.14 *Examples of parts made by printing onto a powder bed: (a) model of an electronic multimeter, made by printing binder onto ceramic powder by the Z-Corp™ process, and (b) stainless steel mesh, sintered after printing onto metal powder by the fcubic™ process.*

process route. It is already widely used industrially for printing text and graphics, and is now being adopted for certain manufacturing processes where its ability to deposit precise volumes of material in well-defined locations under digital control offers special benefits. However, there are important limitations to the use of inkjet processes which must be borne in mind in considering new applications.

The resolution which can be achieved by inkjet printing depends not only on the size of the final printed drop after any solidification, drying or curing has occurred, but also on the precision with which the drop can be deposited onto the substrate. That precision is influenced by the accuracy of movement of the substrate or print-head, by inherent variability in the direction in which the drop leaves the print-head, by aerodynamic and electrostatic influences on the drop in flight, which may themselves depend on the presence of other neighbouring drops, and by other sources of process variability such as variation in drop size and velocity. In practice, drop placement accuracy of better than several micrometres is hard to achieve in direct inkjet printing onto a homogeneous substrate, and $\sim 10\,\mu m$ represents a current lower limit to the sizes of features (circular spots or line widths) which can be printed by DOD methods.

However, the final position of a liquid drop on a substrate can be controlled to some extent by using a heterogeneous substrate, in which different areas are wettable to different degrees by the printed drop. This approach has been very successful, for example in printing thin film transistors, where laser, photolithographic or other methods have been used to pattern the surface energy of the substrate, and thus limit the movement of the deposited drop or steer it away from a hydrophobic region and towards a hydrophilic region (Wang *et al*., 2004). Sub-micrometre features can be fabricated in such ways, and the edge definition which can be achieved is limited largely by the accuracy of the surface energy patterning process. In an ingenious extension of this approach, also used for transistor fabrication, the inkjet printing of materials in sequence (so that the fluid in the second deposit is repelled by the first, which is already on the substrate) can be used to generate extremely narrow channels between the two deposits: gaps of $<100\,nm$ have been demonstrated (Sele *et al*., 2005).

Most research and development on inkjet printing for manufacturing purposes has addressed the formation of thin film deposits, often for electronics applications (Perelaer *et al*., 2010). These films may be tens, hundreds or even thousands of micrometres in lateral extent, but are often sub-micrometre in thickness, and are formed by the printing of small numbers of layers, or even only a single layer, of drops. Their electronic, rather than mechanical, properties have usually been optimised. In comparison there has been limited attention, so far, to building 3D deposits with aspect ratios of one or more by direct material deposition for mechanical applications, with the exceptions of free-standing pillars of metals, ceramics and polymers. Surface energy control of liquid surface curvature has been used in forming optical components such as lenses and waveguides, and in depositing solder drops and bumps.

With the exception of some work on the overprinting of conductors and dielectrics as electronic circuit elements (e.g. Sanchez-Romaguera, Madec and Yeates, 2008), there has so far been little attempt to address the challenges involved in sequential, or even simultaneous, deposition of the droplets of different materials which will be needed to fabricate complex 3D structures from multiple materials. Careful control of surface energies and hence of relative wettabilities will be an essential component of this. Suspensions of small particles are widely used to print both metals and ceramics, but the volume fraction of solids which can be used in the fluid is generally low: there is scope for further development of more heavily loaded colloidal fluids, which still have the rheological properties needed for printing, to extend the range of materials and products which can be achieved. Full exploitation of the undoubted potential of inkjet printing for digital fabrication will require further research into all these aspects.

References

Ainsley, C., Reis, N. and Derby, B. (2002) Freeform fabrication by controlled droplet deposition of powder filled melts. *Journal of Materials Science*, **37**, 3155–3161.

Bathurst, S.P., Lee, H.W. and Kim, S.G. (2008) Ink jet printing of PZT thin films for MEMS applications. Digital Fabrication 2008, Pittsburgh, PA, pp. 897–901.

Bidoki, S.M., Nouri, J. and Heidari, A.A. (2010) Inkjet deposited circuit components. *Journal of Micromechanics and Microengineering*, **20**, 055023.

Calvert, P. (2001) Inkjet printing for materials and devices. *Chemistry of Materials*, **13**, 3299–3305.

Cappi, B., Özkol, E., Ebert, J. and Telle, R. (2008) Direct inkjet printing of Si_3N_4: characterization of ink, green bodies and microstructure. *Journal of the European Ceramic Society*, **28**, 2625–2628.

Chen, C.T., Chiu, C.L., Tseng, Z.F. and Chuang, C.T. (2008) Dynamic evolvement and formation of refractive microlenses self-assembled from evaporative polyurethane droplets. *Sensors & Actuators*, **A147**, 369–377.

Chen, Y.C., Tzeng, Y., Cheng, A.J. *et al*. (2009) Inkjet printing of nanodiamond suspensions in ethylene glycol for CVD growth of patterned diamond structures and practical applications. *Diamond and Related Materials*, **18**, 146–150.

Cheng, S.X., Li, T. and Chandra, S. (2005) Producing molten metal droplets with a pneumatic droplet-on-demand generator. *Journal of Materials Processing Technology*, **159**, 295–302.

Costa, H.L. and Hutchings, I.M. (2008) Ink-jet printing for patterning engineering surfaces. Digital Fabrication 2008, Pittsburgh, PA, Society for Imaging Science and Technology, pp. 256–259.

De Gans, B.J., Duineveld, P.C. and Schubert, U.S. (2004) Inkjet printing of polymers: state of the art and future developments. *Advanced Materials*, **16**, 203–213.

De Gans, B.J., Hoeppner, S. and Schubert, U.S. (2007) Polymer relief structures by inkjet etching. *Journal of Materials Chemistry*, **17**, 3045–3050.

Fuller, S.B., Wilhelm, E.J. and Jacobson, J.M. (2002) Ink-jet printed nanoparticle microelectromechanical systems. *Journal of Microelectromechanical Systems*, **11**, 54–60.

Gallage, R., Matuo, A., Fujiwara, T. *et al*. (2008) On-site fabrication of crystalline cerium oxide films and patterns by ink-jet deposition method at moderate temperatures. *The Journal of the American Ceramic Society*, **91**, 2083–2087.

Hayes, D.J., Wallace, D.B. and Cox, W.R. (1999) Micro jet printing of solder and polymers for multi-chip modules and chip-scale packages. Proceedings of IMAPS International Conference on High Density Packaging and MCMs.

Huber, J., Amgoune, A. and Mecking, S. (2008) Patterning of polymers on a substrate via ink-jet printing of a coordination polymerization catalyst. *Advanced Materials*, **20**, 1978–1981.

Jahn, S.F., Engisch, L., Baumann, R.R. *et al*. (2009) Polymer microsieves manufactured by inkjet technology. *Langmuir*, **25**, 606–610.

James, M. (2004) Photochemical machining by ink jet: a revolution in the making? Proceedings of NIP20, Society for Imaging Science and Technology, pp. 279–283.

Kaydanova, T., Miedaner, A., Perkins, J.D. *et al*. (2007) Direct-write inkjet printing for fabrication of barium strontium titanate-based tunable circuits. *Thin Solid Films*, **515**, 3820–3824.

Keat, Y.C., Sreekantan, S., Hutagalung, S.D. and Ahmad, Z.A. (2007) Fabrication of BaTiO3 thin films through ink-jet printing of TiO2 sol and soluble B salts. *Materials Letters*, **61**, 4536–4539.

Kim, D., Jeong, S., Lee, S. *et al*. (2007) Organic thin film transistor using silver electrodes by ink-jet printing technology. *Thin Solid Films*, **515**, 7692–7696.

Lee, T.M., Kang, T.G., Yang, J.S. *et al*. (2008) Drop-on-demand solder droplet jetting system for fabricating microstructure. *IEEE Transactions on Electronics Packaging Manufacture*, **31**, 202–210.

Lejeune, M., Chartier, T., Dossou-Yovo, C. and Noguera, R.J. (2009) Ink-jet printing of ceramic micro-pillar arrays. *European Ceramic Society*, **29**, 905–911.

Lennon, A.J., Utama, R.Y., Lenio, M.A.T. *et al*. (2008) Forming openings to semiconductor layers of silicon solar cells by inkjet printing. *Solar Energy Materials and Solar Cells*, **92**, 1410–1415.

Liu, Q. and Orme, M. (2001a) High precision solder droplet printing technology and the state-of-the-art. *Journal of Materials Processing Technology*, **115**, 271–283.

Liu, Q. and Orme, M. (2001b) On precision droplet-based net-form manufacturing technology. *Proceedings of the Institution of Mechanical Engineers, Part B: Journal of Engineering Manufacture*, **215**, 1333–1355.

Lok, B.K., Liang, Y.N., Gian, P.W. *et al*. (2007) Process integration of inkjet printing and electroless plating for LTCC substrates. Proceedings of the 9th Electronics Packaging Technology Conference, IEEE, pp. 202–205.

Lu, J.P., Chen, F.C. and Lee, Y.Z. (2009) Ring-edged bank array made by inkjet printing for color filters. *Journal of Display Technology*, **5**, 162–165.

Mäntysalo, M. and Mansikkamäki, P. (2009) An inkjet-deposited antenna for 2.4 GHz applications. *International Journal of Electronics and Communications*, **63**, 31–35.

Mei, J., Lovell, M.R. and Mickle, M.H. (2005) Formulation and processing of novel conductive solution inks in continuous inkjet printing of 3-D electric circuits. *IEEE Transactions on Electronics Packaging Manufacturing*, **28**, 265–273.

Muhl, J. and Alder, G.M. (1995) Direct printing of etch masks under computer control. *International Journal of Machine Tools & Manufacture*, **35**, 333–337.

Park, B.K., Kim, D., Jeong, S. *et al*. (2007) Direct writing of copper conductive patterns by ink-jet printing. *Thin Solid Films*, **515**, 7706–7711.

Perelaer, J., Hendriks, C.E., De Laat, A.W.M. and Schubert, U.S. (2009) One-step inkjet printing of conductive silver tracks on polymer substrates. *Nanotechnology*, **20**, 165303.

Perelaer, J., Hendriks, C.E., Reinhold, I. *et al*. (2008) Inkjet printing of conductive silver tracks in high resolution and the (alternative) sintering thereof. Digital Fabrication 2008, Pittsburgh, PA, Society for Imaging Science and Technology, pp. 697–701.

Perelaer, J., Smith, P.J., Mager, D. *et al*. (2010) Printed electronics: the challenges involved in printing devices, interconnects, and contacts based on inorganic materials. *Journal of Materials Chemistry*, **20**, 8446–8453.

Periccet-Camar, R., Best, A., Nett, S.K. *et al*. (2007) Arrays of microlenses with variable focal lengths. *Optics Express*, **15**, 9877–9882.

Sanchez-Romaguera, V., Madec, M.B. and Yeates, S.G. (2008) Inkjet printing of 3D metal-insulator-metal crossovers. *Reactive & Functional Polymers*, **68**, 1052–1058.

Sele, C.W., Von Werne, T., Friend, R.H. and Sirringhaus, H. (2005) Lithography-free, self-aligned inkjet printing with sub-hundred-nanometer resolution. *Advanced Materials*, **17**, 997–1001.

Smith, P.J., Mager, D., Löffelmann, U. and Korvink, J.G. (2008) Inkjet printing silver-containing inks. Digital Fabrication 2008, Pittsburgh, PA, Society for Imaging Science and Technology, pp. 689–692.

Smith, P.J., Shin, D.Y., Stringer, J.E. *et al*. (2006) Direct ink-jet printing and low temperature conversion of conductive silver patterns. *Journal of Materials Science*, **41**, 4153–4158.

Stringer, J., Xu, B.J. and Derby, B. (2007) Characterisation of photo-reduced silver organometallic salt deposited by inkjet printing. Digital Fabrication 2008, Pittsburgh, PA, Society for Imaging Science and Technology, p. 960.

Tekin, E., Smith, P.J. and Schubert, U.S. (2008) Inkjet printing as a deposition and patterning tool for polymers and inorganic particles. *Soft Matter*, **4**, 703–713.

Tseng, A.A., Lee, M.H. and Zhao, B. (2001) Design and operation of a droplet deposition system for freeform fabrication of metal parts. *Transactions of ASME: Journal of Engineering Materials and Technology*, **123**, 74–84.

Wallace, D., Hayes, D., Chen, T. *et al*. (2006) Ink-jet as a MEMS manufacturing tool. Proceedings of SMTA Pan-Pacific Microelectronics Symposium, Hawaii.

Wang, T. and Derby, B. (2005) Ink-jet printing and sintering of PZT. *Journal of the American Ceramic Society*, **88**, 2053–2058.

Wang, J.Z., Zheng, Z.H., Li, H.W. *et al*. (2004) Dewetting of conducting polymer inkjet droplets on patterned surfaces. *Nature Materials*, **3**, 171–176.

Wilson, S.A., Jourdaina, R.P.J., Zhanga, Q. *et al*. (2007) New materials for micro-scale sensors and actuators: an engineering review. *Materials Science and Engineering*, **R56**, 1–129.

Yokoyama, Y., Endo, K., Iwasaki, T. and Fukumoto, H. (2009) Variable-size droplets by a molten-solder ejection method. *Journal of Microelectromechanical Systems*, **18**, 316–321.

Zapka, W., Voit, W., Loderer, C. and Lang, P. (2008) Low temperature chemical post-treatment of inkjet printed nano-particle silver inks. Digital Fabrication 2008, Pittsburgh, PA, Society for Imaging Science and Technology, pp. 906–911.

2

Fundamentals of Inkjet Technology

Graham D. Martin and Ian M. Hutchings
Inkjet Research Centre, University of Cambridge, United Kingdom

2.1 Introduction

Inkjet printing involves the production of small drops of liquid and their deposition in precise locations on a substrate. Many methods for drop generation have been devised, but currently the most important methods for inkjet printing are described as either continuous inkjet (CIJ) or drop-on-demand (DOD). Inkjet technology has been further categorised in several ways (Le, 1998), but here we shall use the classification shown in Figure 2.1, which also includes the *electrospray technique* described in Section 2.4.3.

In both the CIJ and DOD methods, the liquid flows through a small orifice (usually called a nozzle). The essential difference between the two lies in the nature of the flow through the nozzle: in CIJ, as the name implies, the flow is continuous, while in DOD it is impulsive. A CIJ system produces a continuous stream of drops, from which those to be printed onto the substrate are selected as required, whereas in DOD printing the ink is emitted through the nozzle to form a short jet, which then condenses into a drop only when that drop is required. Figure 2.2 shows examples of these types of flow. The upper image (Figure 2.2a) shows a continuous jet of ink emerging from a nozzle, travelling to the right and progressively breaking up into drops; in this case, the break-up was stimulated by an external vibration, but an unstimulated jet will also break up in a similar way. Figure 2.2b shows images of the jets ejected from three nozzles in DOD mode and travelling to the right. The main droplet which forms the head of the jet is followed by a long ligament of liquid which eventually detaches from the nozzle and is pulled towards the head of the jet, while at the same time becoming thinner and in this

Inkjet Technology for Digital Fabrication, First Edition. Edited by Ian M. Hutchings and Graham D. Martin.

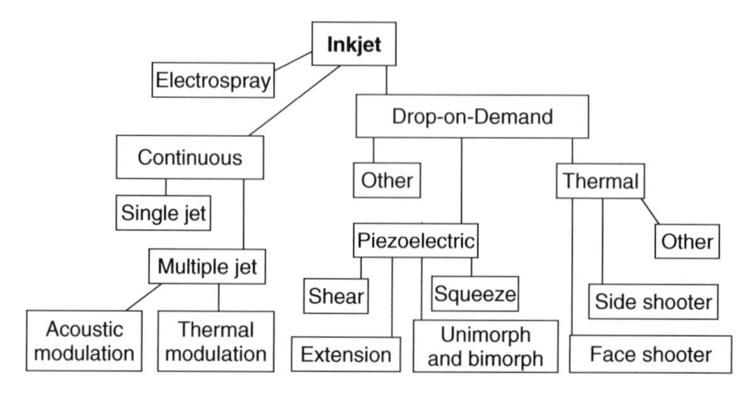

Figure 2.1 *Classification of inkjet technologies.*

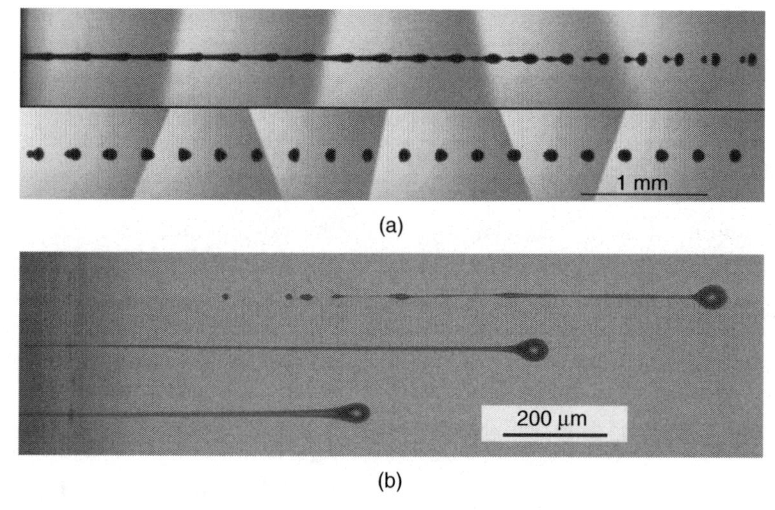

Figure 2.2 *Images of liquid jets moving from left to right, showing (a) the break-up of a continuous jet to form drops (the lower image is a continuation of the upper image) and (b) jet and drop formation in drop-on-demand mode from three nozzles situated towards the left of the frame.*

case (for the upper jet) forming a series of *satellite drops*. The final form of the jet may be a single drop (the ideal case) or, quite commonly, a main drop followed by one or more smaller satellite drops.

Typical drop diameters in inkjet printing lie in the range from 10 to 100 μm, corresponding to drop volumes from 0.5 to 500 pl (see Table 2.1). Drop speeds, at the point at which they strike the substrate, are typically 5–8 $m s^{-1}$ for DOD printing and 10–30 $m s^{-1}$ for CIJ. We shall describe the CIJ and DOD methods more fully in Section 2.4, but we shall first discuss the physical processes involved in the formation of liquid drops.

Table 2.1 *Typical drop diameters and equivalent drop volumes (1 pl = 10^{-12} l).*

Diameter (μm)	Volume (pl)
5.76	0.1
10	0.52
12.41	1
20	4.19
26.73	10
30	14.14
40	33.51
50	65.45
57.59	100
60	113.10
70	179.59
80	268.08
100	523.60
124.07	1000

2.2 Surface Tension and Viscosity

Two physical properties of the liquid dominate the behaviour of the jets and drops involved in inkjet printing: *surface tension* and *viscosity*. We shall remind ourselves briefly of their definitions.

The surface tension of a liquid reflects the fact that atoms or molecules at a free surface have a higher energy than those in the bulk. There is a cost in terms of energy in creating new surface area. For a free droplet of liquid the shape with the lowest surface area, and therefore the lowest surface energy, is a sphere, and in the absence of other influences such as electrostatic or aerodynamic forces that is the shape which a free drop will adopt. If the liquid is in contact with a solid surface, for example after it has been printed onto a substrate, then we need to consider not only the energy of its free surface (which is usually in contact with air) but also the energy of the interface between the liquid and the solid substrate. In that case (and if, as in inkjet printing, the effect of gravity is negligible), the equilibrium shape of the drop becomes a spherical cap, as discussed in Chapter 5 and shown in Figure 5.3.

The surface tension in a liquid causes a force to act in the plane of the free surface perpendicularly to a free edge in that surface, which can be measured directly by various experimental methods. The force is proportional to the length of the edge, and the surface tension σ can therefore be defined as the force per unit length. Since by moving the edge in the direction of the force we will increase the area of the surface and also do work on the system against the surface tension, we can also treat the force per unit length as a *surface energy*: the work done in creating unit area of new surface. The two are equivalent, as are the SI units of $\mathrm{N\,m^{-1}}$ and $\mathrm{J\,m^{-2}}$. Most liquids of practical interest for inkjet deposition have surface tensions σ of the order of tens of $\mathrm{mN\,m^{-1}}$ (or $\mathrm{mJ\,m^{-2}}$). For pure water at 20 °C, $\sigma = 72.5\,\mathrm{mN\,m^{-1}}$, while for many organic liquids (which have smaller intermolecular energies than water), σ lies in the range from 20 to 40 mN m^{-1}. Liquid metals, in contrast, have much higher values, typically several hundred millinewtons per metre for those with low melting points.

The tendency of a liquid to form the shape with the lowest total energy, which in the case of a free drop causes it to become a sphere, is crucial to the processes of inkjet printing. A continuous stream of liquid emerging from a circular nozzle will initially be cylindrical, but that shape is unstable. Under the action of surface tension, disturbances in the shape of the cylinder will grow and the jet will eventually break up into a series of spherical drops. This is the classical Rayleigh–Plateau instability. Lord Rayleigh showed in 1879 that for a liquid cylinder of radius r, the wavelength λ of the disturbance which grows most rapidly is given by $\lambda \approx 9r$. The break-up shown in Figure 2.2a is an example of this phenomenon, which forms the basis of CIJ printing.

The forces which resist the contraction of a liquid jet through the action of surface tension have two origins: the inertia of the liquid and its viscosity. Inertial forces are those associated with a change in a body's momentum: in the case of a moving liquid, they are proportional to its density and the rate of change of velocity. Viscous forces arise from the interactions between molecules of the liquid and act between regions of liquid moving relative to each other. A simple example, which forms the basis for the definition of viscosity, is given by a shear flow as shown in Figure 2.3a. A region of liquid is defined by two parallel surfaces a distance d apart. The lower surface is stationary and the upper surface moves relative to it at a constant velocity v. We assume that this generates a linear gradient of velocity through the liquid that is normal to the upper and lower surfaces. The shear strain rate $\dot{\gamma}$ is given by:

$$\dot{\gamma} = v/d$$

And for many liquids, the shear stress τ acting on the upper and lower surfaces is proportional to $\dot{\gamma}$:

$$\tau = \eta\dot{\gamma}$$

The constant η is the *dynamic viscosity* of the liquid, and if η is independent of $\dot{\gamma}$ then this behaviour is termed *linear* or *Newtonian*. The simple term viscosity, without further qualification, usually means the dynamic viscosity as defined in this section for a shear flow. The SI unit for dynamic viscosity is the pascal second (Pa s). The older (under the centimetre–gram–second system, or cgs) unit of viscosity, poise (P) or more commonly centipoise (cP), is still in widespread use, and conversion is straightforward: 1 mPa s = 1 cP. Water has a viscosity of almost exactly 1 mPa s at 20 °C, while fluids

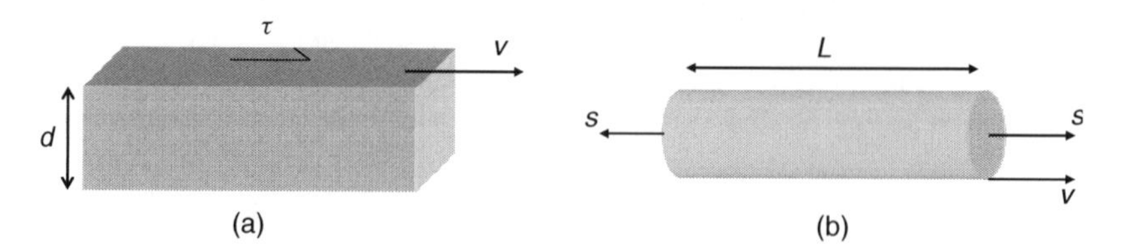

Figure 2.3 *Geometry used to define (a) shear flow and (b) extensional flow.*

which are used in inkjet printing typically have viscosities in the range from ~2 to ~50 mPa s. The viscosities of most liquids fall rapidly with increasing temperature, and this is often exploited in inkjet printing: by varying the temperature of the print-head, it is possible to optimise the viscosity of the ink for drop generation.

The shear flow shown in Figure 2.3a represents a particular pattern of liquid flow, but it does not represent well the flow which occurs in the formation or collapse of a jet. We may also need to consider the extensional or elongational flow which is shown in idealised form in Figure 2.3b. If a cylindrical column of liquid of length L is stretched along its axis at a velocity v, the uniaxial strain rate $\dot{\varepsilon}$ is given by:

$$\dot{\varepsilon} = \frac{v}{L}$$

and the ratio η_T between the stress and the strain rate is then a measure of the *extensional viscosity*:

$$s = \eta_T \dot{\varepsilon}$$

For a Newtonian liquid, η_T is three times the shear viscosity η. The ratio η_T/η is called the Trouton ratio, and it can be significantly greater than 3 for non-Newtonian liquids which are *viscoelastic*. Viscoelastic liquids exhibit a time-dependent elastic response as well as viscosity, due for example to the presence of polymer molecules in solution. Organic solvents are usually Newtonian, as is water, but practical inkjet inks often exhibit some degree of viscoelasticity.

2.3 Dimensionless Groups in Inkjet Printing

Surface tension, inertia and viscosity play key roles in the formation and behaviour of liquid jets and drops. Two important dimensionless numbers can be used to characterise the relative importance of these: the *Reynolds number* and *Weber number*. The Reynolds number Re represents the ratio between inertial and viscous forces in a moving fluid, and is defined by:

$$\mathrm{Re} = \frac{\rho d V}{\eta}$$

where ρ is the density of the fluid, V is its velocity, η is its viscosity and d is a characteristic length, typically the diameter of the jet, nozzle or drop. The Weber number We depends on the ratio between inertia and surface tension:

$$\mathrm{We} = \frac{\rho d V^2}{\sigma}$$

where σ is the surface tension.

The influence of velocity in these two dimensionless groups can be removed by combining them to form a further group, the *Ohnesorge number* Oh, defined by:

$$\mathrm{Oh} = \frac{\sqrt{\mathrm{We}}}{\mathrm{Re}} = \frac{\eta}{\sqrt{\sigma \rho d}}$$

The value of the Ohnesorge number, which reflects only the physical properties of the liquid and the size scale of the jet or drop, but is independent of the driving conditions (which control the velocity), turns out to be closely related to the behaviour of a jet emerging from a nozzle, and thus to the conditions in DOD printing. If the Ohnesorge number is too high ($Oh > \sim 1$), then viscous forces will prevent the separation of a drop, while if it is too low ($Oh < \sim 0.1$), the jet will form a large number of satellite droplets. Satisfactory performance of a fluid in DOD inkjet printing thus requires an appropriate combination of physical properties, which will also depend on the droplet size and velocity (through the value of the Reynolds or Weber number) as shown in Figure 2.4 (Derby, 2010; McKinley and Renardy, 2011). Some authors use the symbol Z for the inverse of the Ohnesorge number ($Z = 1/Oh$). The range over which liquids can be printed is often quoted as $10 > Z > 1$ (e.g. Reis and Derby, 2000), although other work has suggested that $14 > Z > 4$ may be more appropriate (Jang, Kim and Moon, 2009).

The ranges of the Ohnesorge number noted in this section provide some bounds to the 'printability' of the liquid, but other factors must also be considered: the jet must possess enough kinetic energy to be ejected from the nozzle (leading to the solid diagonal line in Figure 2.4 for which $Re = 2/Oh$), and it is also desirable to avoid splashing of the drop on impact with the substrate (which leads to the broken diagonal line for which $OhRe^{5/4} = 50$) (Derby, 2010).

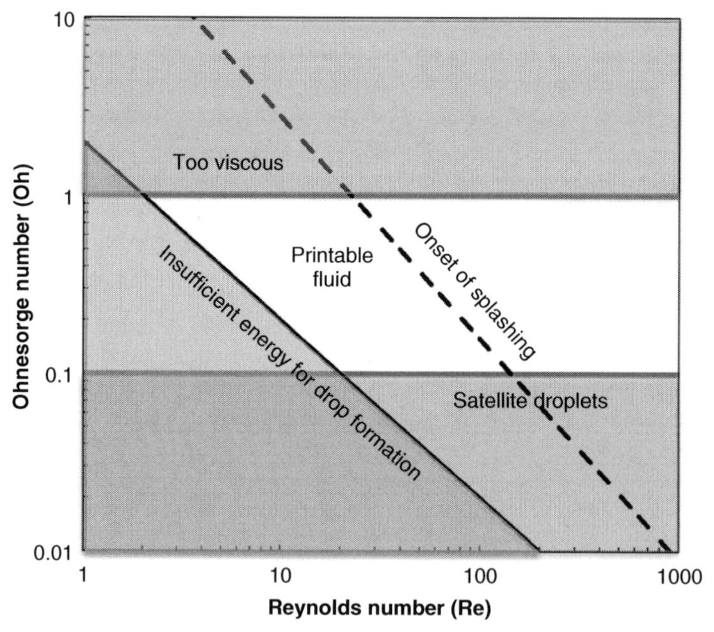

Figure 2.4 *Schematic diagram showing the operating regime for stable operation of drop-on-demand inkjet printing, in terms of the Ohnesorge and Reynolds numbers (Reproduced with permission from McKinley and Renardy (2011) Copyright (2011) American Institute of Physics) (See plate section for coloured version).*

2.4 Methods of Drop Generation

2.4.1 Continuous Inkjet (CIJ)

As outlined in Section 2.1, in this method of printing a continuous jet of liquid is formed by forcing it under pressure through a nozzle. It will tend to break up into drops, driven by surface tension forces, as described in Section 2.2. Normally in a CIJ printer, this natural behaviour is controlled by imposing a disturbance on the jet so that the jet breaks up into a series of drops with well-controlled size and spacing. As drops are produced continually, some means is needed to select those to be printed and to remove (and re-use) the drops not required for printing. This means that CIJ systems tend to be more complex than other inkjet technologies as they require a method to produce drops, a method to select drops and a method to recover and control the liquid. We shall consider both single-jet and multiple-jet variants.

2.4.1.1 Single-Jet CIJ

One of the very first inkjet systems to be developed used a single continuous jet with electrostatic charging and deflection to select the drops and control their placement (Sweet, 1965). Figure 2.5 shows a diagram of such a system. The jet emerges from the nozzle with a velocity typically between 10 and 30 m s^{-1}. A regular disturbance on the jet imposed, for example, by the vibration of a piezoelectric structure causes the jet to break up into a stream of drops as shown in Figure 2.2a. If not selected for printing, drops are collected in a catcher or gutter and the liquid is returned to the supply system for reconditioning and re-use. To select drops, an electrode is placed close to the point where drops are forming from the continuous jet. A potential on this electrode

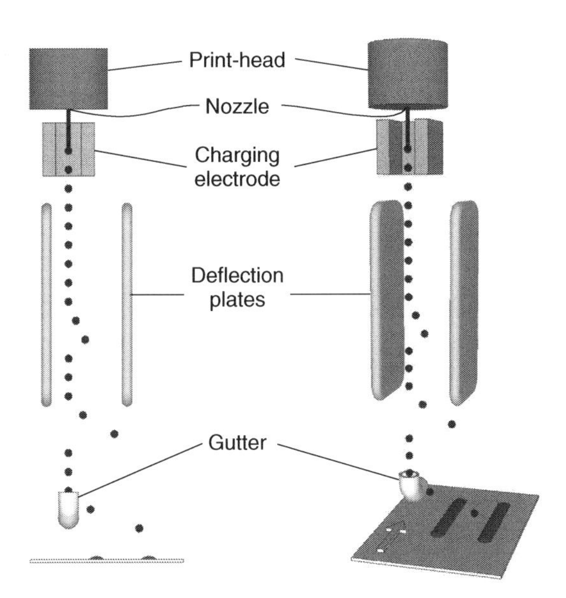

Figure 2.5 *Principles of operation of a single-jet continuous inkjet system.*

will induce an opposite electrical charge on the surface of the forming drop: for this to happen, the liquid must be sufficiently conductive. When the drop breaks away, the charge is trapped on that particular drop and the potential of the electrode can be changed to put a different charge on the next drop (or set to zero for no charge). In this way, each successive drop can be charged as required. The selection is completed by causing the drop stream to move through a fixed electric field, maintained between two more electrodes, before reaching the catcher. Uncharged drops will not deviate, while charged drops will be deflected by the field in proportion to the charge they carry.

If the substrate surface is moving perpendicularly to both the jet and the electric field, then by selecting the time and the level of drop charging, successive drops can be deflected to build up an image on the surface. For text printing, this image would typically be one or a few characters high (in the direction of the deflecting field) and many characters long (in the direction of substrate movement) to form the desired text.

Common applications for these types of inkjet systems include marking and date coding of foods and other products, and addressing and personalising direct mail. Printing is possible on substrates moving at high speeds (several metres per second), capable of keeping up with most production lines, but the print quality can best be described as 'utility'. Because the jet is produced continuously (so that the nozzles do not dry out) and the ink system can be designed to make adjustments to the ink (e.g. by replacing solvent lost by evaporation), this method is able to print volatile liquids.

Despite a high drop generation rate, the narrow print swath and the complexity of the method mean that the scope for application of this technology to digital fabrication is limited.

2.4.1.2 Multiple-Jet CIJ

CIJ nozzles can be arranged in an array so that many jets or drop streams can be produced from a single print-head. Each jet is associated with its own charging electrode (or other means of drop selection), but the jets can share a common deflection electrode and drop catcher. Normally each drop is either selected to print or not print and the substrate surface moves in a direction that is perpendicular to the jets but, in this case, parallel to the deflection field as shown in Figure 2.6. The print swath is determined by the width of the array.

An alternative means to impose a disturbance on the jet, used by Kodak in their recent multi-jet continuous jet systems (e.g. Hawkins and Pond, 2007), is to arrange a resistive heating element close to or at the nozzle. This heater is modulated to change the jet temperature and hence the surface tension of the liquid. This in turn causes a flow which initiates a pinch point in the jet to form drops. Kodak have explored alternative selection techniques, for example by using air flows to separate differently sized drops (which can be created from one stream using the thermal pinch-off method) or by switching the jet direction by using asymmetric heating with more complex heater arrangements at the nozzle.

2.4.2 Drop-on-Demand (DOD)

Except for those used for experimental or research purposes, nearly all DOD print-heads contain multiple nozzles. Early commercial examples typically contained a few tens of

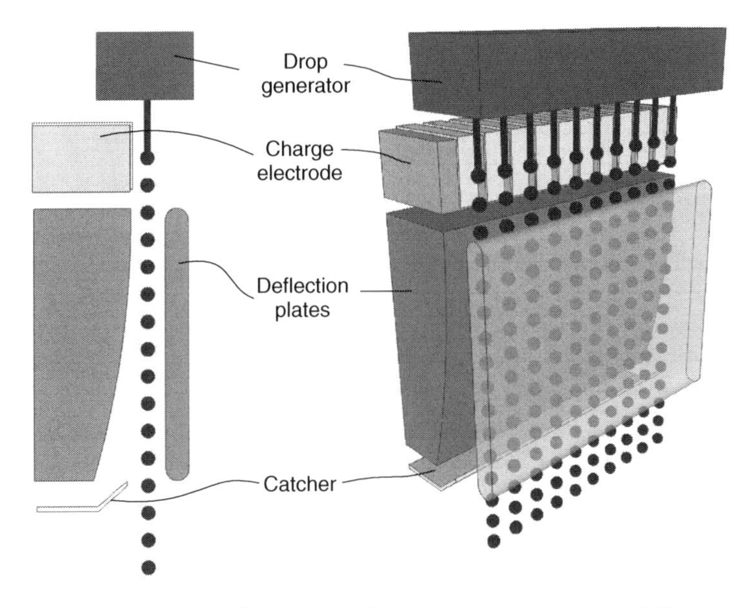

Figure 2.6 *Principles of operation of a multiple-jet continuous inkjet system.*

nozzles, while modern heads contain hundreds or thousands of nozzles and are often used in combination. Each nozzle will contribute to one or a few lines (depending on print configuration) of printed spots making up the complete image.

Unlike continuous jet systems, there are no drop selection or deflection systems and so the nozzles can be placed close to the substrate surface.

As each nozzle fires only as required, the nozzles are inactive for much of the time. This means that changes in the ink at the open nozzle (e.g. by evaporation) can, after a time, affect the performance of that nozzle when it is next fired. In the worst case, the nozzle can fail to fire. Clearly this is to be avoided and various strategies are employed to mitigate the effect, ranging from ink formulation constraints to spitting and purging into waste collectors during operation and capping when shut down.

DOD depends on having an actuation mechanism which will transfer enough energy to a suitable volume of liquid to enable it to form a drop (creating new surface area) and travel to the substrate at a reasonable velocity. Various actuation mechanisms have been proposed and patented, but the two most common techniques are thermal (or bubble) actuation and piezoelectric actuation (PA).

2.4.2.1 Thermal DOD

Thermal inkjet was invented and developed concurrently by Canon and Hewlett Packard in the late 1970s and early 1980s (Vaught *et al*., 1984; Endo *et al*., 1988). This method is still used in a large proportion of home and office inkjet printers. Within a cavity behind the nozzle is a small resistive heater in good thermal contact with the ink. Rapid heating (within a few micro-seconds) causes superheating and vaporisation of a thin layer of ink adjacent to the heater. A vapour bubble quickly expands, producing the fluid displacement and energy necessary to force a drop from the nozzle. Once the drop is ejected and the heat source is switched off the bubble rapidly collapses, drawing in fresh ink, and the cycle is ready to start again once any residual pressure fluctuations and meniscus movements have died away. One particular advantage of this technique is that the actuator is simply a resistive track which can be fabricated in a number of ways consistent with multi-layer fabrication processes such as photolithography. This is particularly suitable for large-quantity manufacture, reducing or eliminating the gluing or assembly stages associated with structures involving PA (discussed in Section 2.4.2.2). To mitigate damage to the resistive elements during repeated operation, the elements are covered by passivation and/or protection layers. Clearly the ink itself needs to be able to support and withstand the heating and vaporisation process, and this will limit the types of solvent which can be used. One problem to be avoided is the deposition of ink components onto the heater or elsewhere within the nozzle ('kogation'), so again attention is required for ink formulation. As only a very thin layer of ink next to the heater is heated and vaporised, the bulk of the ink is heated by only a few degrees Celsius so that it is perfectly possible to print heat-sensitive material such as bio-molecules and living cells. Nevertheless these constraints on ink formulation have made thermal DOD technology a less popular choice than the piezoelectric method discussed in Section 2.4.2.2 for printing the wide range of materials required for fabrication applications.

A wide variety of different thermal DOD designs have been described. Some of the differences concern the position of the heater relative to the nozzle. In *face shooter* designs, the heater is sited directly behind the nozzle as illustrated in Figure 2.7a. The *side shooter* configuration has the heater on a wall beside the nozzle (Figure 2.7b). Other proposed designs include a heater on the nozzle face (Figure 2.7c) or suspended within the ink chamber (Figure 2.7d).

2.4.2.2 *Piezoelectric DOD*

Drop ejection using PA was one of the first DOD techniques to be described (Zoltan, 1972; Kyser and Sears, 1976). PA-based inkjet print-heads are used in home and office

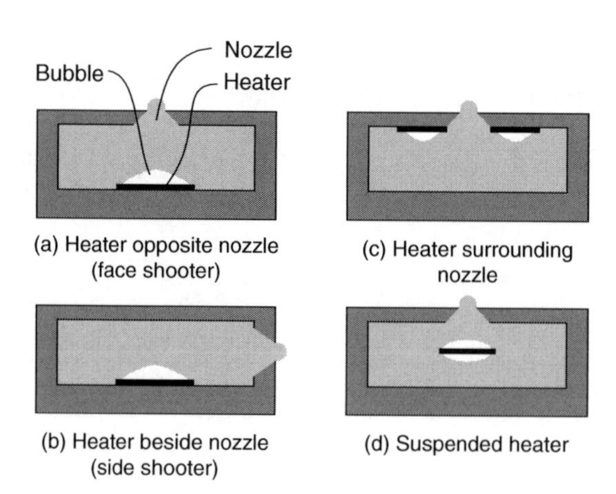

Figure 2.7 *(a–d) Possible configurations of a drop-on-demand thermal inkjet print-head.*

inkjet systems, and PA has predominated in those print-heads developed for industrial application for either printing or digital fabrication.

Lead zirconate titanate (PZT) is a ceramic material that exhibits a strong piezoelectric effect and is used as the active component in many PA systems. The piezoelectric effect is the ability of certain materials to generate an electric field in response to mechanical strain or, conversely, change shape in the presence of an electric field. In PZT, the piezoelectric effect is established during manufacture by exposure to heat and a strong electric field which aligns previously randomly orientated groups of electric dipoles ('poling'). Hence the PZT piezoelectric effect is anisotropic, and the material will change shape in a way which depends on the relative orientation of the applied field to the poling direction. If a field is applied in the same (or opposite) direction as the poling, then the material will expand (or contract) along that direction and contract (or expand) in orthogonal directions. If the field is perpendicular to the poling direction, then the ceramic will shear (Figure 2.8).

Several different actuation mechanisms have been developed. Push-mode actuators rely directly on the expansion and contraction of a PZT rod pushing a membrane, which changes the volume of an ink-filled reservoir behind a nozzle. A unimorph structure in which a thin layer of PZT is bonded to a thin passive sheet will flex when a voltage is applied across the PZT sheet. A similar arrangement uses two active layers (bimorph). A cylinder of PZT with a field directed from the outside to the inside will contract radially (squeeze mode). Shear-mode structures have been used to actuate inkjets. Figure 2.9 shows an arrangement where the fields are arranged so as to cause two regions of

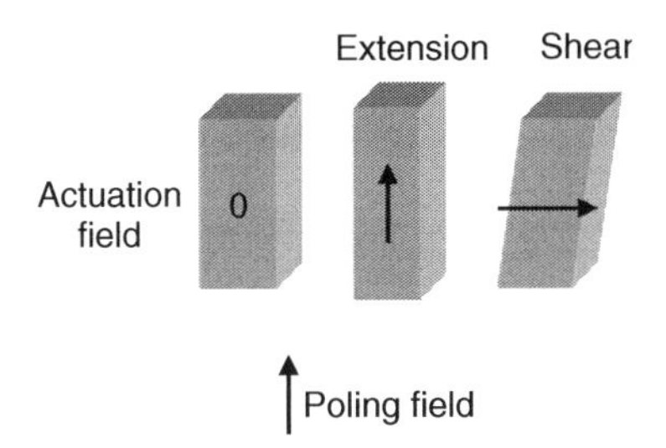

Figure 2.8 *The effect of electric field relative to poling direction in piezoelectric material.*

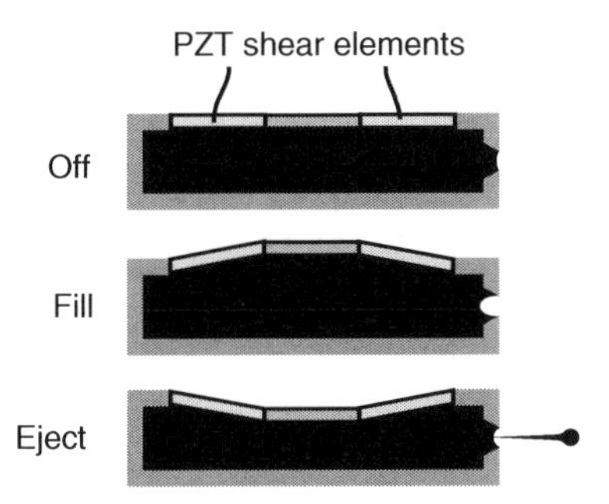

Figure 2.9 *Phases of wall movement in the actuation of a drop-on-demand print-head by shear-mode piezoelectric actuation.*

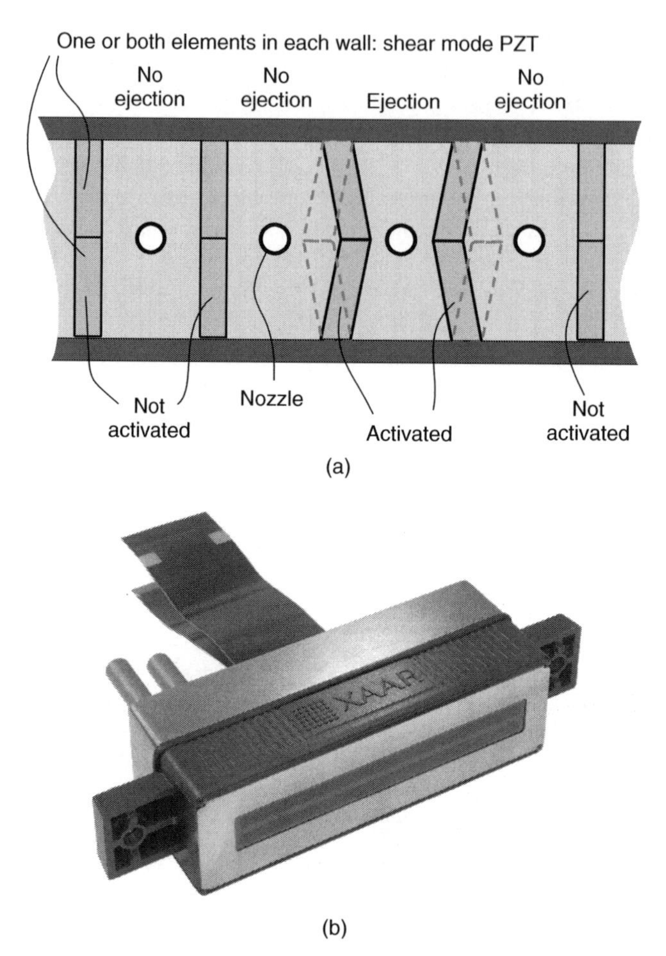

Figure 2.10 *(a) Principles of shared-wall shear-mode actuation in a multiple-nozzle drop-on-demand (DOD) print-head and (b) an example of a modern industrial DOD print--head (Reproduced with kind permission from http://www.xaar.com/printheads-gallery.aspx Copyright (2012) Xaar PLC).*

shear, creating a wall movement above an ink channel. Figure 2.10a illustrates another arrangement in which the ink channels behind the nozzle have walls made with one or two shear-mode PZT elements. Electrodes are placed so that the field direction is perpendicular to the PZT poling direction, and so a shearing action occurs when a voltage is applied. Each ink channel shares these walls with its neighbours. To fire a nozzle, both walls must act in unison and hence while one channel is firing, adjacent channels cannot. A modern industrial DOD print-head based on this technology is illustrated in Figure 2.10b.

Unlike thermal actuation, PA allows low-pressure (i.e. below atmospheric pressure) pulses to be achieved if desired, by arranging the movement of the actuator to increase the volume of the nozzle chamber. Some print-head manufacturers use this to improve performance. For example, an initial low-pressure pulse when reflected and inverted from the end of the chamber can be combined with a later positive pressure pulse to increase the maximum drive pressure (Wijshoff, 2006). The dynamics of piezoelectric DOD print-heads are discussed in more detail in Chapter 3. Drive waveforms can be modified to accommodate printing fluids with different characteristics (e.g. speed of sound, viscosity and surface tension).

2.4.2.3 *Other DOD Methods*

A variety of mechanisms have been proposed and/or demonstrated for DOD actuation other than thermal or PA but have been used so far only in experimental, trial or specialist applications. For example, Figure 2.11 illustrates drop ejection using the electrostatic force between two plates to move an actuator in order to fill an ink cavity and eject ink drops (Kamisuki *et al*., 1998). Silverbrook (2007) has described the use of a thermal bimorph structure in a patent application. Here, a silicon structure includes a moveable nozzle which can be used to provide a pressure transient in the ink behind the nozzle by using electrically heated bimorph strips connected between the nozzle and the print-head structure to move the nozzle relative to the structure. Focussed acoustic energy has also been proposed as a mechanism to eject drops from a free surface (Lovelady and Toye, 1981), which has the advantage that a nozzle is not required.

2.4.3 Electrospray

Drops and jets of liquid can be drawn from liquid surfaces in the presence of electric fields. G.I. Taylor was the first to analyse some of these effects properly (Taylor, 1964), and the raised cone of liquid which is often present is known as a Taylor cone. This technique (also sometimes called electrohydrodynamic inkjet) can produce drops and jets significantly smaller than the nozzle from which they emerge. Producing drops of a well-controlled and reproducible size by this method is challenging. Different modes of operation are seen, depending on the geometry, materials and electric field (Cloupeau and Prunet-Foch, 1994). The liquid can be ejected in a succession of relatively large drops

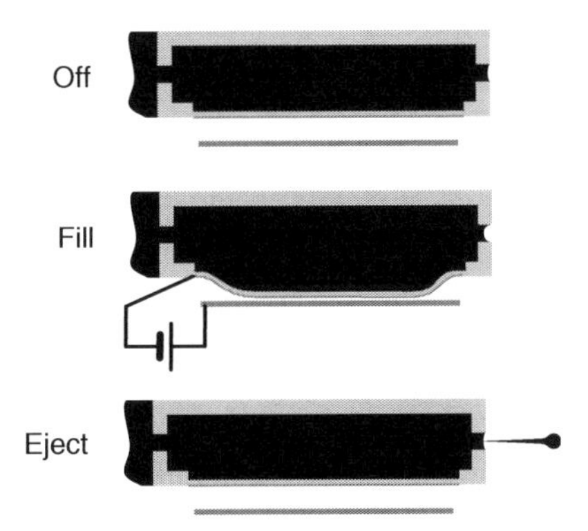

Figure 2.11 *Phases of wall movement in the electrostatic actuation of a drop-on-demand print-head.*

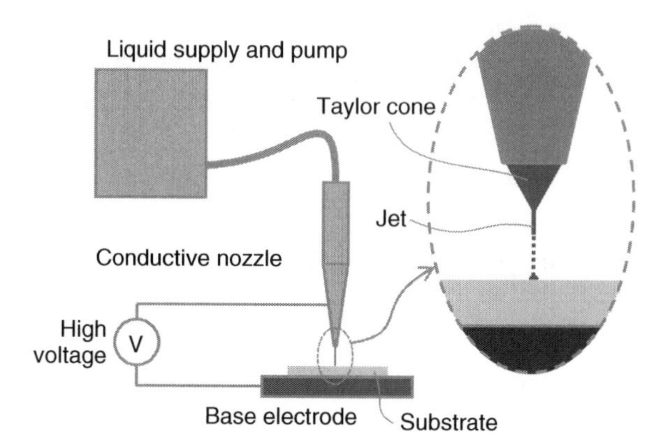

Figure 2.12 *Schematic diagram of electro-spray deposition.*

or, at higher field strengths, as a continuous jet which can break up into a continuous stream of drops which, as they are electrically charged, will repel each other and diverge into a cone. Figure 2.12 shows a typical arrangement.

A variant of this technique has been developed by Tonejet (Newcombe, 2008). The print-head structure has no nozzles but instead an array of points or needles which sit just below the surface of a recirculating ink pool and form one electrode. The ink contains charged pigment particles dispersed in a non-polar solvent. An electric field between the points and another electrode outside the ink causes the charged particles to be ejected. In this method, the concentration of particles is much higher in the ejected drops than in the bulk ink.

2.5 Resolution and Print Quality

Print quality in the context of a graphical inkjet image is often a subjective attribute which is determined by a variety of factors. It is influenced by not only the resolution (i.e. the inverse of spot spacing) but also many other things such as spot size (or sizes), spot shape, ink density, ink colour, substrate colour, substrate reflectivity, substrate texture, drop position accuracy, image processing and type of image. When printing images there is a limit in, say, spot size or print accuracy beyond which the unaided eye can perceive no difference. It can be argued that some office and home inkjet systems have reached this limit. In digital fabrication, for example in printing a functional pattern, then these limits may no longer apply and smaller drops, more accurately placed, may be required.

Given that inkjet-printed spots may be round or have an irregular edge and that there will inevitably be some error in spot placement, it is normal to choose a print resolution where there will be some overlap between adjacent spots to ensure complete coverage and to obscure spot placement errors. This is true for images and is also necessary for printing, for example, a conductive track which needs to provide a continuous conduction path.

It is worth considering the limitations on minimum feature size given a particular resolution, spot size and drop placement accuracy. Clearly it is possible to print lines which have a width similar to the diameter of a single printed spot. The morphology of the line will be determined by the drop spacing s, the print accuracy, the texture and surface energy of the surface, the surface tension of the ink and the timing of printing and drying. The line edge may be straight, have regular variations or sometimes have other shapes (Duineveld, 2003). This type of line can be achieved only parallel or perpendicular to the printing direction (and even then there may be differences). Compare these with a line at 45° to the print direction. In that line the drop spacing will be $\sqrt{2}s$, and therefore the overlap will be less and the line structure will be different. The pattern will also be more prone to discontinuities arising from drop placement. A straight or curved edge at some arbitrary angle can be considered to be made up from segments perpendicular or parallel to the print direction, or at 45° to these. Printing two lines with a gap between them means missing (at least) one row of drops. Given the required spot size as discussed in this chapter, the gap width will be smaller than the track width and here drop placement errors could cause overlap (leading e.g. to a short circuit between conductive tracks). Given the quantised nature of the drop placement there will be constraints on what can be reproduced, and problems can arise when the original 'artwork' was described in a vector-based file or at a different resolution. Multiple-drop grey-scale techniques (discussed in Section 2.6) can help to improve edge quality by printing small (or, in some cases, larger) drops to fill in edges or curves to make them less 'jagged' (sometimes referred to as anti-aliasing). This is a different effect from that obtained by using a grey level at the edge of a character on a computer monitor, which exploits an optical illusion.

2.6 Grey-Scale Printing

Inkjet printers producing drops of one size only are referred to as binary printers. Early commercial DOD printers were all of this type. Each nozzle could print drops of one fixed volume only, although this could typically be varied by up to 20% by varying the printing conditions, for example by using a different ink or changing the drive waveform. However, this was not done dynamically, and it would not have been much of an advantage to do so.

Print quality can be improved by using a larger number of smaller drops at a higher resolution. In the context of image printing, a grey or colour level can be reproduced by using the correct density of drops per unit area. Smaller drops at high resolution present more opportunity to fool the eye. With a higher resolution, edges can be formed more precisely. Higher resolutions require a higher nozzle density (and smaller nozzles) which is achieved by packing more nozzles into each print-head, by overlapping print-heads or by overprinting print swaths. Moving to higher resolution provides the most flexibility in placing drops but it does have disadvantages, including the need for a higher nozzle density and a possible reduction in print speed.

An alternative technique, sometimes called grey-scale printing, improves print quality and flexibility without changing the print resolution. In this case each nozzle is used to print a range of different drop sizes. A variety of techniques have been proposed. Changes in the shape of the driving pulse can produce drops with different volumes, but this can also produce unwanted drop velocity changes and is very dependent on the nozzle geometry and ink properties. It has been shown (Burr *et al*., 1996) that drops of

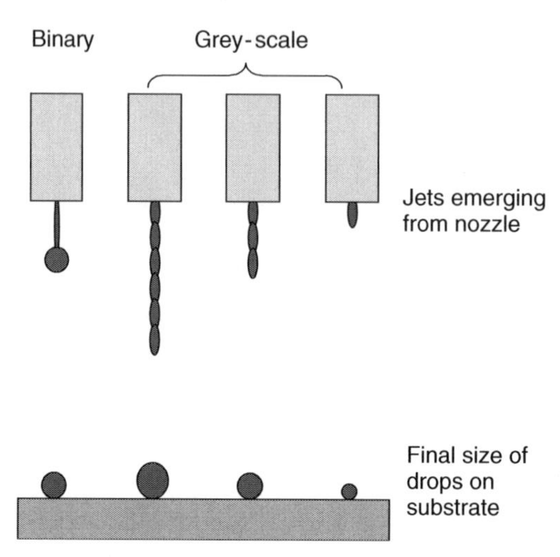

Figure 2.13 Schematic diagram showing the difference between binary and grey-scale printing.

different sizes can be produced by exciting different meniscus modes. This technique is somewhat limited in the number of unique drop sizes which can be obtained, but it can produce drops smaller than those commonly produced from a given nozzle size. A technique used in several recent print-heads was first patented by Lee *et al*. (1985). Here several drops are produced in quick succession which merge together to form one larger drop. Commonly, drops following the initial drop would emerge from the nozzle before the ligament of the first drop had broken off, and so on for successive drops. The series of pulses used to produce these drops can be tuned so that the following drops would quickly catch up with the initial drop to form a single drop before reaching the substrate, as shown in Figure 2.13. Hence each nozzle can produce several sizes of drops, with the size being selected dynamically to suit the image being printed. Normally one limitation on the maximum drop rate, and hence maximum print speed, is the need for the nozzle and cavity to refill and reach equilibrium before the generation of the next drop. If this is not done, then the next drop may well be different in size and velocity in an unpredictable way. However, when a combined drop is produced, a fixed complex waveform can be used (i.e. independent of the print speed) for each combination of drops, and this can be optimised to create these drops at a higher rate than is normally possible.

This technique provides a useful compromise between low- and high-resolution binary printing. It maintains the nozzle densities, potential ink coverage and print speeds of the low-resolution binary arrangement, but combines them with an ability to considerably improve print quality.

2.7 Reliability

Reliable and consistent operation is the key to any printing process, and it can be especially important for digital fabrication. The level of reliability required, and what constitutes a failure, will depend on the application. For example, large-format graphical poster printing may well be able to accommodate jet misdirection and a level of satellite drops (see Section 2.8) which would be totally unacceptable for printing active or passive electronic components. While failure will often be manifested in misdirected or missing drops, there can be many root causes. Clearly particles in the ink can be disruptive. If a particle is large enough, then it can completely or partially block a nozzle. Normally, of course, the ink will be appropriately filtered (both during its manufacture and close to the print-head) to remove such particles. Smaller particles, not large enough to block a nozzle, can still cause disruption (e.g. with one or two drops misdirected). This transient disruption may lead to lasting failure if air bubbles are introduced into the nozzle (discussed later in this section) or if stray ink causes build-up on the nozzle plate or other printer components.

When the print-head is not being used, it is possible that the composition of the ink in the nozzle and/or the state of the nozzle surface will change over time. The nature of the DOD process means that, even during operation, some nozzles may not fire for long periods. In the worst case, solvent evaporation or some chemical change can leave higher viscosity or even solid material in the nozzle, resulting in an inability to restart the nozzle without some operator intervention. It is common to have some means to mitigate this effect through a nozzle wiping or cleaning procedure or by regularly jetting ink into a waste container or absorbent pad while the system is idle. Even without a gross change to the ink composition, the meniscus position or the recent history of the nozzle area wetted during drop ejection can influence the performance. This can result in the first few drops that are printed after a pause being significantly different in velocity, volume and even direction from the drops produced during more regular printing. These effects can sometimes be seen after fairly short pauses of just a few minutes or even seconds. Again, these effects can be mitigated by spitting when idle, or by 'wobbling' the meniscus (by using lower oscillating voltages than the jetting voltages) to sweep the nozzle surfaces which are normally wetted during firing.

Each print-head design will have its own characteristic response, and the drop velocity and volume will depend on the firing frequency. With most print-heads, drop velocity and volume will be reasonably constant up to some operating frequency and then they will start to vary. Although this is not strictly a failure, it may determine the maximum frequency of operation and can be ink-dependent. The demands of the application will establish the range of acceptable drop velocities and volumes.

Another failure mode sometimes seen is termed nozzle-plate flooding. This normally occurs at high print rates (and hence drop frequencies) and is, as the name implies, a failure caused by excess ink on the nozzle plate blocking printing from some or even all the nozzles. Once started, this is normally a catastrophic failure as the flooding spreads and more nozzles feed the flooded area. Recovery usually requires at least a pause in printing and possibly a nozzle cleaning procedure. When the print-head is not printing, or is firing at low frequencies, any excess ink on the nozzle plate would be drawn back into the nozzle by the negative pressure maintained within the nozzle (in a normal non-firing nozzle, the negative pressure is balanced by the surface tension forces of the ink meniscus). However, at some frequency more ink is ejected than can be drawn back by the negative pressure and at this point there is potential for failure, perhaps started by a transitory disruption by a particle or bubble causing local flooding.

DOD inkjets are susceptible to air bubbles within the cavity behind the nozzle. These absorb energy and change the acoustic characteristics, causing a reduction in drop velocity or complete nozzle failure. Worse, very small air bubbles causing little problem can grow over time through a process of rectified diffusion (Fyrillas and Szeri, 1994) to be large enough to disrupt jetting. Air bubbles can be drawn in with the ink supply or introduced through the nozzle during the temporary disruption caused by particles passing through the nozzle or by ink build-up on the nozzle surface (De Jong *et al*., 2006). Manufacturers attempt to reduce these effects, for example by degassing the ink before use or by employing a degassing mechanism in the fluid supply system. Recent alternatives (e.g. the Xaar 1001 print-head shown in Figure 2.10b) use a continuous flow of ink across the inside of the nozzle face to remove bubbles (and other contamination) which would otherwise cause problems.

2.8 Satellite Drops

Satellites are additional drops, smaller than the intended drop, often produced during the inkjet printing process (Hoath *et al*., 2007). These small drops result from the collapse of the liquid column by surface tension, as discussed in Section 2.2, and typically originate when the conditions for the ligament or continuous jet to collapse into single large drops are not fully met.

In continuous jet printers, satellite drops can form from the column which necks down before the drop breaks off from the jet. Different jetting conditions (including the amplitude of the driving disturbance) can result in no satellites, satellites which can remain between the neighbouring main drops as they move or those which can merge with either the preceding or following main drop. The formation of CIJ satellites tends to be regular and repeatable and they can be observed, like the main drops, by strobe illumination. Problems will arise if the satellites have not merged with the main drops by the time they reach the deflecting field. Charged satellites will be deflected onto the substrate or parts of the inkjet equipment, sooner or later causing problems.

Another species of satellites produced by continuous jets are sometimes known as micro-satellites. These are small (sub-micrometre) satellite drops probably produced at the ligament tip just at the moment of main drop break-off. With no forcing disturbance, these will have a distribution of sizes. They will also become charged when the main drop is being charged and, like the larger satellites, will be directed by the deflection field to parts of the equipment, typically a high voltage electrode. Over time, these deposits can build up and cause equipment failure.

A drop emerging from the nozzle of a DOD printer will extrude from the nozzle as a jet which necks down and breaks off from the ink remaining in the nozzle. Ideally, the detached jet will move towards the substrate with surface tension forces pulling all the liquid into a single spherical drop; surface tension will also cause any ink still attached to the nozzle to be drawn back into it. Often, after the initial break-off, the jet consists of a head containing most of the ink, and a tail or ligament which extends behind the head, as seen in Figure 2.2b. Once detached, this ligament will start to merge into the head of the jet, drawn by surface tension forces but opposed by inertia and viscosity. It is from this trailing ligament that satellites can form by the Rayleigh–Plateau instability discussed in Section 2.2. The fluid which originally formed the ligament may, or may not, merge with the head before it reaches the substrate. Unlike CIJ satellites, the sizes and velocities of DOD satellites are not regular and any one nozzle will typically create a distribution of satellite sizes and velocities while producing regular main drops.

Satellites cause a number of problems. If they do not land on top of the main drop, they will contribute to a degradation of print quality, often most clearly seen in the fuzzing or blurring of the trailing edge of a printed area. This can be a very serious problem when printing functional materials. For example, inkjet-printed conductive tracks could be shorted by satellite deposits. Satellite drops can also be caught up in air flows and so find their way onto the equipment, leading to contamination and even eventual failure.

Satellites can be reduced or even eliminated in some circumstances. The detail of the nozzle design and the transducer drive waveform can change the satellite behaviour. Simply reducing the amplitude of the initial impulse to reduce the drop velocity will tend to limit ligament and satellite production. All else being equal, the ligament volume is normally related to the size of the drop, so those grey-scale systems which generate a swift succession of small drops can benefit to some extent as the final ligament is associated with the small drop size and not the final larger drop. There is a tendency to attain higher resolutions by using smaller drops which of course have smaller satellites. Spots resulting from individual 1 pl sized drops (the current smallest drops from commercially available print-heads) are hard to see with the naked eye, and their satellites impossible. That is not to say that small satellites cause no problems. Smaller satellites are more likely to be entrained in air flows. Non-Newtonian components in the ink can change the drop formation process and change the size and number of the satellites produced.

2.9 Print-Head and Substrate Motion

Although much of this section also applies to CIJ, this discussion assumes that we are dealing with a DOD system.

In the simplest case, an image is printed by choosing to place, or not to place, spots of ink on the points of a grid determined by the resolution of the image (the term spot is used here to distinguish the deposit on the substrate from the drop in free flight). Resolution is the inverse of the closest spacing at which individual spots are placed (i.e. spots per unit length) and may be different between the printing direction and the perpendicular direction. Resolution is conventionally described in terms of dots per inch (dpi), and typical values with corresponding dot spacings in micrometres are listed in Table 2.2. Sometimes the grid is not rectangular or aligned with the print direction, so that the resolution is different in the printing direction and perpendicular to it, but we will ignore those possibilities here. Some print-heads, as described in Section 2.6, can print one of several sizes of drops on any grid position so that the resulting image can contain spots of different sizes. This grey-scale capability can be used to improve the quality of the image or structure being printed over that obtained with a binary (single-drop-size) print-head, printing at the same resolution.

The image resolution is determined by a number of factors. When designing an application, one would ideally start from the print quality required and then choose or design a printing system with the drop size, nozzle spacing and head and substrate movement

Table 2.2 Typical resolutions (dots per inch) and equivalent spot spacings.

Resolution (dpi)	Spacing (μm)
50	508.0
100	254.0
200	127.0
300	84.7
400	63.5
600	42.3
800	31.8
1000	25.4
1200	21.2
1600	15.9

necessary to achieve the resolution required. Most applications require that full coverage will be achieved when printing all the spots. As spots are approximately circular this means that for full coverage there will be significant overlap of adjacent spots. Usually there are several trade-offs to consider between printer specifications such as drop size, resolution, print speed, drop placement accuracy and print-head reliability. These will be determined by the nature of the printing, the ink, the substrate and the structures to be created.

In the manufacturing context, there are a number of issues which need to be considered which can determine the success or otherwise of the printing system. The following is a partial list:

- the suitability of the ink for the function intended
- the short- and long-term compatibility of the ink with the materials of the printer
- the 'printability' of the ink in the intended print-head: that is, whether it will form satisfactorily printable drops at the rate required
- the reliability of the ink in the system: different inks vary greatly in their ability to print consistently
- the adhesion between ink and substrate
- the consistency of the substrate
- any substrate pre-treatment required to improve adhesion or consistency
- post-treatment required to provide drying and/or curing
- the local environment: for example chemical contaminants, dust, temperature and humidity variations.

Although print-heads with a substantial two-dimensional array of nozzles have been proposed (e.g. covering the area of an A4 or letter-sized sheet), all current print-heads contain one or a few rows of nozzles, as illustrated in Figure 2.10b. One or two rows of nozzles are common as these structures can be easier to make; however, some manufacturing techniques enable more than two rows to be incorporated in a single head, which can be an advantage.

A particular resolution along the direction of printing is achieved by printing spots at appropriate intervals depending on the relative speed of the head and substrate. Print-head manufacturing constraints may mean that the inter-nozzle gap is larger than the intended resolution perpendicular to the direction of printing. In that case, the required resolution can be achieved in a number of ways:

- **by tilting the print-head to the direction of motion**: in combination with appropriate timing, this will enable spots to be printed closer together in a line perpendicular to the print direction;
- **by overprinting using the same print-head**: that is, print one or more subsequent passes over the same area to 'fill in the gaps';
- **by using more than one print-head for the same area**: that is, position one or more extra print-heads which interleave their spots to print at the required resolution.

Clearly the spot size will need to be appropriate for the intended resolution and application. The exact drop size will depend on many factors, but by far the most influential is the size of the nozzle. Hence a particular print-head design will produce approximately the same size of drop in most circumstances. The grey-scale technique described in Section 2.6 starts with the smallest drop, again dictated by the nozzle size, and then builds drops which have volumes which are larger or smaller multiples of the minimum drop volume. However, once the drop reaches the substrate the size of the resulting spot can vary greatly, influenced by the physical and chemical interaction between the ink and the substrate. One of the keys to producing appropriate functional devices is to be able to control these interactions to create the desired structure, for example by controlling the physical and chemical properties of the ink and substrate, pre-treating the substrate to influence morphology or surface energy and post-treating the printed result to dry, fix or cure the print.

In printing systems, print-heads are often grouped together and rigidly fixed within a framework, which may also be movable. It may be feasible to remove and replace individual print-heads (e.g. should one fail), and it would be common to have a means to mechanically (and/or electronically) fine-tune the alignment.

With one or a small number of print-heads, printing an extended area requires that the print-head (or group of heads) is moved relative to the substrate.

A common configuration involves moving the head in one direction and the substrate (step by step) in a perpendicular direction, and hence printing the substrate in print-head-wide swaths. This arrangement lends itself to overprinting strategies which can be used to increase resolution and mask print defects. This scheme is seen in most home and office, personal-use inkjet printers and is dictated by at least one of three factors:

- the print-head group is not wide enough to cover the substrate in one pass;
- the nozzle spacing is wider than the intended resolution and
- the print-head may have faulty or non-functioning nozzles.

Configurations in which the print-head or group of heads span the width of printing clearly offer several advantages over scanning head arrangements:

- there is no need for a print-head moving mechanism;
- the substrate can move continuously past the print-heads at a fixed velocity;
- print speed is higher because there are more nozzles;

- print speed is also higher because there is no time lost while the substrate or print-heads are indexed to the next print position and
- there are fewer issues related to alignment and drying variations at the edges of adjacent print swaths.

Substrate-wide arrays require:

- a means to stitch together perhaps many individual print-heads in a way which provides good alignment but avoids printing artefacts at head–head boundaries;
- nozzles and print-heads which are reliable enough not to fail when used in large numbers: it may be possible to provide some redundancy by using additional nozzles and/or having a printing scheme in which jets adjacent to faulty jets can be used and
- sufficient data processing power to drive the print-heads and provide the data to them at the rate needed.

2.10 Inkjet Complexity

Here and in other chapters in this book, we see the diversity of techniques and applications associated with modern inkjet printing. To achieve reliable operation which meets the operational requirement of any given application requires a combination of engineering and scientific skills unmatched in many other areas of technology. These include, as a minimum:

- the electronic hardware and software required to transport and manipulate large quantities of data at extremely high rates;
- the chemistry to develop inks which are compatible with the materials of the printer, which reproducibly form drops which behave correctly during impact and spreading on the substrate and which also meet the functional requirements for the product when printed;
- the manufacturing processes for electromechanical systems at the micrometre or even sub-micrometre scale needed to fabricate the very precise actuators, nozzles and other components of the print-head;
- an understanding of the fluid mechanics associated with the transport of liquids, the formation of drops and their flight and spreading on the substrate;
- an appreciation of the rheological factors that can affect drop formation and
- an understanding of the surface chemistry and physics which determine how ink drops interact with the substrate.

If these skills can be brought to bear, then the potential benefits of inkjet technology can be substantial, as illustrated in the remaining chapters of this book.

References

Burr, R.F., Tence, D.A., Le, H.P. *et al.* (1996) Method and apparatus for producing dot size modulation ink jet printing. US Patent 5,495,270, Tektronix Inc.

Cloupeau, M. and Prunet-Foch, B. (1994) Electrohydrodynamic spraying functioning modes: a critical review. *Journal of Aerosol Science*, **25**, 1021–1036.

De Jong, J., De Bruin, G., Reinten, H. *et al*. (2006) Air entrapment in piezo-driven inkjet printheads. *Journal of the Acoustical Society of America*, **120**, 1257–1265.

Derby, B. (2010) Inkjet printing of functional and structural materials: fluid property requirements, feature stability, and resolution. *Annual Review of Materials Research*, **40**, 395–414.

Duineveld, P.C. (2003) The stability of inkjet printed lines of liquid with zero receding contact angle on a homogeneous substrate. *Journal of Fluid Mechanics*, **477**, 175–200.

Endo, I., Sato, Y., Saito, S. *et al*. (1988) Bubble jet recording method and apparatus in which a heating element. US Patent 4,723,129, Canon Kabushiki Kaisha.

Fyrillas, M.M. and Szeri, A.J. (1994) Dissolution or growth of soluble spherical oscillating bubbles. *Journal of Fluid Mechanics*, **277**, 381–407.

Hawkins, G.A. and Pond, S.F. (2007) High speed, high quality liquid pattern deposition apparatus. US Patent 7,249,829, Eastman Kodak Company.

Hoath, S.D., Martin, G.D., Castrejon-Pita, J.R. and Hutchings, I.M. (2007) Satellite formation in drop-on-demand printing of polymer solutions. NIP23: 23rd International Conference on Digital Printing Technologies, Digital Fabrication 2007 DF2007, Society for Imaging Science and Technology, pp. 331–335.

Jang, D., Kim, D. and Moon, J. (2009) Influence of fluid physical properties on ink-jet printability. *Langmuir*, **25**, 2629–2635.

Kamisuki, S., Hagata, T., Tezuka, C. *et al*. (1998) A low power, small, electrostatically-driven commercial inkjet head. Proceedings of the Eleventh Annual International Workshop on Micro Electro Mechanical Systems, 1998, MEMS 98.

Kyser, E.L. and Sears, S.B. (1976) Method and apparatus for recording with writing fluids and drop projection therefor. US Patent 3,946,398, Siliconics Inc.

Le, H.P. (1998) Progress and trends in inkjet printing technology. *Journal of Imaging Science and Technology*, **42**, 49–62.

Lee, F.C., Mills, R.N., Payne, R.N. and Talke, F.E. (1985) Spot size modulation using multiple pulse resonance drop ejection. US Patent 4,513,299, IBM.

Lovelady, K.T. and Toye, L.F. (1981) Liquid drop emitter. US Patent 4,308,547, Recognition Equipment Incorporated.

McKinley, G.R. and Renardy, M. (2011) Wolfgang von Ohnesorge. *Physics of Fluids*, **23**, 127101.

Newcombe, G. (2008) Tonejet: delivering digital printing to the mass market. International Conference on Digital Printing Technologies.

Reis, N. and Derby, B. (2000) Ink jet deposition of ceramic suspensions: modelling and experiments of droplet formation. *Materials Research Society Symposium Proceedings*, **625**, 117–122.

Silverbrook, K. (2007) Nozzle arrangement with movable ink ejection structure. US Patent 2007/0139473, Silverbrook Research Pty Ltd.

Sweet, R.G. (1965) High frequency recording with electrostatically deflected ink jets. *Review of Scientific Instruments*, **36**, 131–136.

Taylor, G.I. (1964) Disintegration of water drops in an electric field. *Proceedings of the Royal Society of London: Series A: Mathematical and Physical Sciences*, **280**, 383–397.

Vaught, J.L., Cloutier, F.L., Donald, D.K. and Meyer, J.D. (1984) Thermal ink jet printer. US Patent 4,490,728, Hewlett-Packard Company.

Wijshoff, H. (2006) Manipulating drop formation in piezo acoustic inkjet. NIP22: 22nd International Conference on Digital Printing Technologies, Society for Imaging Science and Technology, pp. 79–82 .

Zoltan, S.I. (1972) Pulsed droplet ejecting system. US Patent 3,683,212, Clevite Corporation.

3

Dynamics of Piezoelectric Print-Heads

J. Frits Dijksman[1,2] *and Anke Pierik*[1]
[1]*Philips Research Europe, High Tech Campus 11, The Netherlands*
[2]*University of Twente, Faculty of Science and Technology, Physics of Fluids, The Netherlands*

3.1 Introduction

The process of depositing, on demand, droplets with a well-defined volume at a precise given location on a substrate can be split up into several unit operations as shown in Figure 3.1 (Lee, 2003; Dijksman *et al.*, 2007; Dijksman and Pierik, 2008; Wijshoff, 2008).

Here we will consider the ink to be a solvent with a small percentage of solid material, either dissolved or dispersed. Certain inks are solid at room temperature and have to be heated above their melting temperature. Droplets are considered as free-flying volumes of fluid, which after landing form dots. After drying, solidifying and/or permeation the solid content of the ink forms a spot, this being the final result of the process.

Droplets are emitted from a nozzle and fly towards the substrate, possibly losing some mass by evaporation (Bird, Stewart and Lightfoot, 2002b) and losing some speed due to air friction (Bird, Stewart and Lightfoot, 2002a). Upon landing on the substrate, the droplet changes shape and starts spreading (Yarin, 2006). The amount of spreading is dependent on the speed and volume of the droplets and the properties of both the fluid and the substrate (He *et al.*, 2003). On a non-absorbing substrate, the kinetic energy of the droplet is transferred into excess surface energy and a tiny amount of heat during a decaying oscillatory motion of the fluid in the dot (viscous dissipation). Finally, the

Inkjet Technology for Digital Fabrication, First Edition. Edited by Ian M. Hutchings and Graham D. Martin.

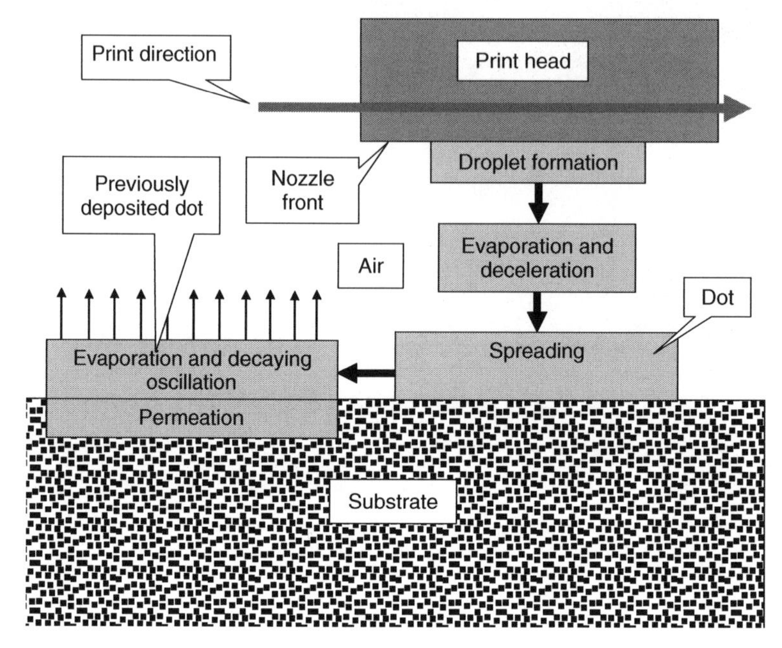

Figure 3.1 Schematic diagram of the inkjet printing process. Note that permeation takes place only in an ink absorbing substrate.

solvent evaporates and the spot is formed. In the case of an absorbing medium, spreading, permeation and evaporation occur in parallel (Dijksman and Pierik, 2008). Finally, the solvent has evaporated and the solid content is partly in and partly on the substrate. For a wax-like material, solidification has to be taken into account as well. A fuller discussion of the processes involved in droplet impact and spreading is presented in Chapter 5.

The nozzles in an inkjet print-head are arranged either along a straight line (linear array print-head) or in a two-dimensional array (matrix array print-head). Since print-heads have limited dimensions, to cover a complete substrate the head is often mounted on a carriage that moves over the substrate. Usually the print-head is moved along one axis whilst another mechanism moves the substrate along an axis perpendicular to the axis of the print-head. As the pitch between the nozzles of a linear array print-head in most cases does not provide the resolution needed, the print-head is placed at an angle with respect to the print direction, as shown in Figure 3.2a. In matrix array print-heads, the print resolution is achieved by displacing the different rows of nozzles with respect to each other as depicted in Figure 3.2b.

Piezoelectrically driven print-heads consist of a multitude of individually addressable, miniature valveless pumps. Each pump is a flow-through arrangement of small-sized ducts without any valves. Such an arrangement consists of a connection to an ink supply, a pump chamber, a connecting duct to the nozzle and finally the nozzle itself. The wall of the pump chamber is partly covered with a small piezoelectric plate, the actuator. When the voltage applied to the piezoelectric actuator changes in a pulse-wise manner,

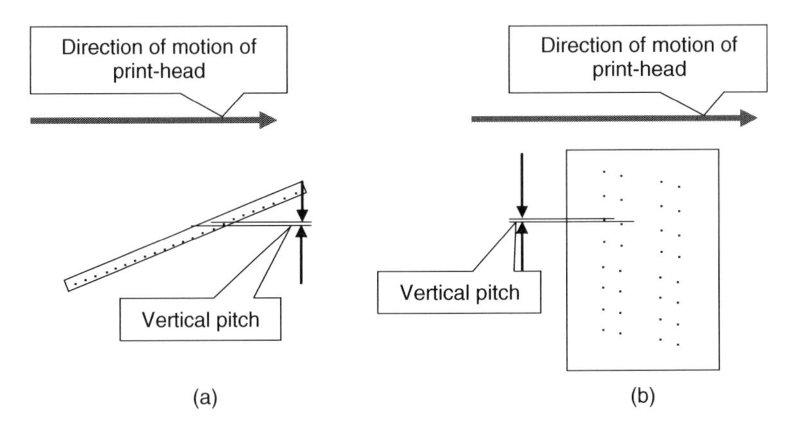

Figure 3.2 *(a) Linear array print-head. (b) Matrix array print-head.*

acoustic waves are induced and one or more droplets, each with a precisely defined volume and speed, are emitted from the nozzle. A piezo-driven inkjet pump is basically an acoustic cavity set in motion by the actuator, the characteristics of which can be presented in both the time domain and the frequency domain. Droplet formation takes place at the exit of the nozzle and is a phenomenon with its own characteristic timescales (Wijshoff, 2008).

Droplet impact is a highly dynamic phenomenon, as discussed in Chapter 5. If the impact speed is too high, it can lead to splashing and consequently to poor printing quality (Yarin, 2006). The impact speed may have to be limited.

In this chapter, we aim to couple the characteristics of droplet formation to the acoustics of the fluidics of the print-head behind the nozzle all the way up into the ink supply. In Section 3.2, we start with an overview of basic designs of multi-nozzle print-heads. This leads to the definition of the basic unit (acoustic cavity) of a multi-nozzle print-head that will be used in this chapter's subsequent sections. To explain the different acoustic features of the basic unit (acoustic cavity), in Section 3.3 we model such a unit as an oscillator with a single degree of freedom. With the results of this modelling we explain the concepts of resonance frequency, damping, pulse shape (positive and negative) and response in the time domain. Using the results of this single-degree-of-freedom oscillator, droplet formation dynamics, maximum jetting frequency and refilling are then analysed in Section 3.4. Finally, we discuss two extensions of the single-degree-of-freedom model: a multi-cavity model to understand the behaviour of different pumps acting in parallel (Section 3.5) and a long duct model that describes the waves travelling back and forth through the pump duct upon activation (Section 3.6).

3.2 Basic Designs of Piezo-Driven Print-Heads

In this section, we discuss two basic designs of multi-nozzle print-heads, referred to as the Helmholtz resonator design and the open-end design (Stemme and Larsson, 1973;

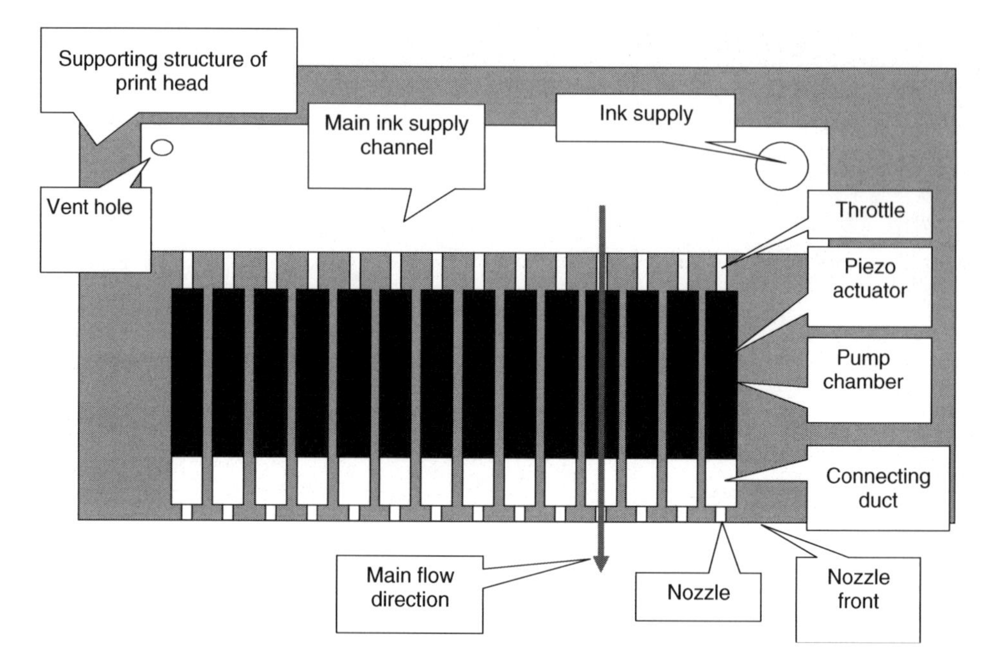

Figure 3.3 *Piezo-driven print-head according to the Helmholtz resonator design.*

Beasley, 1977; Kurz, 1980; Döring, 1982; Rosenstock, 1982; Bogy and Talke, 1984; Dijksman, 1984, 1999, 2003; Lee, Mills and Talke, 1984; Bentin *et al*., 1986; Shield, Bogy and Talke, 1987; Kitahara, 1995; Burr, Tence and Berger, 1996; McDonald, 1996; Usui, 1996; Badie and de Lange, 1997; de Jong *et al*., 2005, 2006; Groot Wassink *et al*., 2005; de Jong, 2007; Dijksman *et al*., 2007; Groot Wassink, 2007; Wijshoff, 2008).

The Helmholtz design is shown in Figure 3.3.

The pump chamber and the connecting duct form a cavity that at one end is connected via a throttle (a relatively long duct with a small cross-section) to the ink supply and at the other end connected to ambient via a nozzle (a duct with a short length and a small-sized exit opening). Ink is fed into the print-head through the ink supply opening. To facilitate filling of the print-head without allowing air bubbles, there is a vent hole at the other end of the supply channel. The arrangement shown is also referred to as the front-shooter design. The main flow direction is a straight line coinciding with or running parallel to the centrelines of the different ducts: throttle, pump chamber, connecting duct and nozzle. Other arrangements are possible as well. In Figure 3.4, a so-called side shooter is shown. Here the centreline of the nozzle is perpendicular to the main directions of the other channels.

Figure 3.5 shows the open-end arrangement. Characteristic of this design is the open connection to the ink supply. It is in principle a $\lambda/4$ resonator with an open connection to the ink supply and a small but open restriction (the nozzle) to ambient.

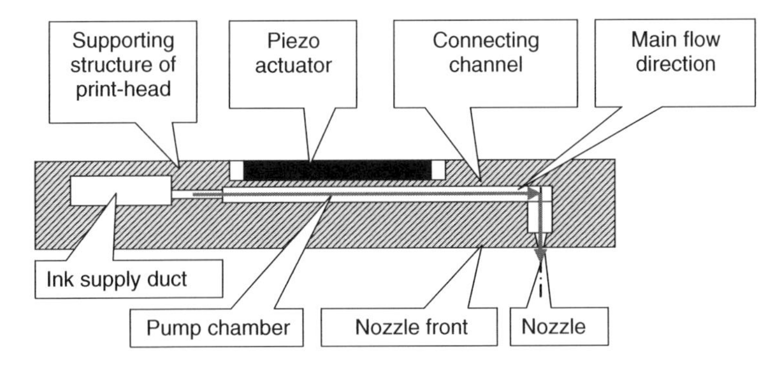

Figure 3.4 *Side-shooter Helmholtz resonator design.*

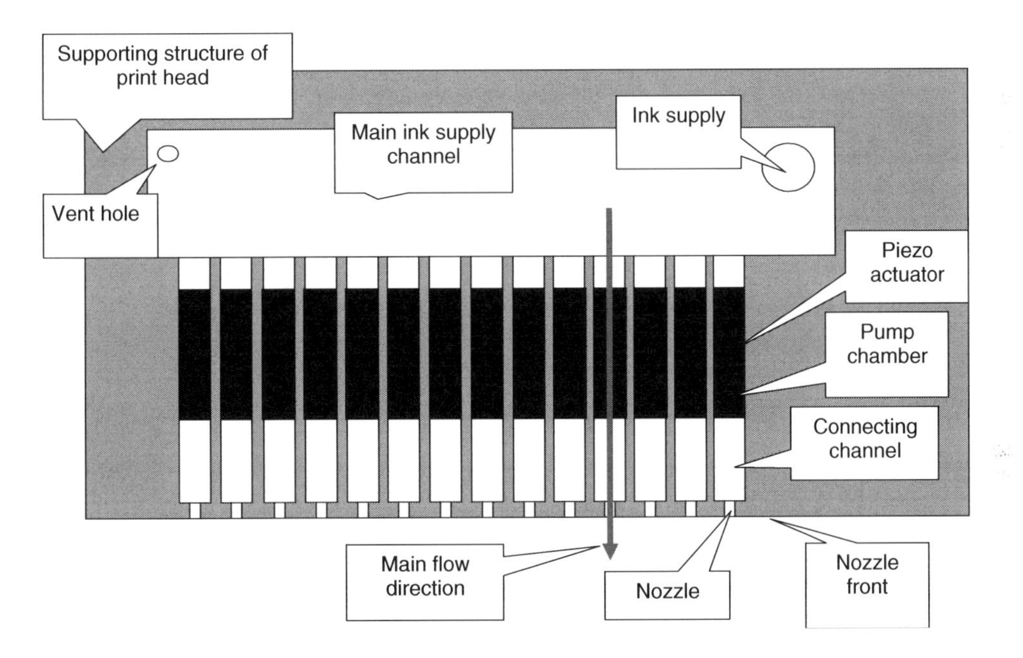

Figure 3.5 *Print-head according to the open-end design.*

3.3 Basic Dynamics of a Piezo-Driven Inkjet Print-Head (Single-Degree-of-Freedom Analysis)

To explain the acoustical properties of a piezo-driven inkjet printer head, we start by considering the most basic arrangement, namely, a small reservoir with volume V_c connected to the external environment by a small hole, the nozzle, with cross-section A_1 and length L_1 (see Figure 3.6). The other side of the pump chamber is connected to

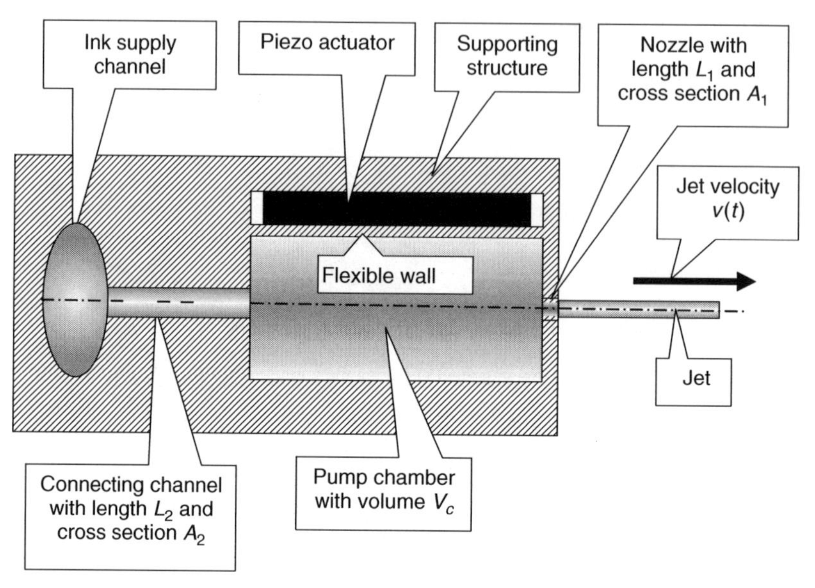

Figure 3.6 Basic arrangement of an inkjet printer pump. The volume of the pump chamber can be changed stepwise by charging the piezo actuator. A jet starts to issue from the nozzle. The mass of the fluid in the connecting channel is large compared to the mass contained in the nozzle. It is therefore assumed that there is hardly any motion in the connecting channel. The jet leaves the nozzle with a velocity: $\dot{x}_1 = v(t)$.

an ink supply channel via a connecting channel which acts as a throttle. Although the cross-sectional dimensions of throttle A_2 and nozzle A_1 are usually chosen to be roughly equal, the length of the throttle L_2 is much larger than the length of the nozzle L_1. Consequently, the mass of the fluid in the connecting channel is much larger than the mass of the fluid contained in the nozzle. For our analysis of the acoustics of the basic arrangement, we will neglect the motion of the fluid in the connecting channel.

The viscosity of the fluid is η, with density ρ_0 and speed of sound c. Upon a sudden change in volume ΔV_0, at $t = 0$ the pressure change with respect to ambient in the pump chamber is given by (with K_B the bulk modulus):

$$p = K_B \frac{\Delta V_0}{V_c}; K_B = \rho_0 c^2$$

The volume change ΔV_0 is considered positive when it generates a positive pressure change. In the following, pressure is used as a shorthand for pressure change with respect to ambient. A flow through the nozzle and a flow in the opposite direction through the throttle initiate, decreasing the volume change according to:

$$\Delta V = \Delta V_0 - \int_0^t Q_1(t)dt + \int_0^t Q_2(t)dt = \Delta V_0 - A_1 \int_0^t \dot{x}_1(t)dt + A_2 \int_0^t \dot{x}_2(t)dt$$
$$= \Delta V_0 - A_1 x_1 + A_2 x_2$$

where $x_1(t)$ denotes the mean volume displacement in the nozzle and $x_2(t)$ the mean volume displacement in the throttle, both being defined positive in the main flow direction. The equilibrium equations (force balances) in nozzle and throttle are:

$$\rho_0 A_1 L_1 \ddot{x}_1 + 8\pi \eta L_1 \dot{x}_1 = K_B \frac{\Delta V_0 - A_1 x_1 + A_2 x_2}{V_c} A_1$$

$$\rho_0 A_2 L_2 \ddot{x}_2 + 8\pi \eta L_2 \dot{x}_2 = K_B \frac{\Delta V_0 - A_1 x_1 + A_2 x_2}{V_c} A_2$$

After dividing the first equation by L_1 and the second equation by L_2, we see that so long as $L_1 \ll L_2$ and $A_1 \approx A_2$, the displacements in the nozzle will be much larger than the displacements in the throttle. Therefore, we neglect the motions in the throttle and we continue with the approximated equation of motion for the fluid contained in the nozzle:

$$\rho_0 A_1 L_1 \ddot{x}_1 + 8\pi \eta L_1 \dot{x}_1 \cong K_B \frac{\Delta V_0 - A_1 x_1}{V_c} A_1$$

After rewriting we get:

$$\rho_0 A_1 L_1 \ddot{x}_1 + 8\pi \eta L_1 \dot{x}_1 + K_B \frac{A_1^2}{V_c} x_1 = K_B \frac{\Delta V_0}{V_c} A_1$$

With $M = \rho_0 A_1 L_1, K = 8\pi \eta L_1, C = K_B A_1^2 / V_c$ and $F_0 = K_B \Delta V_0 A_1 / V_c$, we arrive at the standard second-order non-homogeneous linear differential equation describing the forced motion of a damped oscillator (mass–spring–damper set-up):

$$M \ddot{x}_1 + K \dot{x}_1 + C x_1 = F_0$$

The solution of the homogeneous part of this equation can be found by substitution of:

$$x_1 = Be^{\lambda}$$

The value of λ follows from:

$$\lambda_{1,2} = \left(-\zeta \pm \sqrt{\zeta^2 - 1}\right) \omega_n$$

with:

$$\zeta = \frac{K}{2\sqrt{MC}}$$

$$\omega_n = \sqrt{\frac{C}{M}}$$

where ζ is the damping ratio and ω_n the natural frequency of the system. We have to distinguish between several cases depending on the value of the damping ratio:

$\zeta > 1$: over–damped or aperiodic

$\zeta = 1$: critically damped

$\zeta < 1$: oscillatory motion

In the over-damped or aperiodic case (high viscosity) the motion decays very rapidly, while for the oscillatory case (low viscosity) it takes several cycles before the system comes to rest again after actuation. We first discuss the oscillatory case and write the homogeneous solution in the form:

$$x_1 = e^{-\zeta\omega_n t}(B_1 \sin\omega_n t\sqrt{1-\zeta^2} + B_2 \cos\omega_n t\sqrt{1-\zeta^2})$$

Note that due to damping, the actual resonance frequency has been reduced with respect to the natural frequency of the system by a factor: $\sqrt{1-\zeta^2}$.

The particular solution is:

$$x_1 = \frac{F_0}{C} = \frac{\Delta V_0}{A_1}$$

With the initial conditions of $x_1(t=0)=0$; $\dot{x}_1(t=0)=0$, we find:

$$B_1 = -\frac{\zeta}{\sqrt{1-\zeta^2}}\frac{\Delta V_0}{A_1}, \qquad B_2 = -\frac{\Delta V_0}{A_1}$$

The total solution in terms of displacement and velocity is given by:

$$x_1 = \frac{\Delta V_0}{A_1}\left[1 - e^{-\zeta\omega_n t}\left(\frac{\zeta}{\sqrt{1-\zeta^2}}\sin\omega_n t\sqrt{1-\zeta^2} + \cos\omega_n t\sqrt{1-\zeta^2}\right)\right]$$

$$\dot{x}_1 = \frac{\Delta V_0}{A_1}\omega_n e^{-\zeta\omega_n t}\frac{1}{\sqrt{1-\zeta^2}}\sin\omega_n t\sqrt{1-\zeta^2}$$

A pulse is made out of a sudden stepwise volume decrease as discussed in this section, followed by a volume increase of the same magnitude at the pulse time t_p later. As the governing equation of motion is linear, the solution after $t=t_p$ simply follows by adding the solutions belonging to the two steps (for $t > t_p$):

$$x_1 = \frac{\Delta V_0}{A_1}\left[1 - e^{-\zeta\omega_n t}\left(\frac{\zeta}{\sqrt{1-\zeta^2}}\sin\omega_n t\sqrt{1-\zeta^2}\,t + \cos\omega_n t\sqrt{1-\zeta^2}\right)\right]$$
$$-\frac{\Delta V_0}{A_1}\left[1 - e^{-\zeta\omega_n(t-t_p)}\left\{\frac{\zeta}{\sqrt{1-\zeta^2}}\sin\omega_n(t-t_p)\sqrt{1-\zeta^2} + \cos\omega_n(t-t_p)\sqrt{1-\zeta^2}\right\}\right]$$

$$\dot{x}_1 = \frac{\Delta V_0}{A_1}\omega_n e^{-\zeta\omega_n t}\frac{1}{\sqrt{1-\zeta^2}}\sin\omega_n t\sqrt{1-\zeta^2}$$
$$-\frac{\Delta V_0}{A_1}\omega_n e^{-\zeta\omega_n(t-t_p)}\frac{1}{\sqrt{1-\zeta^2}}\sin\omega_n(t-t_p)\sqrt{1-\zeta^2}$$

Likewise we find for the case of $\zeta > 0$ (the over-damped or aperiodic case):

$$x_1 = \frac{\Delta V_0}{A_1}\left(1 + \frac{\lambda_2}{\lambda_1 - \lambda_2}e^{\lambda_1 t} - \frac{\lambda_1}{\lambda_1 - \lambda_2}e^{\lambda_2 t}\right)$$

$$\dot{x}_1 = \frac{\Delta V_0}{A_1}\frac{\lambda_1\lambda_2}{\lambda_1 - \lambda_2}(e^{\lambda_1 t} - e^{\lambda_2 t})$$

In the same manner as discussed in this section, for a pulse the solution after $t = t_p$ follows by addition of the solutions belonging to the two steps (for $t > t_p$):

$$\begin{aligned} x_1 &= \frac{\Delta V_0}{A_1}\left(1 + \frac{\lambda_2}{\lambda_1 - \lambda_2}e^{\lambda_1 t} - \frac{\lambda_1}{\lambda_1 - \lambda_2}e^{\lambda_2 t}\right) \\ &\quad - \frac{\Delta V_0}{A_1}\left(1 + \frac{\lambda_2}{\lambda_1 - \lambda_2}e^{\lambda_1 (t-t_p)} - \frac{\lambda_1}{\lambda_1 - \lambda_2}e^{\lambda_2 (t-t_p)}\right) \\ \dot{x}_1 &= \frac{\Delta V_0}{A_1}\frac{\lambda_1\lambda_2}{\lambda_1 - \lambda_2}\left(e^{\lambda_1 t} - e^{\lambda_2 t}\right) - \frac{\Delta V_0}{A_1}\frac{\lambda_1\lambda_2}{\lambda_1 - \lambda_2}\left(e^{\lambda_1 (t-t_p)} - e^{\lambda_2 (t-t_p)}\right) \end{aligned}$$

We shall consider two cases as examples of typical print-heads: a pump (basic unit) representative of a large-sized print-head, and one of a small-sized one.

The large pump has a pump chamber with volume $V_c = 4 \times 10^{-9}\,\mathrm{m}^3$ (e.g. with the length of the pump chamber $L = 0.02\,\mathrm{m}$ and cross-sectional area $A = 2 \times 10^{-7}\,\mathrm{m}^2$). The nozzle has a length $L_1 = 150\,\mu\mathrm{m}$ and a diameter such that $R_1 = 25\,\mu\mathrm{m}$ ($A_1 = 1.96 \times 10^{-9}\,\mathrm{m}^2$).

The data for the small-sized print-head are $V_c = 2 \times 10^{-10}\,\mathrm{m}^3$ (e.g. length 0.01 m and cross-section $2 \times 10^{-8}\,\mathrm{m}^2$), $L_1 = 75\,\mu\mathrm{m}$ and $R_1 = 15\,\mu\mathrm{m}$ ($A_1 = 7.07 \times 10^{-10}\,\mathrm{m}^2$).

The fluid used has a density of $1000\,\mathrm{kg\,m^{-3}}$ and a speed of sound of $1400\,\mathrm{m\,s^{-1}}$. The bulk modulus of the fluid follows from $K_B = \rho_0 c^2 = 1.96 \times 10^9\,\mathrm{Pa}$.

In order to compare different cases, we look for settings for which the maximum underpressure is about equal to and does not exceed 1 bar, in order to avoid cavitation in the ink.

Before we proceed, we must distinguish between two cases: the response of the system to a positive pulse and the response of the system to a negative pulse.

When a positive pulse is applied, the fluid in the nozzle immediately starts to flow out. In case of a negative pulse, the fluid is first sucked in and then starts to flow out.

For the positive pulse the length of the pulse, the pulse time, is chosen so that after the velocity has passed through a maximum the voltage is switched off in order to avoid large negative velocities and too large a recoil of the meniscus inside the nozzle after the release of a droplet. Excessive recoil causes air entrapment into the pump chamber (de Jong *et al*., 2005; de Jong, 2007): air bubbles in the pump chamber immediately destroy the action of the print-head. So for the case of a positive pulse we use, for the oscillatory case:

$$\zeta < 1 : t_p = \frac{1}{\omega_n\sqrt{1-\zeta^2}}\arctan\frac{\sqrt{1-\zeta^2}}{\zeta}$$

For small damping the pulse time is given by:

$$\zeta \ll 1 : t_p \approx \frac{\pi}{2\omega_n}$$

For the over-damped or aperiodic case we find, using the same argument (positive pulse):

$$\zeta > 1 : t_p = \frac{\ln(\lambda_1/\lambda_2)}{\lambda_2 - \lambda_1}$$

For a negative pulse, we use the concept of constructive interference. The leading edge of the pulse sets the fluid in motion, and the pulse is switched off at the moment the velocity goes through zero. In that way the effects of the leading edge and trailing edge of the pulse are added optimally. This makes sense only where the damping is rather small, because constructive interference would be prevented by excessive damping. So, for the negative pulse, the pulse time is given by:

$$\zeta < 0.5 : t_p = \frac{\pi}{\omega_n\sqrt{1-\zeta^2}}$$

The results for the small-sized print-head, for an ink with a viscosity of 2 mPa s, are shown in Figures 3.7 and 3.8. Figure 3.7 gives the mean velocity in the nozzle and the length of the jet issuing from the nozzle, both as functions of time. The pressure is a function of the volume change induced by the actuator. The value of the volume change ΔV_0 or relative volume change $\Delta V_0/V_0$ is tuned such that the maximum underpressure does not exceed 1 bar.

The pressure as a function of time in the pump chamber for the case depicted in Figure 3.7 is given by Figure 3.8. The pressure in the pump chamber follows from:

$$p = K_B \frac{\Delta V_0 - A_1 x_1}{V_c}$$

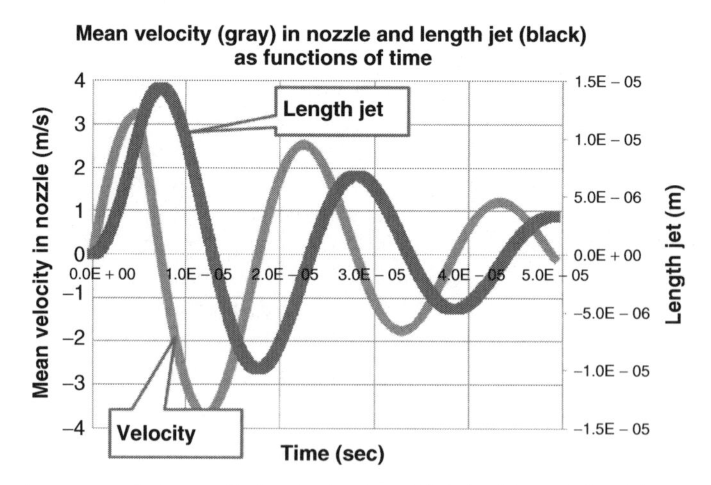

Figure 3.7 *Response of the small-sized print-head filled with an ink with viscosity 2 mPa s driven by a pulse of 4.82 µs and a positive relative volume change of 4.5×10^{-5}. Velocity and length of jet are shown as functions of time. The system is underdamped ($\zeta = 0.117$ and $\omega_n = 303912\ \text{rad s}^{-1} = 48.369$ kHz).*

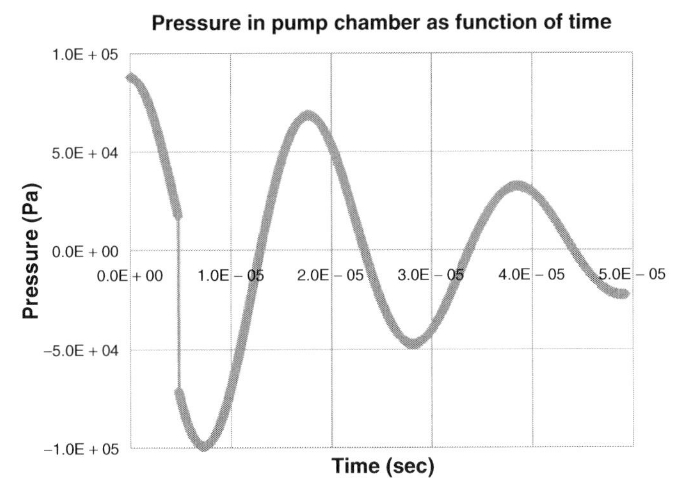

Figure 3.8 Pressure in pump chamber as a function of time caused by a positive relative volume pulse of 4.5×10^{-5} and a duration of 4.82 μs. The underpressure does not exceed 1 bar to avoid cavitation in the liquid.

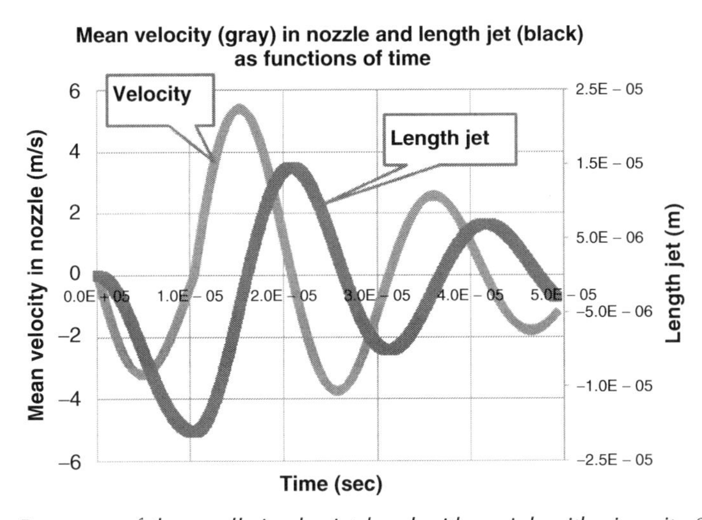

Figure 3.9 Response of the small-sized print-head with an ink with viscosity 2 mPa s with a pulse of 10.4 μs and a negative relative volume change of -4.36×10^{-5}. Velocity and length of jet are shown as functions of time. The system is underdamped ($\zeta = 0.117$ and $\omega_n = 303912\ \text{rad s}^{-1} = 48.369\ \text{kHz}$).

Figures 3.9 and 3.10 show the results for the small-sized print-head with an ink with a viscosity of 2 mPa s, now actuated by a negative pulse.

The results shown in Figures 3.11 and 3.12 relate to the small-sized print-head with an ink with a viscosity of 20 mPa s actuated by a positive pulse.

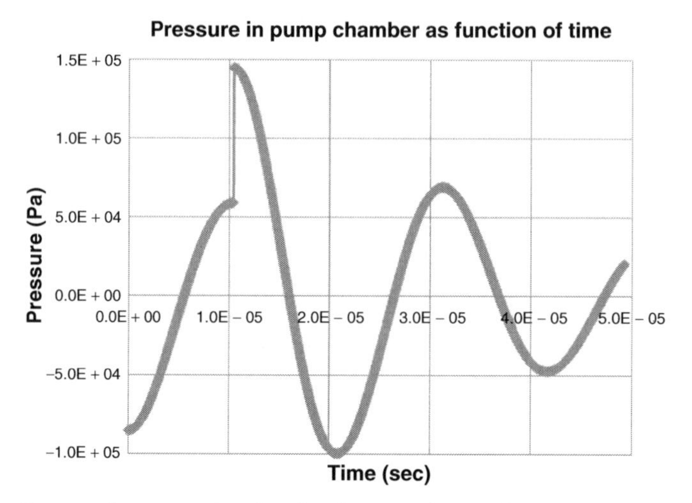

Figure 3.10 Pressure in pump chamber for the case depicted in Figure 3.9 as a function of time caused by a negative relative volume pulse of -4.36×10^{-5} and a duration of 10.4 μs. The underpressure does not exceed 1 bar.

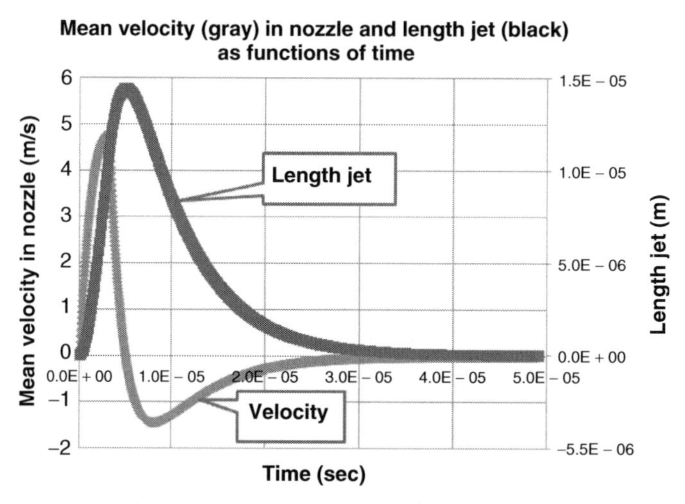

Figure 3.11 Response of the small-sized print-head with an ink with viscosity 20 mPa s with a pulse of 3.12 μs and a positive relative volume change of 1.68×10^{-4}. Velocity and length of jet are shown as functions of time. The system is overdamped or aperiodic ($\zeta = 1.17$ and $\omega_n = 303912\ \text{rad s}^{-1} = 48.369\ \text{kHz}$).

For the high-viscosity case the value of the relative volume displacement has been increased almost fourfold to achieve comparable velocity and displacement to those for the low-viscosity case. Note that the underpressure still does not exceed 1 bar.

A typical result for the pump representative of a large print-head is shown in Figure 3.13.

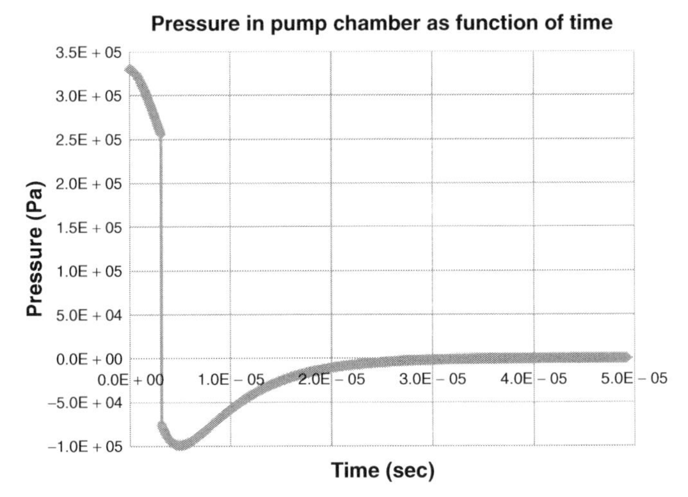

Figure 3.12 Pressure in pump chamber for the case of Figure 3.11 as a function of time caused by a positive relative volume pulse of 1.68×10^{-4} and a duration of 3.12 μs. The pulse height is chosen such that the underpressure does not exceed 1 bar.

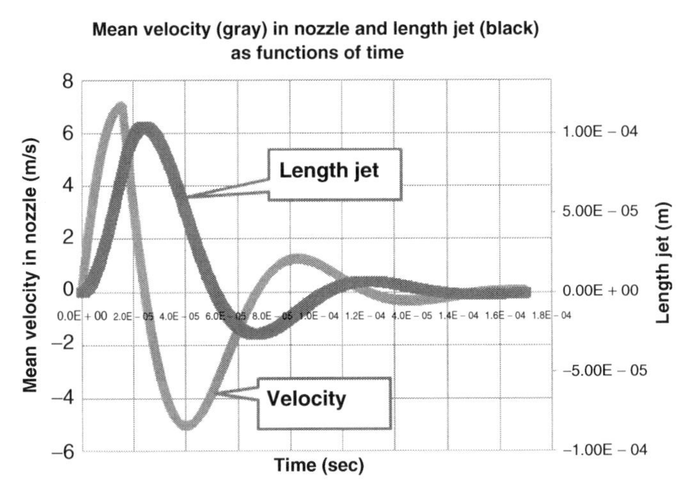

Figure 3.13 Response of the large-sized print-head with an ink with viscosity 5 mPa s with a pulse of 15.8 μs and a positive relative volume change of 7.14×10^{-5}. Velocity and length of jet are shown as functions of time. The system is under-damped or oscillatory ($\zeta = 0.4$ and $\omega_n = 80088\ \text{rad}\,\text{s}^{-1} = 12.746\ \text{kHz}$). The pulse height is chosen such that the underpressure does not exceed 1 bar.

Up to now we have considered the case in which the response of the system immediately follows the electrical pulse. The response cannot in fact occur immediately. The solid structure around the pump chamber has its dynamic characteristics, and the electronic circuitry needed to drive the piezo actuator has its own RC time (where R is the resistance of the connecting wires and C the capacitance of the piezo element).

To account for the fact that it takes some time to apply the pulse and to shut it off, we assume that the volume is changed according to the following ramp function:

$$\Delta V = \Delta V_0 \frac{t}{t_{p1}} - \int_0^t Q(t)dt = \Delta V_0 \frac{t}{t_{p1}} - A_1 \int_0^t \dot{x}_1(t)dt$$

Following the same procedure as above, the equation describing the dynamics of the system is given by:

$$M\ddot{x}_1 + K\dot{x}_1 + Cx_1 = F_0 \frac{t}{t_{p1}}$$

The particular solution is:

$$x_1 = \frac{F_0}{C}\frac{1}{t_{p1}}(-\frac{K}{C} + t)$$

For the oscillatory case ($\zeta < 1$), the total solution is made out of the sum of the homogeneous solution and particular solution just derived:

$$x_1 = \frac{F_0}{C}\frac{1}{t_{p1}}\left(-\frac{K}{C} + t\right) + e^{-\zeta\omega_n t}(B_1 \sin \omega_n t\sqrt{1-\zeta^2} + B_2 \cos \omega_n t\sqrt{1-\zeta^2})$$

The constants B_1 and B_2 are found by evaluation of the initial conditions: $x_1(t=0)=0$; $\dot{x}_1(t=0)=0$:

$$B_1 = -\frac{F_0}{C}\frac{1}{t_{p1}}\frac{1-2\zeta^2}{\sqrt{1-\zeta^2}}, B_2 = -\frac{F_0}{C}\frac{1}{t_{p1}}\frac{K}{C}$$

The response of a ramp function characterised by a volume displacement ΔV_0, applied to the system in t_{p1} seconds, is given by:

$$x_1(t, t_{p1}) = \frac{\Delta V_0}{A_1}\frac{1}{\omega_n t_{p1}}\left[-2\zeta + \omega_n t + e^{-\omega_n t}\left\{-\frac{1-2\zeta^2}{\sqrt{1-\zeta^2}} \sin \omega_n t\sqrt{1-\zeta^2} + 2\zeta \cos \omega_n t\sqrt{1-\zeta^2}\right\}\right]$$

$$\dot{x}_1(t, t_{p1}) = \frac{\Delta V_0}{A_1}\frac{1}{t_{p1}}\left[1 - e^{-\zeta\omega_n t}\left\{\frac{\zeta}{\sqrt{1-\zeta^2}} \sin \omega_n t\sqrt{1-\zeta^2} + \cos \omega_n t\sqrt{1-\zeta^2}\right\}\right]$$

For the over-damped or aperiodic case, we find:

$$x_1(t, t_{p1}) = \frac{\Delta V_0}{A_1}\frac{1}{\omega_n t_{p1}}\left[-2\zeta + \omega_n t \frac{\omega_n}{\lambda_n - \lambda_2}\left\{-\left(1 + 2\frac{\zeta\lambda_2}{\omega_n}\right) e^{\lambda_1 t} + \left(1 + 2\frac{\zeta\lambda_1}{\omega_n}\right) e^{\lambda_2 t}\right\}\right]$$

$$\dot{x}_1(t, t_{p1}) = \frac{\Delta V_0}{A_1}\frac{1}{t_{p1}}\left[1 + \frac{1}{\lambda_n - \lambda_2}\left\{-\lambda_1\left(1 + 2\frac{\zeta\lambda_2}{\omega_n}\right) e^{\lambda_1 t} + \lambda_2\left(1 + 2\frac{\zeta\lambda_1}{\omega_n}\right) e^{\lambda_2 t}\right\}\right]$$

Using these equations, we can construct the response of the system to a pulse in the time domain as a sum of different ramp functions applied to the actuator after each other as depicted in Figure 3.14.

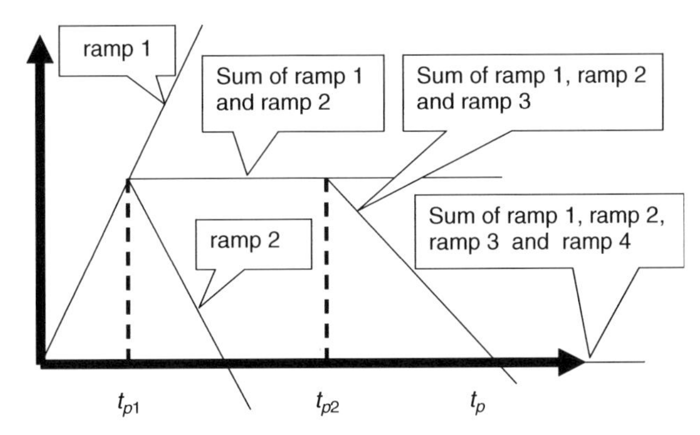

Figure 3.14 *Positive pulse made out of different ramp functions. The first two ramp functions have as base pulse time* t_{p1}*, and the third and fourth* $t_p - t_{p2}$*. In this way, the ramp-up time of the leading edge of the pulse can be chosen differently from the ramp-down time (trailing edge).*

The response of the system to a pulse as shown in Figure 3.14 is built up from different solutions according to the scheme given here:

$$0 < t < t_{p1} : x_1 = x_1\left(t, t_{p1}\right), \quad \dot{x}_1 = \dot{x}_1\left(t, t_{p1}\right)$$

$$t_{p1} < t < t_{p2} : x_1 = x_1\left(t, t_{p1}\right) - x_1\left(t - t_{p1}, t_{p1}\right),$$
$$\dot{x}_1 = \dot{x}_1\left(t, t_{p1}\right) - \dot{x}_1\left(t - t_{p1}, t_{p1}\right)$$

$$t_{p2} < t < t_p : x_1 = x_1\left(t, t_{p1}\right) - x_1\left(t - t_{p1}, t_{p1}\right) - x_1\left(t - t_{p2}, t_p - t_{p2}\right)$$
$$\dot{x}_1 = \dot{x}_1\left(t, t_{p1}\right) - \dot{x}_1\left(t - t_{p1}, t_{p1}\right) - \dot{x}_1\left(t - t_{p2}, t_p - t_{p2}\right)$$

$$t > t_p : x_1 = x_1\left(t, t_{p1}\right) - x_1\left(t - t_{p1}, t_{p1}\right) - x_1\left(t - t_{p2}, t_p - t_{p2}\right)$$
$$+ x_1\left(t - t_p, t_p - t_{p2}\right),$$
$$\dot{x}_1 = \dot{x}_1\left(t, t_{p1}\right) - \dot{x}_1\left(t - t_{p1}, t_{p1}\right) - \dot{x}_1\left(t - t_{p2}, t_p - t_{p2}\right) + \dot{x}\left(t - t_p, t_p - t_{p2}\right)$$

This scheme allows us to choose a ramp-up time (leading edge) different from the ramp-down time (trailing edge). To show the effect of this, we consider the case of a fast positive ramp-up in 1 μs followed by a slow ramp-down in about 25 μs for the small print-head configuration (ink viscosity 10 mPas).

As the pulse is switched off slowly, the negative pressure is much smaller (see Figure 3.15). This means that we can apply a larger volume change and end up with a much higher positive velocity in the nozzle.

This method can be extended to construct the response of dynamical systems to arbitrarily shaped pulses.

In this section we have explained basic concepts of the design of the basic unit of a print-head (acoustic cavity) in terms of its natural frequency, damped frequency, damping ratio, over-damping and oscillatory motion, positive pulse, negative pulse, pulse time,

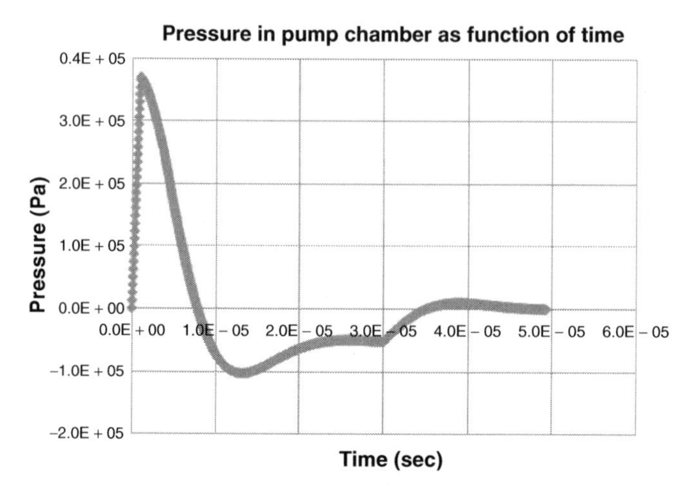

Figure 3.15 *Pressure in pump chamber of the small sized print-head filled with a 10 mPas ink, as a function of time, caused by a positive relative volume pulse of 1.91×10^{-4} and a pulse time of 30 μs. The ramp-up time is 1 μs, $t_{p2} = 3.84$ μs. The pulse height is chosen such that the maximum underpressure does not exceed 1 bar (see also Figure 3.17).*

ramped pulse and the responses in the time domain of fluid velocity in the nozzle and pressure in the pump chamber. To summarise:

- In order to avoid cavitation in the liquid, the underpressure should not exceed 1 bar. This condition determines the maximum volume change the actuator is allowed to generate.
- For positive pulsing the pulse time is determined such that the recoil of the fluid in the nozzle is minimal, in order to prevent air entrapment.
- Only where the damping is small can negative pulsing be used. The pulse time is then chosen so that optimal constructive interference can be exploited. The recoil during negative pulsing, however, is large.
- By carefully choosing the ramp-up time and ramp-down time of a sloped pulse, large fluid velocities in the nozzle can be achieved without cavitation.

3.4 Design Considerations for Droplet Emission from Piezo-Driven Print-Heads

3.4.1 Droplet Formation

Droplet formation is a dynamic process, with the different stages depicted schematically in Figure 3.16 (see also Stemme and Larsson, 1973; Bentin *et al.*, 1986; Shield, Bogy and Talke, 1987; Kitahara, 1995; Usui, 1996; de Jong, 2007; Dijksman *et al.*, 2007; Hutchings *et al.*, 2007; Wijshoff, 2008).

Upon the application of charge to the piezoelectric actuator, the fluid starts to flow out of the nozzle (positive pulsing). Along with the flow, kinetic energy is transported outwards. Droplet formation involves the exchange of kinetic energy that has been accumulated in the volume outside the nozzle, on the one hand, and the surface energy and

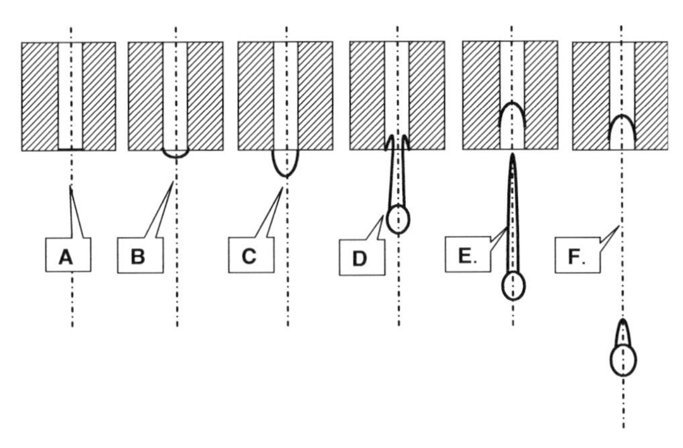

Figure 3.16 *Different stages of single droplet formation caused by a positive pulse. (A) equilibrium; (B) outflow starts; (C) droplet formation starts; (D) elongational phase with retracting meniscus; (E) free flying droplet with filament tail and retracted meniscus and (F) filament almost in main droplet and refilling of nozzle started.*

kinetic energy of the freely moving droplet, on the other. Above a certain threshold value for the maximum velocity in the nozzle, after some time there is enough momentum for the droplet to escape from the nozzle. For some time the droplet is connected with the fluid inside the nozzle by an extending filament of fluid. Surface tension forces lead to the ligament to pinch off, and the remnants of the filament attached to the droplet condense into it, consumed by the main droplet. These processes have been discussed further in Chapter 2. The meniscus retracts inside the nozzle, and refilling starts. In this section, we confine our discussion to the ejection of a single droplet. This means that the maximum velocity of the fluid in the nozzle is slightly above the threshold value.

3.4.1.1 Relation between Droplet Formation and Characteristic Frequency

As discussed in Chapter 2, the characteristic dimensionless numbers defining droplet formation are (de Gennes, Brochard-Wyart and Quéré, 2004):

$$\mathrm{We} = \frac{2\rho_0 R v^2}{\sigma}, \mathrm{Oh} = \frac{\eta}{\sqrt{2R\sigma\rho_0}}, \quad \mathrm{Re} = \frac{2\rho_0 v R}{\eta}$$

where speed, size and material parameters relate to the droplet. The Weber number We is a measure of the influence of the kinetic energy relative to the surface energy. The Ohnesorge number Oh compares the viscous forces to the surface tension forces, and the Reynolds number Re compares inertia forces to viscous forces. For the droplet sizes commonly used in inkjet printing with volumes up to 200 pl, inertial effects dominate over influences associated with surface tension and viscosity, because both the We and Re numbers are large compared to unity. The Oh number is small, so damping due to viscous dissipation plays a minor role. It is the exchange between kinetic and surface energy which matters (Dijksman, 1984).

As shown in Section 3.3, for all cases discussed when the velocity in the nozzle is initially positive, it is approximately a sinusoidal function of time (approximately because the waveform can be strongly distorted by the pulse shape, especially for the cases where damping is important):

$$v(t) = \dot{x}_1(t) = v_{max} \sin \omega_0 t$$

Here v_{max} denotes the amplitude of the velocity in the nozzle. The time t measures the time after the instant the velocity is positive for the first time (for a positive pulse this is the case immediately, whilst for a negative pulse this holds true after the velocity has changed its sign from negative to positive). Additional explanation is needed to define ω_0.

In the case of a positive pulse, the fluid starts to flow out immediately. As shown in Section 3.3, for a positive pulse the pulse time is chosen such that the pulse is switched off after the velocity has passed through a maximum. Therefore, for the analysis of droplet formation in this case we take:

$$\omega_0 \approx \frac{\pi}{2t_p}$$

The larger the damping, the higher ω_0 will be compared to ω_n.

If the damping is low, a negative pulse can also be used. In that case, ω_0 is given by:

$$\omega_0 = \omega_n \sqrt{1 - \zeta^2}$$

3.4.1.2 Analysis of Droplet Formation (Positive Pulse)

The displacement of the fluid in the nozzle for the positive pulse case follows from:

$$x_1(t) = \int_0^t v_{max} \sin \omega_0 t' \, dt'$$

$$x_1(t) = \frac{v_{max}}{\omega_0}(1 - \cos \omega_0 t)$$

With $x_1(t)$, we measure the length of the jet leaving the nozzle, with respect to the nozzle front. At $t = 0$, the pulse is applied to the piezo actuator followed by an immediate rise of the pressure in the pump chamber, causing the fluid to start to flow out at increasing speed. We assume that the jet issues from the nozzle like a cylinder with the same cross-sectional dimension as the nozzle. For a cross-sectional area of the jet A_1, we have for a circular jet $A_1 = \pi {R_1}^2$, where R_1 is the radius of the nozzle. Then the kinetic energy passing the nozzle front is given by:

$$T(t) = \frac{1}{2}\rho_0 A_1 \int_0^5 v^3(t') dt'$$

The fluid portion outside the nozzle moves as a mass for which the kinetic energy is defined by:

$$T_d = \frac{1}{2}\rho_0 A_1 x_1(t) v^2(t)$$

During the outflow, extra free surface is created. As our treatment is based on the fact that the jet almost moves as a solid cylinder, the increase in surface energy W can be expressed as:

$$W = \sigma P_1 x_1(t)$$

We note that if P_1 is the perimeter of a cylindrical jet, for a circular jet we have $P_1 = 2\pi R_1$.

Droplet formation is a dynamic phenomenon, and if the ink contains surfactants the actual surface tension may change because of the sudden enlargement of the free surface (Stückrad *et al*., 1993; Grigorieva *et al*., 2004, 2005). A depletion of the local surfactant concentration at the surface occurs which has to be replenished by diffusion from the bulk. For a water-based ink, this means that the surface tension during droplet formation is generally higher than that measured by a static method (e.g. the pendant drop technique).

The condition for the formation of a droplet is reached as soon as the kinetic energy transported along with the fluid issuing from the nozzle equals the kinetic energy of the cylinder (droplet) plus the extra surface energy:

$$T(t) = T_d + W$$

On substituting the expressions given here, we get:

$$\int_0^t v(t')\{v^2(t') - v^2(t)\}dt' = \frac{2\sigma P_1}{\rho_0 A_1} x_1(t)$$

Note that this condition can be fulfilled only after the velocity has passed through a maximum. Carrying out all the integrations and reducing the equation from cubic to quadratic by factoring out $(1 - cos\ \omega_0 t)$, we arrive at:

$$2\cos^2 \omega_0 t - \cos\omega_0 t - \left(\frac{6\sigma P_1}{\rho_0 A_1 v_{max}^2} + 1\right) = 0$$

The solutions in $\cos\omega_0 t$ are:

$$\cos\omega_0 t^* = \frac{1}{4} \pm \frac{3}{4}\sqrt{1 + \frac{16}{3}\frac{\sigma P_1}{\rho_0 A_1 v_{max}^2}}$$

The positive root delivers a value larger than unity and is therefore invalid. The negative root is valid as long as v_{max} is above a certain threshold value given by:

$$v_{max} > \sqrt{3\sigma\frac{P_1}{\rho_0 A_1}}$$

Where the amplitude of the velocity in the nozzle exceeds the threshold value, the volume and speed of the droplet are given by:

$$V = A_1 x_1(t^*) = A_1 \frac{v_{max}}{\omega_0}(1 - \cos\omega_0 t^*)$$

$$\dot{x}_1(t^*) = v_{max} \sin\omega_0 t^*$$

At the threshold velocity we have $\omega_0 t^* = \pi$, and the droplet velocity is zero. For values of the velocity amplitude in the nozzle much larger than the threshold frequency, we find $\omega_0 t^* = 2\pi/3$. To calculate the final droplet speed, we take into account that the droplet must separate from the fluid in the nozzle by creating twice the meniscus surface area A_1. One surface belongs to the droplet, and the other is the new meniscus in the nozzle. This surface tension effect reduces the final droplet kinetic energy according to:

$$\frac{1}{2}\rho_0 V \dot{x}_1(t^*)^2 - \sigma A_1 = \frac{1}{2}\rho_0 V v^2$$

The droplet velocity is given by:

$$v = \dot{x}_1(t^*)\sqrt{1 - 2\sigma \frac{A_1}{\rho_0 V} \frac{1}{\dot{x}_1(t^*)^2}}$$

This expression defines another threshold value for the velocity of the fluid in the nozzle:

$$\dot{x}_1(t^*) > \sqrt{2\sigma \frac{A_1}{\rho_0 V}}$$

As discussed in this section, the velocity has passed through a maximum before a droplet is released. Close to the threshold value of the fluid velocity in the nozzle, slow droplets are created. For amplitudes far above the threshold value, a high-speed stream is jetted that will break up into a number of small droplets. The analysis described here holds true for fluid velocities in the nozzle close to the threshold value; at higher velocities, it must be established whether more than one droplet will leave the nozzle (Bogy and Talke, 1984; Dijksman *et al*., 2007).

The droplet leaving the nozzle has about the same dimensions as the nozzle from which it emerges. In other words, droplet formation is controlled by the size of the nozzle. This statement leads to the following condition:

$$V = A_1 x_1(t^*) = A_1 \frac{v_{max}}{\omega_0}(1 - \cos\omega_0 t^*) \approx \frac{4}{3}\pi R_1^3$$

$$\frac{v_{max}}{\omega_0 R_1} \approx 1$$

Thus the maximum velocity during droplet formation and the frequency ω_0 are related through the radius of the nozzle. This basic, but important, equation couples the characteristics of the droplet formation process to the acoustic properties of the print-head.

By evaluating the droplet formation characteristics for the small-sized print-head, we find the threshold value $v_{max} = 4.47\,\mathrm{m\,s^{-1}}$ for an ink with $\sigma = 50\,\mathrm{mN\,m^{-1}}$. For the low-viscosity case ($\eta = 2$ mPa s, as shown in Figures 3.7 and 3.8), no droplet formation is possible. For the high-viscosity case ($\eta = 20$ mPa s, Figures 3.11 and 3.12), droplets can barely be generated, because the maximum fluid velocity in the nozzle is about equal to the threshold value.

By using a higher viscosity ink ($\eta = 10$ mPa s) and adjusting the pulse shape as shown in Figure 3.17, the fluid velocity in the nozzle can be as high as $8\,\mathrm{m\,s^{-1}}$ before cavitation occurs. For $v_{max} = 8\,\mathrm{m\,s^{-1}}$, the print-head delivers droplets with a volume of 23.3 pl and a speed of $5.56\,\mathrm{m\,s^{-1}}$ (threshold $x_1(t^*) > 1.74\,\mathrm{m\,s^{-1}}$, pulse positive, rise time 1 µs, $t_{p2} = 3.84$ µs, $t_p = 30$ µs, $\omega_0 = 409306\,\mathrm{rad\,s^{-1}}$, $v_{max}/\omega_0 R_1 = 1.3$).

For the large print-head ($\eta = 5$ mPa s) (Figure 3.13), the threshold value for amplitude of the fluid motion in the nozzle v_{max} is $3.46\,\mathrm{m\,s^{-1}}$ ($\dot{x}_1(t^*) > 1.1\,\mathrm{m\,s^{-1}}$). With

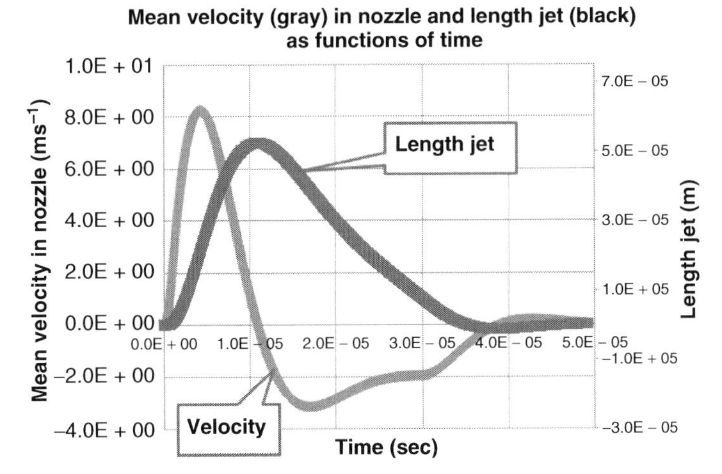

Figure 3.17 *Response of a small-sized print-head with ink with a viscosity of 10 mPa s with a positive slope pulse with a total pulse length of 30 µs, in terms of velocity and length of the jet as functions of time. The ramp-up time is 1 µs,* t_{p2}*= 3.84 µs. The positive relative volume displacement is* 1.91×10^{-4}*. The system is underdamped (*$\zeta = 0.585$ *and* $\omega_n = 303912\ \text{rad s}^{-1} = 48.369\ \text{kHz}$*). See Figure 3.15.*

$v_{max} = 4.5\ \text{m s}^{-1}$, the droplet speed and volume are $2.29\ \text{m s}^{-1}$ and 162 pl, respectively (pulse positive with pulse time 15.8 µs and $\omega_0 = 99437\ \text{rad s}^{-1}$, $v_{max}/\omega_0 R_1 = 1.81$).

3.4.1.3 Analysis of Droplet Formation (Negative Pulse)

With a negative pulse, the fluid in the nozzle first retracts and then after some time the fluid velocity in the nozzle changes sign and the nozzle will be filled again. Droplet formation starts only after the meniscus has passed the nozzle front. For a negative pulse, the velocity and displacement of the fluid in the nozzle are given by (note that t starts after the velocity has changed sign from negative to positive):

$$\dot{x}_1(t) = v_{max} \sin \omega_0 t$$

$$x_1(t) = x_0 + \frac{v_{max}}{\omega_0}(1 - \cos \omega_0 t)$$

where x_0 measures the distance of the meniscus from the nozzle front at the moment the velocity changes sign from negative to positive (note that $x_0 < 0$ for a retracted meniscus). The moment the meniscus passes the nozzle follows from:

$$\cos \omega_0 t_0 = 1 + \frac{x_0 \omega_0}{v_{max}}$$

Using a negative pulse makes sense only when the viscosity is low. This means that velocity and displacement of the fluid in the nozzle are almost 90° out of phase (see Figure 3.9) and $\cos \omega_0 t_0 \approx 0$. On substituting the expressions for the different energies given above, we obtain:

$$\int_{t_0}^{t} v(t')\{v^2(t') - v^2(t)\}dt' = \frac{2\sigma P_1}{\rho_0 A_1}\int_{t_0}^{t} v(t')dt' = \frac{2\sigma P_1}{\rho_0 A_1}x_1(t)$$

Again, this condition can be fulfilled only after the velocity has gone through a maximum. Carrying out the integrations and reducing the equation from cubic to quadratic by factoring out $(1-\cos\omega_0 t)$, we arrive at:

$$2\cos^2\omega_0 t - \cos\omega_0 t_0 \cos\omega_0 t - \left(\frac{6\sigma P_1}{\rho_0 A_1 v_{max}^2} + \cos^2\omega_0 t_0\right) = 0$$

Because $\cos\omega_0 t_0 \approx 0$, this equation simplifies to:

$$\cos^2\omega_0 t \approx \frac{3\sigma P_1}{\rho_0 A_1 v_{max}^2}$$

with solutions:

$$\cos\omega_0 t^* = \pm\sqrt{\frac{3\sigma P_1}{\rho_0 A_1 v_{max}^2}}$$

As the velocity has gone through a maximum, the negative root applies. The negative root is valid so long as v_{max} is above a certain threshold value given by:

$$v_{max} > \sqrt{3\sigma\frac{P_1}{\rho_0 A_1}}$$

This value for the threshold velocity is equal to the corresponding value found for the positive pulse. When the amplitude of the velocity in the nozzle exceeds the threshold value, the volume and $\dot{x}_1(t^*)$ are given by:

$$V = A_1 x_1(t^*) = A_1\left[x_0 + \frac{v_{max}}{\omega_0}(1 - cos\,\omega_0 t^*)\right] \approx A_1\frac{1}{\omega_0}\sqrt{3\sigma\frac{P_1}{\rho A_1}}$$

$$\dot{x}_1(t^*) = v_{max}\sin\omega_0 t^*$$

The final droplet velocity is found as for the case of a positive pulse. It is remarkable that for this case the droplet volume does not depend on the characteristics of the pulse such as v_{max}, a fact that is supported by experiments.

For the small-sized print-head driven by a negative pulse, we find as a threshold value $v_{max} = 4.47\,\mathrm{m\,s^{-1}}$, for an ink with $\sigma = 50\,\mathrm{mN\,m^{-1}}$. In the low-viscosity case ($\eta = 2\,\mathrm{mPa\,s}$, Figure 3.9), we find that $v_{max} = 5.5\,\mathrm{m\,s^{-1}}$. With this setting the print-head delivers droplets with a volume of 10 pl and a speed of $1.87\,\mathrm{m\,s^{-1}}$ (negative pulse with pulse time 10.4 μs and $\omega_0 = 301825\,\mathrm{rad\,s^{-1}}$, $v_{max}/\omega_0 R_1 = 1.2$).

3.4.2 Damping

After droplet emission, the meniscus retracts and the ink in the nozzle is still in motion. Due to damping, these motions decay with time. Before charging the actuator for the next droplet, the fluid in the system must be returned almost to the equilibrium position. We take here as a measure that 5% of the initial motion of the previous droplet is allowed to be present to ensure that there is almost no influence on the formation of the next droplet. From this statement, the minimum time between droplets (or the maximum droplet frequency) is given by:

$$t_{\min} = \frac{\pi}{\zeta \omega_n}, f_{max} = \frac{\zeta \omega_n}{\pi}$$

For the example of the small print-head with an ink of viscosity 2 mPa s, the maximum frequency for stable droplet emission is thus 11.3 kHz.

3.4.3 Refilling

After droplet formation, the meniscus retracts in the nozzle over a distance ΔL_1 related to the volume of the droplet:

$$\Delta L_1 = \frac{V}{A_1}$$

In order to end up with stable droplet emission, the meniscus must return to the equilibrium position before the next droplet can be launched. The volume of the droplet must be replenished first. This mechanism is called refilling and is controlled by surface tension and inertia-related phenomena in the nozzle. We will discuss here three mechanisms for refilling:

1. Surface tension–driven flow
2. Flow generated by Bernoulli underpressure
3. Flow caused by the asymmetric filling of the nozzle during damping of the motion of the fluid in the nozzle after releasing a droplet.

3.4.3.1 Capillary-Driven Refilling

The fluid retracts in the nozzle after firing a droplet, and refilling starts due to surface tension. The situation is visualised in Figure 3.18. The droplet has separated, and the meniscus has retracted inside the nozzle.

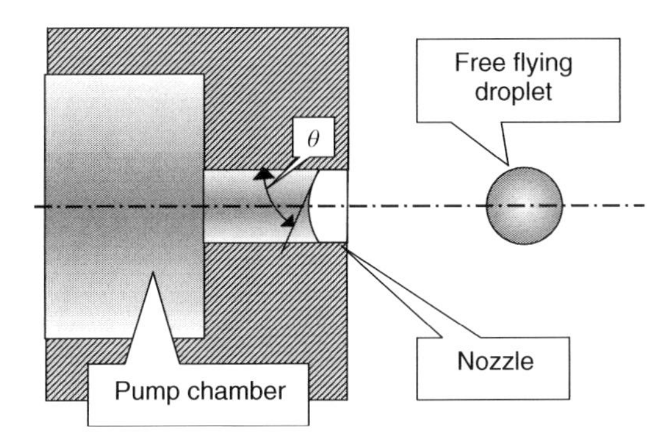

Figure 3.18 *Situation after a droplet has separated from the fluid inside the nozzle.*

The capillary pressure is given by de Gennes, Brochard-Wyart and Quéré (2004):

$$p_{cap} = \frac{2\sigma \cos\theta}{R_1}$$

A simple estimate is used for calculating the surface tension–driven filling time based on combining the Poiseuille law and the capillary underpressure given above:

$$\tau_{filling,\ surface\ tension} = \frac{4\pi\eta R_1\left(\frac{L_1}{A_1^2}+\frac{L_2}{A_2^2}\right)}{\sigma\cos\theta}V$$

where θ is the contact angle of the ink on the inside surface of the nozzle (usually in the nozzle, θ is small). Note that for refilling, the dimensions of the throttle are included in the time involved in refilling due to surface tension. Capillary filling is active only when the fluid inside the nozzle is behind the nozzle front. At the moment the liquid approaches the rim of the nozzle, filling due to capillary underpressure starts to slow down until the meniscus is flat or given by the underpressure control of the print-head. For the small-sized system (with $L_2 = 0.001$ m and $A_2 = 6.36 \times 10^{-9}\ \text{m}^3$) filled with an ink with viscosity $\eta = 10$ mPa s, surface tension $\sigma = 50\ \text{mN m}^{-1}$ and contact angle $\theta = 20°$, driven such that $v_{max} = 8\ \text{m s}^{-1}$ (Figure 3.17), we find a droplet volume and speed of 23.3 pl and $5.56\ \text{m s}^{-1}$, respectively, and a capillary-driven refilling time of 0.16 ms. Taking into account capillary-driven refilling only, this would allow for a maximum droplet frequency of almost 6 kHz.

3.4.3.2 *Bernoulli Underpressure-Driven Refilling*

Due to fact that the meniscus faces ambient pressure and the fluid in the nozzle moves back and forth, Bernoulli pressure is also present (Dijksman *et al*., 2007). This pressure is active only at the moment the fluid is in motion; the higher the velocity, the higher its value. There is also a strong dependence on viscosity: the higher the viscosity, the higher the damping and the shorter the Bernoulli underpressure-driven replenishing time. Therefore we confine ourselves here to the underdamped or oscillatory case ($\zeta \ll 1$). The value of the Bernoulli pressure depends on time according to:

$$p_{Bernoulli} = \frac{1}{2}\rho_0 v_{max}^2 e^{-2\zeta\omega_n t}\sin^2(\omega_n t\sqrt{1-\zeta^2}) \approx \frac{1}{2}\rho_0 v_{max}^2 e^{-2\zeta\omega_n t}\sin^2(\omega_n t)$$

As we have to consider a longer period of time, we use here the damped resonance frequency rather than ω_0. The volume rate of flow as a function of time is given by:

$$Q(t) = \frac{\rho_0 v_{max}^2}{16\pi\eta\left(\frac{L_1}{A_1^2}+\frac{L_2}{A_2^2}\right)}e^{-2\zeta\omega_n t}\sin^2(\omega_n t)$$

The characteristic time $\tau_{Bernoulli}$ for replenishing the volume of the droplet after release of the droplet (at $t = t^*$) follows from:

$$V = \frac{\rho_0 v_{max}^2}{16\pi\eta\left(\frac{L_1}{A_1^2}+\frac{L_2}{A_2^2}\right)}\int_{t*}^{\tau_{Bernoulli}} e^{-2\zeta\omega_n t}\sin^2(\omega_n t)dt$$

An estimate of the refilling time due to the velocity-related underpressure can be found from:

$$\zeta \ll 1 : \tau_{Bernoulli} = -\frac{\ln\left[e^{-2\zeta\omega_n t^*} - \frac{64\pi\eta\left(\frac{L_1}{A_1^2}+\frac{L_2}{A_2^2}\right)\zeta\omega_n}{\rho_0 v_{max}^2}V\right]}{2\zeta\omega_n}$$

The maximum volume that can be delivered by the Bernoulli underpressure is given by:

$$V_{max} = \frac{\rho_0 v_{max}^2}{64\pi\eta\omega_n\zeta\left(\frac{L_1}{A_1^2}+\frac{L_2}{A_2^2}\right)}e^{-2\zeta\omega_n t^*}$$

For the small pump discussed in Section 3.4.3.1, with an ink viscosity of 10 mPa s, there is very little Bernoulli-driven refilling ($V_{max} = 0.06$ pl). For $\eta = 2$ mPa s, however, this contribution to refilling is of the same order of magnitude as that from capillary pressure ($V_{max} = 8.33$ pl).

For $\eta = 1$ mPa s, the volume delivered by the Bernoulli underpressure becomes so large that flooding of the nozzle plate may occur ($V_{max} = 43$ pl, $\tau_{Bernoulli} = 31$). Note that for these short times, damping of the fluid motion becomes the determining factor (see Section 3.4.2).

3.4.3.3 *Refilling Associated with Partial Filling of the Nozzle after Release of the Droplet*

After the droplet has separated from the fluid in the nozzle, the meniscus retracts and the nozzle is not completely filled with fluid. During the damped oscillations, the length of the fluid column in the nozzle changes (Wijshoff, 2008). This effect influences the mass and the damping according to (with $L_1^* = L_1 - \Delta L_1$):

$$\rho_0 A_1 (L_1^* + x_1)\ddot{x}_1 + 8\pi\eta(L_1^* + x_1)\dot{x}_1 + K_B\frac{A_1^2}{V_c}x_1 = 0$$

When ΔL_1 is small compared to L_1, we assume that the solution of the homogeneous non-linear second-order differential equation can be written as:

$$x_1 = x_{1,linear} + \varepsilon f(t), \qquad \varepsilon = \frac{\Delta L_1}{L_1^*} \ll 1$$

Note that the change in length of the fluid column in the nozzle is in phase with the displacement, and because of damping it holds that $|x_1| < |\Delta L_1|$. Upon substitution in the equation of motion and subtracting the linear solution, we get:

$$M^*\ddot{f} + K^*\dot{f} + Cf = C\frac{x_{1,linear}^2}{\Delta L_1}$$

with $M^* = \rho_0 A_1 L_1^*$, $K^* = 8\pi\eta L_1^*$, $C = K_B A_1^2/V_c$ and ω_n^* and ζ^* defined accordingly. To estimate the effect on refilling, we take into account that the condition for droplet formation is fulfilled at $t = t^*$. At that moment the meniscus of the fluid in the nozzle jumps back. We start with a new timescale defined by $t_R = t - t^*$ and rewrite the expressions for velocity and displacement:

$$v(t_R) = \dot{x}_{1,linear}(t_R) = v_{max} e^{-\zeta^* \omega_n^* tR} \cos \omega_n^* t_R \sqrt{1-\zeta^{*2}}$$

$$x_{1,linear}(t_R) = \frac{v_{max}}{\omega_n^*}$$

$$\left[\zeta^* - e^{-\zeta^* \omega_n^* t_R}\left(\zeta^* \cos \omega_n^* t_R \sqrt{1-\zeta^{*2}} - \sqrt{1-\zeta^{*2}} \sin \omega_n^* t_R \sqrt{1-\zeta^{*2}}\right)\right]$$

$$x_{1,linear}(t_R) = \frac{v_{max}}{\omega_n^*} \zeta^* \left[1 - e^{-\zeta^* \omega_n^* t_R} \cos(\omega_n^* t_R \sqrt{1-\zeta^{*2}} + \phi)\right]$$

The result is:

$$M^* \ddot{f} + K^* \dot{f} + Cf$$
$$= \frac{C}{\Delta L_1} \frac{v_{\max}^2}{\omega_n^{*2}} \zeta^{*2} \left[1 - e^{-\zeta^*} \omega_n^* t_R \cos\left(\omega_n^* t_R \sqrt{1-\zeta^{*2}} + \phi\right)\right]^2$$

For our analysis, the particular solution is of interest (harmonic terms fall to zero):

$$f = \frac{1}{\Delta L_1} \frac{v_{max}^2}{\omega_n^{*2}} \zeta^{*2} [1 + e^{-\zeta^* \omega_n^* t_R}(\ldots\ldots) + e^{-2\zeta^* \omega_n^* t_R}(\ldots\ldots)] \approx \frac{1}{\Delta L_1} \frac{v_{max}^2}{\omega_n^{*2}} \zeta^{*2}$$

$$x_{1f} \approx \frac{1}{L_1} \frac{v_{max}^2}{\omega_n^{*2}} \zeta^{*2}$$

It should be mentioned that this effect becomes stronger for higher viscosities as it is the viscous drag that keeps the fluid from moving backwards. With the small-sized pump for a viscosity of 2 mPa s, hardly any displacement is generated, whilst for viscosities above 20 mPa s, part of the volume of the droplet can be replenished by this effect. For even higher viscosities, this effect becomes small again because the motion of the fluid in the nozzle is damped quickly.

3.4.4 Deceleration Due to Elongational Effects Prior to Pinching Off

Before the droplet separates from the fluid in the nozzle, it is connected for a short time to the fluid in the nozzle by a thin filament (Dijksman, 1984). In order to stretch this filament starting from $t = t^*$, a force is needed that decelerates the droplet. During stretching we assume that the volume of the droplet is contained in the filament with radius R_f and length L_f, and so we have:

$$\pi R_f^2(t_R) L_f(t_R) = V$$

To derive the force balance we assume that all the mass is contained in the tip of the filament:

$$\rho_0 V \frac{dv}{dt_R} = -3\eta \frac{v(t_R) - \dot{x}_1(t_R)}{L_f(t_R)} \pi R_f^2(t_R)$$

The term 3η represents the elongational viscosity of a cylindrical filament of fluid (Bird, Armstrong and Hassager, 1987a). The difference between the velocity of the droplet and the velocity of the fluid in the nozzle equals the time rate of change of the length of the filament. From this, we obtain:

$$\frac{dv}{dt_R} = -\frac{3\eta}{\rho_0}\frac{1}{L_f^2}\frac{dL_f}{dt_R}$$

As the filament becomes very thin at the moment of pinching off, the approximate solution of this equation is:

$$\Delta v = -\frac{3\eta}{\rho_0}\frac{1}{L_f(t_R = 0)} = -\frac{3\eta}{\rho_0}\frac{1}{x_1(t^*)}$$

This effect is most prominent for larger droplets and high-viscosity inks. If, however, the ink is slightly non-Newtonian (visco-elastic), the elongational viscosity becomes a function of the elongation rate: the larger the elongational rate, the larger the elongational viscosity. This effect is strongly related to the molecular weight of polymer solute in the ink. Above a certain molecular weight of polymer (and at a certain concentration), droplet formation becomes impossible.

3.4.5 Summary

We have discussed in this section a number of aspects of droplet formation and chamber refilling. The main result is given by the equation:

$$\frac{v_{max}}{\omega_0 R_1} \approx 1$$

which shows the conditions for droplet formation, with the main argument that the droplet size is controlled by the nozzle dimensions. The maximum value of the velocity during the first time the velocity is positive (v_{max}) and the characteristic frequency given by the pulse time (ω_0) are determined by the dimensions of the basic unit of the print-head and the pulse shape.

Droplet formation is possible only above a certain threshold velocity. Our analysis to estimate droplet velocity and volume is valid as long as the nozzle velocity is slightly higher than the threshold value.

Refilling is a complicated process potentially involving at least three phenomena: capillary-driven refilling, filling associated with Bernoulli underpressure, and an inertia effect caused by partial filling of the nozzle upon retraction of the meniscus caused by the release of the droplet. The Bernoulli effect is most important for low-viscosity inks, while the inertia effect is most pronounced for high-viscosity inks.

Before pinching off, the droplet is decelerated by an extending fluid filament between the droplet and meniscus in the nozzle.

3.5 Multi-Cavity Helmholtz Resonator Theory

We now return to consider the set-up shown in Figure 3.3, the Helmholtz arrangement. The print-head contains n pumps all connected to the same supply channel

(Dijksman, 1999). By charging one or more pumps, pressure waves are generated in the supply channel. These pressure waves influence the motion in the charged pumps as well as in the non-activated pumps. In order to derive the governing equations of motion, we follow the way of thinking developed by Helmholtz (1885) and Rayleigh (1896). All pumps have the same dimensions and are filled with the same liquid. In Figure 3.19, one out of the ensemble of pumps is shown together with the relevant dimensions.

The fluid displacements in the nozzles and throttles of the activated pumps are denoted by x_{1i} and x_{2i} ($i = 1, \ldots, k$), respectively. In the non-activated pumps, the corresponding fluid displacements induced by the activated pumps are given likewise by y_{1j} and y_{2j} ($j = 1, \ldots, n-k$). The fluid motion in the connection to the ink supply is given by x_3. Motions in the main direction of the flow are considered positive. Per volume, whether an activated pump chamber V_{ci}, a non-activated pump chamber V_{cj} or the ink supply V_{supply}, we can calculate the excess volumes, being defined as the difference between volume in and volume out:

$$\Delta V_{supply} = A_3 x_3 - A_2 \sum_{i=1}^{k} x_{2i} - A_2 \sum_{j=1}^{n-k} y_{2j}$$

$$\Delta V_{ci} = A_2 x_{2i} - A_1 x_{1i} + \Delta V_0 \; (i = 1, \ldots k)$$

$$\Delta V_{cj} = A_2 x_{2j} - A_1 x_{1j} \; (j = 1, \ldots n - k)$$

The volume change caused by the actuator is denoted by: ΔV_0.

The relations between excess volumes and excess pressures are given by:

$$p_{supply} = \frac{\rho_0 c_{supply}^2}{V_{supply}} \Delta V_{supply}, \quad p_{ci} = \frac{\rho_0 c_c^2}{V_c} \Delta V_{ci}, \quad p_{cj} = \frac{\rho_0 c_c^2}{V_c} \Delta V_{cj}$$

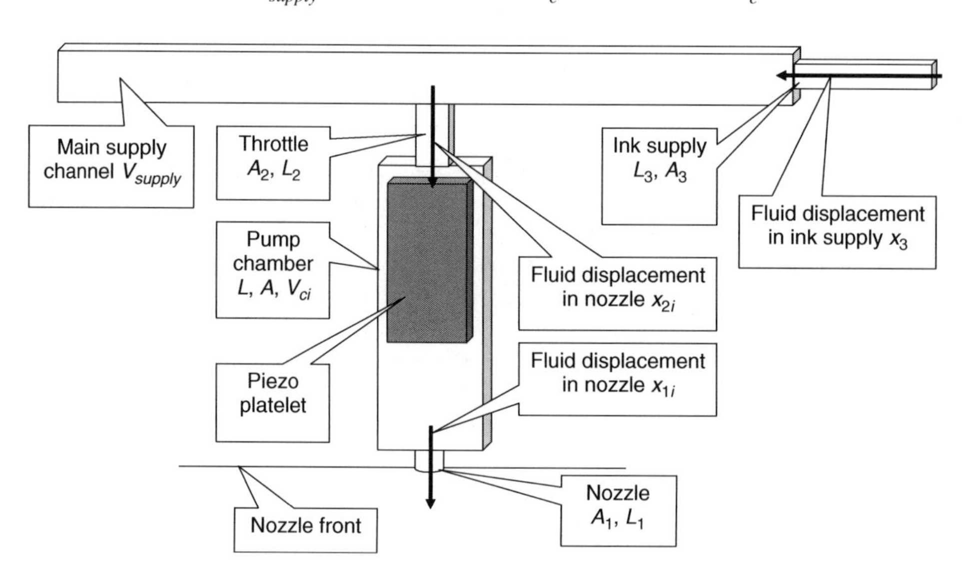

Figure 3.19 *Geometrical data of activated pump i (i = 1, ..., k) out of an ensemble of n pumps integrated in a print-head. For a non-activated pump j (j = 1, ..., n − k), the fluid displacements in the nozzle and throttle are denoted by* y_{1j} *and* y_{2j}*.*

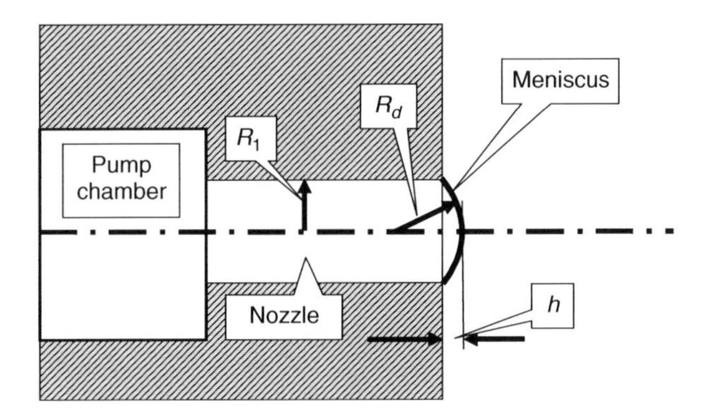

Figure 3.20 *Geometric details of meniscus shape.*

Per element the velocity of sound can be entered. This allows for taking into account that the velocity of sound may depend on the compliance of the environment, e.g. in a duct with a flexible wall the velocity of sound can be considerably lower than the isentropic value.

In the equilibrium situation the fluid at the end of the nozzle forms a meniscus. The solid, fluid and air contact line pins at the rim of the nozzle. The geometric details are shown in Figure 3.20.

When the fluid flows out of the nozzle, the meniscus is deformed in such a way that the curvature increases and as a consequence the capillary pressure increases as well. The capillary pressure inside the domed meniscus opposes the motion of the fluid column in the nozzle. When the fluid retracts into the nozzle, a concave meniscus is formed. The capillary pressure is again opposing the retracting motion of the fluid. We have:

$$A_1 x_1 = \frac{1}{6}\pi h(3R_1^2 + h^2)$$

The radius of curvature of the dome follows from:

$$R_d = \frac{R_1^2 + h^2}{2h}$$

From the Laplace equation (de Gennes, Brochard-Wyart and Quéré, 2004), we obtain an expression for the capillary pressure as a function of the fluid displacement x_1:

$$p_{cap} = \frac{2\sigma}{R_d} \cong \frac{8\pi\sigma}{A_1} x_1 = \frac{8\sigma}{R_1^2} x_1$$

The approximate expression shows that the pressure depends linearly on the fluid displacement x_1. In the following, we will use this dependence of the capillary pressure on fluid displacement. We derive the equilibrium equations for the fluid contained in the nozzles, throttles and ink supply in a similar way to that used in Section 3.3:

$$\rho_0 L_1 \ddot{x}_{1i} = p_{ci} - \frac{8\pi\sigma}{A_1} x_{1i}, \quad \rho_0 L_1 \ddot{y}_{1j} = p_{cj} - \frac{8\pi\sigma}{A_1} y_{1j}$$

$$\rho_0 L_2 \ddot{x}_{2i} = p_{supply} - p_{ci}, \quad \rho_0 L_2 \ddot{y}_{2j} = p_{supply} - p_{cj}, \quad \rho_0 L_3 \ddot{x}_3 = -p_{supply}$$

All activated pumps are charged in exactly the same way. Consequently they induce equal fluid motions in the activated pumps, and the same holds true for the fluid motions in the non-activated pumps set in motion by the pressure waves in the supply channel. Using this as a starting point, we can skip the indices i and j, and instead of n separate equations of motion we end up with five equations of motion:

$$\ddot{x}_1 + \left(\frac{c_c^2}{V_c}\frac{A_1}{L_1} + \frac{8\pi\sigma}{\rho_0 A_1 L_1}\right)x_1 - \frac{c_c^2}{V_c}\frac{A_2}{L_1}x_2 = \frac{c_c^2}{V_c L_1}\Delta L_0$$

$$\ddot{x}_2 + \left(\frac{c_c^2}{V_c}\frac{A_2}{L_2} + k\frac{c_{supply}^2}{V_{supply}}\frac{A_2}{L_2}\right)x_2 + (n-k)\frac{c_{supply}^2}{V_{supply}}\frac{A_2}{L_2}y_2 - \frac{c_c^2}{V_c}\frac{A_1}{L_2}x_1 - \frac{c_{supply}^2}{V_{supply}}\frac{A_3}{L_2}x_3$$

$$= -\frac{c_c^2}{V_c L_2}\Delta V_0$$

$$\ddot{x}_3 + \frac{c_{supply}^2}{V_{supply}}\frac{A_3}{L_3}x_3 - k\frac{c_{supply}^2}{V_{supply}}\frac{A_2}{L_3}x_2 - (n-k)\frac{c_{supply}^2}{V_{supply}}\frac{A_2}{L_3}y_2 = 0$$

$$\ddot{y}_1 + \left(\frac{c_c^2}{V_c}\frac{A_1}{L_1} + \frac{8\pi\sigma}{\rho_0 A_1 L_1}\right)y_1 - \frac{c_c^2}{V_c}\frac{A_2}{L_1}y_2 = 0$$

$$\ddot{y}_2 + \left(\frac{c_c^2}{V_c}\frac{A_2}{L_2} + (n-k)\frac{c_{supply}^2}{V_{supply}}\frac{A_2}{L_2}\right)y_2 + k\frac{c_{supply}^2}{V_{supply}}\frac{A_2}{L_2}x_2 - \frac{c_c^2}{V_c}\frac{A_1}{L_2}y_1 - \frac{c_{supply}^2}{V_{supply}}\frac{A_3}{L_2}x_3 = 0$$

To solve this set of five linear non-homogeneous second-order differential equations, we proceed as outlined in Section 3.3.

Firstly, we specify a ramp function for the volume change according to:

$$\Delta V = \Delta V_0 \frac{t}{t_{p1}}$$

We pose a particular solution:

$$x_1 = y_1 = y_2 = 0,\ x_2 = -\frac{\Delta V_0}{A_2}\frac{t}{t_{p1}},\ x_3 = -k\frac{\Delta V_0}{A_3}\frac{t}{t_{p1}}$$

Geometrically, this solution means that in each pump chamber, the volume displacement of the actuator is pushed through the throttle. As k pumps are actuated, the sum of the volume displacements of all activated pumps is pushed through the ink inlet. The set of equations of motion describes a system with five degrees of freedom.

When the k actuators are switched on to start the gradual decrease of the volume of the pump chamber, five different eigen (resonance) modes are excited, each with its own frequency ω_j $(j = 1, \ldots, 5)$ and amplitude. The resonant frequencies associated with the modes follow from the solution of the homogeneous set of equations. The different vibrational modes are characterised by the associated eigenvectors ξ_{ij}. The index i refers to the different channels involved: $i = 1$ is the nozzle of an activated pump, and $i = 2$ refers to the corresponding throttle. The ink supply is denoted by $i = 3$. The nozzle and throttle of a non-activated pump are indicated by $i = 4$ and 5, respectively. The index j

lists the different modes. With known ω_j and ξ_{ij}, the solution of the non-homogeneous set of equations can be written as:

$$x_1 = \sum_{j=1}^{5} D_j \xi_{1j} \sin \omega_j t, \ \dot{x}_1 = \sum_{j=1}^{5} D_j \xi_{1j} \omega_j \cos \omega_j t,$$

$$x_2 = \sum_{j=1}^{5} D_j \xi_{2j} \sin \omega_j t - \frac{\Delta V_0}{A_2} \frac{t}{t_{p1}}, \ \dot{x}_2 = \sum_{j=1}^{5} D_j \xi_{2j} \omega_j \cos \omega_j t - \frac{\Delta V_0}{A_2} \frac{t}{t_{p1}},$$

$$x_3 = \sum_{j=1}^{5} D_j \xi_{3j} \sin \omega_j t - k \frac{\Delta V_0}{A_3} \frac{t}{t_{p1}}, \ \dot{x}_3 = \sum_{j=1}^{5} D_j \xi_{3j} \omega_j \cos \omega_j t - k \frac{\Delta V_0}{A_3} \frac{t}{t_{p1}},$$

$$y_1 = \sum_{j=1}^{5} D_j \xi_{4j} \sin \omega_j t, \ \dot{y}_1 = \sum_{j=1}^{5} D_j \xi_{4j} \omega_j \cos \omega_j t,$$

$$y_2 = \sum_{j=1}^{5} D_j \xi_{5j} \sin \omega_j t, \ \dot{y}_2 = \sum_{j=1}^{5} D_j \xi_{5j} \omega_j \cos \omega_j t.$$

For $t = 0$, all displacements and velocities are zero. This is already true for the displacements given here. The condition that all velocities are zero for $t = 0$ gives five linear algebraic equations for the five constants D_j $(j = 1, \ldots, 5)$.

In order to estimate the damping, we calculate the decrease in kinetic energy caused by viscous dissipation. This method is applicable provided it takes some oscillations for the system to come to rest (low damping). The eigen (resonance) modes are independent of each other. Therefore we can calculate the damping per mode. With the results obtained so far, the amplitude of the kinetic energy is given by:

$$E_{kin,i} = \frac{1}{2} \rho_0 \omega_i^2 D_i^2 [k(A_1 L_1 \xi_{1i}^2 + A_2 L_2 \xi_{2i}^2) + A_3 L_3 \xi_{3i}^2 + (n-k)(A_1 L_1 \xi_{4i}^2 + A_2 L_2 \xi_{5i}^2)]$$

As it takes a few cycles for the system to be damped, the decrease in amplitude of the kinetic energy can be approximated by a Taylor expansion around $t = t$:

$$\begin{aligned} E_{kin,i} = \frac{1}{2} \rho_0 \omega_i^2 [& k(A_1 L_1 \xi_{1i}^2 + A_2 L_2 \xi_{2i}^2) + A_3 L_3 \xi_{3i}^2 \\ & + (n-k)(A_1 L_1 \xi_{4i}^2 + A_2 L_2 \xi_{5i}^2)] \left\{ D_i^2(t) + 2 D_i(t) \frac{dD_i}{dt} \Delta t \right\} \end{aligned}$$

The viscous damping per mode per half period $\Delta t = \pi / \omega_i$ is given by:

$$P_{dis,i} = \frac{4\pi^2 \eta}{\rho_0} D_i^2 [k(L_1 \xi_{1i}^2 + L_2 \xi_{2i}^2) + L_3 \xi_{3i}^2 + (n-k)(L_1 \xi_{4i}^2 + L_2 \xi_{5i}^2)]$$

The decrease in kinetic energy per half cycle just equals the power dissipated by the viscous forces per half cycle. From that, it can be shown that the damping coefficient ζ_i of the i-th mode equals:

$$D_j(t) = D(t=0) e^{-\zeta_j \omega_j t}, \quad \zeta_j = \frac{P_{dis.j}}{2\pi E_{kin,j}}$$

To determine the response of the multi-compartment system to an arbitrarily shaped pulse, we use the method developed in Section 3.3.

We show here the results of an analysis for a print-head consisting of 24 small-sized pumps with the dimensions $V_c = 2 \times 10^{-10}\,\text{m}^3$ (e.g. length 0.01 m and cross-section $2 \times 10^{-8}\,\text{m}^2$), $L_1 = 75\,\mu\text{m}$ and $R_1 = 15\,\mu\text{m}$ ($A_1 = 7.07 \times 10^{-10}\,\text{m}^2$). The throttle and ink supply connection dimensions are $A_2 = 6.36 \times 10^{-9}\,\text{m}^2$, length $L_2 = 1 \times 10^{-3}$ m and $A_3 = 7.78 \times 10^{-7}\,\text{m}^2$, length $L_3 = 4 \times 10^{-3}$ m, respectively. The supply channel volume is $3 \times 10^{-9}\,\text{m}^3$. The fluid has a density of $1000\,\text{kg}\,\text{m}^{-3}$ and a speed of sound of $1400\,\text{m}\,\text{s}^{-1}$. The bulk modulus of the fluid follows from $K_B = \rho_0 c^2 = 1.96 \times 10^9$ Pa. In the main supply channel, the speed of sound is $1000\,\text{m}\,\text{s}^{-1}$. For the sample calculation, we have taken $\eta = 2$ mPa s.

Table 3.1 lists the five resonance frequencies with associated eigenvectors normalised with respect to the displacement in the activated nozzle(s). The results are shown for the case of 12 pumps activated out of the total of 24.

These different eigenmodes represent different motion patterns. In the first, the fluid columns in the activated and non-activated pumps and the fluid in the fluid connection move in phase against the surface tension springs in all the nozzles. The second mode is also a surface tension spring-driven motion, but now the fluid columns in the activated pumps move in the opposite direction with respect to the fluid columns in the non-activated pumps. There is no motion induced in the fluid connection to the ink reservoir. In the third mode, the fluid columns in both the activated and non-activated pumps move against the motion of the fluid in the connection to the ink reservoir with the spring action of the compressible fluid in the supply channel in between. The last two modes are the Helmholtz modes. In the fourth mode the motions in the activated and non-activated pumps run in opposite directions, while in the fifth mode they all move in the same direction. The eigenmodes and associated frequencies depend on the number of activated pumps.

Although the dimensions used here are the same as for the one-degree-of-freedom model (see Figures 3.7, 3.8 and 3.9), the result is quite different. Because of the surface tension spring there are low-frequency modes, whilst because of the presence of a throttle the frequency associated with droplet formation is higher. Some results for negative pulsing are shown in Figure 3.21 (see also Figure 3.9).

We have used here the same pulse setting as those used for Figure 3.9. At the moment, the velocity induced by the leading edge of the pulse changes sign from negative to positive, if the pulse is switched off. In that way, due to constructive interference a high

Table 3.1 *Resonance frequencies and eigen vectors for the five modes for the case of 12 pumps activated out of a total of 24.*

Mode number j	Resonance frequency (Hz)	ξ_{1j}	ξ_{2j}	ξ_{3j}	ξ_{4j}	ξ_{5j}
1	11 823	1	0.133	0.028	1	0.133
2	14 851	1	0.129	0	−1	−0.129
3	46 519	1	0.037	−0.023	1	0.037
4	65 561	1	−0.065	0	−1	0.065
5	71 735	1	−0.105	0.01	1	−0.105

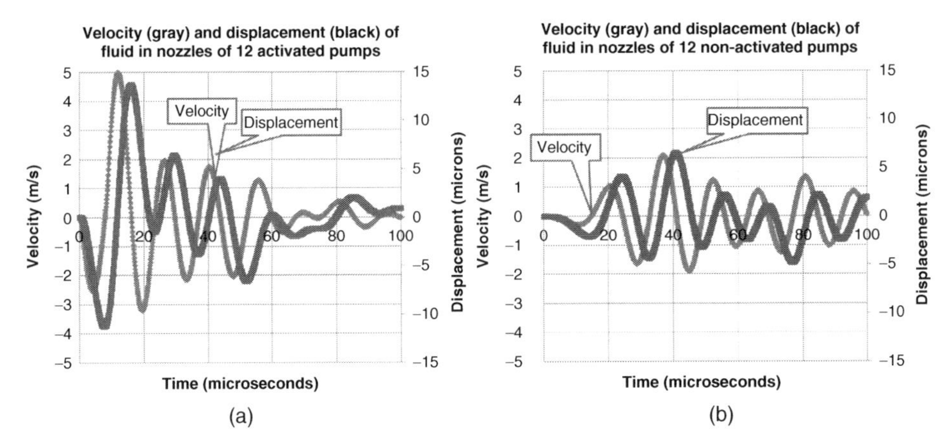

Figure 3.21 *Results for the case of 12 pumps that are activated out of a total of 24. (a) Response in terms of the velocity and displacement of the fluid in the nozzle of an activated pump on a negative pulse, with a relative volume displacement of 0.45×10^{-5}, ramp-up time and switch-down time equal to 1 μs and a pulse length of 9 μs. (b) Shows the fluid motion in the non-activated pumps (see also Figure 3.9).*

positive value of the velocity of the fluid in the nozzle is obtained. The compliance of the wall of the supply channel is modelled to be stiff (i.e. with a high value of the speed of sound). Figure 3.21b shows that in the non-activated pumps, there is considerable fluid motion. These motions can be reduced by reducing the stiffness of the environment of the supply channel. In such a channel the effective speed of sound is much lower. For example, taking a value of $100\,\mathrm{m\,s^{-1}}$ as the velocity of sound in the ink supply channel reduces the induced motions in the non-activated pumps by more than a factor of 10. A very low velocity of sound in the main supply channel can be obtained by covering the channel with a thin and flexible foil.

What is also seen in Figure 3.21 is that the motion of the fluid in the nozzle is a superposition of a low-frequency mode and a high-frequency mode. We conclude by showing the difference between driving only a single nozzle and driving 23 nozzles at the same time. There are two main features that contribute to the difference between activating either one or 23 pumps (see Figures 3.21 and 3.22):

- for 23 activated pumps, there is clearly interference of two oscillatory motions with nearby frequencies, changing the motion of the fluid in the nozzle drastically;
- with 23 activated pumps, the amplitude of the low-frequency mode is greater.

3.6 Long Duct Theory

In a multi-nozzle piezo-electrically driven print-head, a large number of miniature valveless pumps are integrated. In order to obtain a design with a small nozzle pitch, the pumps are placed close to each other. This means that the length of the pump chamber has to be long compared with the cross-sectional dimensions in order to

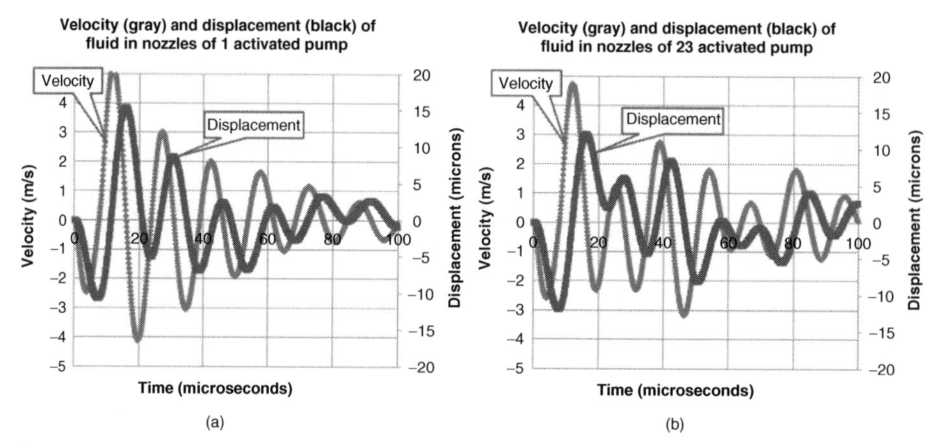

Figure 3.22 *Results for cases in which either a single pump (a) or 23 pumps (b) are activated out of a total of 24. Response in terms of fluid velocity in the nozzle and position of tip of jet of an activated pump for a negative pulse with a relative volume displacement of* 0.45×10^{-5}*, ramp-up and ramp-down times of* $1\,\mu s$ *and a pulse length of* $9\,\mu s$*.*

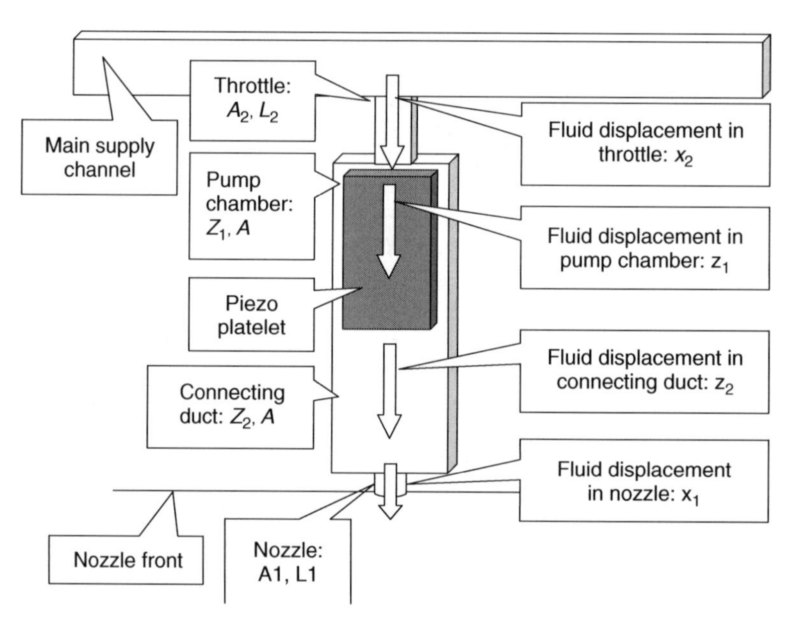

Figure 3.23 *Geometrical data used in the model of the valveless pump, together with the local co-ordinates in the pump chamber and connecting part to the nozzle.*

generate enough volume displacement by the piezo actuator. Rather than a single cavity resonator, we have here a wave tube (Dijksman, 1984, 2003). The basic layout of a wave tube type of pump is shown in Figure 3.23. It consists of a pump chamber, a connecting channel, a nozzle and a restriction. The restriction communicates with the main supply channel.

If the restriction is absent, we have a pump with an open-end design (Wijshoff, 2008) (see also Figure 3.5). The pump chamber is partly surrounded by a piezo-electric actuator. Upon charging the actuator, the volume of the pump chamber changes, causing pressure waves to travel back and forth through the wave tube. At the nozzle, these pressure waves are transformed to fluid velocity and ultimately into a freely flying droplet or a series of freely flying droplets.

The pump section is partly surrounded by the piezo-electric actuator. Apart from end effects, the motion of the actuator is uniform over the active length. To make the calculation simple for the time being, it is assumed that the pump chamber is circularly cylindrical and that the actuator surrounds the pump chamber completely. The continuum equation in cylindrical co-ordinates is (Bird, Stewart and Lightfoot, 2002c):

$$\frac{\partial \rho}{\partial t} + \frac{1}{r}\frac{\partial}{\partial r}(\rho r v_r) + \frac{\partial}{\partial z_1}(\rho v_z) = 0$$

Integration of the continuum equation over the cross-section of the fluid column contained in the pump section gives (for v_r radial velocity, v_z velocity in z_1 direction and v_1 the mean velocity in z_1 direction):

$$AZ_1\left\{\frac{\partial \rho}{\partial t} + \frac{\partial}{\partial z_1}(\rho v_1)\right\} + O\rho_0 Z_1 v_r|_{wall} = 0$$

with:

$$OZ_1 v_r|_{wall} = -\frac{d\Delta V_0}{dt}$$

where O is the perimeter of the pump chamber. Note that ΔV_0 is defined as positive for a volume decrease (see Section 3.3). The linearised equations of motion and continuity and the equation of state are:

$$-\frac{\partial p_1}{\partial z_1} = \rho_0\frac{\partial v_1}{\partial t}$$

$$\frac{\partial \rho}{\partial t} + \rho_0\frac{\partial v_1}{\partial z_1} - \frac{\rho_0}{AZ_1}\frac{d\Delta V_0}{dt} = 0$$

$$\frac{\partial p_1}{\partial \rho} = c_1^2$$

We end up with:

$$\frac{1}{c_1^2}\frac{\partial^2 p_1}{\partial t^2} - \frac{\partial^2 p_1}{\partial z_1^2} = \frac{\rho_0}{AZ_1}\frac{d^2\Delta V_0}{dt^2}$$

This equation can be generalised in the sense that it holds for a pump section with arbitrary cross-sectional dimensions and with actuators only partly deforming the wall of the pump chamber.

The equations governing the motion of the fluid inside the valveless pump are listed below:

$$A_2\dot{x}_2 = A\dot{z}_1(0)$$

$$\rho_0 L_2 A_2 \ddot{x}_2 = A_2\{p_{supply} - p_1(0)\}$$

$$\frac{1}{c_1^2}\frac{\partial^2 p_1}{\partial t^2} - \frac{\partial^2 p_1}{\partial z_1^2} = \frac{\rho_0}{AZ_1}\frac{d^2 \Delta V}{dt^2}$$

$$\dot{z}_1(Z_1) = \dot{z}_2(0)$$

$$p_1(Z_1) = p_2(0)$$

$$\frac{1}{c_2^2}\frac{\partial^2 p_2}{\partial t^2} - \frac{\partial^2 p_2}{\partial z_2^2} = 0$$

$$A\dot{z}_2(Z_2) = A_1\dot{x}_1$$

$$\rho_0 L_1 A_1 \ddot{x}_1 = A_1\left\{p_2(Z_2) - \frac{8\pi\sigma}{A_1}x_1\right\}$$

In summary, these equations represent:

- the continuity equation for volume rate of flow through the restriction into the pump chamber;
- the equation of motion of the fluid contained in the restriction;
- the wave equation in the pump section with the forcing term;
- the continuity equations for pressure and volume rate of flow at the transition from pump section to connecting part of the pump chamber to the nozzle;
- the wave equation in the connecting duct between pump section and nozzle;
- the continuity equation for the volume rate of flow at the nozzle;
- the equation of motion of the fluid contained in the nozzle including the capillary pressure.

To solve the homogeneous part of this set of equations, we assume ($i = 1, 2$):

$$p_i = \left(C_{1i}\cos\frac{\omega}{c_i}z_i + C_{2i}\sin\frac{\omega}{c_i}z_i\right)\sin\omega t$$

With the equation of motion in both the pump chamber and the connecting duct ($i = 1, 2$):

$$-\frac{\partial p_i}{\partial z_i} = \rho_0\frac{\partial v_i}{\partial t}$$

We find:

$$-\frac{\partial p_i}{\partial z_i} = \frac{\omega}{c_i}\left(-C_{1i}\sin\frac{\omega}{c_i}z_i + C_{2i}\cos\frac{\omega}{c_i}z_i\right)\sin\omega t = \rho_0\frac{\partial v_i}{\partial t}$$

$$v_i = \frac{1}{\rho_0 c_i}\left(-C_{1i}\sin\frac{\omega}{c_i}z_i + C_{2i}\cos\frac{\omega}{c_i}z_i\right)\cos\omega t + F_i(z_i)$$

$$\int v_i\,dt = \frac{1}{\rho_0\,\omega c_i}\left(-C_{1i}\sin\frac{\omega}{c_i}z_i + C_{2i}\cos\frac{\omega}{c_i}z_i\right)\sin\omega t + F_i(z_i)t + G_i(z_i)$$

For standing waves, the integration constants F_i and G_i are equal to zero. Substitution of the general solutions into the boundary and connection conditions delivers a set of linear equations in the constants C_{ij}:

$$-C_{11} + C_{21}\frac{A}{A_2}\frac{\omega L_2}{c_1} = 0$$

$$-C_{11}\sin\frac{\omega Z_1}{c_1} + C_{21}\cos\frac{\omega Z_1}{c_1} - C_{22}\frac{c_1}{c_2} = 0$$

$$C_{11}\cos\frac{\omega Z_1}{c_1} + C_{21}\sin\frac{\omega Z_1}{c_1} - C_{12} = 0$$

$$\begin{aligned} C_{12}&\left(\frac{A}{A_1}\frac{\omega L_1}{c_2}\sin\frac{\omega Z_2}{c_1} - \cos\frac{\omega Z_2}{c_1} - \frac{8\pi\sigma}{\rho_0\omega c\,2}\frac{A}{A_1^2}\sin\frac{\omega Z_2}{c_1}\right) \\ +C_{22}&\left(-\frac{A}{A_1}\frac{\omega L_1}{c_2}\cos\frac{\omega Z_2}{c_1} - \sin\frac{\omega Z_2}{c_1} + \frac{8\pi\sigma}{\rho_0\omega c_2}\frac{A}{A_1^2}\cos\frac{\omega Z_2}{c_1}\right) = 0 \end{aligned}$$

This set of linear algebraic equations has a solution where C_{11}, C_{21}, C_{12} and C_{22} are all equal to zero. Only if the determinant of the set is zero do non-zero solutions for C_{11}, C_{21}, C_{12} and C_{22} exist. The determinant or characteristic equation in ω is solved by means of a simple scanning method. In that way an infinite number of resonance frequencies ω_j is found. At a certain resonance frequency ω_j apart from a common multiplication factor, the constants C_{1ij} and C_{2ij} can be determined.

A linear ramp-up of the volume is represented by (see Figure 3.14):

$$\Delta V = \Delta V_0\frac{t}{t_{p1}}$$

The particular solutions in the pump section and the reservoir are given by:

$$v_1 = \frac{\Delta V_0}{AZ_1}\frac{Z_1 - z_1}{t_{p1}}$$

$$v_2 = 0$$

By the sudden start of volume increase as shown in Figure 3.14, an infinite number of eigenmodes will be excited and set in motion. The solutions for the pressure and velocity distributions inside the print-head pump are represented by:

$$p_i = \sum_{j=1}^{\infty} B_j\left(C_{1ij}\cos\frac{\omega_j}{c_i}z_i + C_{2ij}\sin\frac{\omega_j}{c_i}z_i\right)\sin\omega_j t, \quad i = 1, 2$$

$$v_1 = \frac{\Delta V_0}{AZ_1}\frac{Z_1 - z_1}{t_{p1}} + \sum_{j=1}^{\infty}\frac{B_j}{\rho_0 c_1}\left(-C_{11j}\sin\frac{\omega_j}{c_1}z_1 + C_{21j}\cos\frac{\omega_j}{c_1}z_1\right)\cos\omega_j t$$

$$v_2 = \sum_{j=1}^{\infty}\frac{B_j}{\rho_0 c_2}\left(-C_{12j}\sin\frac{\omega_j}{c_2}z_2 + C_{22j}\cos\frac{\omega_j}{c_2}z_2\right)\cos\omega_j t$$

The initial condition that $p = 0$ everywhere for $t = 0$ is automatically fulfilled. The initial condition that the velocities are zero for $t = 0$ leads to an infinite set of non-homogeneous and transcendental equations in the constants B_j, which cannot be solved directly. Firstly, it is an infinite set. Secondly, the set is transcendental in the sense that trigonometric functions are combined with terms that are constant or linear in the length co-ordinate z_1.

In order to arrive at an approximate solution, only the first few terms of the series will be taken into account. That means that the condition that the velocity is zero everywhere for $t = 0$ cannot be fulfilled for all z_i $(i = 1, 2)$. This will hold true only at a number of equidistant points along the axis in the pump chamber and the connecting channel. Between these points, the velocities at $t = 0$ are only approximately zero.

In order to estimate the effect of damping (low damping), the same procedure as outlined at the end of Section 3.5 can be used.

To determine the response of the multi-compartment system to an arbitrarily shaped pulse, we use the method developed in Section 3.3. As an example, we show in Figure 3.24 the results for a small-sized pump with a pump chamber of 10 mm length covered over 6 mm with a piezo platelet. The connecting duct to the nozzle measures 4 mm. For the other dimensions, we refer to the other sections of this chapter.

The volume displacement takes place so fast that it takes some time for the pressure wave to reach the nozzle. The first five resonance frequency modes are given in Table 3.2.

To place these values in perspective, we have also calculated the slosh mode frequency, the Helmholtz frequency and the $\lambda/4$ and $\lambda/2$ mode frequencies (with $\lambda = Z_1 + Z_2$) (Table 3.3):

$$f_{slosh\ mode} = \frac{1}{2\pi}\sqrt{8\pi\frac{\sigma}{\rho_0}\frac{A_2}{A_1^2 L_2}}, \quad f_{Helmholtx} = \frac{c}{2\pi}\sqrt{\frac{1}{A(Z_1 + Z_1)}\left(\frac{A_1}{L_1} + \frac{A_2}{L_2}\right)}$$

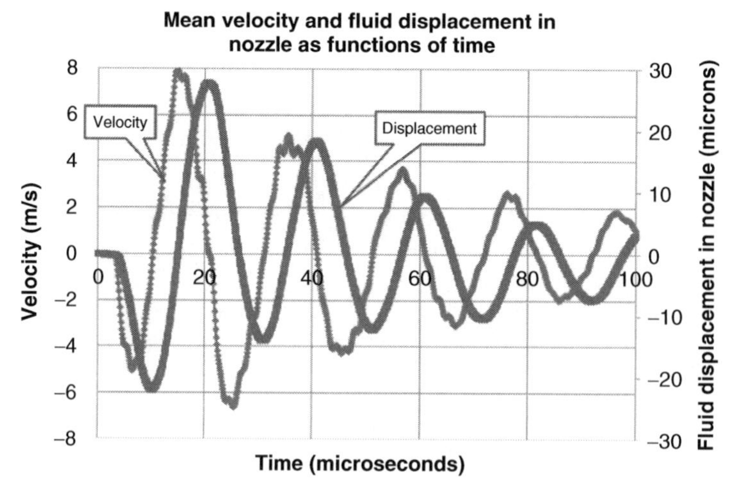

Figure 3.24 *Response of a small-sized inkjet pump in terms of fluid velocity in the nozzle and the position of the tip of the jet for a negative relative volume displacement of* 1.67×10^{-4} *(defined for the pump section) using 12 modes (viscosity of ink 2 mPa s).*

Table 3.2 *First five resonance frequencies of the small-sized pump calculated from the long duct theory.*

Mode number	Resonance frequency (Hz)
1	8772
2	49 014
3	102 012
4	162 214
5	226 528

Table 3.3 *Slosh mode, Helmholtz and λ/2 and λ/4 frequencies.*

Mode type	Resonance frequency (Hz)
Slosh mode	18 905
Helmholtz mode	62 600
λ/4 mode	35 000
λ/2 mode	70 000

3.7 Concluding Remarks

In this chapter we have developed a detailed acoustic model of a piezoelectric-driven valveless inkjet pump, the key building block of a print-head. We started by considering such a pump as a one-degree-of-freedom system to explain the resonance behaviour, damping, cavitation and method of analysis of the response to an arbitrarily shaped driving pulse.

Using the results of this analysis, we outlined an approximate method to predict droplet volume and speed. Before the droplet pinches off from the fluid in the nozzle, it is decelerated by a stretching fluid filament. Different mechanisms to refill the nozzle have been discussed, such as capillary-driven refilling and the effects of Bernoulli underpressure and asymmetric filling of the nozzle.

The key relationship connecting single droplet formation characteristics, actuation and the acoustics of the pump is given by:

$$\frac{v_{max}}{\omega_0 R_1} \approx 1$$

where v_{max} is the maximum velocity in the nozzle generated by the piezo actuation, ω_0 is the frequency of the fluid motion in the nozzle that depends on the acoustic characteristics of the pump design, the pulse shape (positive or negative driving) and the viscosity of the ink and R_1 is the radius of the nozzle.

A multi-nozzle print-head contains a large number of miniature valveless pumps, integrated so that they communicate with a common ink supply channel. In order to predict what happens when a number of pumps is activated simultaneously, we derived a multi-cavity Helmholtz resonator model. This model shows that droplet properties depend on the number of pumps activated, as the fluid velocity in the nozzle and the displaced volume depend on the number of activated pumps.

In order to integrate a large number of pumps at a small pitch, the cross-sectional dimensions of the pump chamber have to be small. To end up with sufficient volume displacement generated by the piezo actuator, the pump chamber must be relatively long. Rather than a single cavity resonator, the set-up is shaped like a long duct in which sound waves travel back and forth. This effect has been analysed in Section 3.6. Higher order waves influence the velocity of the fluid in the nozzle and the fluid jet tip displacement over the course of time.

References

Badie, R. and de Lange, D.F. (1997) Mechanism of drop constriction in a drop-on-demand inkjet system. *Proceedings of the Royal Society of London, Series A: Mathematical, Physical and Engineering Sciences*, **453**, 2573–2581.

Beasley, J.D. (1977) Model for fluid ejection and refill in an impulse drive jet. *Photographic Science and Engineering*, **21**, 78–82.

Bentin, H., Döring, M., Radtke, W. and Rothgordt, U. (1986) Physical properties of micro-planar ink-drop generators. *Journal of Imaging Technology*, **12** (3), 152–155.

Bird, R.B., Armstrong, R.C. and Hassager, O. (1987a) *Dynamics of Polymeric Liquids Fluid Mechanics*, 2nd edn, vol. 1, Wiley-Interscience, Chichester, pp. 14–16.

Bird, R.B., Armstrong, R.C. and Hassager, O. (1987b) *Dynamics of Polymeric Liquids Fluid Mechanics*, 2nd edn, vol. 1, Wiley-Interscience, Chichester, pp. 39–39.

Bird, R.B., Stewart, W.E. and Lightfoot, E.N. (2002a) *Transport Phenomena*, 2nd edn, John Wiley & Sons, Inc., Hoboken, NJ, pp. 185–188.

Bird, R.B., Stewart, W.E. and Lightfoot, E.N. (2002b) *Transport Phenomena*, 2nd edn, John Wiley & Sons, Inc., Hoboken, NJ, pp. 682–683.

Bird, R.B., Stewart, W.E. and Lightfoot, E.N. (2002c) *Transport Phenomena*, 2nd edn, John Wiley & Sons, Inc., Hoboken, NJ, Appendix A.

Bogy, D.B. and Talke, F.E. (1984) Experimental and theoretical study of wave propagation phenomena in drop-on-demand ink jet devices. *IBM Journal of Research and Development*, **28** (3), 314–321.

Burr, R.F., Tence, D.A. and Berger, S.S. (1996) Multiple dot size fluidics for phase change piezoelectric ink jets. NIP12: International Conference on Digital Printing Technologies, The Society for Imaging Science and Technology, pp. 12–18.

Dijksman, J.F. (1984) Hydrodynamics of small tubular pumps. *Journal of Fluid Mechanics*, **139**, 173–191.

Dijksman, J.F. (1999) Hydro-acoustics of piezoelectrically driven ink-jet print heads. *Flow, Turbulence and Combustion*, **61**, 211–237.

Dijksman, J.F. (2003) Long duct analysis of miniature piezoelectrically driven ink-jet printer pumps, in *From Physics to Devices: A Survey of Materials Research at Philips Research Laboratories Eindhoven* (eds B.H. Huisman and J.F. Dijksman), Philips, Eindhoven, The Netherlands, pp 211–221.

Dijksman, J.F., Duineveld, P.C., Hack, M.J.J. *et al.* (2007) Precision ink jet printing of polymer light emitting displays. *Journal of Materials Chemistry*, **17**, 511–522.

Dijksman, J.F. and Pierik, A. (2008) Fluid dynamical analysis of the distribution of ink jet printed biomolecules in microarray substrates for genotyping applications. *Biomicrofluidics*, **2**, 044101.

Döring, M. (1982) Ink-jet printing. *Philips Technical Review*, **40** (7), 192–198.

de Gennes, P-G., Brochard-Wyart, F. and Quéré, D. (2004) *Capillarity and Wetting Phenomena*, Springer-Verlag, Berlin.

Grigorieva, O.V., Grigoriev, D.O., Kovalchuk, N.M. and Vollhardt, D. (2005) Auto-oscillation of surface tension: heptanol in water and water/ethanol systems. *Colloids and Surfaces A: Physicochemical and Engineering Aspects*, **256**, 61–68.

Grigorieva, O.V., Kovalchuk, N.M., Grigoriev, D.O. and Vollhardt, D. (2004) Spontaneous non-linear surface tension oscillations in the presence of a spread surfactant monolayer at the air/water interface. *Colloids and Surfaces A: Physicochemical and Engineering Aspects*, **250**, 141–151.

Groot Wassink, M.B. (2007) Inkjet printhead performance enhancement by feedforward input design based on two-port modeling. PhD dissertation, Delft University of Technology.

Groot Wassink, M.B., Bosch, N.J.M., Bosgra, O.H. and Koekebakker, S. (2005) Enabling higher jet frequencies for an inkjet printhead using iterative learning control. Proceedings of the IEEE Conference on Control Applications, pp. 791–796.

He, B., Neelesh, A., Patankar, N.A. and Lee, J. (2003) Multiple equilibrium droplet shapes and design criterion for rough hydrophobic surfaces. *Langmuir*, **19**, 4999–5003.

Helmholtz, H. (1885) *On the Sensations of Tone*, Longman & Co, New York, pp. 43, 372–374. (Republished by Dover Publications, New York (1954))

Hutchings, I.M., Martin, G.D. and Hoath, S.D. (2007) High speed imaging and analysis of jet and drop formation. *Journal of Imaging Science and Technology*, **51**, 438–444.

de Jong, J. (2007) Air entrapment in piezo inkjet printing. PhD dissertation, University of Twente.

de Jong, J., de Bruin, G., Reinten, H. *et al*. (2005) Acoustical and optical characterisation of air entrapment in piezo driven inkjet printheads. *Proceedings IEEE Ultrasonics*, **2**, 1270–1271.

de Jong, J., Jeurissen, R., Borel, H. *et al*. (2006) Entrapped air bubbles in piezo-driven inkjet printing: their effect on the droplet velocity. *Physics of Fluids*, **18**, 121511.

Kitahara, T. (1995) Ink jet head with multi-layer piezoelectric actuator. Eleventh International Congress on Advances in Non-Impact Printing Technologies, The Society for Imaging Science and Technology, pp. 346–349.

Kurz, H. (1980) *Tintenstrahldrucker. Philips Unsere Forschung in Deutschland Band III*, Philips Forschungslabor Aachen, Aachen, The Netherlands, pp. 194–196.

Lee, E.R. (2003) *Mirodrop Generation*, CRC Press, Boca Raton, FL, pp. 9–10.

Lee, F.C., Mills, R.N. and Talke, F.E. (1984) The application of drop-on-demand ink jet technology to color printing. *IBM Journal of Research and Development*, **28** (3), 307–313.

McDonald, M. (1996) Scaling of piezoelectric drop-on-demand jets for high resolution applications. NIP12: International Conference on Digital Printing Technologies, Society for Imaging Science and Technology, pp. 53–56.

Rayleigh, J.W.S. (1896) *The Theory of Sound*, MacMillan Company, New York, pp. 170–172. (Republished by Dover Publications, New York (1945))

Rosenstock, G. (1982) Erzeugung schnell fliegender Tropfen für Tintendrucker mit Hilfe von Druckwellen. PhD thesis, University of München.

Shield, T.W., Bogy, D.B. and Talke, F.E. (1987) Drop formation by DOD ink-jet nozzles: a comparison of experiment and numerical simulation. *IBM Journal of Research and Development*, **31** (1), 96–110.

Stemme, E. and Larsson, S.G. (1973) The piezoelectric capillary injector – a new hydrodynamic method for dot pattern generation. *IEEE Transactions on Electron Devices*, **20**, 14–19.

Stückrad, B., Hiler, W.J. and Kowalewski, T.A. (1993) Measurement of dynamic surface tension by oscillating droplet method. Experiments in Fluids, **15**, 332–340.

Usui, M. (1996) Development of the new multilayer actuator head (MACH with multilayer ceramic with hyper-integrated piezi segments). NIP12: International Conference on Digital Printing Technologies, Society for Imaging Science and Technology, pp. 50–53.

Wijshoff, H. (2008) Structure-and fluid-dynamics in piezo inkjet print-heads. PhD thesis, University of Twente.

Yarin, A.L. (2006) Drop impact dynamics: splashing, spreading, receding, bouncing .... *Annual Review of Fluid Mechanics*, **38**, 159–192.

4

Fluids for Inkjet Printing

Stephen G. Yeates, Desheng Xu, Marie-Beatrice Madec, Dolores Caras-Quintero, Khalid A. Alamry, Andromachi Malandraki and Veronica Sanchez-Romaguera
Organic Materials Innovation Centre, School of Chemistry, University of Manchester United Kingdom

4.1 Introduction

In this chapter, we will address the various components and their function within an inkjet fluid. At the simplest level the components within an inkjet fluid fulfil one of two primary roles. Firstly there are components such as solvent, humectants and surfactants which are present to enable a stable ink to be formulated which is capable of being jetted reliably under standard operating conditions, and secondly there are components such as dyes, pigments and functional materials whose role is to give an effect to the printed substrate. It is not possible to give prescriptive formulation guidance in a text such as this but instead to make the reader aware of the important design rules as well as the exciting opportunities which exist. It should be recognised from the outset that the reliability and printing characteristics of any ink are not inherent properties of the formulation but are strongly dependent on the print-head and process architecture, thus requiring a holistic approach to ink design.

Therefore in the first part of this chapter we focus upon the constraints imposed by the different head technologies on the types of inks which can be successfully printed. Since inkjet digital fabrication is dominated by piezoelectric drop-on-demand (DOD) technology, we will then focus on the fundamental principles of droplet formation and its relation to successful fluid formation for DOD printing, before considering the different ink technologies available, their specific formulation attributes and examples of application in digital fabrication.

Inkjet Technology for Digital Fabrication, First Edition. Edited by Ian M. Hutchings and Graham D. Martin.

4.2 Print-Head Considerations

As described in Chapter 2, there are two major mechanisms through which drops are generated and positioned: continuous inkjet printing (CIJ) and Drop On Demand (DOD) inkjet printing. For CIJ and DOD the individual drop generation process and ink management systems present different challenges and constraints on ink design, making them more or less suitable for the specific application in mind.

4.2.1 Continuous Inkjet (CIJ)

In CIJ the liquid ink passes through a chamber where a vibrating piezoelectric crystal creates an acoustic wave causing the emerging stream of liquid to break into droplets via the Rayleigh instability (Cameron, 2006). The ink droplets, which have conductivity typically between 50 and 2000 Ω cm, are electrostatically charged by induction as they break off and then directed (deflected) by electrostatic deflection plates to print on the receptor material (substrate), or allowed to continue on undeflected to a collection gutter for re-use. Only a small fraction of the droplets generated are used to print, the majority being recycled. Since the non-printed ink is recycled, the ink system requires active solvent regulation to counteract solvent evaporation during the time of flight (the time between nozzle ejection and gutter recycling) and from the venting process whereby air that is drawn into the gutter along with the unused drops is vented from the reservoir. Viscosity is monitored, and a solvent (or solvent blend) is added in order to counteract any solvent loss.

Specific ink formulation challenges for CIJ include:

- ink stability at high drop generation frequency with elongation shear rates up to $1\,000\,000\ s^{-1}$;
- the ability to recycle ink without loss of jetting performance or final product performance;
- the generation of stable charge through the use of additives such as potassium thiocyanate (0.5–2.0 wt%) which may have a negative effect on final device performance, especially in electronic and biological print applications.

4.2.2 Thermal Inkjet (TIJ)

The majority of DOD TIJ printers use print cartridges with a series of tiny electrically heated chambers constructed by photolithography behind each nozzle (Dawson, 2004; Cameron, 2006). To produce a droplet an electrical current is passed through the heating elements causing a vapour bubble to form in the chamber. When the ink droplet is ejected the heat pulse is removed, causing the bubble to collapse which pulls a further charge of ink into the chamber through a narrow channel attached to an ink reservoir. TIJ requires the fluid to have a high vapour pressure and low boiling point, and consequently water-based inks containing dyes or pigments are exclusively used.

TIJ has found wide utility in both small office and home office (SOHO) and industrial graphics applications, but the effective limitation to water-based inks for simple head architectures has limited its use in digital fabrication applications. Specific formulation challenges include the avoidance of the ink drying at the nozzle plate and the formulation of inks which are stable with respect to 'kogation' (Shirota *et al*., 1996), the phenomenon whereby ink ingredients are thermally decomposed and deposited on the surface of the heating element. More recent print-head designs have been proposed where direct contact between the heater and the ink is eliminated, thus minimising issues of kogation and extending the range of potential fluids (Sen and Darabi, 2007).

4.2.3 Piezoelectric Drop-on-Demand (Piezo-DOD)

Piezo-DOD inkjet printers use a piezoelectric material adjacent to the ink chamber behind each nozzle. When a voltage is applied, the piezoelectric material changes shape or size, which generates a pressure pulse in the fluid, forcing a droplet of ink with volume typically in the range 1–100 pl from the nozzle (Dijksman, 1984; Fromm, 1984). This is essentially the same mechanism as thermal inkjet but generates the pressure pulse by using a different physical principle. It is discussed in more detail in Chapters 2 and 3.

Because the drop generation process imposes fewer design constraints on the ink than either CIJ or thermal DOD, piezo-DOD has found wide utility in SOHO, commercial and industrial applications. The ability to print a wide range of different fluid formulations is the reason why the majority of new digital fabrication applications utilise piezo-DOD inkjet head technology (Dijksman *et al*., 2007; Sanchez-Romaguera, Madec and Yeates, 2008).

4.3 Physical Considerations in DOD Droplet Formation

As outlined in Chapter 1 and discussed in more detail in Chapters 2, 3 and 5, the printing process consists of three distinct stages: drop generation, drop flight and drop impact. Each of these operations imposes physical limitations on both the ink formulation and the printing process.

In a DOD print-head, the pressure pulse that drives drop formation can be generated (as we have seen in Section 4.2) either by a piezoelectric actuator or by a thermal pulse creating a vapour bubble. Liquid is contained at the orifice by surface tension and if the pressure pulse exceeds some critical value, a liquid column protrudes from the orifice with surface tension leading to an unstable neck forming and a drop separating. The time scale for the whole process is of the order of 5–250 μs and is shown schematically in Figure 4.1.

The important physical properties of a fluid forming droplets are its density ρ, surface tension σ and viscosity η, commonly grouped in dimensionless form using the Weber number $We = v^2 d\rho/\sigma$ and Reynolds number $Re = vd\sigma/\eta$, where v and d are the fluid velocity and a characteristic length respectively (see Chapter 2). The Weber number is important in the early stage of drop generation where the interplay of forward inertia and surface tension is important (period I in Figure 4.1), and the Reynolds number in the middle to latter stages where the interplay with the viscoelasticity of the ligament is critical (period II).

Fromm (1984) used the dimensionless Ohnesorge number $Oh = We^{1/2}/Re$ to remove the velocity dependence and calculated that DOD printing should be possible if $Oh < 0.5$, with the drop volume increasing as Oh decreases. Reis, Ainsley and Derby (2005)

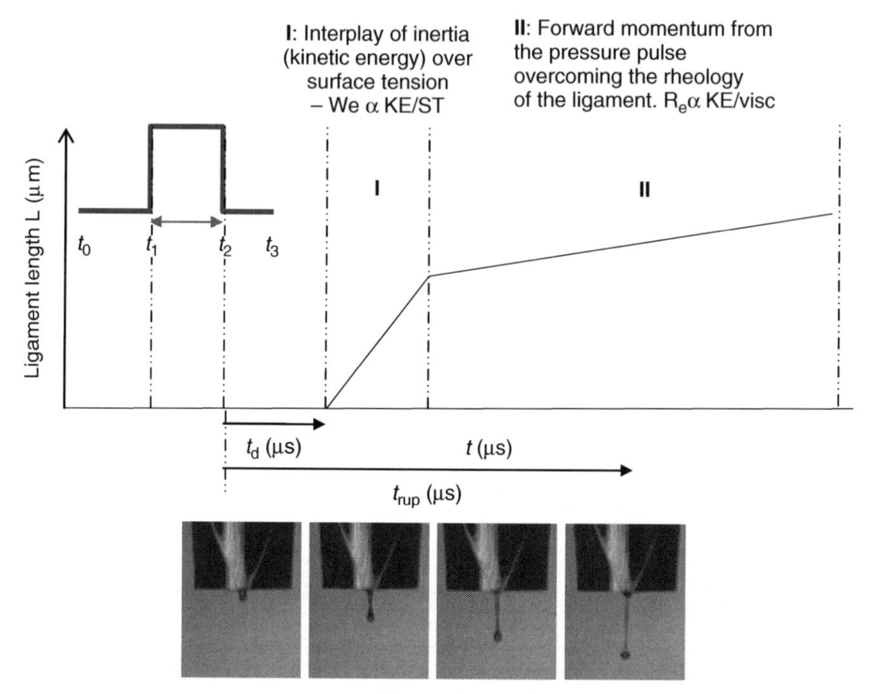

Figure 4.1 *Schematic representation of pressure pulse generation and droplet ejection for DOD inkjet.*

investigated the inkjet printing conditions for a range of different fluids and further explored the influence of Oh through fluid dynamics simulations to define a range $0.1 < \mathrm{Oh} < 1$, within which DOD printing is possible. When $\mathrm{Oh} > 1$ viscous dissipation within the liquid prevents drop formation, and if $\mathrm{Oh} < 0.1$ the balance between surface tension and viscosity results in the liquid breaking into a series of satellite drops rather than a single drop as desired. They found that, for a range of fluids of different physical properties, printing was possible within these limits and that, for a number of different fluids ejected through the same print-head, the droplet volume was controlled by the Ohnesorge number (Dijksman *et al*., 2007), consistent with the prediction of Fromm (1984). In practice, systems in which Oh is much less than 0.1 are printable so long as the satellites merge with the main droplet.

A further consideration for inkjet printing is aerodynamic drag on the droplet in flight. This can slow the drop's velocity significantly before impact and also allow random fluctuations in the surrounding atmosphere to deflect a drop away from its desired trajectory, which is especially important where absolute drop placement is critical. Stringer and Derby (2009) showed that the critical value of Re will vary from 19.5 for a liquid of viscosity 1 mPa s to 0.48 for a liquid of viscosity 40 mPa s.

The final stage of the printing process is drop impact, discussed in detail in Chapter 5. During the impact process, the drop loses kinetic energy through viscous dissipation and by surface extension. A balance of surface forces defines the final equilibrium drop shape, and substantial oscillations may occur before this is achieved. However, if the impact velocity is too high the drop may splash, defining the limiting drop velocity for printing which is strongly dependent on the roughness and porosity of the surface (Hoath *et al*., 2009).

These considerations can be used to define the limits of drop generation, drop flight and drop impact in a parameter space, in terms of the Reynolds and Weber numbers, within which DOD inkjet printing is feasible under normal atmospheric conditions at standard temperature and pressure (Stringer and Derby, 2009). From these considerations, the optimal ranges of static values for DOD inkjet printing are typically with viscosity (η) lying in the range of 2–20 mPa s and surface tension (γ) in the range of 20–50 mN m^{-1}. It is broadly against these criteria that inks are initially formulated and specified, although narrower ranges may be applicable to specific print-heads.

However, given the high elongational shear rates experienced by the ink upon drop generation, typically greater than $10^5\ s^{-1}$, it is the dynamic properties which are the most important. It is either difficult or impossible to measure these directly in the laboratory, and consequently direct visualisation of the drop generation process becomes a key aspect in successful ink design (Hoath *et al*., 2009). For many digital fabrication opportunities, it is necessary to print high-molecular-weight polymer at high concentration in order to maximise either material throughput and/or functionality. The addition of polymer to an ink has a strong effect on the nature of the drop generation and ejection processes (Christianni and Walker, 2001; De Gans *et al*., 2004; Store and Harrison, 2005).

The influence of added polymer on drop formation and filament break-up has been studied as a function of concentration, typically in the dilute regime for polyacrylamide (Meyer, Bazileyky and Rozkhov, 1997), polyethylene oxide (Store and Harrison, 2005), polystyrene (Dawson, 2004), poly[2-methoxy-5-(2′-ethylhexyloxy)-1,4-phenylenevinylene] (MEH–PPV) (Tekin *et al*., 2007) and cellulose ester (Xu *et al*., 2007). Four different regimes have been observed in inkjet drop generation behaviour as a combined function of concentration and molecular weight as illustrated in Table 4.1 (Christianni and Walker, 2001; De Gans *et al*., 2004; Store and Harrison, 2005), with regime 3 offering the desired result.

Rather than consider the polymer concentration within an ink in terms of mass per unit volume (e.g. g/dl), it is better to define the reduced concentration as $[\eta]$.c or c/c*, where c* is the coil overlap concentration, the point at which individual polymer chains in solution are just in contact, and $[\eta]$ is the polymer intrinsic viscosity in dl/g and $c^* = 1/[\eta]$ (Hamley, 2000). Regime 3 in Table 4.1 is typically observed for reduced concentrations between 0.2 and 1.0 in which relative viscosity approximately scales as c^1 and above which it scales as $c^{15/4}$, as shown in Figure 4.2 for cellulose ester in a good solvent γ-butyrolactone over the range $M_w = 6–155$ kDa (Xu, 2009).

Drop break-up behaviour of polymer-containing inks is not simply a function of the effect of polymer concentration on the low-shear, static, solution viscosity typically measured by standard laboratory techniques. It has been found that drop break-up behaviour is strongly related to the strain hardening resulting from the presence of polymer at high strain rate (De Gennes, 1974; Rabin, 1986; De Gans *et al*., 2004). The micro-rheological explanation for strain hardening is the sudden transition of the polymer chain from a coiled to a stretched state, which is accompanied by a strong increase of the hydrodynamic drag which must relax over the timescale of the drop formation process as shown schematically in Figure 4.3.

Table 4.1 *Details of the four different regimes observed in polymer solution inkjet drop generation behaviour.*

Regime	Characteristic behaviour	Typical picture
1	The first regime occurs at either very low concentrations and/or very low molecular weight, where a long tail is formed that simultaneously disintegrates along its axis to form several satellite droplets. This regime can often be a highly chaotic and irreproducible in nature. Characteristic of the dilute polymer regime.	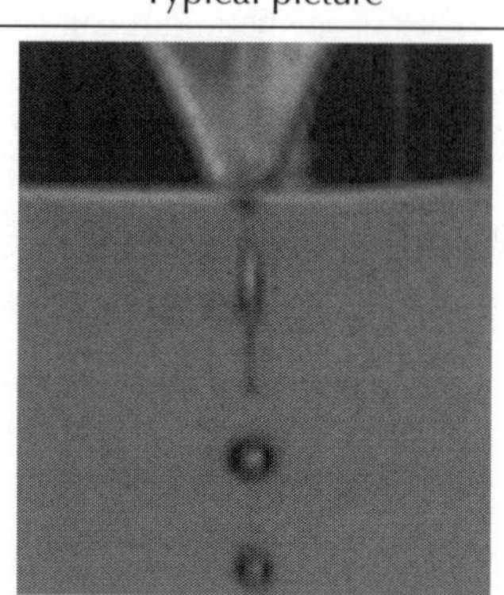
2	The second regime occurs upon increasing concentration or molecular weight when only a few satellites appear at the tail's end.	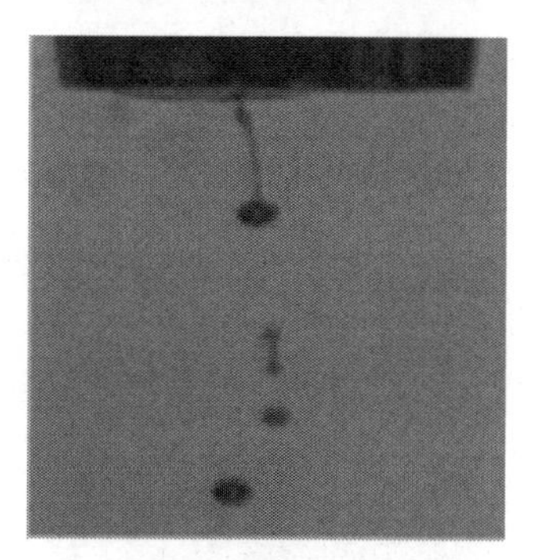
3	Raising concentration or molecular weight further yields a single droplet without a tail. Characteristic behaviour of dilute to semi-dilute solutions.	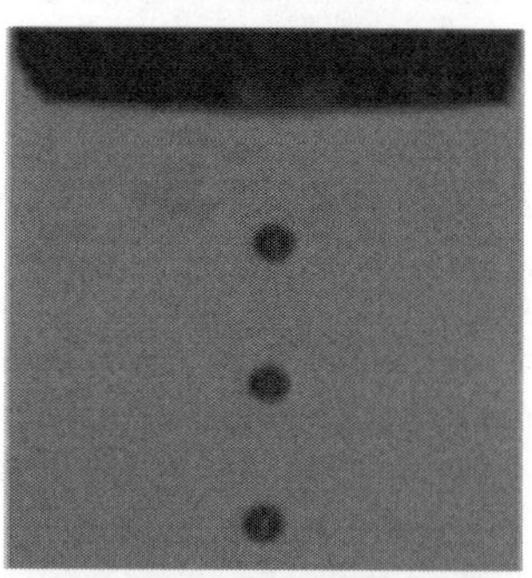
4	At high concentration or molecular weight, the polymer solution becomes highly viscoelastic and the droplet does not detach and returns into the nozzle. Characteristic of the concentrated regime.	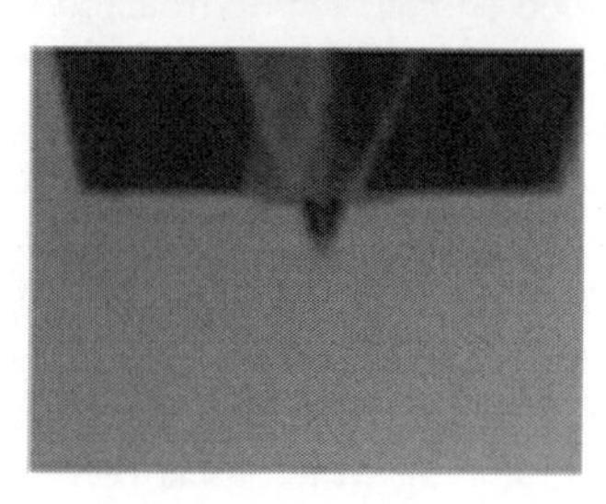

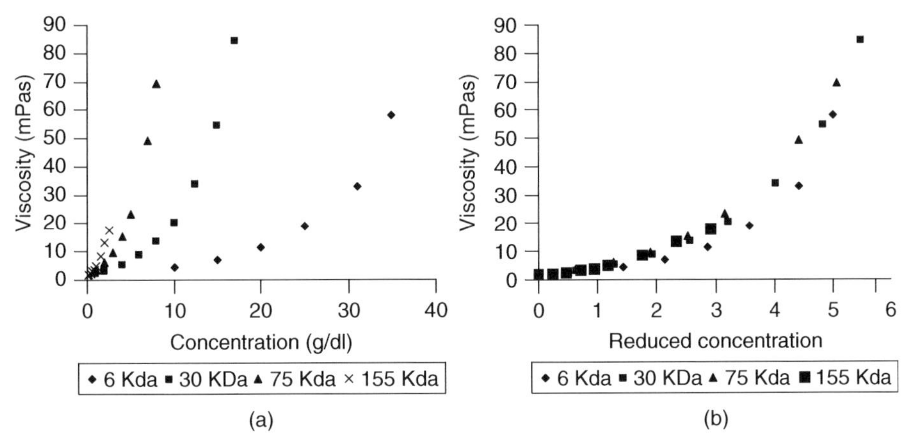

Figure 4.2 *Effect of polymer molecular weight and concentration on solution viscosity expressed as (a) concentration in g/dl and (b) reduced concentration c/c* for cellulose ester in γ-butyrolactone (Xu, 2009).*

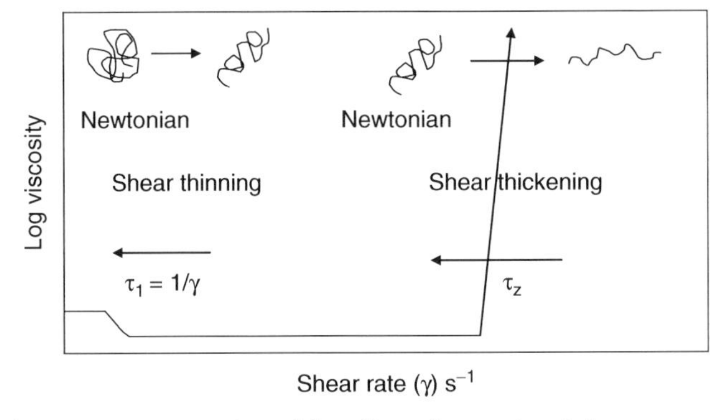

Figure 4.3 *Schematic representation of the effect of extensional shear rate on polymer chain conformation and solution viscosity.*

The coil–stretch transition occurs for linear polymers at a critical strain rate (ε_{crit}) where the rate of deformation of the chain exceeds its rate of relaxation so that it passes from a slightly distorted random coil to an extended state. This critical condition is achieved when the critical Weissenberg number ($Wi_{ecrit} = \varepsilon_{crit.}\tau_z$) > 0.5, where τ_z denotes the longest relaxation time (De Gennes, 1974; De Gans *et al*., 2004). The longest polymer chain relaxation time is typically described by the Zimm non-free-draining relaxation time $\eta_s[\eta]M_w/RT$, where η_s is the viscosity of the solvent, $[\eta]$ is the intrinsic viscosity of the polymer solution and M_w is the weight average molecular weight. The apparent universality of Zimm behaviour is believed to arise from either the fact that elongational flow experiments probe only the dynamics of the partially stretched coil, or the fact that the coil–stretch transition is essentially non-equilibrium since molecules experience only a finite residence time in the flow field (Rabin, 1986). For the inkjet process the strain rate at the nozzle tip is such (>50 000 s^{-1}) that this critical condition is typically exceeded for all polymers having M_w > 20 000 Da at the

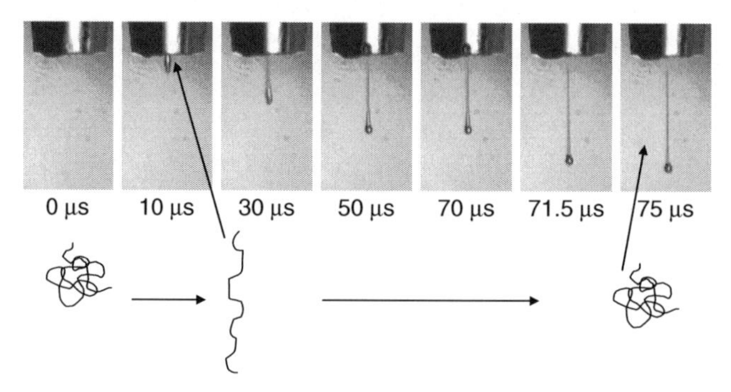

Figure 4.4 Change in polymer coil confirmation during inkjet drop generation for cellulose ester in γ-butyrolactone at c/c* = 0.6 using a MicroFab print head (Xu et al., 2007; Xu, 2009).

pinch region, the point at which the ligament is attached to the nozzle tip meniscus (Christianni and Walker, 2001), as illustrated in Figure 4.4.

It should be noted, however, that the inkjet process can give rise under certain conditions to degradation in molecular weight. Model studies on the DOD inkjet printing of poly(methylmethacrylate) and polystyrene in good solvent showed that polymers having a weight average molecular weight (M_w) either less than 100 kDa or greater than ~1000 kDa show no evidence of molecular weight degradation (Figure 4.5). The lower boundary condition is a consequence of a low Deborah number imposed by the print-head geometry, and the upper boundary condition due to the short residence time of the polymer in the extensional flow field (Alamry *et al*., 2011). For intermediate molecular weights the effect is greatest at a high elongational strain rate and low solution concentration, with higher polydispersity polymers being most sensitive to molecular weight degradation. For low-polydispersity samples, PDi $\leq$ 1.3 chain breakage is essentially centro-symmetric, induced either by overstretching when the strain rate increases well beyond a critical value (i.e. the stretching rate is high enough to exceed the rate of relaxation) or by turbulence. For higher polydispersity samples chain breakage is consistent with almost random scission along the chain, inferring that the forces required

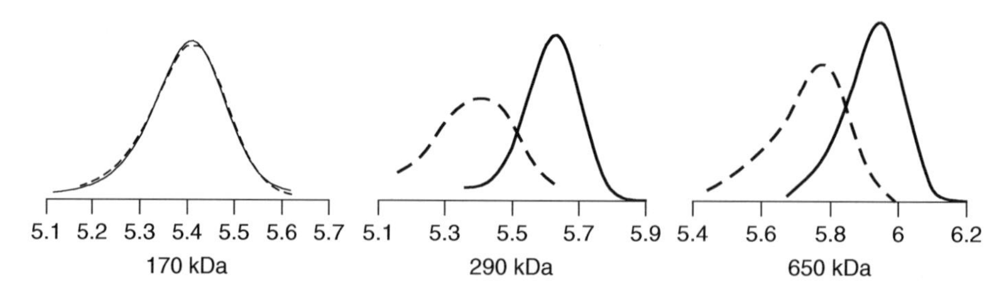

Figure 4.5 (WF/dLogMwt against log mol wt) before (—) and after (– –) single-pass jetting for low-polydispersity polystyrene in tetralin at c/c* = 0.15 jetted at 18–26 V using Dimatix 10 pl DMP Printhead at 25 °C. WF: weight flow (Alamry et al., 2011).

to break the chain are additionally transmitted by valence bonds (i.e. network chains and junctions or discrete entanglements) rather than solely by hydrodynamic interaction. These observations have implications with respect to the printing of functional and biological materials where retention of polymer functionality is critical.

4.4 Ink Design Considerations

The main differences between SOHO and industrial inkjet printers relate to the materials of construction of the print-head, the mode, frequency and size of the generated droplet, the absolute and relative accuracy of droplet placement on the substrate and the overall speed of production. In formulating a robust inkjet ink, consideration must be given to the following aspects:

- **ink chemistry**: ability to formulate stable inks which do not detrimentally interact with the materials of construction of the inkjet printer;
- **ink flow within the print-head**: ability to wet the different surfaces throughout the head allowing ease of priming and avoidance of bubble nucleation;
- **create a stable meniscus at the nozzle**: dependent upon whether a wetting or non-wetting faceplate technology is used, and on the need to balance the surface tension forces of the meniscus and gravity feed from the ink reservoir;
- **droplet formation and jet stability**: stable droplet formation with control of ligament length, drop size, satellite formation and directionality;
- **drop–substrate interaction**: interaction of the droplet with the surface to enable the correct spatial deposition of the functional material in the x, y and z planes;
- **in-use functionality**: dried printed ink must have the correct topology, morphology and physical properties to perform the required function.

In any ink development programme it is critical that the print-head technology is carefully specified with respect to the needs of the print process followed by definition of the type of substrate to be printed, which will enable the initial definition of the appropriate surface tension and viscosity limits of the ink. Carefully attention is paid at all times to component purity as well as filtration and degassing of the ink as these all help successful ink formulation. Once the operating windows for these parameters have been determined, preliminary inks can be printed and initial print evaluations carried out. An iterative process will then follow where various parameters, defined by the specific application, are varied in order to improve the printing characteristics, with the capability to visualise droplet formation from the print-head and print on the substrates of interest being important. Once candidate inks are identified, they are subjected to a more comprehensive series of tests looking at aspects such as kogation, degassing and long-term stability.

4.5 Ink Classification

A variety of different ink technologies have been developed for inkjet applications. These range from aqueous inks, which are found in most SOHO-based inkjet printers, to oil-based and UV curing inks which are used in industrial and packaging-type applications. Digital fabrication applications have begun to extend the formulation envelope of existing inks whilst requiring new ink chemistries to be developed.

Inkjet inks can be classified into five generic classes based upon the nature of the solvent or carrier vehicle:

- water-based or aqueous inks, which often contain up to 20% of organic co-solvents;
- 'oil'-based inks based on very-high-boiling-point long-chain glycols and hydrocarbons;
- phase-change or hot-melt inks based on waxes and long-chain organics which are jetted at elevated temperature above the melting point of the 'solvent';
- UV-curable inks which can be either 100% reactive (solvent-free) or diluted with solvent or water;
- solvent-based inks typically based on methyl ethyl ketone (MEK), alcohols, short-chain glycols, lactates and aromatics.

4.5.1 Aqueous Ink Technology

The main advantage that the water-based inks developed over the last 30 years for both thermal and piezo inkjet applications have over solvent-based inks is that they contain little or no volatile organic components (VOCs). The initial application for these inks was in SOHO, where it was desirable to be able to print text with some colour capability on porous media. As the use of the technology has expanded, the requirements placed on the ink have also grown, for example the output from inkjet printers for offices is now required to match laser printing text quality, or to match silver halide permanence for digital photography. To cope with the greater demands, dyes and pigment dispersions have been specifically designed for inkjet applications where precise requirements are placed on the ink properties of the colorants, as well as on the image properties. As the speed and flexibility of inkjet print-head technologies have increased, the number of different potential applications has also increased. A typical aqueous ink formulation is given in Table 4.2, which also indicates the function of each component.

Whilst the subject of this text is focused upon the use of inkjet for digital fabrication, it is useful to review the body of knowledge used in the successful development of aqueous inks based upon dyes and pigments for SOHO.

Table 4.2 *Water-based inkjet ink composition (Hue, 1998).*

Component	Function	Concentration (wt%)
Deionised water	Aqueous carrier medium	60–90
Water soluble co-solvent	Humectant, viscosity control and enhance dye solubility	5–30
Dye or pigment	Provides colour	1–10
Surfactant	Wetting and penetration	0.1–10
Biocide	Prevents biological growth	0.05–1
Buffer	Controls the pH of the ink	0.1–0.5
Other additives	Chelating agents, defoamers, solubilisers, binders and charging agent	>1

4.5.1.1 Inkjet Colorants

Although, as discussed in Chapter 1, inkjet printing was initially developed during the 1960s and 1970s, it was in the mid-1980s – when Canon and Hewlett-Packard successfully introduced printers for the SOHO – that the real growth in the use of inkjet technology was observed. Initially, the demands placed on the ink were associated with getting colorants onto paper using this new technology. The inks were originally based on water-soluble textile dyes, which proved to be extremely unreliable with respect to nozzle blockage and this was subsequently overcome through further processing and purification. Today the colorant can be either dye or pigment, the choice being determined by the particular application for which the ink is being designed (Gregory, Katrizky and Sabongi, 1991).

The choice of dye depends on the ink used, on whether it is aqueous, solvent or hot melt, and on the type of printer (thermal or non-thermal). However, irrespective of the solvent system, all inkjet dyes have to satisfy a number of stringent criteria listed in Table 4.3.

Many SOHO and graphics inkjet applications are dominated by aqueous ink technology where the aqueous colorant is made water soluble or dispersible through the introduction of an ionised anionic group such as a carboxylate or sulfonate salt. Cationically stabilised colorants are not preferred since the acidity of the resultant ink gives rise to corrosion issues within the print-head. Purely non-ionically stabilised colorants, through the introduction of polyethylene oxide functionality, similarly are not preferred since water sensitivity persists after printing. In the case of anionic aqueous dyes the chromophores are usually based on azo or phthalocyanine chemistry (Gregory, Katrizky and Sabongi, 1991), as illustrated in Figure 4.6.

The salt form of the dyes varies depending on the dye structure and the solubility of the complex, but is typically sodium, lithium or ammonium. The main advantages of dyes over pigments include the large number of chromophores available that can be fine-tuned to deliver the specific colour needs of particular applications, together with the colour gamut and the formulation latitude that these chromophores can supply. In addition, dyes have very good special media (photographic) performance with no gloss issues. However, dyes do not generally display good light or ozone fastness compared with pigments, and there tends to be a compromise between dyes that are bright but not very light or ozone fast, and dyes that have a lower colour gamut but generate images that have extremely good permanence. It must also be stressed that the permanence of

Table 4.3 *Property requirements of inkjet dyes.*

Colour	Yellow, cyan, magenta and black
Colour strength	High optical density at low concentration
Solubility	5–20 wt%
Insolubles	<0.5 μm
Electrolytes and metals	Cl^-, SO_4^{2-}, Ca^{2+} ... (ppm)
Fastness	Light, water, smear and ozone
Shade	Same on different substrates; print definition
Toxicology	Ames −ve
Thermal stability (TIJ)	Avoid build-up of insoluble deposits (kogation)

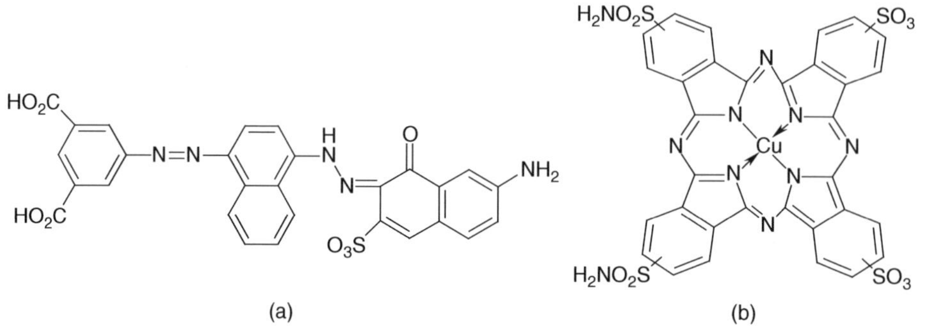

(a)

(b)

Figure 4.6 *Examples of (a) azo and (b) phthalocyanine inkjet dyes.*

each dye is directly dependent upon the structure of the substrate on which it is printed, and the conditions under which the image is stored. In recent years there has been a drive towards the use of micro-porous media in preference to swellable polymer materials for photographic inkjet applications, where the use of alumina and silica provides substrates that look and feel like traditional photographs, and deliver rapid drying characteristics enabling fast printing. Unfortunately, it would appear that in most cases the permanence of dyes is reduced when placed on substrates of this nature.

The fade resistance of dyes can be improved by the use of specific additives to the ink. The exact mechanism of fade is unknown in most cases, but it is believed that most fading processes involve oxidation of the chromophore via either free radical processes or the interaction of the dye with species such as singlet oxygen. In some cases, depending upon the substrate composition and the nature of the dye, the fade mechanism could also involve a reduction process. There are a variety of additive classes that can be included in the ink in an attempt to prevent fading. As UV light is believed to be more damaging than visible light, UV absorbers or quenchers such as 2-hydroxyphenyl benzophenones or metal salts are used. Antioxidants such as hindered phenols could also be added to intercept the oxidative processes. This approach is extremely difficult to deliver as the effectiveness of the additives will depend on the fading mechanisms for each dye and media type as discussed here, but will also rely upon the location of the additive relative to the dye once deposited on the substrate.

Developments in dye chromophore technology are widely reported in the patent literature and are outside the scope of this chapter, but highlight is made of recent work on polymeric dyes where the chromophore is in the main chain, chain pendant or terminal as a method for designing specific media interaction properties into the dye (Avecia Ltd., 2002). Dye encapsulation, however, has been less widely reported.

The use of pigments has now become widely established in both SOHO and industrial applications with advances in pigment stabilisation technology. To ensure good operability it is necessary that the pigment particles are colloidally stable and have a desired typical particle size in the region of 100–200 nm with no large particle shoulder, so that they do not flocculate and cause nozzle blockages. Either a dispersant is included in the ink or the pigment particles are chemically modified so that they are self-dispersing (Spinelli, 1998). The main advantages of pigment-based inks are related to the colour density that can be generated, particularly for black text, and the permanence of the image, where good water fastness and excellent resistance to fading by light and ozone are observed. However, pigments are by their nature particulate, and it is not always straightforward to achieve the correct particle size distribution or acceptable stability of the dispersion. If the colloidal dispersion is made too stable, re-peptisation of the pigments (re-dispersion of the coagulated colloidal particles in the presence of water) may occur leading to less water-fast images. In addition, images formed from pigments do not generally display the colour gamut available with dye-based inks, and as the particles are deposited on the surface of the substrate, they can be easily smudged or smeared. The performance of many pigmented inks is also unacceptable on glossy photographic media as the particles do not tend to penetrate the media surface, leading to unevenness of the gloss.

Recent developments in pigment technology have seen the use of polymer encapsulation in order to give better control of particle size and distribution as well as to build in better substrate interaction properties, particularly with respect to use in photographic media (Leelajariyakul, Noguchi and Kiatkamjornwong, 2008).

4.5.1.2 Colorant–Substrate Interactions

In the case of aqueous colorants such as water-soluble dyes or functionalised pigments, both the water fastness and the quality of the image generated on plain papers can often be poor. Papers specially designed for inkjet applications tend to contain mordants or fixing agents such as polyvalent metal ions or quaternary amines. A mordant is a substance used to fix anionic stabilised aqueous colorants onto a surface by forming an insoluble complex. Polyvalent metal ions such as calcium acetate are effective at improving water fastness but may have a negative effect on dye chroma and shade. Polymeric quaternary amines such as polyethyleneimine, poly(diallyldimethyl ammonium chloride) and poly(hexamethylene biguanide) have been reported demonstrating good water fastness and improved optical density, although performance is critically dependent upon the net charge between the dye and/or pigment and complexing agent if resolubility of the dye complex is to be overcome. Figure 4.7 shows the effect of mordant–pigment interaction for a sulfonate functionalised carbon black pigment on plain and treated paper. Here we see that in the presence of pigment–fixer interaction, the pigment is localised at the

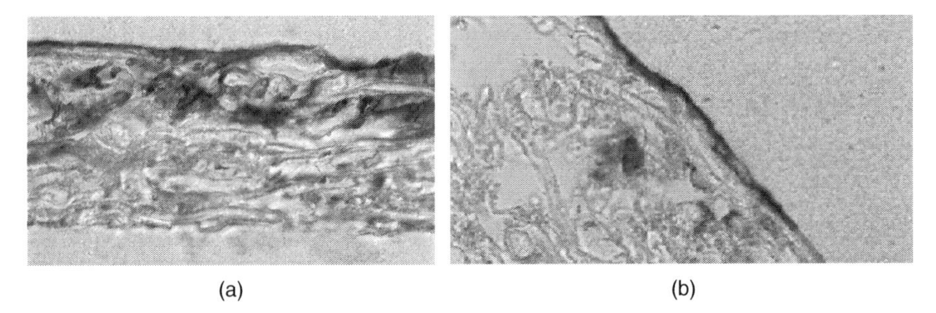

Figure 4.7 *Effect of polymeric cationic fixer – sulfonic acid functionalised pigment interaction on pigment distribution in plain paper (a) with no fixer and (b) with fixer.*

paper surface resulting in an increase in optical density as well as better print edge acuity and colour-to-colour bleed.

Alternatively the fixing agent can be applied to the substrate during the printing process from an additional print-head either just before or just after the colorant ink. This approach, sometimes referred to as the fifth pen approach, can be very effective if the extra complexity and cost of an additional print-head can be incorporated into the printer.

In the absence of specific mordant–colorant interactions, inks need to be developed that display good substrate versatility and as such will provide good print performance on a variety of substrates, including cheap non-inkjet papers. In order to provide prints with good resistance to water, the ink can be adapted specifically to reduce the solubility of the colorant once the ink has been printed. Water-fast prints can be achieved through the use of aqueous colorants whose solubility is extremely pH dependent: at the pH of the ink, usually pH 8–9, the dyes tend to be very water soluble; whereas on the paper, where the pH tends to be less than 7, the aqueous solubility is minimal. Aqueous colorants that can display this pH switching behaviour generally contain aromatic carboxylic acids or contain amines that can form insoluble salts with sulfonic acids. It is also possible to improve the water fastness of aqueous colorants by the inclusion of certain water-soluble amines and ammonium carboxylate salts in the ink. In the ink, the amine is unprotonated, but once printed ammonia can evaporate from the surface leaving the free acid which protonates the amine. If correctly chosen, this protonated amine can then form an insoluble salt with, for example, the sulfonic acid groups on the dye.

4.5.1.3 Polymeric Additives

The use of polymeric additives in aqueous ink formulation is widely reported. Examples include:

- use of low concentrations (100 ppm) of high-molecular-weight water-soluble polymers such as polyethylene oxide and polyacrylamide both as drag reduction agents and to control droplet generation;
- use of low-molar-mass water-soluble and -dispersible polymers such as styrene acrylates (Merrington *et al*., 2008), aqueous polyurethanes (Avecia Ltd., 1999), cellulose esters (Xu *et al*., 2007) and polyesters (Avecia Ltd., 2003) at 0.1–5 wt% as binders to improve image permanence on the substrate especially with respect to water and humidity. An example of the use of reactive polymer binders is in the inkjet fabrication of liquid crystal display colour filters (Shin and Smith, 2008);
- as pigment dispersants and stabilisers (Merrington, Hodge and Yeates, 2006).

4.5.2 Non-aqueous Ink Technologies

4.5.2.1 Oil-Based Pigment Inks

In this technology, the pigments are dispersed in a low-viscosity, non-volatile oil. Again, the particle size of the pigments is below 1 μm to ensure good operability, and to maintain good colloidal stability polymeric dispersants may be employed. The inks typically have a viscosity in the region of 10 mPa s, have a surface tension below 30 mN m^{-1} and tend to give excellent operability and long open nozzle times due to the lack of volatile components. The use of these inks is largely for printing onto plain and coated papers, as well as coated vinyls as the substrate needs to be able to absorb the ink vehicle. The images that are formed have excellent light and water fastness, and as the inks contain no water there are no issues with cockle when printing onto paper substrates. These inks are used in wide-format applications, as well in commercial printing, for example receipts and lottery tickets as well as several digital fabrication applications.

4.5.2.2 Phase Change Inks

Phase change inks are solid inks at room temperature, consisting mainly of a transparent mixture of synthetic waxes. Tackifiers and plasticisers are added to impart improved adhesion and flexibility, and antioxidants give improved heat stability. Both dyes and pigments are used to give colour. The solid ink is heated within the piezo-DOD print-head at around 60 °C, and the hot wax ejected onto the substrate. As the droplet hits the substrate, it freezes immediately with little penetration into the substrate. When printing conventional text and image, the printed drop needs to be cold pressure-fused into the substrate in order to improve adhesion, prevent light scattering owing to the lens effect of the hemispherical droplets and minimise surface texture. As an example, the 3D nature of the drops has been exploited for the printing of masks for metal layer etching in a transistor using an 80 °C melting wax with a 20–40 μm spot size and drop placement accuracy of 1 μm (Hue, 1998).

4.5.2.3 100% Solids UV Cure

These are typically either clear or pigmented inks containing a blend of monomeric and oligomeric acrylates or epoxies that are polymerised by UV light in the presence of a photo-initiator (Wicks, Jones and Pappas, 2007). The printed film is instantly hardened on the substrate, as all the ink components are chemically cross-linked on exposure to the irradiation. The inks give excellent print performance across a range of non-porous substrates, for example metals and plastics, and can be used for printing applications such as beverage can labelling and credit cards. The images formed have very high durability, with good resistance to chemicals and physical abrasion. As the inks typically contain no VOCs, the nozzles can be left uncapped for long periods in the absence of light, and as the rate of curing is fast, good line speeds can be achieved for production printing. Ink viscosity at the operating temperature is typically around 10 mPa s, and the surface tension is in the region of 23–29 $mN\,m^{-1}$. To achieve good operability in the inkjet print-head, the pigment particle size is typically smaller than 1 μm.

Typical ink formulations contain a mixture of mono-, di-, tri-, tetra- and penta-functional monomers, with respect to the polymerising group; examples are shown in Figure 4.8. Alkoxylated monomers are often favoured because this can reduce sensitisation and odour, and by balancing the degree of ethoxylation and/or propoxylation, speed of cure, film flexibility and the ability to wet pigment can be finely tuned.

The final composition is chosen so as to give the appropriate balance of cure speed, film shrinkage, adhesion, flexibility, hardness and solvent resistance (Table 4.4). Mono-functional monomers are typically characterised by low viscosity and low cure speed, giving films with good flexibility and poor hardness and solvent resistance, whereas higher functional monomers typically have higher viscosity and are faster curing, giving films with good hardness and solvent resistance but lower flexibility. Photo-initiator is

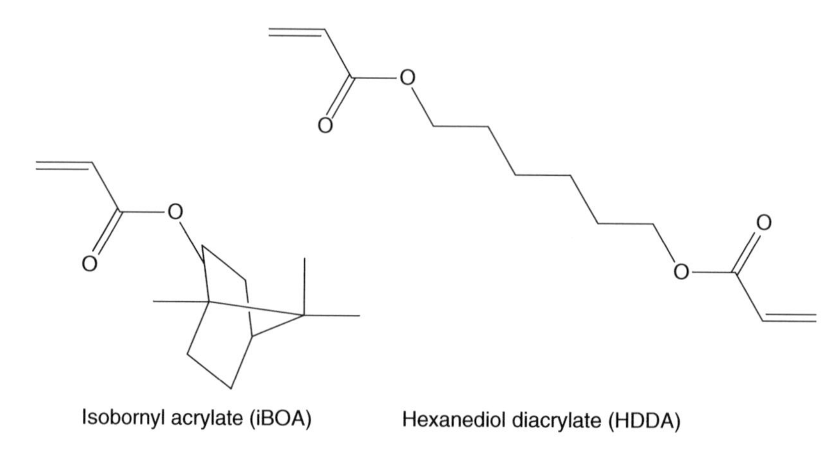

Figure 4.8 *Typical acrylate monomers used in 100% UV cure inkjet.*

Table 4.4 *Typical UV-curable monomer properties as a function of acrylate functionality.*

Property	Functionality of acrylate molecule	
	Mono	Penta
Cure speed	Slow	Fast
Flexibility	Flexible	Brittle
Hardness	Soft	Hard
Solvent resistance	Less	Best
Shrinkage	Low	High

present, typically at around 1 wt%, chosen on the basis of the cross-linking chemistry used, UV high-source spectral output and the required cure speed (Wicks, Jones and Pappas, 2007). Whilst epoxy-based systems are typically slower to cure than free radical, with thermal post-curing possible, they give good wetting and adhesion on low-energy substrates with lower volume shrinkage, 3–5% compared with 10–15% for free radical.

One hundred percent UV-curable inks have many advantages in digital fabrication related to their high operability and low maintenance and their ability to crosslink on demand. These attributes are exemplified by their application in printed circuit board (PCB) manufacture (as discussed more fully in Chapter 8) and the printing of dielectric or insulating layers (see Chapter 7). Inkjet printing has been suggested as a technique for manufacturing PCBs in a fast and efficient way, augmenting conventional screen printing and photolithographic methods. This technology has now advanced to a point where primary track imaging, solder mask application and legend printing can be achieved commercially. The benefits of digital imaging methods have been recognised as a way to simplify, speed up and add flexibility to board production by eliminating the need for a screen or photo-tool. Laser direct imaging (LDI) was developed in part to address this problem but has not been widely used for solder mask application to date.

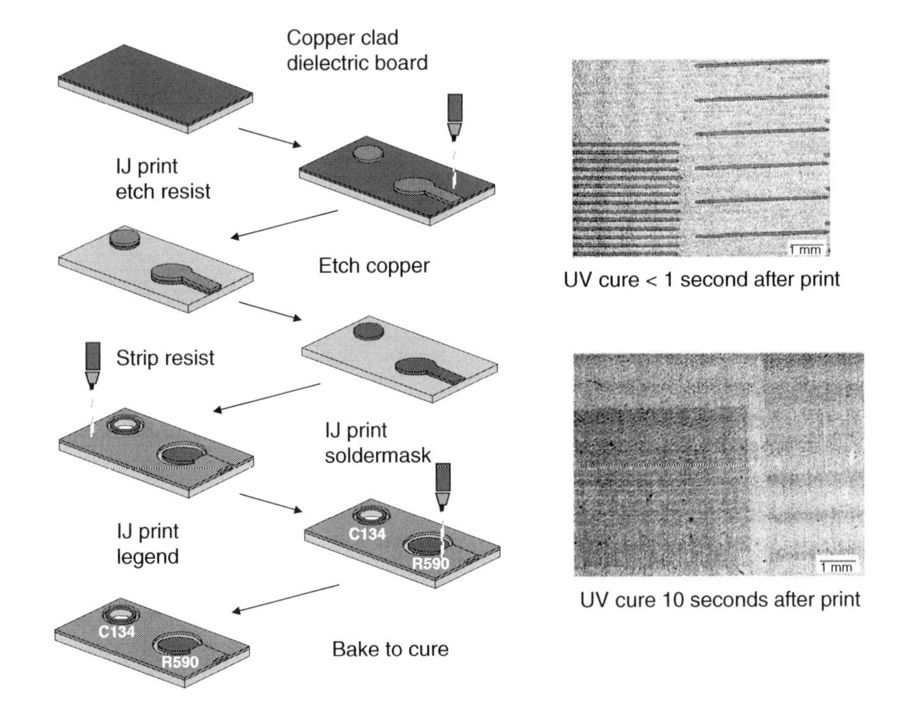

Figure 4.9 *PCB production using IJ printing (See plate section for coloured version).*

Simplified inkjet processes can be designed for many of today's board requirements, and Figure 4.9 depicts a schematic inkjet-based process to manufacture single-sided PCBs. It also highlights the importance of the print process whereby narrow line features are obtained if the ink is cured rapidly after printing before spreading across the substrate leads to line broadening as highlighted for the ink cured after 10 seconds.

Inkjet-printed UV-curable epoxies have also been reported for use as thin-film, 1 μm, pinhole-free dielectric layers in thin-film transistor manufacture (Sanchez-Romaguera, Madec and Yeates, 2008). Figure 4.10a shows the characteristic line profile of an inkjet-printed UV-cured ink with an absence of the characteristic coffee stain typical of solvent containing inks with low solute content, and Figure 4.10b shows the line profile of a subsequently printed silver ink as it traverses the glass–dielectric step. It is useful to note that pinhole-free films can be printed, but in designing such a multi-layer process the differential wetting of subsequent inks on energetically different surfaces needs to be taken into account.

4.5.2.4 Solvent-Based Inks

Solvent-based inks based on MEK, short-chain glycols, lactates and alcohols are widely used, for example in industrial marking by CIJ (Dawson, 2004; Cameron, 2006). However, the increasing realisation and utilisation of inkjet printing as a fabrication tool, primarily driven by the electronics industry, have necessitated the use of increasingly aggressive solvents such as toluene, anisole, tetralin and other

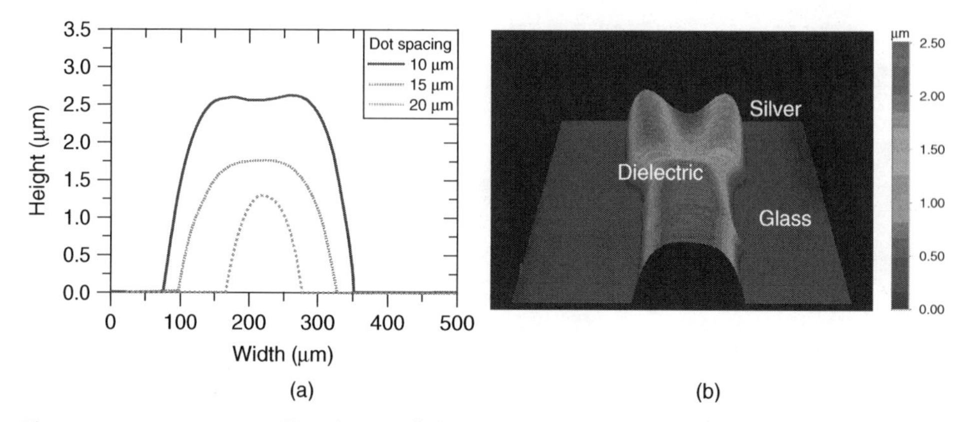

Figure 4.10 *(a) Line profile of SU-8 dielectric layer on glass after UV cure and (b) image of SU-8 dielectric traversing a glass–silver film interface (Sanchez-Romaguera, Madec and Yeates, 2008) (See plate section for coloured version).*

substituted aromatics (Gregory, Katrizky and Sabongi, 1991). This has been driven by the need to form low-viscosity solutions of high-molecular-weight polymers such as polyphenylene-vinylenes, polyfluorenes (Steiger, Heun and Tallant, 2003) and poly 3-hexylthiophene (Speakman *et al*., 2001), structured carbons including carbon nanotubes (Hopkins, Kruk and Lipeles, 2007) as well as low-molar-mass conjugated molecules such as phenyl-C61-butyric acid methyl ester (PCBM) (Hoth *et al*., 2009) and substituted polyacenes (Lee *et al*., 2008). The move to more aggressive solvents has not been without its challenges as these can interact in a negative manner with the materials of construction of the print-head. Coupled with the additional challenges of exceptional drop uniformity and directionality, this has brought about the development of new families of print-heads. The formulations are generally 'simple', comprising solute (functional material) and a mixture of solvents. Solvent selection is critical and must fulfil a number of primary functions:

- must be a good solvent for the solute and stable to precipitation or gelation over extended periods;
- must not have any unfavourable interactions with the materials of construction of the print-head;
- stable to drying out of the ink at the nozzle, leading to print failure. For this reason, solvents with boiling points higher than 130 °C tend to be preferred, with anisole (154 °C), mesitylene (164 °C), ortho dichlorobenzene (180 °C) and tetralin (207 °C) frequently being cited;
- dry in such a way that no solvent is retained within the printed feature which can result in downgraded performance, and avoid undesired print phenomena such as pinholing and/or coffee staining For these reasons there tends to be an upper boiling point limit of around 210 °C, with the use of binary mixtures of low and high boiling point solvents to eliminate the coffee staining effect (Tekin, De Gans and Schubert, 2004).

The use of thermo-cleavable solvents (Figure 4.11) for printing conjugated polymers in solar cell applications has recently being reported, which combine the benefits of low

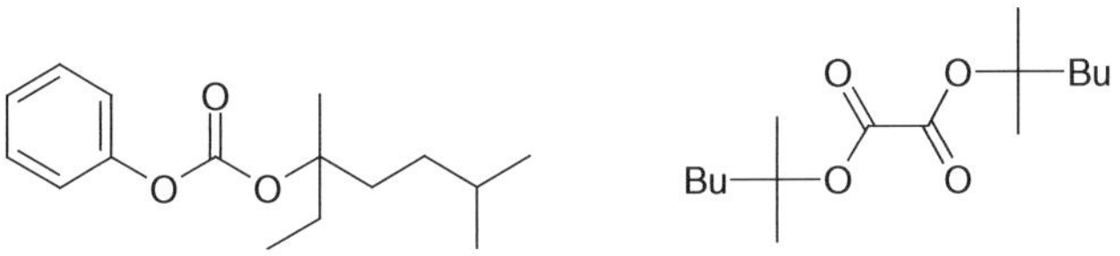

Figure 4.11 *Illustration of potential thermo-cleavable solvents (Jorgensen* et al.*, 2009).*

volatility at room temperature but decompose thermally between 130 and 180 °C to yield low boiling, highly volatile products (Jorgensen *et al.*, 2009).

4.6 Applications in Electronic Devices

Patterning is a vital part in the development and future production of organic electronic devices, and as such inkjet printing is ideally suited for a number of process steps because it is additive and maskless in nature. These applications are further discussed in Chapters 7, 9, 10 and 11. Here we discuss a number of the material sets used with specific reference to the relationship between material properties and inkjet performance.

4.6.1 Organic Conducting Polymers

The use of conductive poly(4-ethylenedioxythiophene)-poly(styrenesulfonate) (PEDOT: PSS) blends is widespread in the fabrication of organic thin-film transistors (OTFTs), organic solar cells, organic light emitting displays (OLEDs) and sensors. The sulfonyl groups on PSS are deprotonated and carry a negative charge, whilst PEDOT carries a positive charge, with the charged polymers forming a polymeric salt which is dispersible in water (Yuka and Jabbour, 2007; Figure 4.12).

Dependent upon the specific formulation, typically through the use of high boiling solvents like methylpyrrolidone, dimethylsulfoxide, sorbitol and glycerol, the conductivity of the resultant transparent polymer film can vary over several orders of magnitude from 10^{-4} to 10^{3} S cm^{-1} making it suitable both as a replacement for indium tin oxide (ITO) and for printing conducting features such as interconnects (Ouyang *et al.*, 2005). In order to spatially modify the conductivity still further, the inkjet printing of hydrogen peroxide ink onto a PEDOT:PSS anode, in order to alter its oxidation state, has been

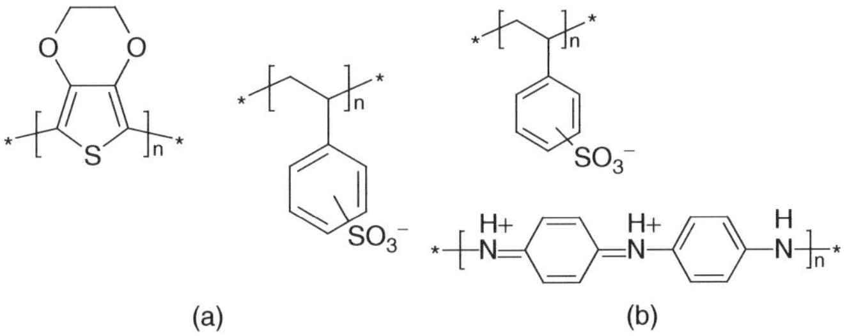

Figure 4.12 *Structure of (a) PEDOT:PSS and (b) PANI–PSS emeraldine salt form.*

reported (Ouyang *et al*., 2004). This ability to spatially modify the local conductivity of the PEDOT:PSS layer opens up the possibility of electrical greyscale imaging capability (Yoshioka *et al*., 2005; Yoshioka and Jabbour, 2006). PEDOT:PSS layers have been deposited by inkjet for a number of sensor applications exploiting the change in conductivity as a function of external stimuli including humidity and strain (Mabrook, Pearson and Petty, 2006).

The use of PEDOT:PSS as printed source, drain and gate electrodes in OTFTs has been reported (Zhang *et al*., 2008), although direct inkjet printing leads to large-scale features where the critical dimensions are typically $>10\,\mu m$. An innovative way of increasing both resolution and drop placement is the sequential printing of a PEDOT:PSS line, hydrophobically modified by either surfactant or carbon tetrafluoride plasma, followed by a hydrophilic PEDOT:PSS ink which de-wets from the first feature to give a channel down to 250 nm wide (Burns *et al*., 2003). PEDOT:PSS as the conductive anode in OLED displays has been extensively reported, both to modify the ITO work function and to planarise the anode surface to avoid shorts and black spots (Shimoda *et al*., 2003).

Reports have demonstrated the use of inkjet in the fabrication of organic solar cells (Hoth *et al*., 2007) and highlighted the importance of obtaining the correct PEDOT:PSS layer morphology for enhanced performance. PEDOT:PSS inks with additives of glycerol and surfactant show improved surface morphology and high conductivity resulting in enhanced photovoltaic performance. Using optimised ink formulation and print patterns, solar cell efficiencies greater than 4.0% have been reported (Steirer *et al*., 2009).

Whilst there is a large literature on the inkjet printing of PEDOT:PSS, there have been proportionately less reports on the use of water-dispersible polyaniline (PANI) which may be due to either the lack of facile and reliable procedures for preparing high-quality nanometre-size particles required for robust inkjet printing or the need for highly corrosive and acidic dopants such as dodecylbenzenesulfonic acid (DBSA) which with the advent of more robust print-heads is only now becoming feasible. The use of PANI–PSS has been reported as a chemical sensor for ammonia, showing enhanced response when compared with conventional PANI-based chemical sensors (Jang, Ha and Cho, 2007).

The inkjet printing of structured carbons including carbon nanotubes and graphene, both as conductors and as the semiconductor in an OTFT (Beecher *et al*., 2007), has been reported. These inks consist of a dispersion of carbon nanotubes in solvents such as N-methyl-2-pyrrolidone and dimethylformamide which have been shown to partially break up the strong van der Waals forces of attraction between individual carbon nanotubes under ultrasonic treatment, enabling the formulation of stable inks. The use of additional agents such as single-strand DNA has been reported to further aid dispersion.

4.6.2 Conjugated Organic Semiconductors

Many early developments in the high-resolution inkjet printing of solvent-based inks were driven by the desire to fabricate OLEDs. The desire to print high-molecular-weight conjugated polymers such as polyphenylene-vinylenes and polyfluorenes required print-heads to be developed which had tolerance to aggressive solvents and, at the same time, exceptional drop-to-drop uniformity coupled with high absolute print accuracy. These approaches have now been extended to the printing of both polymeric (Barret, Sanaur and Collot, 2008) and small-molecule organic semiconductors (Madec *et al.*, 2010). In all cases once robust droplet generation is obtained and high accuracy drop placement achieved, uniform pinhole-free film formation is required, which can be achieved both through formulation with solvent mixtures (Tekin, De Gans and Schubert, 2004) and by considering droplet–substrate interaction.

When printing conjugated materials and in particular polymers, inter-chain interactions in solution are important as these can lead to aggregation in solution and an increase in solution viscosity at low concentration coupled with undesirable viscoelastic behaviour on printing, as reported for MEH–PPV (Tekin *et al.*, 2007). This can be overcome using mild ultrasonic treatment which can partially break down the physical aggregates, but care needs to be observed if main-chain degradation is to be avoided (Alamry *et al.*, 2011). Recently the inkjet printing of water-based dispersions of MEH–PPV with high-molecular-weight polyethylene oxide has been reported in order to overcome some of the issues of printing solutions.

High-performance small-molecule semi-conductors are currently being developed for OTFT applications. These materials are generally soluble in strong aromatic solvents such as dichlorobenzene, up to a few weight percent. These materials form highly crystalline thin films, and control of the evaporation rate of the solvent(s) is critical in ensuring uniformity across a given device (Madec *et al.*, 2010). Blends of 1,2,4-trichlorobenzene and cyclohexanol have been shown to be particularly effective at controlling viscosity, but often low viscosity and device uniformity are issues. Similarly, when printing the active p-n heterojunction layer in an organic solar cell by use of a mixture of ortho-dichlorobenzene and mesitylene, it is possible to print PCBM:P3HT having the correct morphology giving device efficiency of around 4%.

4.6.3 Inorganic Particles

The formulation challenges relating to metal particles in inkjet fluids are a consequence of the high density of the disperse phase relative to the dispersion medium. For dilute suspensions of small spheres in a fluid, either air or water, Stokes' law predicts the settling velocity by:

$$w = \frac{2\left(\rho_p - \rho_f\right) g r^2}{9\eta}$$

where w is the settling velocity, ρ is density (the subscripts p and f indicate particle and fluid respectively), g is the acceleration due to gravity, r is the radius of the particle and η is the dynamic viscosity of the fluid. Therefore to formulate stable, metal-containing inks, two different approaches have been pursued.

- Use of metal salts such as silver neodecanoate in xylene and copper hexanoate in either isopropanol or chloroform (Dearden *et al.*, 2005) or
- Use of metal nanoparticles having a diameter typically much smaller than 200 nm. Small particle size is important since the settling velocity scales as r^2, but also the sintering temperature decreases rapidly as the surface area–volume ratio decreases (Buffat and Burrel, 1976). The synthesis and stabilisation of metal nanoparticles have been extensively reviewed elsewhere (Bonmemann and Nagabhushana, 2008).

The direct inkjet printing of metal nanoparticles such as gold and silver is attractive because of the ability to directly print electrodes, conductive tracks and vertical interconnects at low temperature such that the overall processes are compatible with temperature-sensitive substrates such as polyethylene terephthalate (PET), and this is discussed in detail in Chapter 7. Use of metal salts which decompose under the application of heat and/or light has been reported (Dearden *et al*., 2005), although examples such as silver neodecanoate in xylene typically convert to bulk silver at temperatures greater than 150 °C for tens of minutes. More recent studies have shown that the conversion temperature can be reduced to 130 °C if the ink is printed directly onto a pre-heated substrate, giving resistivity eight times that of bulk silver (Perelaer *et al*., 2009). Silver nanoparticles can sinter at lower temperatures, although the conductivities typically approach about 10% of bulk metal (Perelaer *et al*., 2008). An alternative approach is to print either a palladium (II) solution onto a substrate, followed by reduction to catalytic palladium (0) and the subsequent electroless deposition of copper (Busato, Belloli and Ermani, 2007); or direct inkjet printing of palladium (0) nanoparticles onto the substrate, followed by the electroless deposition of either copper or nickel (Tseng *et al*., 2009). The advantage of such approaches is that conductive tracks can be printed having near-bulk metal conductivities with processing temperatures well below 100 °C, although a wet processing step is required. Inkjet printing is not restricted to purely conductive features, with recent reports on the inkjet printing of light emitting quantum dots based on cadmium selenide (Wood *et al*., 2009).

References

Alamry, K.A., Nixon, K., Hindley, R. *et al*. (2011) Flow-induced polymer degradation during ink-jet printing. *Macromolecular Rapid Communications*, **32** (3), 316–320.

Avecia Ltd. (1999) Composition based on water-dissipatable polyurethane. International Patent Application 9950364.

Avecia Ltd. (2002) Coloured, water-dissipatable polyurethanes. International Patent Application 050197.

Avecia Ltd. (2003) Ink-jet compositions comprising a water-dissipatable polymer. United Kingdom Patent Application 2351292 A1.

Barret, M., Sanaur, S. and Collot, P. (2008) Inkjet-printed polymer thin-film transistors: Enhancing performances by contact resistances engineering. *Organic Electronics*, **9** (6), 1093–1100.

Beecher, P., Servati, P., Rozhin, A. *et al*. (2007) Ink-jet printing of carbon nanotube thin film transistors. *Journal of Applied Physics*, **102** (4), 043710/1–043710/7.

Bonmemann, H. and Nagabhushana, K.S. (2008) Metal nanoclusters: synthesis and strategies for their size control, in *Metal Nanoclusters in Catalysis and Materials Science: The Issue of Size Control*, 1st edn (eds B. Corain, G. Schmid and N. Toshima), Elsevier, Amsterdam, pp. 21–49.

Buffat, P. and Burrel, J-P. (1976) Size effect on the melting temperature of gold particles. *Physical Review A*, **13** (6), 2287–2298.

Burns, S.E., Cain, P., Mills, J. *et al*. (2003) Inkjet printing of polymer thin-film transistor circuits. *MRS Bulletin*, **28** (11), 829–834.

Busato, S., Belloli, A. and Ermani, P. (2007) Inkjet printing of palladium catalyst patterns on polyimide film for electroless copper plating. *Sensors and Actuators B*, **123** (2), 840–846.

Cameron, N.L. (2006) Ink-jet printing, in *Coatings Technology Handbook*, 3rd edn (ed. A.A. Tracton), CRC Press, Boca Raton, FL, pp. 25/1–25/4.

Christianni, Y. and Walker, L.M. (2001) Surface tension driven jet break up of strain-hardening polymer solutions. *Journal of Non-Newtonian Fluid Mechanics*, **100** (1–3), 9–26.

Dawson, T.L. (2004) Inkjet printing of textiles: an overview of its developments and the principles behind ink drop formation and deposition. *Colourage*, **51** (10), 75–82.

De Gans, B.J., Kazancioglu, E., Meyer, W. and Schubert, U.S. (2004) Ink-jet printing polymers and polymer libraries using micropipettes. *Macromolecular Rapid Communications*, **25** (1), 292–296.

De Gennes, P.G. (1974) Coil-stretched-transition of dilute flexible polymer under ultra-high velocity gradients. *Journal of Chemical Physics*, **60**, 5030–5042.

Dearden, A.L., Smith, P.J., Shin, D.Y. *et al*. (2005) A low curing temperature silver ink for use in ink-jet printing and subsequent production of conductive tracks. *Macromolecular Rapid Communications*, **26** (4), 315–318.

Dijksman, J.F. (1984) Hydrodynamics of small tubular pumps. *Journal of Fluid Mechanics*, **139**, 173–191.

Dijksman, J.F., Duineveld, P.C., Hack, M.J.J. *et al*. (2007) Precision ink jet printing of polymer light emitting displays. *Journal of Materials Chemistry*, **17** (6), 511–522.

Fromm, J.E. (1984) Numerical calculations of the fluid dynamics of drop-on-demand jets. *IBM Journal of Research and Development*, **28** (3), 322–333.

Gregory, P, Katrizky A.R and Sabongi, G.J. (eds) (1991) *High-Technology Applications of Organic Colorants*, Plenum Press, New York.

Hamley, I.W. (2000) *Introduction to Soft Matter: Polymers, Colloids, Amphiphiles and Liquid Crystals*, Wiley-Interscience, West Sussex.

Hoath, S.D., Hutchings, I.M., Martin, G.D. *et al*. (2009) Links between ink rheology, drop-on-demand jet formation, and printability. *Journal of Imaging Science and Technology*, **53** (4), 041208/1–041208/8.

Hopkins, A.R., Kruk, N.A. and Lipeles, R.A. (2007) Macroscopic alignment of single-walled carbon nanotubes (SWNTs). *Surface and Coatings Technology*, **202** (4–7), 1282–1286.

Hoth, C.N., Choulis, S.A., Schilinsky, P. and Brabec, C.J. (2007) High photovoltaic performance of inkjet printed polymer: fullerene blends. *Advanced Materials*, **19** (22), 3973–3978.

Hoth, C.N., Choulis, S.A., Schilinsky, P. and Brabec, C.J. (2009) On the effect of poly(3-hexylthiophene) regioregularity on inkjet printed organic solar cells. *Journal of Materials Chemistry*, **19** (30), 5398–5404.

Hue, P.L. (1998) Progress and trends in ink-jet printing technology. *Journal of Imaging Science and Technology*, **42** (1), 49–62.

Jang, J., Ha, J. and Cho, J. (2007) Fabrication of water-dispersible polyaniline-poly(4-styrenesulfonate) nanoparticles for inkjet-printed chemical-sensor applications. *Advanced Materials*, **19** (13), 1772–1775.

Jorgensen, M., Hagemann, O., Alstrup, J. and Krebs, F.C. (2009) Thermo-cleavable solvents for printing conjugated polymers: application in polymer solar cells. *Solar Energy Materials and Solar Cells*, **93** (4), 413–421.

Lee, S.H., Choi, M.H., Han, S.H. *et al*. (2008) High-performance thin-film transistor with 6,13-bis(triisopropylsilylethynyl) pentacene by inkjet printing. *Organic Electronics*, **9** (5), 721–726.

Leelajariyakul, S., Noguchi, H. and Kiatkamjornwong, S. (2008) Surface-modified and macro-encapsulated pigmented inks for inkjet printing on textile fabrics. *Progress in Organic Coatings*, **62** (2), 145–161.

Mabrook, M.F., Pearson, C. and Petty, M.C. (2006) Inkjet-printed polymer films for the detection of organic vapors. *IEEE Sensors Journal*, **6** (6), 1435–1444.

Madec, M-B., Smith, P.J., Malandraki, A. *et al*. (2010) Enhanced reproducibility of inkjet printed organic thin film transistors based on solution processable polymer-small molecule blends. *Journal of Materials Chemistry*, **20** (41), 9155–9160.

Merrington, J., Hodge, P. and Yeates, S.G. (2006) A high-throughput method for determining the stability of pigment dispersions. *Macromolecules Rapid Communications*, **27** (11), 835–840.

Merrington, J., Yeates, S.G., Hodge, P. and Christian, P. (2008) High-throughput screening of polymeric dispersants to accelerate the development of stable pigment dispersions. *Journal of Materials Chemistry*, **18** (2), 182–189.

Meyer, J.D., Bazileyky, A.A. and Rozkhov, A.N. (1997) Effects of polymeric additives on thermal ink jets. Proceedings of IS&T's NIP13: International Conference on Digital Printing Technologies, Seattle, November 2–7, 1997, Seattle (IS&T, Springfield, VA), p. 675.

Ouyang, B.Y., Chi, C.W., Chen, F.C. *et al*. (2005) High-conductivity poly (3,4-ethylenedioxythiophene): poly(styrene sulfonate) film and its application in polymer optoelectronic devices. *Advanced Functional Materials*, **15** (2), 203–208.

Ouyang, J., Xu, Q.F., Chu, C.W. *et al*. (2004) On the mechanism of conductivity enhancement in poly (3,4-ethylenedioxythiophene): poly(styrene sulfonate) film through solvent treatment. *Polymer*, **45** (25), 8443–8450.

Perelaer, J., De Laat, A.W.M., Hendriks, C.E. and Schubert, U.S. (2008) Inkjet-printed silver tracks: low temperature curing and thermal stability investigation. *Journal of Materials Chemistry*, **18** (27), 3209–3215.

Perelaer, J., Hendriks, C.E., De Laat, A.W.M. and Schubert, U.S. (2009) One-step inkjet printing of conductive silver tracks on polymer substrates. *Nanotechnology*, **20** (16), 165303/1–165303/5.

Rabin, Y. (1986) *Polymer-Flow Interaction: La Jolla Institute 1985*, AIP Press, New York.

Reis, N., Ainsley, C., Derby, B. (2005) Ink-jet delivery of particle suspensions by piezoelectric droplet ejectors. *Journal of Applied Physics*, **97**, 094903/1–094903/6.

Sanchez-Romaguera, V., Madec, B-M. and Yeates, S.G. (2008) Inkjet printing of 3D metal-insulator-metal crossovers. *Reactive and Functional Polymers*, **68**, 1052–1058.

Sen, A.K. and Darabi, J. (2007) Droplet ejection performance of a monolithic thermal inkjet print head. *Journal of Micromechanics and Microengineering*, **17** (8), 1420–1427.

Shimoda, T., Morii, K., Seki, S. and Kiguchi, H. (2003) Inkjet printing of light-emitting polymer displays. *MRS Bulletin*, **28** (11), 821–827.

Shin, D-Y. and Smith, P.J. (2008) Theoretical investigation of the influence of nozzle diameter variation on the fabrication of thin film transistor liquid crystal display color filters. *Journal of Applied Physics*, **103** (11), 114905/1–114905/11.

Shirota, K., Shioya, M., Suga, Y. *et al*. (1996) Kogation of inorganic impurities in bubble jet ink, in *Recent Progress in Inkjet Technologies* (eds I. Rezanka and R. Eschbach), Society for Imaging Science and Technology, Washington, DC, pp. 218–219.

Speakman, S.P., Rozenberg, G.G., Clay, K.J. *et al*. (2001) High performance organic semiconducting thin films: ink jet printed polythiophene [rr-P3HT]. *Organic Electronics*, **2** (2), 65–73.

Spinelli, H.J. (1998) Polymeric dispersants in ink jet technology. *Advanced Materials*, **10** (15), 1215–1218.

Steiger, J., Heun, S. and Tallant, N. (2003) Polymer light emitting diodes made by ink jet printing. *Journal of Imaging Science and Technology*, **47** (6), 473–478.

Steirer, K.X., Berry, J.J., Reese, M.O. *et al*. (2009) Ultrasonically sprayed and inkjet printed thin film electrodes for organic solar cells. *Thin Solid Films*, **517** (8), 2781–2786.

Store, H.J. and Harrison, G.M. (2005) The effect of added polymers in the formation of drops ejected from a nozzle. *Physics of Fluids*, **17** (3), 033104/1–033104/7.

Stringer, J. and Derby, B. (2009) Limits to feature size and resolution in inkjet printing. *Journal of the European Ceramic Society*, **29** (5), 913–918.

Tekin, E., De Gans, B-J. and Schubert, U.S. (2004) Ink-jet printing of polymers – from single dots to thin film libraries. *Journal of Materials Chemistry*, **14**, 2627–2632.

Tekin, E., Holder, E., Kozodaev, D. and Schubert, U.S. (2007) Controlled pattern formation of poly [2-methoxy-5(2′-ethylhexyloxyl)-1,4-phenylenevinylene] (meh-ppv) by ink-jet printing. *Advanced Functional Materials*, **17** (2), 277–284.

Tseng, C-C., Chang, C-P., Sung, Y. *et al*. (2009) A novel method to produce PD nanoparticle ink for ink-jet printing technology. *Colloids and Surfaces A-Physicochemical and Engineering Aspects*, **339** (1–3), 206–210.

Wicks, Z.W., Jones, F.N. and Pappas, S.P. (2007) *Organic Coatings: Science and Technology*, 3rd edn, Wiley-Interscience, Chichester, pp. 575–595.

Wood, V., Panzer, M.J., Chen, J. *et al*. (2009) Inkjet-printed quantum dot-polymer composites for full-color AC-driven displays. *Advanced Materials*, **21** (21), 2151–2155.

Xu, D. (2009) Inkjet deposited conductive tracks via electroless deposition. PhD thesis, University of Manchester.

Xu, D., Sanchez-Romaguera, V., Barbosa, S. *et al*. (2007) Inkjet printing of polymer solutions and the role of chain entanglement. *Journal of Materials Chemistry*, **17** (46), 4902–4907.

Yoshioka, Y., Calvert, P.D. and Jabbour, G.E. (2005) Simple modification of sheet resistivity of conducting polymeric anodes via combinatorial ink-jet printing techniques. *Macromolecules Rapid Communication*, **26** (4), 238–246.

Yoshioka, Y. and Jabbour, G.E. (2006) Inkjet printing of oxidants for patterning of nanometer-thick conducting polymer electrodes. *Advanced Materials*, **18** (10), 1307–1312.

Yuka, Y. and Jabbour, G.E. (2007) Inkjet printing and patterning of PEDOT:PSS: application to optoelectronic devices, in *Handbook of Conductiong Polymers*, 3rd edn, vol. 1 (eds T.A. Skotheim and J.R. Reynolds), CRC Press, Boca Raton, FL, p. 3–1.

Zhang, X.H., Lee, S.M., Domercq, B. and Kippelen, B. (2008) Transparent organic field-effect transistors with polymeric source and drain electrodes fabricated by inkjet printing. *Applied Physics Letters*, **92** (24), 24307/1–24307/3.

5

When the Drop Hits the Substrate

Jonathan Stringer[1] *and Brian Derby*[2]
[1]*Department of Mechanical Engineering, University of Sheffield, United Kingdom*
[2]*School of Materials, University of Manchester, United Kingdom*

5.1 Introduction

Inkjet printing is a method for the generation and precise placement of liquid drops on a substrate. The final desired image or pattern is a solid object on the substrate. Thus an important stage in the generation of the image or pattern is the impact of the liquid drop on a substrate and the subsequent phase change that transforms the liquid into a solid. The liquid-to-solid phase change can occur by a number of mechanisms including: solvent evaporation, cooling through a transition temperature, gelling of a polymer precursor and chemical reaction. In all these cases, solidification occurs post-deposition and the printed pattern must retain some stability in the liquid state prior to solidification. In order to fully understand the processes that occur between the printed drop and the substrate prior to attaining the final structure, we must identify the interactions that occur between the substrate and the fluid drop prior to solidification.

Whilst droplet impact and its subsequent interaction with a substrate have been studied extensively for a number of years (Worthington, 1876; Armster *et al.*, 2002; Yarin, 2006), a complete understanding of the possible mechanisms involved in the process has not yet been obtained. This is due to the complex nature of the flow as the initial droplet kinetic energy drives droplet spreading across a substrate until this energy has been lost through viscous dissipation or converted into surface energy. The outcome of a droplet deposition event will therefore be influenced by a number of factors, such as the initial mass and velocity of the impinging droplet, the viscosity and surface tension of the fluid,

Inkjet Technology for Digital Fabrication, First Edition. Edited by Ian M. Hutchings and Graham D. Martin.

the nature of the substrate and any phase change undergone by the fluid. To help quantify the impact conditions, it is useful to use a series of dimensionless numbers that express order-of-magnitude ratios between different energies experienced by the droplet. These were introduced in Chapter 2, but their definitions will be briefly repeated here as we shall use them extensively.

The impact Reynolds number (Re) represents the ratio between inertial and viscous forces, and is defined as:

$$\mathrm{Re} = \frac{\rho U_0 D_0}{\eta} \tag{5.1}$$

where ρ is the fluid density, U_0 is the impact velocity, D_0 is the initial droplet diameter and η is the fluid dynamic viscosity. The impact Weber number (We) represents the ratio between inertial and surface energies, and is defined as:

$$\mathrm{We} = \frac{\rho U_0^2 D_0}{\sigma} \tag{5.2}$$

where σ is the fluid surface tension. The Bond number (Bo) represents the ratio between gravitational and surface energies, and is defined as:

$$\mathrm{Bo} = \frac{\rho g D_0^2}{\sigma} \tag{5.3}$$

where g is the acceleration due to gravity. The impact capillary number (Ca) represents the ratio of viscous and surface energies and is defined as:

$$\mathrm{Ca} = \frac{\eta U_0}{\sigma} \tag{5.4}$$

The Ohnesorge number (Oh) is another dimensionless number that shows the ratio of viscous forces to inertia and surface forces independent of impact velocity and is defined as:

$$\mathrm{Oh} = \frac{\eta}{(\rho \sigma D_0)^{1/2}} \tag{5.5}$$

5.2 Stable Droplet Deposition

5.2.1 Deposition Maps

The range of impact conditions and fluid properties that are of interest for various industrial applications covers several orders of magnitude of the appropriate physical parameters. It is therefore useful to use the dimensionless groups defined in Section 5.1 to construct a map that will classify the type of impact expected for a given set of conditions. This was originally proposed by Schiaffino and Sonin (1997b), who identified four regions of interest (Figure 5.1) using a parameter space defined by axes of Oh and We, producing a map with one axis independent of impact velocity. This is of particular interest at low We because the impact velocity will have little or no influence on the driving force of spreading.

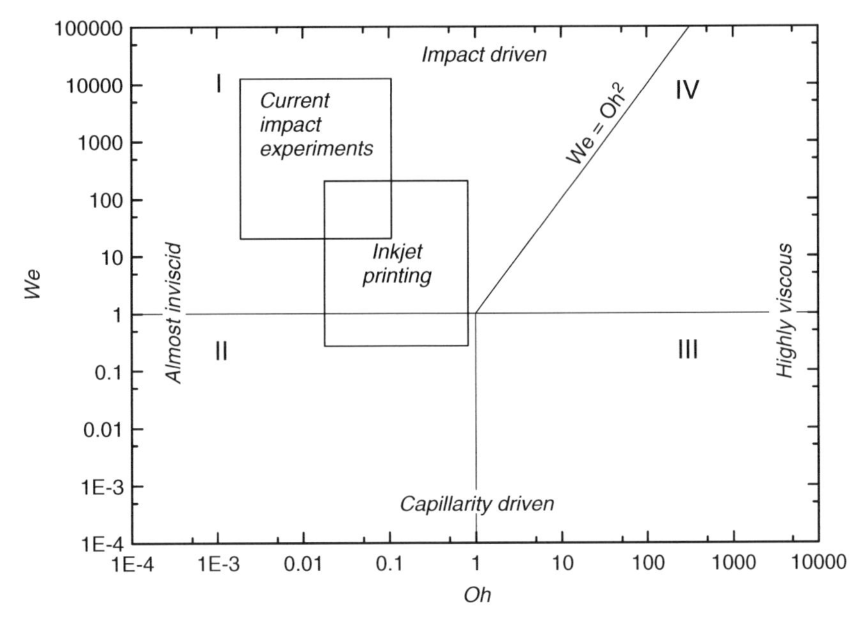

Figure 5.1 *We–Oh map based on that initially constructed by Schiaffino and Sonin (1997b) to demonstrate the balance of capillarity, inertia and viscous dissipation involved in different droplet impact regimes.*

The boundaries between the four regions in Figure 5.1 are determined by balances between the different variables that influence the flow, namely, inertia, viscosity and surface energy. This approach, whilst useful, tends to over-simplify the characterisation of droplet behaviour, particularly in the transition regions close to the boundaries.

Problems can arise initially in defining where the boundaries should be, and this was realised when the map was first devised (Schiaffino and Sonin, 1997b). The transition between groups II and III could occur at Oh $\sim$0.013 as this would be indicated by the Hoffman–Tanner–Voinov relationship between dynamic contact angle and velocity (Hoffman, 1975; Tanner, 1979). It should also be noted that the value of We where the kinetic and surface energy of an impacting droplet will be equal is 12, which would present a reasonable argument for this to be the location of the boundary between the regions I–II and III–IV.

Another over-simplification caused by these deposition maps is the assumption that the impact conditions accurately represent the conditions experienced by the droplet throughout deposition. During the spreading process, the velocity of the droplet normal to the surface will decrease from the initial impact velocity to zero as spreading concludes. This reduction in velocity will reduce We, which could change the balance between surface energy, viscous dissipation and inertia that governs the droplet dynamics. The axes used in the deposition map take no account of changing fluid properties dependent upon shearing of the fluid (i.e. non-Newtonian viscosity) or any change due to evaporation during the spreading process. Even with these drawbacks considered, however, the deposition map similar to that shown in Figure 5.1 is a useful starting point when defining the nature of a given impact and the relevance of previous work to these circumstances.

As can be seen in Figure 5.1, the impact conditions typically experienced in inkjet printing are situated primarily in region I (inviscid and impact driven), but are reasonably close to all of the other regions. It can also be seen that droplet impact conditions investigated previously with millimetre-size droplets are situated more definitely in region I, meaning that previous models derived from these studies will not necessarily apply to inkjet printing. It is therefore appropriate to review previous work on capillarity-driven spreading, in addition to impact-driven spreading, to gain a more complete understanding of the spreading of drops during inkjet printing. Indeed, it is likely that during inkjet printing there is a transition from impact-driven spreading to that controlled by capillarity as time increases during the spreading process (Derby, 2010), as illustrated schematically in Figure 5.2.

5.2.2 Impact of Millimetre-Size Droplets

The initial energy state of a droplet can be separated into two components: kinetic energy (E_{K0}) and surface energy (E_{S0}) (Madejski, 1983). Generally the droplet is assumed to be spherical in shape, although there is usually a slight eccentricity to the droplet that has been observed (Stow and Hadfield, 1981). For inkjet printing, it should also be noted that the liquid may well not be spherical if the characteristic ligament that is often produced (as discussed in Chapters 2 and 3) does not spheroidise completely before impact. However, by assuming that the droplet is spherical in shape, it is possible to

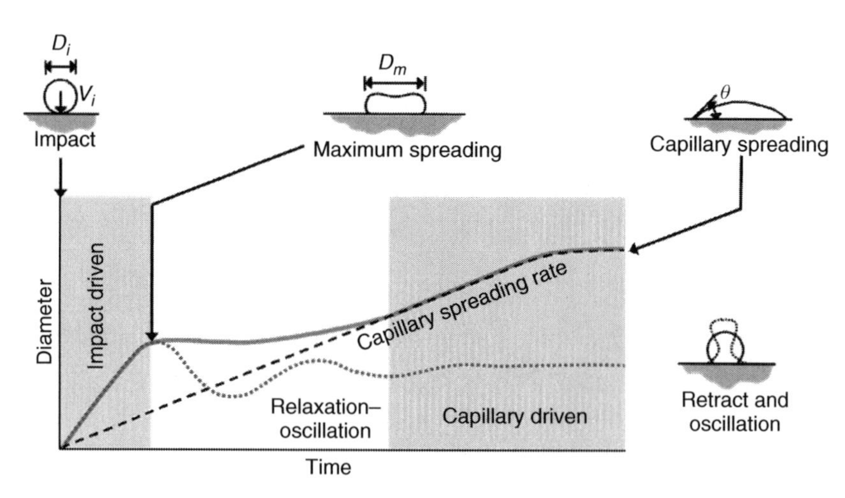

Figure 5.2 *Schematic of the time evolution of the driving force for liquid drop spreading on a flat surface after impact; timescale is non-linear (Reproduced with permission from Inkjet Printing of Functional and Structural Materials: Fluid Property Requirements, Feature Stability, and Resolution by Brian Derby, Annual Review of Materials Research, 40, 395–414 Copyright (2010) Annual Reviews. With thanks to Dr. W.K. Hsiao, Institute for Manufacturing, University of Cambridge, UK).*

define both of the component energies in terms of the properties of the fluid and the size and velocity of the droplet:

$$E_{K0} = \frac{\pi \rho D_0^3 U_0^2}{12} \tag{5.6}$$

$$E_{S0} = \pi D_0^2 \sigma \tag{5.7}$$

It is therefore possible to define the initial energy of the droplet E_0 as the sum of Equations 5.6 and 5.7:

$$E_0 = E_{K0} + E_{S0} = \pi D_0^2 \left(\frac{\rho D_0 U_0^2}{12} + \sigma \right) \tag{5.8}$$

The role of potential energy has not been considered in previous energy balance models. For gravitational potential energy this is sensible because Bo is small for drop sizes of the millimetre scale or smaller. The potential energy of a droplet due to its pressure would seem to merit more consideration for inclusion, but as yet this has not been included in any model of droplet deposition. This has partly been because it constitutes only one-sixth of the surface energy and therefore is considered negligible compared to the other energies involved (Madejski, 1983). It is also because it is difficult to evaluate the change in potential energy upon impact; the change in pressure is very slight in all but the highest impact speeds, whereupon the change in energy is negligible compared to the change in kinetic energy.

Upon contact with the substrate, the droplet deforms, with liquid being pushed radially outwards from the point of contact onto the substrate. This deformation encompasses changes in energy due to viscous dissipation, the change in surface area of the droplet and the interaction between the droplet and the substrate. Previous attempts at modelling the spreading of a droplet have done so by trying to predict the maximum extent of fluid spreading (β_{max}) defined here:

$$\beta_{max} = \frac{D_{max}}{D_0} \tag{5.9}$$

where D_{max} is the maximum footprint diameter reached by the droplet due to impact-driven spreading. Many models have been proposed to predict β_{max}, but their applicability to certain conditions is often limited due to the method by which they were derived. Models devised for high-velocity impact (and concurrent high Re and We) such as those used for thermal spraying of metals tend to discount any influence of surface tension due to this being largely insignificant when considering purely impact-driven spreading (Watanabe *et al*., 1992). These models are therefore not applicable to lower velocity impacts such as those encountered in inkjet printing.

Most predictions of β_{max} are based around an energy balance (Madejski, 1983; Chandra and Avedisian, 1991; Pasandideh-Fard *et al.*, 1996; Mao, Kuhn and Tran, 1997). These models have the general form given here:

$$E_0 = E_{K0} + E_{S0} = E_{K1} + E_{S1} + E_D \tag{5.10}$$

where E_{K1} is the droplet kinetic energy at maximum spreading, E_{S1} is the surface energy at maximum spreading and E_D is the energy loss due to dissipation. In all cases, the value of E_0 is taken as that shown in Equation 5.8. Observation has shown that the shape of a droplet at maximum spreading extent can vary from a spherical cap at low Re and We, to a toroidal shape as Re and We are increased (Roisman, Rioboo and Tropea, 2002; Son *et al.*, 2008). Despite this discrepancy, most models have estimated the value of E_{SI} by approximating the spread droplet to be a cylinder of diameter $\beta_{\max} D_0$ with the same volume as the initial droplet. The surface energy of the spread droplet is therefore given by:

$$E_{S1} = \sigma \left(\frac{\pi}{4} \beta_{\max} D_0 \left(1 - \cos\theta_{\text{eqm}}\right) + \frac{2D_0^3}{\beta_{\max} D_0} \right) \tag{5.11}$$

At the point of maximum spreading, the value of E_{KI} is assumed to be zero. Whilst this is the case in terms of the velocity of the advancing contact line, other parts of the drop may still have kinetic energy (Roisman, Rioboo and Tropea, 2002). The difference in the various energy balance models for calculating $\beta_{\max}$ is caused by the different methods of determining the value of E_D. This value represents the energy loss due to viscous dissipation in deforming the droplet from its initial spherical shape to its maximum diameter.

The E_D term has been approximated by different researchers using different boundary conditions for the flow. Madejski (1983) assumed a simplified average radial velocity profile and was therefore able to predict $\beta_{\max}$ as follows:

$$\frac{3.629}{Re} \beta_{\max}^5 + \frac{3}{We} \beta_{\max}^2 - 1 = 0 \tag{5.12}$$

The E_D term could not be determined by an analytical expression and was therefore evaluated numerically. The predicted values agreed with experimental values at higher values of Re and We. This model takes no account of the surface energy interactions between the droplet and substrate due to their insignificance at higher values of Re and We.

The work of Chandra and Avedisian (1991) evaluated E_D by integration of a simple dissipation function over droplet volume and deformation time. This approach led to the following solution for the energy balance:

$$\frac{2We}{3Re} \beta_{\max}^4 + (1 - \cos\theta)\, \beta_{\max}^2 - \frac{We + 12}{3} \approx 0 \tag{5.13}$$

where θ_a is the dynamic advancing contact angle of the spreading lamella over the substrate. This model gave better agreement with experimental data, particularly at lower impact velocities due to the inclusion of a term associated with the droplet–substrate surface energy interaction (the contact angle). The model, however, still tended to over-predict the spreading of the droplet because it assumes that the viscous dissipation upon spreading occurs throughout the entire thickness of the spreading lamella. This model was refined further to allow for the dissipation to occur only in a boundary layer region of the spreading droplet close to the substrate (Pasandideh-Fard *et al.*, 1996).

$$\beta_{\max} = \sqrt{\frac{We + 12}{3\left(1 - \cos\theta_a\right) + 4We/\sqrt{Re}}} \tag{5.14}$$

This equation gave good agreement with experimental results within the region in which the assumptions necessary to define the boundary layer were valid (Re $\gg$ 1). Outside this range, as would be expected, agreement is not so good. Another issue with this model is the use of the dynamic advancing contact angle rather than the equilibrium contact angle, although the difference in results obtained in the region where the boundary layer assumptions are valid is negligible. This is because the second term in the denominator of Equation 5.14 is far larger than the first term at higher impact velocities. Whilst there is a clear link between contact line velocity, fluid properties and equilibrium contact angle as discussed in Section 5.4.2, modelling of this as a function of time during the spreading process introduces a great deal of complexity. For this reason, the value of θ_a was assumed to have a constant average value of 110° throughout the spreading process, meaning that any link between the actual droplet–substrate surface energy balance and the model was lost.

The most rigorous model of the droplet spreading process was conducted by Roisman, Rioboo and Tropea (2002), in which the dissipation of the impact kinetic energy was separated into two distinct stages. This model is quite similar in principle to that first devised by Scheller and Bousfield (1995), but with the addition of a term for the surface energy interactions between the droplet fluid and substrate. In principle, this should make the model more applicable to lower kinetic energy impacts. The first stage is taken as the time when there is liquid flowing perpendicular to the substrate due to the impact, with it being possible to approximate the flow during this period as the squeeze flow encountered between two parallel plates being brought together. The modelling of this first stage gives a radial mass and momentum distribution, which is used as an input for the second stage of the model where this mass and momentum distribution is balanced against the increasing surface energy and dissipation at the contact line of the spreading lamella. This produces a model that gives the temporal evolution of the spreading droplet footprint during both spreading and receding of the substrate, which can be used to predict the maximum spreading diameter. The model agrees well with both experimental results and numerical models of the droplet impingement process.

5.2.3 Impact of Inkjet-Sized Droplets

As shown in Figure 5.1, the range of We encountered in inkjet printing tends to be lower than in previous impact-driven spreading studies. Models for low-We impact are neither as numerous nor as well developed as for the higher We impacts discussed in this chapter, primarily due to the overwhelming influence of surface energy interactions. To eliminate the influence of wetting, impact experiments have been performed on non-wetting and super-hydrophobic surfaces (Clanet *et al.*, 2004). This work showed that at low We, the maximum extent of droplet spreading strongly correlated with $\mathrm{We}^{0.25}$. A transition between the low-We and high-We behaviour was derived to occur at $\mathrm{WeRe}^{-0.8} = 1$, with impact conditions below this transition exhibiting low-We behaviour. The printing conditions experienced during inkjet printing are typically close to this transition (Derby and Reis, 2003), meaning that great care must be taken when relating any of the aforementioned models to the inkjet printing process.

Due to the stringent temporal and spatial resolutions required for droplet impact studies in the inkjet regime, few studies have been performed. The work of Van Dam and Le

Clerc (2004) imaged the impact of water droplets on glass at varying velocities produced by inkjet printing. They found that there was poor agreement between the observed maximum spreading and any models derived for larger droplets, with all models tending to predict greater spreading. They suggested that this was due to oscillation of the droplet (once impacted) being an important means of kinetic energy dissipation, and demonstrated this by showing that the oscillation time of the droplet compared well with analytical models based upon a balance of inertia and surface energy. Dong, Carr and Morris (2006) also showed the oscillation of printed water droplets, with the oscillations being more pronounced at higher contact angles. The overriding influence of surface energy on the final droplet diameter was also shown, with drops on hydrophilic surfaces continuing to spread at a much slower rate due to capillarity after the impact kinetic energy has been dissipated.

Son *et al.* (2008) presented work on the impact of printed water droplets on substrates of varying surface energy. Again it was found that existing models for larger droplets predicted higher spreading than observed, and the deviation was found to increase as the contact angle was increased. This was attributed to the dissipation at the contact line not being included in previous models of larger droplets where it is of negligible significance. To allow for the contact line dynamics in a simple manner, a correlation between the spreading and We modified to reflect the equilibrium surface energy of a sessile droplet (We*) was found:

$$\ln \beta = 0.09 \ln \mathrm{We}^* + 0.151 \tag{5.15}$$

This correlation bears reasonable resemblance to that derived by Clanet *et al.* (2004) in that it shows a power law dependence with We, although in this case the surface energy of the substrate is included as super-hydrophobic surfaces are not used. This correlation has also been found to agree well with experimental results using non-Newtonian fluids containing polyethylene oxide (Son and Kim, 2009), suggesting that any influence of extensional viscosity upon the impact process is minimal.

5.3 Unstable Droplet Deposition

When a droplet impacts upon a substrate, it does not necessarily spread in a stable manner. Depending upon the impact conditions, the fluid properties and the condition of the substrate, the impact process can lead to an instability that disturbs the advancing liquid front. This disturbance manifests itself either as a series of regular perturbations to the contact line or as secondary droplets ejected from the spreading bulk. The occurrence of splashing appears to follow a threshold behaviour, with the threshold being a parameter of impact and fluid conditions. The most prevalent form of threshold used to describe this behaviour is one that takes the form of a dimensionless formulation with $K_c^* = \mathrm{Re}^{0.25}\mathrm{We}^{0.5}$ (Stow and Hadfield, 1981).

The splashing threshold therefore takes into account all salient impact and fluid conditions for a Newtonian fluid. Non-Newtonian fluids have been found to exhibit different dependencies of splashing behaviour (Crooks and Boger, 2000). The parameter $K_c{}^* = \mathrm{Re}^{0.25}\mathrm{We}^{0.5}$ was initially determined empirically, and so far has been explained by means of a kinematic discontinuity (Yarin and Weiss, 1995) and propagation of a Rayleigh–Taylor instability along the spreading contact line (Bhola and Chandra, 1999). The splashing threshold is also influenced by the surrounding environment, with removal of the surrounding atmosphere suppressing splash formation on a smooth surface (Xu, Zhang and Nagel, 2005). Splashing, however, has been observed in vacuum plasma spraying (Montavon *et al.*, 1997), indicating the importance of other factors (e.g. surface roughness and phase change of the fluid) in the splashing mechanism.

The nature of the substrate is of critical importance when determining the splashing behaviour of a droplet. Previous work has primarily looked into the influence of surface roughness on the splashing behaviour, with the impact velocity required to induce a splash found to decrease with increasing surface roughness amplitude (Stow and Hadfield, 1981). The exact mechanism describing the influence of surface roughness on splashing, much like that for splashing in general, is not well understood. Most attempts at explaining how surface roughness alters the splashing mechanism have been qualitative explanations based upon the generation of a Rayleigh–Taylor instability. The most mathematically complete explanation considers the break-up of the spreading liquid rim by this instability (Wu, 2003); however, this explanation is for a gradual break-up over the whole spreading process of an inviscid fluid. Although some experimental results have suggested a negligible influence of viscosity on the splashing process (Range and Feuillebois, 1998), the bulk of experimental observations show that viscosity is a key parameter (Stow and Hadfield, 1981; Vander Wal, Berger and Mozes, 2006).

Previous observation has shown that perturbations of the contact line associated with splashing initiate very early in the impact process on a rough surface, and the number of perturbations stays constant over the whole spreading process. This would suggest that any attempt to explain the splashing mechanism must do so over an initial period of time shorter than the bulk spreading time. The introduction of surface roughness to the substrate acts to reduce the number of observed perturbations, with the number of perturbations decreasing as the surface roughness increases (Range and Feuillebois, 1998).

Attempts to quantify the influence of surface roughness on $K_c{}^*$ has yielded the following empirical expression based on experimentally obtained data (Cossali, Coghe and Marengo, 1997):

$$K_c^* = 57.23 + \frac{2.288}{\bar{R}_a^{0.39}} \tag{5.16}$$

where $\bar{R}_a$ is the substrate roughness R_a made dimensionless with respect to D_0 ($\bar{R}_a = R_a/D_0$). This correlation shows good agreement at both very small values of $\bar{R}_a$ (Stow and Hadfield, 1981; Cossali, Coghe and Marengo, 1997) and at values approaching unity (Mundo, Sommerfeld and Tropea, 1995), but fails to show such agreement at intermediate values of $0.001 \le \bar{R}_a \le 0.1$ (Crooks and Boger, 2000). This range of roughness is of particular interest for processes involving micrometre-size droplets, such as inkjet printing, as this corresponds to the roughness most likely encountered on common substrates such as paper (Moss *et al.*, 1993).

5.4 Capillarity-Driven Spreading

5.4.1 Droplet–Substrate Equilibrium

When a liquid is put into contact with a solid, the liquid will tend towards an equilibrium shape if no other forces act upon it. This manifests itself in the form of an equilibrium contact angle (θ_{eqm}) that is dependent upon the balance of the three salient surface energies (as shown in Figure 5.3): the surface energy between the liquid and the surrounding environment (σ_{LV}), the surface energy between the solid and the surrounding environment (σ_{SV}) and the surface energy between the liquid and the solid (σ_{LS}). The balance between these three surface energies and their relation to θ_{eqm} is given by the Young equation (Young, 1805):

$$\sigma_{SV} = \sigma_{LS} + \sigma_{LV} \cos\theta_{\text{eqm}} \tag{5.17}$$

This equation is valid in the immediate vicinity of the contact line at equilibrium; however, gravitational forces may influence the shape of the droplet at distances far from the contact line. The relative magnitude of the surface energy and gravitational forces is given by Bo (defined by Equation 5.3), with Bo > 1 indicating that gravitational forces are significant. If gravitational forces are significant, the profile of the deposited liquid will need to be calculated by solving the Young–Laplace equation over the whole surface of the liquid (Laplace, 1806):

$$\Delta P_0 + \Delta\rho g z = \sigma_{LV}\left(\frac{1}{R_1} + \frac{1}{R_2}\right) \tag{5.18}$$

where P_0 is the pressure at a fixed reference point, $\Delta\rho$ is the density difference between the liquid and the surrounding vapour, g is the acceleration due to gravity, z is the distance from the fixed reference point and R_1 and R_2 are the two principal radii of curvature of the liquid. For inkjet printing, the Bond number of an individual droplet is typically orders of magnitude less than 1, and as such this calculation is unnecessary and the liquid profile can be calculated on the assumption that the liquid will tend to minimise surface energy by forming truncated sphere geometries (Figure 5.3).

In the case of a finite contact angle, it is possible to model the deposited droplet as a spherical cap if gravitational forces are negligible, that is Bo ≪ 1. If we assume volume conservation between the deposited drop and the spherical cap, the following expression can be derived (Van Dam and Le Clerc, 2004) where β_{eqm} is the diameter of

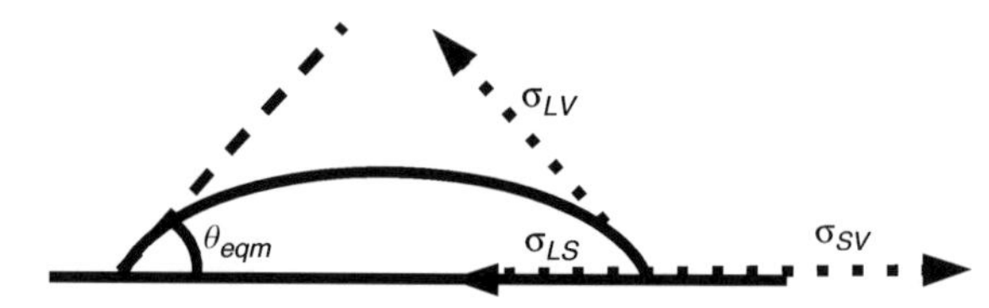

Figure 5.3 *Schematic diagram illustrating the truncated sphere geometry adopted by sessile droplets when Bo ≪ 1, as encountered in inkjet printing. The relevant surface energies and consequent equilibrium contact angle are also illustrated.*

the spherical cap on the substrate, D_{eqm}, normalised to D_0:

$$\beta_{\text{eqm}} = \frac{D_{\text{eqm}}}{D_0} = \sqrt[3]{\frac{8}{\tan\frac{\theta_{\text{eqm}}}{2}\left(3+\tan^2\frac{\theta_{\text{eqm}}}{2}\right)}} \tag{5.19}$$

The equilibrium wetting of a droplet on a substrate can be further influenced by the use of heterogeneous and structured substrates. This is due to the added energetic barriers placed in the way of the contact line that may lead to a metastable energy state different from equilibrium. Without a means of supplying additional energy, a droplet on a rough surface will therefore not always form a truncated sphere geometry that would be expected on a smooth surface. This pinning of the contact line in a metastable state has been demonstrated by use of additional vibrational energy to overcome the energy barrier, at which point the droplet adopts the expected truncated sphere geometry (Meiron, Marmur and Saguy, 2004). Due to the increased area of contact between droplet and substrate caused by the rough surface, the surface energy balance at equilibrium is also altered. This is expressed in the Wenzel relation (Wenzel, 1936):

$$\sigma_{LV}\cos\theta_W = r_W\left(\sigma_{SV} - \sigma_{LS}\right) \tag{5.20}$$

where θ_W is the apparent equilibrium contact angle with the rough substrate and r_W is the ratio between the actual area of contact between droplet and substrate and the equivalent area on a smooth surface. This relation predicts that any hydrophobic or hydrophilic character of the droplet is accentuated by the surface roughness, as has been shown experimentally (Palasantzas and de Hosson, 2001). The Wenzel relation described here assumes that the liquid fully penetrates in between the points of roughness. Another possibility is that the surface energy interactions between solid, liquid and atmosphere mean that a lower energy state exists when roughness asperities are bridged and a series of air pockets formed. This can be modelled using a particular form of the Cassie–Baxter relation, which more generically describes wetting on a chemically heterogeneous surface (Cassie and Baxter, 1944):

$$\cos\theta_{CB} = f_1\cos\theta_{e1} + f_2\cos\theta_{e2} \tag{5.21}$$

where θ_{CB} is the apparent equilibrium contact angle on the surface, f_1 and f_2 are the fraction of components 1 and 2 respectively, and θ_{e1} and θ_{e2} are the equilibrium contact angles with component 1 and 2 respectively. In the case of air pockets forming when spreading on a rough surface, f_1 is the fraction of contact made with roughness asperities, $f_2 = 1 - f_1$ and $\theta_{e2} = 180°$. By minimising the area of contact with roughness asperities whilst maintaining this form of wetting, it is possible to produce a surface with a contact angle approaching 180°. These surfaces are known as super-hydrophobic, and have been fabricated by means of nanosphere lithography (Shiu *et al.*, 2004) and ordered carbon nanotube arrays (Lau *et al.*, 2003).

Equation 5.21 represents the wetting behaviour of a liquid on a chemically heterogeneous substrate where the variation of surface energy is significantly smaller than the size of the droplet. Using patterned surfaces with heterogeneities in either surface energy or topography of a similar order of magnitude to the droplet enables the controlled wetting of only predefined areas of substrate. This can be particularly useful for inkjet printing of functional materials (Sirringhaus *et al.*, 2000; de Gans, Hoeppener and Schubert, 2006) because it both enables a reduction in feature size on the substrate and supplies a self-correcting mechanism for the placement of droplets. This has been exploited to produce features with dimensions much smaller than the diameter of a spread droplet, for example gate channels in printed field effect transistors (Sirringhaus *et al.*, 2000).

5.4.2 Capillarity-Driven Contact Line Motion

The spontaneous spreading of a liquid upon a substrate entails the redistribution of the liquid volume such that it ends up in equilibrium with the substrate at the equilibrium contact angle. To achieve this, it is necessary to both move the contact line from an initial position and redistribute the liquid volume across the contact area with the substrate (De Gennes, 1985). The driving force for this contact line motion is provided by the lower energy obtained by incrementally tending towards the lowest energy, the equilibrium state.

For an individual droplet placed upon a substrate, this spreading process will have an initial contact angle of 180°, which will decrease with time and tend towards θ_{eqm} as the liquid spreads. Experimental observations of advancing interfaces have shown that there is a correlation between the dynamic contact angle (θ_t) of the spreading drop and the contact line velocity (U_{CL}), as shown by the Hoffman–Tanner–Voinov relation (Hoffman, 1975; Tanner, 1979):

$$U_{CL} = \frac{\kappa\sigma}{\eta}\left(\theta_t^3 - \theta_{\mathrm{eqm}}^3\right) \tag{5.22}$$

where κ is a constant, found experimentally to be 0.013. As noted here, spontaneous spreading of a droplet occurs due to incremental lowering of the surface energy towards equilibrium. This necessitates that during the spreading process, there is a mechanism by which this energy can be dissipated during the flow of fluid. There have been two approaches to this taken in the literature, one in which the energy is dissipated within the core of the droplet (Cox, 1986) and one in which the energy is dissipated in a region surrounding the moving contact line (Blake and Haynes, 1969).

In the first case, the dissipation mechanism follows standard hydrodynamic theory of viscous laminar flow, with a no-slip condition on the substrate and free flow on the droplet surface. As there is a three-phase interface at the contact line, it is also necessary to impose a cut-off distance some length from the contact line so as to avoid a non-integrable stress singularity at the contact line (Huh and Scriven, 1971); the cut-off distance used is typically of the order 10^{-8}–10^{-9} m. A model of this form proposed by Cox (1986) and developed by de Gennes, Hua and Levinson (1990) gives the following expression for U_{CL} under the assumption of a finite equilibrium contact angle below 135°:

$$U_{CL} = \left(\frac{\sigma}{9\eta\left[\ln\left(R_m/\lambda_c\right) + Q_e\right]}\right)\left(\theta_t^3 - \theta_{\mathrm{eqm}}^3\right) \tag{5.23}$$

where R_m is a macroscopic length scale, λ_c is the cut-off distance and Q_e is the dissipation in length λ. This expression is similar in form to Equation 5.22, and may well provide the physical basis of the empirical relation.

In the second case, the dissipation in the region adjacent to the contact is achieved by means of the molecular kinetic theory of wetting first proposed by Blake and Haynes (1969). Here, the motion of the contact line is modelled as an activated hopping process of molecules from the droplet onto adsorption sites on the substrate in the region of the contact line. The movement of the contact line is assumed to be due to a net flux of molecules from the droplet to the surface adsorption sites, with the velocity of the contact line given by the product of the net frequency of molecular jumps in the direction of motion and the separation between adsorption sites. By assuming that this process is driven by the imbalance in surface energy between the current state and equilibrium, the following equation for U_{CL} is derived:

$$U_{CL} = 2\xi_w \lambda_m \sinh\left[\frac{\sigma}{2nkT}\left(\cos\theta_{\text{eqm}} - \cos\theta_t\right)\right] \tag{5.24}$$

where ξ_w is the equilibrium jump frequency, λ_m is the molecular displacement length, k is the Boltzmann constant, T is the absolute temperature and n_a is the number of adsorption sites per unit area of substrate ($n_a \sim \lambda_m{}^{-2}$). This model does not include a viscosity term, which is of importance as shown by the empirical relationship derived in Equation 5.22. Such a term can be included by expressing the dissipation at the contact line as a function of both the retarding influence of the solid and the liquid molecular interactions. By expressing the excess Gibbs activation energy of both of these retarding influences on the basis of absolute reaction rates, it is possible to obtain the following expression for ξ_w, with

$$\xi_w = \xi_s\left(\frac{h_p}{\eta v}\right) \tag{5.25}$$

where ξ_s is the solid–liquid interaction parameter, h_p is Planck's constant and v is the specific volume of the droplet fluid. When Equation 5.25 is substituted into 5.24, the following formula is obtained:

$$U_{CL} = \frac{2\xi_s h_p \lambda_m}{\eta v}\sinh\left[\frac{\sigma}{2nkT}\left(\cos\theta_{\text{eqm}} - \cos\theta_t\right)\right] \tag{5.26}$$

A large number of experiments (Schwartz and Tejada, 1972; Foister, 1990; Brochard-Wyart and de Gennes, 1992; Hayes and Ralston, 1993; Sedev *et al.*, 1993) have been conducted to investigate the effectiveness of both these models. Of the studies that have compared the two (Brochard-Wyart and de Gennes, 1992; Hayes and Ralston, 1993; Sedev *et al.*, 1993), it is generally found that the molecular kinetic approach shows better agreement at higher equilibrium contact angles, whilst at lower contact angles and contact line velocities the hydrodynamic model shows better agreement.

5.4.3 Contact Angle Hysteresis

Contact angle hysteresis is caused by an energy barrier inhibiting the movement of the contact line and leads to there being two contact angles measured on a substrate that are dependent upon the direction of contact line movement: an advancing contact angle and a receding contact angle (de Gennes, 1985). There are a number of possible sources for this hysteresis, such as surface roughness and chemical heterogeneity of the substrate;

but perhaps the most relevant to inkjet printing is the presence of an additional solid phase at the contact line, such as dust contamination or a solid phase within the droplet. The presence of the additional solid phase can introduce either a mechanical barrier to the receding contact line or a change in the surface energy equilibrium in the vicinity of the contact line.

If the deposition of solid material at the contact line is great enough, this will lead to the receding contact angle being reduced to zero, meaning that it is not energetically favourable for the contact line to retract at any point during the evaporation process. This means that the initial equilibrium state of the droplet on the substrate before phase change will control the size of deposit left after evaporation of the carrier solvent has occurred (Duineveld, 2003).

5.5 Coalescence

5.5.1 Stages of Coalescence

5.5.1.1 Bridge Formation and Broadening

When two droplets come into contact with each other, the radius of curvature of the surface at the initiation point of the liquid bridge is instantaneously inverted. The initiation could therefore be expected to be a singularity, in a similar way to the triple point at a contact line, and has been modelled as such (Diez and Kondic, 2002). Experiments conducted with spherical drops coming into contact have shown, however, that there is a flattening of the droplet curvature that leads to a finite contact area (Thoroddsen, Takehara and Etoh, 2005). This was explained as being due to the compression of the volume between the two approaching droplet surfaces; this increases the pressure and the surface equilibrium accordingly (Equation 5.18). The nature of the flow within the liquid bridge will be dependent upon the properties of the droplet liquid, namely, the density, surface tension and viscosity.

Initially, due to the very small size of the liquid bridge (and consequent $\mathrm{Re} \ll 1$), the flow will be dominated by a balance between the viscous and capillary forces. This will continue to be the case until the bridge width exceeds a viscous length scale that corresponds to $Re = 1$, with the velocity of flow equal to σ/η (Thoroddsen, Takehara and Etoh, 2005). For fluids used in previous experimental investigations such as water (Thoroddsen, Takehara and Etoh, 2005) and mercury (Menchaca-Rocha *et al*., 2001), this length is of the order of nanometres, meaning that this stage of bridge growth is very difficult to observe. Attempts have been made to analyse this stage by using high-viscosity silicone oils (Yao *et al.*, 2005), with the growth rate decreasing linearly with bridge width, which does not match previously derived analytical models (Hopper, 1993). This deviation from the model was expected due to assumptions in the model being valid only over the very short time and length scales experienced with less viscous fluid.

By use of phase-separated binary colloidal suspension, the effective surface tension between the two phases is lowered by many orders of magnitude (Aarts *et al.*, 2005); this greatly extends the viscous length scale, enabling similar experiments to be conducted. It was found that the growth rate of the bridge was constant with time, which was similar to the model proposed by Hopper (1993) but excluding the logarithmic corrections included in the model. When the bridge width exceeds the viscous length scale, the flow can be assumed to be inviscid. This has led to the derivation of a scaling law, with the bridge width being proportional to the square root of the time since bridge formation (Duchemin, Eggers and Josserand, 2003). The constant of proportionality used in this scaling law was identified by fitting experimental data, but was found to be dependent upon droplet diameter and fluid properties. This relationship has been found to agree well with spherical droplets of mercury (Menchaca-Rocha *et al.*, 2001), spherical droplets of water and silicone oil (Aarts *et al.*, 2005) and sessile droplets of silicone oil (Ristenpart *et al.*, 2006). With regard to inkjet printing, the experiments conducted with sessile droplets are the most relevant. In these experiments, it was found that the constant of proportionality for the scaling law was far more dependent upon the droplet size and geometry than with spherical droplets, with the scaling law expressed as:

$$R_b = \sqrt{\frac{\sigma h_{bd}^3}{\eta R_{bd}^2}} \sqrt{t_b} \tag{5.27}$$

where R_b is the width of the meniscus bridge, h_{bd} is the height of the droplet, $\mathrm{R_{bd}}$ is the droplet radius and t_b is the time of bridge growth. By rearranging Equation 5.27, using fluid and droplet geometry typical of inkjet printing and assuming that bridge growth proceeds until $R_b = 2R_{bd}$, the timescale of coalescence is found to be of the order of 10^{-3} s. The situation described previously consists of two droplets with negligible kinetic energy coming into contact and coalescing. This situation may not be directly applicable due to the coalescence taking place when there is a significant inertial force due to the impact of the droplet. Whilst this situation has not been analysed in any depth, an estimate for the timescale of flow has been made, assuming that the flow is a balance between the inertial and capillary forces (Schiaffino and Sonin, 1997a):

$$t_b = \sqrt{\frac{\rho D_0^3}{\sigma}} \tag{5.28}$$

In Equation 5.28, together with parameters typical of inkjet printing, t_b is found to be on the order of 10^{-4} s. Both of these timescale estimates coincide with the separation times between droplet depositions regularly encountered during inkjet printing.

5.5.1.2 *Droplet Relaxation*

Once the droplets have coalesced into a single body, there is a driving force to minimise surface energy to satisfy the Young equation (Equation 5.17). This means that the droplet will tend towards a spherical cap geometry, which requires advancement and retraction of the contact line upon the substrate.

It has been found that the retraction is highly dependent on the environmental conditions experienced by the coalescing droplets, with several orders of magnitude of difference in timescale between ambient and saturated conditions (Andrieu *et al.*, 2002; Beysens and Narhe, 2006). The retraction of the contact line must overcome any contact line hysteresis and the dissipation at the contact line, which, assuming no external force, will be due to evaporation of liquid from the bulk to the surrounding environment in the vicinity of the contact line. The saturation of the environment is therefore critical to the relaxation time, due to the significant influence it has on the evaporation rate. This explanation of the contact line motion bears a large amount of similarity to the molecular kinetic approach to dynamic wetting (see Section 5.4.1).

In the presence of an external force, the relaxation of the droplet displays no such relationship with the environment (Beysens and Narhe, 2006). This is due to the external force being able to overcome the aforementioned hysteresis and dissipation. Under these conditions, the relaxation time is more similar to what would be expected with a bulk hydrodynamic flow (Beysens and Narhe, 2006). Using typical conditions experienced during inkjet printing, this leads to the velocity of contact line motion of 10^{-2}–10^{-3} m s^{-1}. An estimate of the total retraction time is given by the ratio of droplet diameter and velocity of contact line motion, which results in a timescale of 10^{-2}–10^{-1} s. This timescale is significantly larger than the time separation between droplets typically encountered in inkjet printing, although this timescale is for a large retraction equal to the diameter of one droplet. If a smaller retraction of a deposited droplet is sufficient to prevent overlap of a newly deposited droplet, this retraction would lead to the breaking up of a continuous bead into a series of discontinuous regions of single or multiple droplets.

5.5.2 Coalescence and Pattern Formation

To fabricate a two-dimensional pattern, it is necessary for a series of droplets to be deposited upon a substrate. For this pattern to consist of continuous deposits larger than a single droplet, the droplets must coalesce to form a liquid structure on a substrate. Understanding how the droplets coalesce is necessary in understanding what structures can be obtained by coalescence, so that droplet deposition patterns and ink formulations can be designed accordingly. One of the simplest structures that can be formed through droplet coalescence is that of a bead, which is formed by the repeated coalescence of droplets in one direction across a substrate. This structure can be used in a number of applications, such as conductive tracks in printed electronics (Szczech *et al.*, 2002; Stringer and Derby, 2010), and digital micro-fabrication of wax-based solids (Gao and Sonin, 1994).

5.5.3 Stable Bead Formation

Two droplets deposited so that they overlap will tend to coalesce. These droplets will coalesce into a single body of liquid if the first deposited droplet does not form a solid before the next one is deposited. A series of droplets deposited linearly on a substrate will therefore form a liquid bead on the substrate. However, in order to form a continuous bead, the drops must be positioned sufficiently close together to interact and coalesce (Smith *et al.*, 2006). In order to form a stable parallel-sided bead, Soltman and Subramanian observed that the drops need to overlap by a critical amount but if the drops are too close together an irregular track forms (Soltman and Subramanian, 2008), as illustrated in Figure 5.4. Thus there appears to be a limited set of drop spacings over which stable parallel-sided liquid beads form.

Assuming that the contact line of this liquid bead cannot recede due to contact line pinning, ignoring any end effects and the fact that gravitational forces are negligible ($\mathrm{Bo} \ll 1$), the bead will have a constant cross-section of a circular segment. Based on these assumptions and that of volume conservation between the impinging droplets and

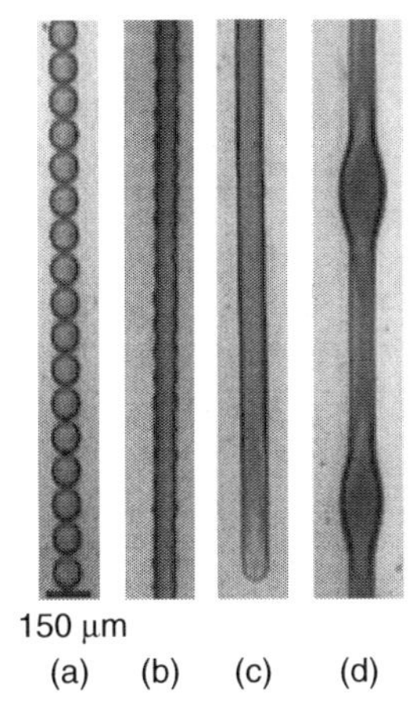

Figure 5.4 *A stable liquid bead is formed through the overlap of printed drops. (a) If the drops are too far apart, they do not coalesce to form a bead. (b) There is a critical droplet spacing above which, although drops coalesce, they do not form a parallel-sided bead. (c) There is a range of drop spacing over which parallel-sided beads form. (d) However, if the drop spacing is too small, unstable bulges form along the length of the liquid bead (Reproduced with permission from Soltman and Subramanian (2008) Copyright (2008) American Chemical Society).*

the liquid bead, it is possible to construct a simple geometric model of the bead as a function of initial droplet diameter D_0, droplet spacing p, bead width w and θ_{eqm} (Stringer and Derby, 2009):

$$w = \sqrt{\frac{2\pi D_0^3}{3p\left(\dfrac{\theta_{\mathrm{eqm}}}{\sin^2\theta_{\mathrm{eqm}}} - \dfrac{\cos\theta_{\mathrm{eqm}}}{\sin\theta_{\mathrm{eqm}}}\right)}} \tag{5.29}$$

The bead width can be normalised with respect to $\beta_{\mathrm{eqm}} D_0$, giving the following dimensionless relationship for all stable bead widths:

$$w^* = \frac{w}{\beta_{\mathrm{eqm}} D_0} = \sqrt{\frac{2\pi D_0}{3p\beta_{\mathrm{eqm}}^2\left(\dfrac{\theta_{\mathrm{eqm}}}{\sin^2\theta_{\mathrm{eqm}}} - \dfrac{\cos\theta_{\mathrm{eqm}}}{\sin\theta_{\mathrm{eqm}}}\right)}} \tag{5.30}$$

where w^* is the dimensionless bead width. This derivation and similar (Schiaffino and Sonin, 1997c; Duineveld, 2003; Stringer and Derby, 2009), have been found to successfully predict bead width for solution-based inks, suspension-based inks and solidifying systems.

Assuming that the contact line of an individual droplet is pinned and cannot retract, the minimum width of the bead is equal to $\beta_{\mathrm{eqm}} D_0$. By substituting this into Equation 5.29 for w and rearranging, the following equation for the maximum droplet spacing $p_{\max}$

is obtained:

$$p_{max} = \frac{2\pi d_0}{3\beta_{eqm}^2 \left(\frac{\theta_{eqm}}{\sin^2 \theta_{eqm}} - \frac{\cos \theta_{eqm}}{\sin \theta_{eqm}} \right)} \tag{5.31}$$

or in dimensionless form using the same normalisation as w^*:

$$p_{max}^* f\left(\theta_{eqm}\right) = \frac{2\pi}{3} \tag{5.32}$$

where

$$f\left(\theta_{eqm}\right) = \beta_{eqm}^3 \left(\frac{\theta_{eqm}}{\sin^2 \theta_{eqm}} - \frac{\cos \theta_{eqm}}{\sin \theta_{eqm}} \right) \tag{5.33}$$

With these equations, it is therefore possible to select appropriate printing parameters to produce linear features with parallel contact lines of a prescribed size.

5.5.4 Unstable Bead Formation

The stability of a liquid bead with parallel contact lines upon a substrate is analogous to the Rayleigh instability of a liquid jet (Rayleigh, 1879), with the additional complexity of a contact line between the liquid bead and the substrate. The contact line can be considered freely moving with a constant contact angle, freely moving with contact angle dependent upon contact line velocity or a pinned contact line (Davis, 1980; Sekimoto, Oguma and Kawasaki, 1987).

A bead with a freely moving contact line is inherently unstable in the presence of a perturbation, as the liquid will minimise surface energy by forming a series of unconnected spherical caps on the substrate. The case where the contact angle is constant is not found in practice, as the contact angle is found to change with contact line velocity; however, the final result given by this condition should be identical to the condition of a velocity-dependent contact angle unless the contact line is arrested by a phase change during the contact line motion.

The case of a pinned contact line is not inherently unstable, as the contact line pinning may present a sufficient energy barrier that prevents a perturbation from growing and causing an instability across the substrate. It was found that, theoretically, a bead with pinned contact lines should remain stable as long as the contact angle with the substrate remains below $\pi/2$ radians. This criterion for stability of a ridge with pinned contact lines was confirmed experimentally for a solidifying wax system (Schiaffino and Sonin, 1997a), with the instability exhibited as a series of bulges. This bulging morphology and $\pi/2$ threshold were also observed when water was condensed onto hydrophilic stripes on a substrate (Gau *et al.*, 1999).

For both the solidifying system and hydrophilic stripes mentioned here, the contact line is constrained over a timescale much shorter than that of coalescence and spreading. For the inkjet printing of inks that experience a much slower phase change (e.g. evaporation), contact line pinning is unlikely to be so dramatic. In these circumstances, the conditions that control the contact line motion could be considered more similar to the freely moving contact line condition described in this section. It is found experimentally (Duineveld, 2003) that bead morphologies are similar to those expected by this condition if the

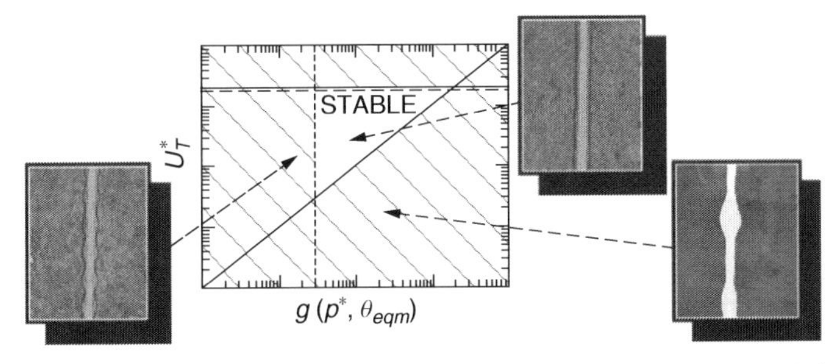

Figure 5.5 *Parameter space map showing a region for stable liquid bead formation as a function of dimensionless substrate velocity* U^*_T *and a dimensionless grouping of droplet spacing* p^* *and advancing contact angle* θ*. The stable triangular region is shown, together with instabilities due to bulge formation and 'scalloping' (Reproduced with permission from Stringer and Derby (2010) Copyright (2010) American Chemical Society).*

contact angle is free to move. The contact line motion of inks that experience evaporation often experiences a hysteresis, with the receding contact angle significantly less than the advancing contact angle. If the receding contact angle is zero or approaching zero, the model of bulge formation described here is no longer appropriate as the contact line is free to advance upon the substrate, but cannot retract and form individual sessile droplets.

Experiments with the deposition of an inkjet-printed polymer solution with a zero receding contact angle have shown a bulging instability (Duineveld, 2003). Due to the ability of the contact line to advance freely, the model for describing the bulging instability for a completely pinned contact line is invalid. A new model was developed that described the bulging instability as a function of the competing capillary flow of a newly deposited droplet with the axial flow within the bead due to a pressure difference. This model was found to successfully predict the bulging instability, and the use of a thermogel ink was found to prevent this bulging by making the axial flow within the bead unfavourable (van den Berg *et al.*, 2007).

The model of Duineveld was further developed by Stringer and Derby (2010) so as to enable the construction of a stability map (Figure 5.5). This produced a parameter space consisting of fluid properties and substrate velocity on one axis and droplet size, droplet spacing and contact angle on the other. Due to this structure, it is possible to plot both the bulging instability and the maximum droplet spacing given in Equation 5.33, with good agreement shown with experimental results.

5.6 Phase Change

In most applications of inkjet printing, it is necessary for the liquid ink to be converted to a solid phase, whether this is by the evaporation of solvent to leave a coloured pigment or dye in graphical applications (Kang, 1991), or the solidification of wax droplets in the formation of 3D ceramic pre-forms (Seerden *et al.*, 2001). The phase change influences the resolution of the printing by interacting with the fluid flow mechanisms discussed in this chapter, as well as determining the distribution of material within the deposit via other, supplementary processes.

5.6.1 Solidification

If a fast and direct phase change can take place before this equilibrium configuration is reached, the liquid will be unable to reach equilibrium and the contact line will be arrested in position. An understanding of the phase change kinetics is therefore necessary in understanding the final contact area produced by a printed droplet solidifying or gelling on a substrate. Work on solidifying systems (Gao and Sonin, 1994; Schiaffino and Sonin, 1997c) has demonstrated that at the low values of We that can be encountered in inkjet printing, the final deposit size is controlled primarily by the temperature of both the molten droplet and the target substrate. Using an order of magnitude analysis of the solidification and spreading times, it can be shown that this deposit size control is due to a highly localised solidification adjacent to the substrate that is primarily dependent upon the temperature difference between the fusion temperature of the droplet and the temperature of the substrate (Schiaffino and Sonin, 1997c).

As the value of We is increased, the size of deposit will show a greater dependence upon the impact conditions as this will have an influence upon the speed of the contact line over the substrate. Work conducted at higher values of We has tended to have been with larger millimetre-size droplets, and has modelled the influence of temperature as another path for energy loss within the energy balance given in Equation 5.10 (Bhola and Chandra, 1999). This produced a prediction of $\beta_{\max}$ similar to Equation 5.14 with the addition of a thermal energy term, although the balance assumes that there is no freezing of the contact line until the maximum spreading is reached. A droplet that undergoes gelation can produce similar phenomena to those observed in solidification, with a freezing of the contact line occurring as the gelation proceeds. This has been demonstrated in inkjet printing with a droplet undergoing a thermally triggered flocculation to rapidly increase viscosity above a threshold temperature (van den Berg *et al*., 2007). This was found again to show a dependence of droplet size with substrate temperature due to contact line arrest, as well as control of solid segregation and line instabilities as discussed in this chapter.

5.6.2 Evaporation

If the phase change involved in the deposition process takes place over a timescale much greater than deposition, contact line pinning (and, therefore, determination of feature resolution) that occurs will be as a result of contact angle hysteresis. There are a number of possible sources for this hysteresis, such as surface roughness and chemical heterogeneity of the substrate as discussed previously; but perhaps the most relevant to inkjet printing is the presence of an additional solid phase at the contact line, such as dust contamination or a solid phase within the droplet. The presence of the additional solid phase can introduce either a mechanical barrier to the receding contact line or a change in the surface energy equilibrium in the vicinity of the contact line.

Contact line pinning results in the contact line retraction of an evaporating droplet proceeding in a two-stage process (Shanahan and Bourges, 1994). In the first stage, the contact area between droplet and substrate remains constant and the contact angle between droplet and substrate is reduced. This stage continues until the contact angle has been reduced to the receding contact angle, at which point the contact line will retract and the contact area shrink. These two stages can be repeated several times over the entire

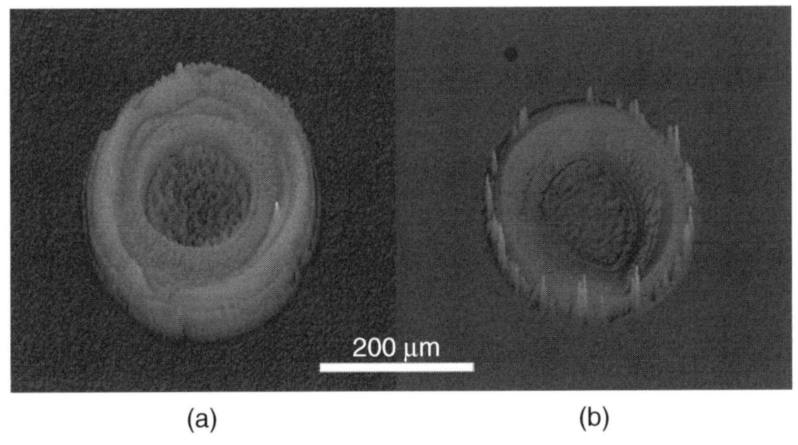

Figure 5.6 Images of coffee staining seen when a suspension of ZrO_2 nanoparticles dries on a glass surface at (a) 20°C and (b) 35°C. In both images coffee staining is evident with all the solid material forming a characteristic ring around the initial contact line and the drop interior being free of deposit. In image (a) the contact line has broken away from its initial pinned position and a secondary drying ring has formed within the original coffee stain (Reproduced with permission from Dou et al. (2011) Copyright (2011) John Wiley & Sons Ltd.).

retraction process, resulting in the droplet having a series of concentric metastable states as the evaporation proceeds. Addition of a small amount of solid particles to the droplet has enabled this to be observed as a series of deposited rings during the evaporation process (Deegan *et al.*, 1997). An example of this is shown in Figure 5.6.

In the inkjet printing of solid-bearing, high-vapour-pressure inks, the evaporation will exacerbate pinning by increasing the amount of additional solid material present at the contact line. This will increase any resistance to the receding of the contact line and reduce the receding contact angle of the liquid across the substrate. If the deposition of solid material at the contact line is great enough, this will lead to the receding contact angle being reduced to zero, meaning that it is not energetically favourable for the contact line to retract at any point during the evaporation process. This means that the initial equilibrium state of the droplet on the substrate before evaporation will control the size of deposit left after evaporation of the carrier solvent has occurred (Duineveld, 2003).

The mass transport of fluid upon evaporation is proportional to the surface area of the liquid, but in a drop with pinned contact lines the droplet cannot uniformly change volume. The extra volume needed to maintain the mass transport due to evaporation must come from the centre of the droplet, and an outward radial flow is induced. This fluid flow possesses a great enough force to overcome gravitational forces acting upon any small suspended particles, and the solid is therefore segregated to the perimeter of the droplet. It has been found that this 'coffee staining' can be overcome by inducing a Marangoni flow within the droplet towards the centre to counteract the flow due to mass transport (Hu and Larson, 2006). A Marangoni flow will be induced due to the temperature gradient caused by evaporation changing the surface tension profile across the droplet, which is found to occur primarily with organic solvents.

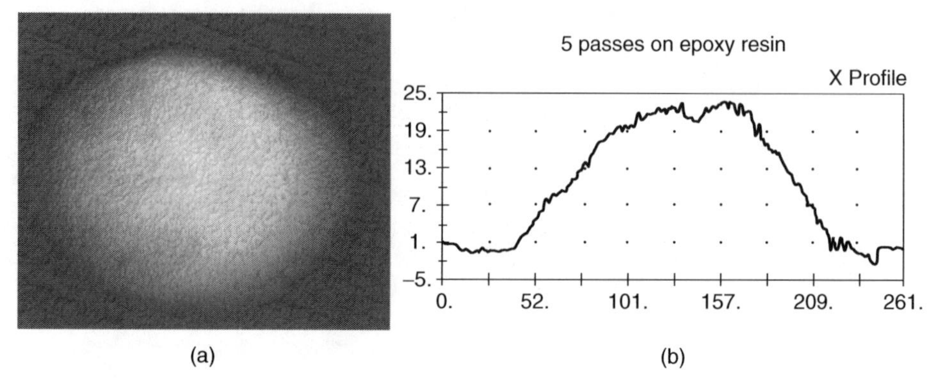

Figure 5.7 *Example of a solvent mixture (water–polyethylene glycol (PEG) solution) that has been used to suppress the formation of coffee stains during the drying of a ZrO_2 suspension. (a) Image of dried drop showing no coffee staining. (b) Contactless profilometer measurement showing a clear height maximum in the centre of the dried drop (Reproduced with permission from Dou* et al. *(2011) Copyright (2011) John Wiley & Sons Ltd.).*

Solute segregation has also been controlled by the use of binary solvent mixtures (de Gans and Schubert, 2004), where two solvents with a significantly different volatility are used. As the more volatile species will be first to evaporate, the concentration of this species will be reduced at the periphery of the droplet compared to the centre due to the differing relative evaporation rates across the droplet. The increased concentration of the volatile species at the centre of the droplet will then compensate for the variation in evaporation rate due to the droplet geometry and prevent solute segregation (Figure 5.7). Although originally devised for dilute suspensions, previous work has shown that segregation also occurs in more concentrated suspensions (Noguera, Lejeune and Chartier, 2005). Rather than complete segregation of the solid to the perimeter, a raised ring of material is present at the perimeter. The incomplete segregation is most probably due to both the magnitude of the induced fluid flow not being sufficient to move all the material within the timescale of evaporation, and the solid material being unable to move outside the constraints of the droplet fluid.

A similar segregation has been found to occur in fluids that contain both a suspended solid phase and a dissolved salt (Yarin *et al.*, 2006), in which there is a definite segregation towards the edge but far less abrupt than in dilute suspensions. This morphology was explained in terms of the precipitating salt forming a 3D network that reduced the mobility of the suspended phase.

5.7 Summary

A rich and complex series of phenomena describe the behaviour of inkjet-printed drops as they impact and solidify on a substrate. Unfortunately, much of the prior work concerning the impact of drops on surfaces has considered drops either with diameter in the range of 10^{-3} m or with velocities that differ from those used in inkjet printing. These are considerably larger than the drops generated during inkjet printing, and have corresponding different dimensionless numbers that can be used to describe the impact conditions. Thus one must be careful when considering the applicability of such observations and models to inkjet printing. Dimensionless plots of fluid properties to identify the important fluid mechanisms that occur during drop impact are useful (Schiaffino and Sonin, 1997b), but these indicate that inkjet printing occurs in a fluid flow regime that is intermediate in behaviour between impact- and capillary-driven flow. It is clear, however, that the final shape of printed droplets is controlled predominantly by capillary forces.

For most applications in graphics the final desired pattern consists of a series of isolated droplets and the processes that lead to this configuration are fairly well understood. For applications in digital fabrication, structures are often fabricated from the coalescence of drops on the substrate to form linear or more complex features. To achieve this the spacing of adjacent droplets is of critical importance and two instabilities occur: at large drop spacings there is a critical spacing above which isolated drops do not interact or coalesce, whilst at small drop centre spacings (in which drops overlap significantly) there is another instability that occurs below a critical spacing.

The physics of drop drying is also of great importance, with fluid flow and solute distribution strongly influenced by capillary and Marangoni flows. There are as yet no good models of the drying process that attempt to predict and include Marangoni flows that are formed through the evaporation of solvent mixtures. Such models would be very useful in predicting and controlling solute or suspension distribution during drying, otherwise known as coffee staining.

References

Aarts, D.G.A.L., Lekkerkerker, H.N.W., Guo, H. *et al*. (2005) Hydrodynamics of droplet coalescence. *Physical Review Letters*, **95** (16), 164503.

Andrieu, C., Beysens, D.A., Nikolayev, V.S. and Pomeau, Y. (2002) Coalescence of sessile drops. *Journal of Fluid Mechanics*, **453**, 427–438.

Armster, S.Q., Delplanque, J.P., Rein, M. and Lavernia, E.J. (2002) Thermo-fluid mechanisms controlling droplet based materials processes. *International Materials Reviews*, **47** (6), 265–301.

van den Berg, A.M.J., de Laat, A.W.M., Smith, P.J. *et al*. (2007) Geometric control of inkjet printed features using a gelating polymer. *Journal of Materials Chemistry*, **17** (7), 677–683.

Beysens, D.A. and Narhe, R.D. (2006) Contact line dynamics in the late-stage coalescence of diethylene glycol drops. *Journal of Physical Chemistry B*, **110** (44), 22133–22135.

Bhola, R. and Chandra, S. (1999) Parameters controlling solidification of molten wax droplets falling on a solid surface. *Journal of Materials Science*, **34** (19), 4883–4894.

Blake, T.D. and Haynes, J.M. (1969) Kinetics of liquid/liquid displacement. *Journal of Colloid and Interface Science*, **30** (3), 421–423.

Brochard-Wyart, F. and de Gennes, P.G. (1992) Dynamics of partial wetting. *Advances in Colloid and Interface Science*, **39**, 1–11.

Cassie, A. and Baxter, S. (1944) Wettability of porous surfaces. *Transactions of the Faraday Society*, **40**, 546–551.

Chandra, S. and Avedisian, C.T. (1991) On the collision of a droplet with a solid-surface. *Proceedings of the Royal Society of London, Series A*, **432** (1884), 13–41.

Clanet, C., Beguin, C., Richard, D. and Quere, D. (2004) Maximal deformation of an impacting drop. *Journal of Fluid Mechanics*, **517**, 199–208.

Cossali, G.E., Coghe, A. and Marengo, M. (1997) The impact of a single drop on a wetted solid surface. *Experiments in Fluids*, **22** (6), 463–472.

Cox, R.G. (1986) The dynamics of the spreading of liquids on a solid-surface. Part 1: viscous-flow. *Journal of Fluid Mechanics*, **168**, 169–194.

Crooks, R. and Boger, D.V. (2000) Influence of fluid elasticity on drops impacting on dry surfaces. *Journal of Rheology*, **44** (4), 973–996.

Davis, S.H. (1980) Moving contact lines and rivulet instabilities. Part 1: the static rivulet. *Journal of Fluid Mechanics*, **98** (5), 225–242.

Deegan, R.D., Bakajin, O., Dupont, T.F. *et al*. (1997) Capillary flow as the cause of ring stains from dried liquid drops. *Nature*, **389** (6653), 827–829.

de Gans, B.J., Hoeppener, S. and Schubert, U.S. (2006) Polymer-relief microstructures by inkjet etching. *Advanced Materials*, **18** (7), 910–914.

de Gans, B.J. and Schubert, U.S. (2004) Inkjet printing of well-defined polymer dots and arrays. *Langmuir*, **20** (18), 7789–7793.

de Gennes, P.G. (1985) Wetting – statics and dynamics. *Reviews of Modern Physics*, **57** (3), 827–863.

de Gennes, P.G., Hua, X. and Levinson, P. (1990) Dynamics of wetting – local contact angles. *Journal of Fluid Mechanics*, **212**, 55–63.

Derby, B. (2010) Inkjet printing of functional and structural materials – fluid property requirements, feature stability and resolution. *Annual Review of Materials Research*, **40**, 395–414.

Derby, B. and Reis, N. (2003) Inkjet printing of highly loaded particulate suspensions. *MRS Bulletin*, **28**, 815–818.

Diez, J.A. and Kondic, L. (2002) Computing three-dimensional thin film flows including contact lines. *Journal of Computational Physics*, **183** (1), 274–306.

Dong, H.M., Carr, W.W. and Morris, J.F. (2006) An experimental study of drop-on-demand drop formation. *Physics of Fluids*, **18** (7), 072102.

Dou, R., Wang, T., Guo, Y. and Derby, B. (2011) Ink-jet printing of zirconia: coffee staining and line stability. *Journal of the American Ceramic Society*, **94** (11), 3787–3792.

Duchemin, L., Eggers, J. and Josserand, C. (2003) Inviscid coalescence of drops. *Journal of Fluid Mechanics*, **487**, 167–178.

Duineveld, P.C. (2003) The stability of inkjet printed lines of liquid with zero receding contact angle on a homogeneous substrate. *Journal of Fluid Mechanics*, **477**, 175–200.

Foister, R.T. (1990) The kinetics of displacement wetting in liquid/liquid/solid systems. *Journal of Colloid and Interface Science*, **136** (1), 266–282.

Gao, F.Q. and Sonin, A.A. (1994) Precise deposition of molten microdrops – the physics of digital microfabrication. *Proceedings of the Royal Society of London, Series A*, **444** (1922), 533–554.

Gau, H., Herminghaus, S., Lenz, P. and Lipowsky, R. (1999) Liquid morphologies on structured surfaces: from microchannels to microchips. *Science*, **283** (5398), 46–49.

Hayes, R.A. and Ralston, J. (1993) Forced liquid movement on low-energy surfaces. *Journal of Colloid and Interface Science*, **159** (2), 429–438.

Hoffman, R.L. (1975) Study of advancing interface. Part 1: interface shape in liquid–gas systems. *Journal of Colloid and Interface Science*, **50** (2), 228–241.

Hopper, R.W. (1993) Coalescence of 2 viscous cylinders by capillarity. 1: theory. *Journal of the American Ceramic Society*, **76** (12), 2947–2952.

Hu, H. and Larson, R.G. (2006) Marangoni effect reverses coffee-ring depositions. *Journal of Physical Chemistry B*, **110** (14), 7090–7094.

Huh, C. and Scriven, L.E. (1971) Hydrodynamic model of steady movement of a solid/liquid/fluid contact line. *Journal of Colloid and Interface Science*, **35** (1), 85–101.

Kang, H.R. (1991) Water-based inkjet ink. Part 1: formulation. *Journal of Imaging Science*, **35** (3), 179–188.

Laplace, P.S. (1806) *Mecanique Celeste*, Suppl. to 10th edn, Paris.

Lau, K-T., Shi, S-Q. and Cheng, H-M. (2003) Micromechanical properties and morphological observation on fracture surfaces of carbon nanotube composites pre-treated at different temperatures. *Composites Science and Technology*, **63**, 1161–1164.

Madejski, J. (1983) Droplets on impact with a solid-surface. *International Journal of Heat and Mass Transfer*, **26** (7), 1095–1098.

Mao, T., Kuhn, D.C.S. and Tran, H. (1997) Spread and rebound of liquid droplets upon impact on flat surfaces. *AIChE Journal*, **43** (9), 2169–2179.

Meiron, T.S., Marmur, A. and Saguy, I.S. (2004) Contact angle measurement on rough surfaces. *Journal of Colloid and Interface Science*, **274** (2), 637–644.

Menchaca-Rocha, A., Martinez-Davalos, A., Nunez, R. *et al*. (2001) Coalescence of liquid drops by surface tension. *Physical Review E*, **63** (4), 046309.

Montavon, G., Sampath, S., Berndt, C.C. *et al*. (1997) Effects of the spray angle on splat morphology during thermal spraying. *Surface and Coatings Technology*, **91** (1–2), 107–115.

Moss, P.A., Retulainen, E., Paulapuro, H. and Aaltonen, P. (1993) Taking a new look at pulp and paper: applications of confocal laser scanning microscopy (CLSM) to pulp and paper research. *Paperi Ja Puu-Paper and Timber*, **75** (1–2), 74–79.

Mundo, C., Sommerfeld, M. and Tropea, C. (1995) Droplet-wall collisions – experimental studies of the deformation and breakup process. *International Journal of Multiphase Flow*, **21** (2), 151–173.

Noguera, R., Lejeune, M. and Chartier, T. (2005) 3D fine scale ceramic components formed by inkjet prototyping process. *Journal of the European Ceramic Society*, **25** (12), 2055–2059.

Palasantzas, G. and de Hosson, J.T.M. (2001) Wetting on rough surfaces. *Acta Materialia*, **49** (17), 3533–3538.

Pasandideh-Fard, M., Qiao, Y.M., Chandra, S. and Mostaghimi, J. (1996) Capillary effects during droplet impact on a solid surface. *Physics of Fluids*, **8** (3), 650–659.

Range, K. and Feuillebois, F. (1998) Influence of surface roughness on liquid drop impact. *Journal of Colloid and Interface Science*, **203** (1), 16–30.

Rayleigh, L. (1879) On the capillary phenomena of jets. *Proceedings of the Royal Society (London)*, **29**, 71–97.

Ristenpart, W.D., McCalla, P.M., Roy, R.V. and Stone, H.A. (2006) Coalescence of spreading droplets on a wettable substrate. *Physical Review Letters*, **97** (6), 064501.

Roisman, I.V., Rioboo, R. and Tropea, C. (2002) Normal impact of a liquid drop on a dry surface: model for spreading and receding. *Proceedings of the Royal Society of London, Series A*, **458** (2022), 1411–1430.

Scheller, B.L. and Bousfield, D.W. (1995) Newtonian drop impact with a solid-surface. *AIChE Journal*, **41** (6), 1357–1367.

Schiaffino, S. and Sonin, A.A. (1997a) Formation and stability of liquid and molten beads on a solid surface. *Journal of Fluid Mechanics*, **343**, 95–110.

Schiaffino, S. and Sonin, A.A. (1997b) Molten droplet deposition and solidification at low Weber numbers. *Physics of Fluids*, **9** (11), 3172–3187.

Schiaffino, S. and Sonin, A.A. (1997c) Motion and arrest of a molten contact line on a cold surface: an experimental study. *Physics of Fluids*, **9** (8), 2217–2226.

Schwartz, A.M. and Tejada, S.B. (1972) Studies of dynamic contact angles on solids. *Journal of Colloid and Interface Science*, **38** (2), 359–374.

Sedev, R.V., Budziak, C.J., Petrov, J.G. and Neumann, A.W. (1993) Dynamic contact angles at low velocities. *Journal of Colloid and Interface Science*, **159** (2), 392–399.

Seerden, K.A.M., Reis, N., Evans, J.R.G. *et al*. (2001) Inkjet printing of wax-based alumina suspensions. *Journal of the American Ceramic Society*, **84** (11), 2514–2520.

Sekimoto, K., Oguma, R. and Kawasaki, K. (1987) Morphological stability analysis of partial wetting. *Annals of Physics*, **176** (2), 359–392.

Shanahan, M.E.R. and Bourges, C. (1994) Effects of evaporation on contact angles on polymer surfaces. *International Journal of Adhesion and Adhesives*, **14** (3), 201–205.

Shiu, J.Y., Kuo, C.W., Chen, P.L. and Mou, C.Y. (2004) Fabrication of tunable superhydrophobic surfaces by nanosphere lithography. *Chemistry of Materials*, **16** (4), 561–564.

Sirringhaus, H., Kawase, T., Friend, R.H. *et al*. (2000) High-resolution inkjet printing of all-polymer transistor circuits. *Science*, **290** (5499), 2123–2126.

Soltman, D. and Subramanian, V. (2008) Inkjet-printed line morphologies and temperature control of the coffee ring effect. *Langmuir*, **24** (5), 2224–2231.

Smith, P., Dearden, A., Stringer, J. *et al*. (2006) Direct inkjet printing and low temperature conversion of conductive silver patterns. *Journal of Materials Science*, **41** (13), 4153–4158.

Son, Y. and Kim, C. (2009) Spreading of inkjet droplet of non-Newtonian fluid on solid surface with controlled contact angle at low Weber and Reynolds numbers. *Journal of Non-Newtonian Fluid Mechanics*, **162** (1–3), 78–87.

Son, Y., Kim, C., Yang, D.H. and Alm, D.J. (2008) Spreading of an inkjet droplet on a solid surface with a controlled contact angle at low Weber and Reynolds numbers. *Langmuir*, **24** (6), 2900–2907.

Stow, C.D. and Hadfield, M.G. (1981) An experimental investigation of fluid-flow resulting from the impact of a water drop with an unyielding dry surface. *Proceedings of the Royal Society of London, Series A*, **373** (1755), 419–441.

Stringer, J. and Derby, B. (2009) Limits to feature size and resolution in inkjet printing. *Journal of the European Ceramic Society*, **29** (5), 913–918.

Stringer, J. and Derby, B. (2010) The formation and stability of lines produced by inkjet printing. *Langmuir*, **26** (12), 10365–10372.

Szczech, J.B., Megaridis, C.M., Gamota, D.R. and Zhang, J. (2002) Fine-line conductor manufacturing using drop-on-demand PZT printing technology. *IEEE Transactions on Electronics Packaging Manufacturing*, **25** (1), 26–33.

Tanner, L.H. (1979) Spreading of silicone oil drops on horizontal surfaces. *Journal of Physics D: Applied Physics*, **12** (9), 1473–1484.

Thoroddsen, S.T., Takehara, K. and Etoh, T.G. (2005) The coalescence speed of a pendent and a sessile drop. *Journal of Fluid Mechanics*, **527**, 85–114.

Van Dam, D.B. and Le Clerc, C. (2004) Experimental study of the impact of an ink-jet printed droplet on a solid substrate. *Physics of Fluids*, **16**, 3403–3414.

Vander Wal, R.L., Berger, G.M. and Mozes, S.D. (2006) The combined influence of a rough surface and thin fluid film upon the splashing threshold and splash dynamics of a droplet impacting onto them. *Experiments in Fluids*, **40** (1), 23–32.

Watanabe, T., Kuribayashi, I., Honda, T. and Kanzawa, A. (1992) Deformation and solidification of a droplet on a cold substrate. *Chemical Engineering Science*, **47** (12), 3059–3065.

Wenzel, R. (1936) Resistance of solid surfaces to wetting by water. *Industrial & Engineering Chemistry Research*, **28** (8), 988–994.

Worthington, A.M. (1876) On the forms assumed by drops of liquids falling vertically on a horizontal plate. *Proceedings of the Royal Society (London)*, **25**, 261–272.

Wu, Z.N. (2003) Approximate critical Weber number for the breakup of an expanding torus. *Acta Mechanica*, **166** (1–4), 231–239.

Xu, L., Zhang, W.W. and Nagel, S.R. (2005) Drop splashing on a dry smooth surface. *Physical Review Letters*, **94** (18), 184505.

Yao, W., Maris, H.J., Pennington, P. and Seidel, G.M. (2005) Coalescence of viscous liquid drops. *Physical Review E*, **71** (1), 016309.

Yarin, A.L. (2006) Drop impact dynamics: splashing, spreading, receding, bouncing. *Annual Review of Fluid Mechanics*, **38**, 159–192.

Yarin, A.L., Szczech, J.B., Megaridis, C.M. *et al*. (2006) Lines of dense nanoparticle colloidal suspensions evaporating on a flat surface: formation of non-uniform dried deposits. *Journal of Colloid and Interface Science*, **294** (2), 343–354.

Yarin, A.L. and Weiss, D.A. (1995) Impact of drops on solid-surfaces – self-similar capillary waves, and splashing as a new-type of kinematic discontinuity. *Journal of Fluid Mechanics*, **283**, 141–173.

Young, T. (1805) An essay on the cohesion of fluids. *Philosophical Transactions of the Royal Society of London*, **95**, 65–87.

6

Manufacturing of Micro-Electro-Mechanical Systems (MEMS)

David B. Wallace
MicroFab Technologies, Inc., USA

6.1 Introduction

Micro-electro-mechanical systems (MEMS) fabrication technology was developed initially in the 1980s with the goal of integrating electro-mechanical sensors and actuators with their conditioning electronics. By adding silicon micromachining and deposition of metals and oxides to silicon-based analog integrated circuit (IC) fabrication technology, very small electro-mechanical sensors and actuators could be fabricated at very low cost and in high volume (Kuehnel and Sherman, 1994; Tseng *et al*., 2000). In addition, these MEMS-based devices were more robust and reliable than their larger conventional counterparts. MEMS-based accelerometers are the most widely used MEMS products. They have enabled the use of airbags in automobiles and can currently be found in cell phones, computers, cameras, golf clubs, and skis. MEMS-based pressure transducers have also been in widespread use for decades, primarily in automobiles, airplanes, process control equipment, and disposable biomedical sensors.

In most applications, the benefits of a high degree of system integration are readily apparent. The success of MEMS-based accelerometers and pressure sensors, along with the general drive toward small, low-cost, high-volume products, has led to an explosion of both process technologies that can be used in MEMS fabrication and applications targeted by these process technologies. Inclusion of digital electronics was

Inkjet Technology for Digital Fabrication, First Edition. Edited by Ian M. Hutchings and Graham D. Martin.

a fairly obvious step. Expansion to optical emitters, detectors, and switches has resulted in micro-optical-electro-mechanical systems (MOEMS) (Liu, 2002). Examples include micro-mirror devices that are used in televisions and projectors, and solid-state laser devices that have enabled the growth of the telecommunication and data communication industries (Sampsell, 1993; Tatum, 2000). Inclusion of biological functions has resulted in bio-MEMS devices (James *et al.*, 2006). The lab-on-a-chip devices for point-of-care diagnostics are one of the few examples of available bio-MEMS products (Oosterbroek and van den Berg, 2003; Daw and Finkelstein, 2006), although a large number of applications are being developed, including implantable sensors (e.g., for glucose) and microneedle-based transdermal drug delivery devices (Kim and Lee, 2007). Finally, by including fluid flow with thermal or piezoelectric actuators, many inkjet print-heads fall into the category of MEMS devices, and inkjet printers have long been "adopted" by analysts when reporting MEMS industry revenue figures (Seto *et al.*, 2008).

Many optical devices now included in the MOEMS category do not have a mechanical function, being more properly categorized as electro-optics (EO) devices. Other devices cited in this section have similar issues when it comes to categorization. This illustrates the difficulty in using the term MEMS in any strict sense at this point in time, since it has been broadened and generalized, partly as a result of its successes and familiarity. In this chapter, we will discuss the fabrication of small, integrated devices and refer to them all as "MEMS devices."

6.2 Limitations and Opportunities in MEMS Fabrication

MEMS fabrication methods grew out of the silicon-based semiconductor industry, and so most rely on photolithography. Photolithographic processes are particularly well suited for large-volume, high-feature-density manufacturing of devices with low fabrication processes and fabricated feature diversity (Elliott, 1989; Jaeger, 2002). The prime example is a dynamic random-access memory (DRAM) memory device with repetition of the same features millions of times using a limited number of fabrication processes. MEMS has successfully built on the huge microelectronics equipment and technology base, adding the feature diversity required to create a "system" through a limited number of additional compatible processes. However, photolithography and other "IC-like" fabrication processes are severely limited in the types of materials that can be used. In addition, there are technical and cost limitations that limit the number of "layers" that can be created in fabricating a MEMS device.

Materials limitations in conventional MEMS fabrication fall into two categories: compatibility and cost. Firstly, materials must be compatible with photolithography, which means that they must survive the application, patterning, and removal of photosensitive masking materials, and in general be compatible with the creation of one or more additional layers. Functional materials such as ones that are biologically active, chemically active or receptive, or optically active (emitters and receivers) are typically difficult or impossible to use in photolithographic or other IC-like fabrication methods. Yet it is these types of materials that could enable a broad range of small, integrated system devices for medicine, security, communications, and so on. Some of the most interesting materials are rare or expensive and thus would be cost-prohibitive to use in the current subtractive processes employed in MEMS fabrication. Many biologically active materials fall into this category.

Finally, there is an inherent conflict between the desire to increase the number of functions in MEMS devices and the cost of each "layer" in a photolithographic or other IC-like fabrication method. Even if the functional materials are compatible with multiple layers of photolithography and are not particularly expensive, each layer has a substantial total cost (equipment, labor, facility, materials, yield, testing, etc.) associated with it. Thus, additional functions always have a strong cost counterweight in MEMS device design based on current manufacturing methods.

6.3 Benefits of Inkjet in MEMS Fabrication

Inkjet printing technology has a number of attributes that can overcome some of the inherent limitations of photolithographically based MEMS fabrication methods. Since it is an additive method, material is deposited only where it is desired. Usage of rare or expensive materials can thus be conserved. The net savings can be significant for even moderately priced materials if the amount of area required to be covered is small compared to the total area of the substrate (low feature density). In addition to the cost savings due to low materials usage, additive methods have little or no waste, so they are much more environmentally friendly. In contrast, subtractive methods waste large amounts of the functional material, plus the photosensitive masking material (if different) and the cleaning and etching solvents.

Since inkjet is noncontact, the interaction between depositions of different materials is eliminated or greatly reduced and the requirement to consider each material to be deposited as an expensive additional "layer" is greatly reduced. Deposition can occur on nonplanar surfaces, eliminating the planarization steps required if using photolithography on a previously formed nonplanar layer. It is possible to print by inkjet even on very fragile or sensitive surfaces such as released layers (thin structures not supported by any material underneath them), because of the extremely low inertial force exerted by a deposited drop. Active layers (detectors, emitters, and biological) can be deposited on without degrading their function. Thick films can be created by overprinting the same location multiple times, and locally layered structures can be created without having to process the entire substrate area.

Lastly, since inkjet is a fundamentally digital, data-driven process, the cost and time associating with producing the masks required by photolithographically based MEMS fabrication methods are eliminated. This removes both the cost and turn-around time associated with fixed tooling. Also accelerating the development time is the ability to deposit onto individual dies and partial wafers, and to perform parametric process development experiments under digital control.

Since inkjet printing technology has the ability to enable a wide range of MEMS device types and functions (again, using "MEMS" to denote small, integrated devices), a comprehensive approach to the use, and potential use, of inkjet in MEMS fabrication would require an inordinate, and uninteresting, amount of time to be spent in organizing and classifying device types and subtypes. Therefore, the approach taken in this chapter will be to present and discuss a number of specific applications in a few application areas.

6.4 Chemical Sensors

Chemical sensors represent a fairly new and broad area of research and development for MEMS devices, driven by the need for large numbers of low-cost sensors for explosives, chemical warfare agents, drugs of abuse, industrial gases, residential gases, and many others. A majority of these sensors use materials that are electrically or photonically active, or more simply have surfaces that cause the molecules of interest to temporarily adhere to them. Not surprisingly, most of these sensing materials are "sensitive" (i.e., delicate) and cannot be photolithographically processed. Also, because they are sensitive, they are applied in the final or nearly final fabrication process, and typically this is onto nonplanar, feature-rich surfaces that can be very fragile. All these factors make MEMS chemical sensor manufacturing an area that is broadly exploring the use of inkjet fabrication technology.

Chemoresistive materials, which change resistance when exposed to specific molecules of interest, are the oldest and most broadly used sensing materials in MEMS sensor devices (Sakai, Sadaoka, and Matsuguchi, 1996). Recent developments in nanomaterials and MEMS structures have expanded the number of materials and sensor structures being developed (Bochenkov and Sergeev, 2007). An example of a MEMS chemoresistive sensor is one that is being developed at Carnegie Mellon University to detect volatile organic compounds (VOCs) in respirators, indicating end of life (Fedder *et al*., 2008). The basic sensor structure, shown in Figure 6.1a, is a pair of spiral electrodes in a 250 μm circle that is in a 350 μm diameter SU-8 well. Multiple sensing and reference elements, which in general could contain multiple sensing materials, are incorporated on a 2.65 mm die that also contains all the required control electronics, as shown in Figure 6.1b. The die is assembled into a TO-5 package (Figure 6.1c) which is commonly used for optical devices.

The sensing materials, gold–thiolate nanoparticles, are suspended in a carrier fluid (5–10 mg/ml) and deposited onto the sensing area. Although not visible in Figure 6.1a, 15 drops of nominally 30 pl volume have been deposited onto the sensor using an inkjet device. Figure 6.1d shows the sensing area after 225 nominally 30 pl drops of solution have been deposited, producing an average film thickness of 1.5 μm. It is interesting to

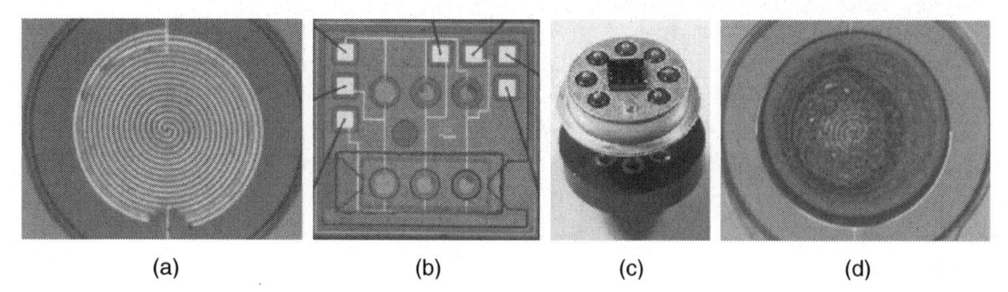

Figure 6.1 Carnegie Mellon chemoresistive sensor: (a) sensing element configuration showing a 250 μm diameter dual-electrode spiral in a 350 μm SU-8 well; (b) multiple sensing and reference elements on a 2.65 mm die; (c) sensor device in a TO-5 package; and (d) sensor element printed with 225 nominally 30 pl drops of solution containing gold–thiolate nanoparticles (Images courtesy of Carnegie Mellon University).

note their use of two wetting "stops" in the sensing area. The SU-8 well contains the initial fluid volume dispensed, preventing undesired wetting onto other areas of the die. In addition, the fluid de-wets from the outer portion of the well during drying so that all of the particles are deposited onto the electrode region. This self-centering behavior results in impedance variation of less than 10%.

The dispensing of the sensing material occurs not only on the individual die but also after it is assembled into the package. This effectively limits the sensing material deposition method to additive dispensing methods, and the requirement to print multiple sensors held in a fixture in production would require a data-driven method unless the fixture is high precision (i.e., expensive). If a contact dispensing method were used, throughput would be limited by the requirement for the dispenser to make a vertical movement for each dispense.

Resonant MEMS structures detect a change in resonant frequency associated with a change in the mass of the resonating structure due to the adsorption of molecules of interest. Detection of changes in resonance can be accomplished to great accuracy with well-known electronic circuitry that can be implemented into integrated circuitry on a MEMS device. MEMS fabrication techniques can produce extremely low-mass, high-Q resonant structures that allow detection of very low concentrations of the molecules of interest (Brand, 2005).

In most cases, the material used in resonance-based MEMS detectors is a polymer that selectively adsorbs the molecules of interest, similar to a capillary gas chromatography column. To form a functioning detector, the vibrating structures in a resonance-based MEMS detector must be coated with the selectively adsorbing polymer and the coating must be thin and uniform to allow the structure to remain low mass and high Q. Since these structures can be on the scale of 1–10 μm or less, inkjet would not appear to be a viable approach as a coating method. However, researchers at Carnegie Mellon have created a detector that has an integrated well structure large enough for a 30–50 μm fluid drop to impact and fill. The well is connected to a 1–10 μm resonant structure via a channel that allows the resonant structure to be coated by means of an inkjet-deposited drop and capillary-induced flow (Bedair and Fedder, 2008). Figure 6.2a shows the top view of this sensor with the well structure taking up the left half of the device. The resonating structure is the center beam on the right half of the device and connects to the apex of the well. The structures surrounding the center beam are used to drive it electrostatically. The center beam is undercut (i.e., not supported underneath) to allow it to vibrate freely. Figure 6.2b shows a close-up, oblique view of the center beam and drive elements. It shows the U-shaped groove in the center beam that causes the polymer to flow along the beam. Figure 6.2c shows the groove at higher magnification. This device has demonstrated the ability to detect gas molecules down to femtogram levels.

Another resonant-mode MEMS device, manufactured by Boston Microsystems, is shown in Figure 6.3. The piezoelectrically driven resonators shown in Figure 6.3a are coated with chemoselective polymers for detection of hazardous chemicals (Tuller, 2005; Mlcak *et al*., 2004). In this image, the coating can be seen because of the interference fringes it produces. The resonator length (horizontal dimension) is 100 μm. The side view in Figure 6.3b shows the undercut that allows the resonator to vibrate about the point attachment to the structure on the left side of this image. Figure 6.3c shows the

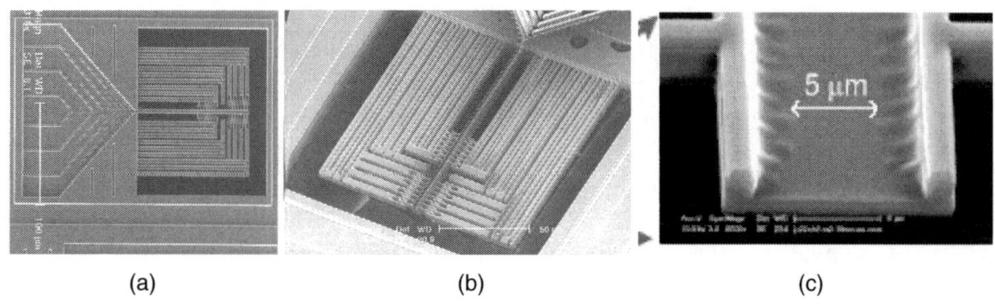

(a) (b) (c)

Figure 6.2 *Carnegie Mellon resonance-based MEMS sensor. (a) The left half contains a well structure for inkjet deposition of selectively adsorbing polymer, and the right half contains a resonant structure in the center surrounded by electrostatic drive elements: polymer flows from the well to the resonant structure due to surface tension; (b) angled view showing detail of a resonant structure and drive elements; and (c) detailed view of a resonant structure showing the channel structure used to create a surface tension–driven flow of polymer, an undercut that allows the structure to resonate, and "wings" that drive the structure electrostatically (Images courtesy of Carnegie Mellon University).*

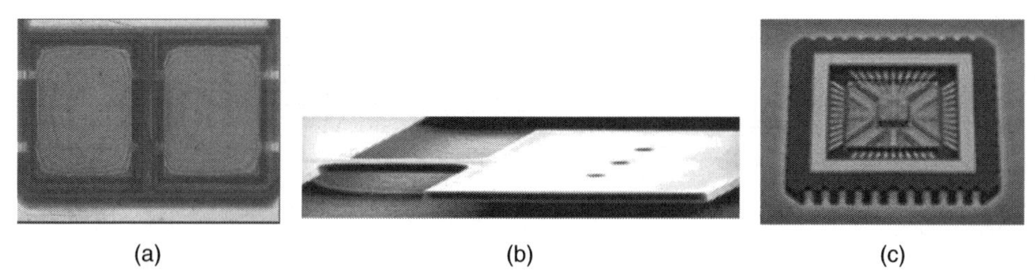

(a) (b) (c)

Figure 6.3 *Piezoelectrically driven MEMS resonators coated with chemoselective polymers for detection of hazardous chemicals. The resonator length (horizontal distance) for the two sensing elements shown at (a) is 100 µm. The side view seen in (b) shows the undercut that allows the resonator to vibrate about the point attachment to the structure on the left side. (c) Shows the MEMS device with several individual sensing elements as assembled into a standard LCC package (Images courtesy of Boston MicroSystems).*

MEMS device with several individual sensing elements as assembled into a standard leadless chip carrier (LCC) package. In contrast to the Carnegie Mellon chemoresistive device discussed in this section, the sensing material, in this case a polymer in solution, is dispensed onto the MEMS device before packaging.

Optically based biosensors have undergone intensive research over the past decade (Epstein and Walt, 2003; Kishen *et al*., 2003; Preejith *et al*., 2003). In these sensors, an optical indicator chemistry is designed to change optical properties in response to the presence of a target molecule. Light of a suitable wavelength is used to illuminate the sensing material, and the absorption or re-emission due to fluorescence is measured with a photodetector. Conventional optical-based chemical sensors are based on optical fibers, allowing only a single sensing material per fiber and making system integration of multiple sensing materials cumbersome. The ability to place multiple sensing materials on the end of a fiber-optic image bundle allows for convenient integration of

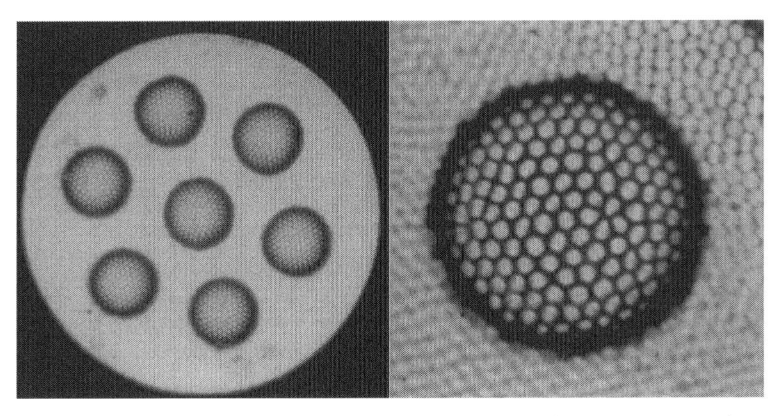

Figure 6.4 *90 μm diameter sensing polymer elements printed onto a 500 μm diameter fiber bundle in a prototype optically based biosensor.*

multiple sensing materials onto a small, integrated structure, and this type of system has been demonstrated by the Lawrence Livermore National Laboratory (LLNL) and Tufts University (Ferguson *et al.*, 1997) for a blood gases sensor (O_2, CO_2, pH, etc.). However, fabrication of the sensing materials onto the end of the fiber bundle using subtractive processes (one sensing material at a time: mask, deposit, remove mask, and repeat) results in deformation and degradation of the sensing materials, and thus variability in performance. Figure 6.4 demonstrates the use of inkjet in this application and shows 90 μm diameter features of fluorescein-doped polymer (pH-sensitive) printed onto a 500 μm diameter fiber bundle (Hayes, Cooley, and Wallace, 2004). Since no masking and mask removal steps are required, the placement of one sensor element does not degrade the previously placed elements.

Although the optical fiber bundle sensor can be considered miniature and integrated at the sensor location, the fiber bundle itself plus the light source, detector, and electronics are macroscopic and discrete components and required to make a complete, functioning system. Given the ability to create integrated arrays of light sources (such as the lasers discussed in Sections 6.5 and 6.7) and detectors on a single MEMS device, one can visualize adding the printing of an optical filter (to separate the excitation light from the emitted light) onto a detector array and using inkjet to print different sensing materials onto individual source–detector pairs to create a fully miniaturized, integrated, optically based sensor array, as is illustrated in Figure 6.5.

6.5 Optical MEMS Devices

Refractive optical components in MEMS devices present a number of difficulties. Refractive lens and waveguide features are highly three-dimensional, have high spatial resolution requirements (location, diameter, and curvature), must be highly transmissive, and in general should be able to tolerate solder reflow temperatures. Diffractive optical components have been widely employed in MEMS devices because their planar structure allows them to be fabricated using traditional MEMS processes. However,

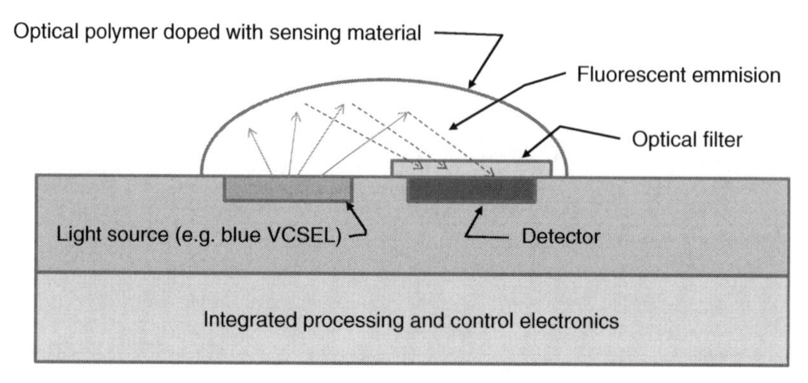

Figure 6.5 Concept for an integrated, optically based MEMS chemical sensor with integrated light source, detector, optical filter, and processing and control electronics.

refractive optics offer higher performance than diffractive optics, making their use in MEMS optical devices highly desirable.

By using pre-polymer solutions that are cross-linked after deposition, it is possible to formulate jettable polymeric solutions that have no solvents; cure into hard, durable microlenses; and can tolerate reflow temperatures. Figure 6.6a shows a portion of a lens array that has been fabricated by inkjet deposition of this type of material, jetted at 120 °C to lower the viscosity (Cox *et al.*, 2000). The lens diameters are 225 μm and the center-to-center spacing 250 μm, both ±1 μm. This tight dimensional control is achieved by patterning the substrate, in this case borosilicate glass, into wetting and nonwetting regions. In this configuration, the substrate thickness would provide the required standoff distance so that the lenses could collimate or focus light from a point source such as a vertical cavity surface-emitting laser (VCSEL) array. Durability, an important consideration in any practical application of microlenses, is indicated by the data shown in Figure 6.6b (data courtesy Agilent Corp.). Here, the focal length is seen to be stable at 85 °C for over 3000 hours. Additional data show similar results for 2 hours at 180 °C, for 1 hour at 200 °C, and for a 96-hour pressure pot test at 120 °C and 100% RH.

Fabrication of lenses onto MEMS devices at the wafer level has advantages in both cost and throughput. Figure 6.7a shows a portion of a gallium arsenide wafer

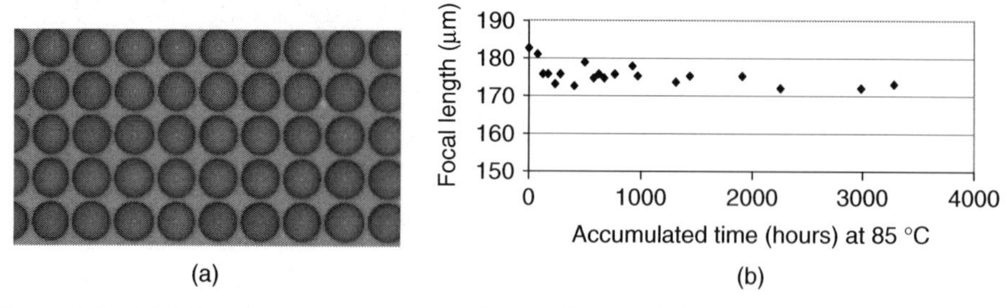

Figure 6.6 (a) Micro-lens array on borosilicate glass with lens diameters of 225 μm and center-to-center spacing of 250 μm, both ±1 μm. (b) Focal length data for polymer micro-lens indicating stability at 85 °C for over 3000 hours (Data courtesy of Agilent Corp).

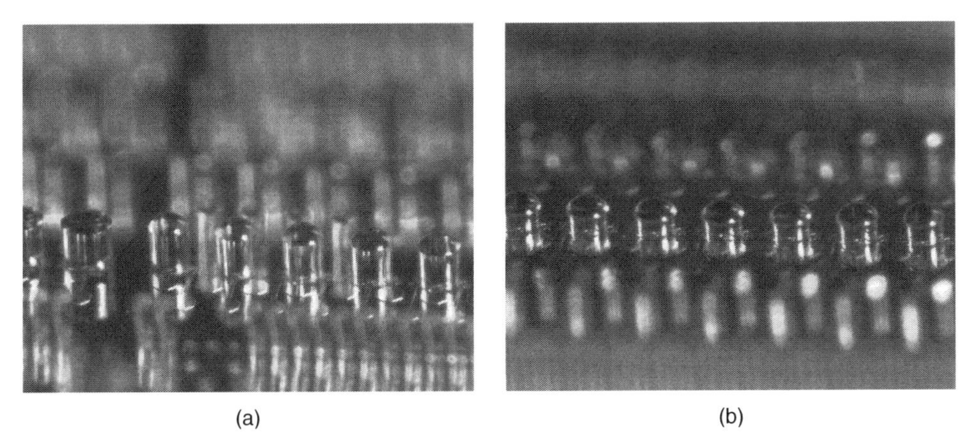

Figure 6.7 *One hundred micrometers high, 125 μm diameter SU-8 posts over VCSELs on a gallium arsenide wafer. The posts are on 225 μm center-to-center spacing with polymer lenses printed onto the posts. (a) Variable height and focal-length lenses created by printing 2–18 18 pl drops per post. (b) 22 × 18 pl drops per post. Wafer with posts (Courtesy of Vixar Corp).*

containing 125 μm diameter SU-8 posts on 225 μm center-to-center spacing (from Vixar Corporation) with polymer lenses printed onto the posts. Variable height and focal-length lenses are created by printing 2–1818 pl drops onto each post. The capability to print variable lens heights can be used to experimentally determine the optimal lens height very rapidly during development, and can be used in production to create lenses of different height on a single wafer and/or device. Figure 6.7b shows a section of the wafer with uniform height lenses printed onto the posts. Figure 6.8 shows the results of divergence angle measurement for VCSELs with and without lenses printed onto SU-8 posts, indicating collimation for lensed posts. Besides optical performance, the lens and postmanufacturing processes must not degrade the performance of the VCSEL in terms of threshold voltage and output power. Previous measurements of these VCSELs with lensed posts fabricated using inkjet have shown no degradation in these performance parameters (Hayes and Chen, 2004).

Lenses on photodetector arrays can be used to increase their collection efficiency and can be printed directly onto the detector device, in contrast to the posts required for VCSEL collimation. Again, the process for creating the lens cannot degrade the performance of the detector, and this also has been demonstrated (Hayes and Chen, 2004).

Released structures in MEMS are ones that are fabricated onto a sacrificial layer that is subsequently removed, leaving the structure suspended over air (or vacuum). The resonance-based MEMS devices discussed in this chapter are examples. They are frequently employed to allow the structure to move with relatively large deflections and are very fragile, making application of thick-film materials directly onto released structures very difficult. Figure 6.9 shows a 100 μm diameter released structure (oscillating micromirror), suspended by 10 μm width and thick supports, onto which a polymer lens has been deposited without breaking or deforming the released structure. A nominally

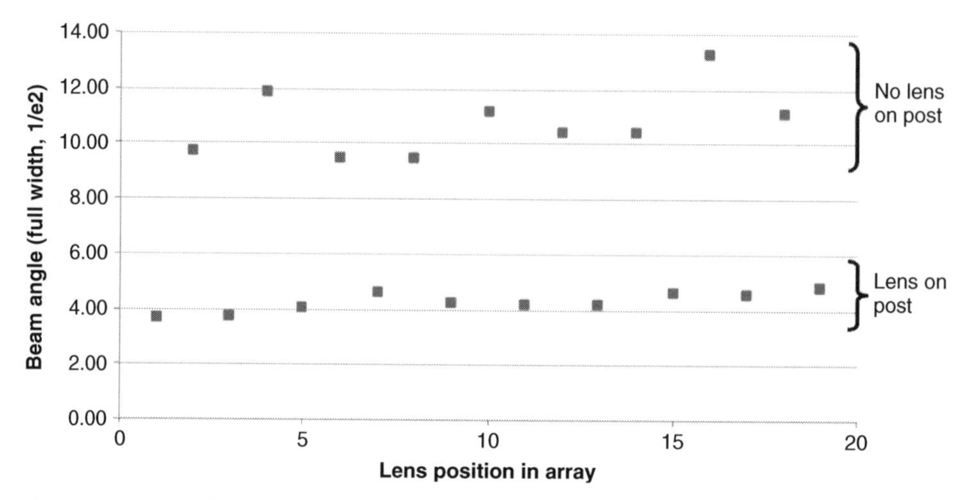

Figure 6.8 Data from VCSELS with and without lenses printed onto SU-8 posts showing collimation for lensed posts (Data courtesy of Vixar Corp).

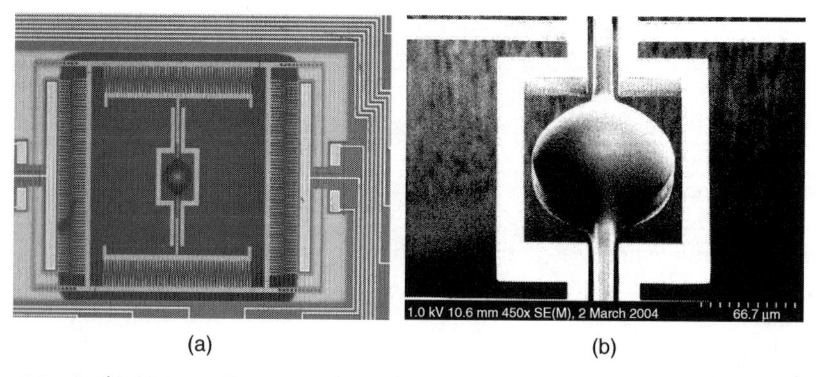

Figure 6.9 (a, b) 100 µm diameter released structure (oscillating micro-mirror), suspended by 10 µm width thick supports, onto which a polymer lens has been deposited.

50 pl (46 µm diameter) drop at 2 m/s was used to form this lens, making the impact momentum ~0.1 µN s.

Deposition of a number of other optical materials and features has been accomplished by inkjet printing technology and could be applied to fabrication of MEMS devices with optical functions. Multimode waveguides have been printed with inkjet using the same polymer used to print the lenses discussed in this section (Cox and Chen, 2001). Figure 6.10 shows nominally 120 µm polymer waveguides printed using inkjet as a branching structure on the left and as parallel waveguides on the right. Figure 6.11 shows 60 × 40 µm waveguides, UV-cure optical epoxy printed into grooves micromachined into silicon.

Light-emitting polymers, phosphors, and color filter materials have been deposited using inkjet technology for display applications (Grove *et al.*, 1999; Shimoda *et al.*, 1999;

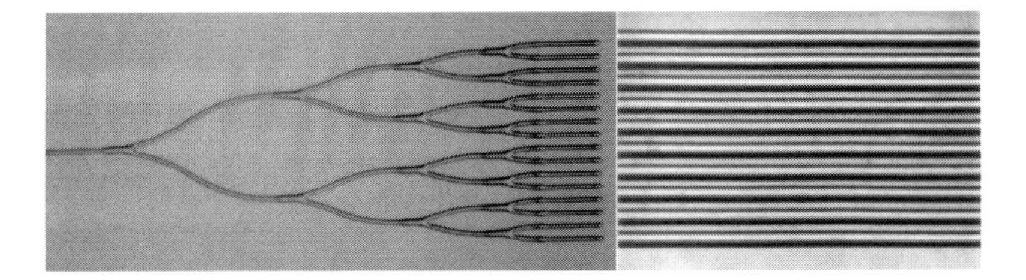

Figure 6.10 *Nominally 120 μm polymer waveguides printed using inkjet.*

Figure 6.11 *60 × 40 μm waveguides, UV-cure optical epoxy printed into grooves micro-machined in silicon.*

Kiguchi *et al*., 2001), and photovoltaic polymers for solar cell applications (Shah and Wallace, 2004). To date these have been used in macroscopic applications, but they also could be used in the fabrication or packaging of a MEMS device to create optical emitter and detector functions. If integrated with printed waveguide and electrical components, they could be used for printed electro-optic circuits on MEMS devices or packages.

6.6 Bio-MEMS Devices

MEMS devices that are implantable in humans have both great potential and significant challenges. Continuous monitoring and adjustment of biological functions create great potential for more effective monitoring and treatment regimens. However, implantable MEMS devices must deal with biocompatibility and biofouling issues (Park and Park, 1996). The implantable microelectrode from the University of Kentucky shown in Figure 6.12 has four 20 × 60 μm electrochemical measurement sites on a ceramic substrate that are used to measure brain activity (Day *et al*., 2006). All four sites and the surrounding regions in Figure 6.12a have been coated with a glutamate oxidase enzyme and overcoated with glutaraldyhyde, a fixative, both using inkjet printing. Since the enzyme coating is thin and transparent, it is not visible in Figure 6.12a. In Figure 6.12b, two of the four electrodes have been coated with a fluorescent dye to illustrate the capability of localized deposition control. Currently this type of probe is used for research into neurological function and diseases such as Parkinson's. Future variants of this type of device could be used as an indwelling sensor in humans that would monitor and report brain function as part of an overall treatment plan, in a

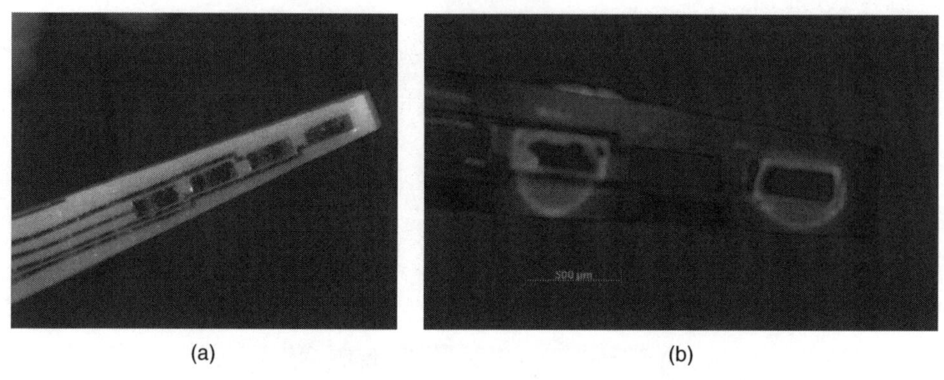

(a) (b)

Figure 6.12 *Implantable (brain) microelectrode with four 20 × 60 μm electrochemical measurement sites on a ceramic substrate. All four sites and the surrounding regions in (a) have been coated with a glutamate oxidase enzyme and overcoated with glutaraldyhyde, a fixative, both by inkjet printing. (b) Two of the four electrodes have been coated with a fluorescent dye to illustrate the capability of localized deposition control (Images courtesy of University of Kentucky Center for Microelectrode Technology) (See plate section for coloured version).*

similar way to blood glucose measurement and insulin injection. Indwelling sensors for measuring cardiac function using a MEMS sensor combined with wireless data reporting have been under development under license from NASA (NASA, 2012).

Delivery of drugs by implantable devices (Staples *et al*., 2006) is of as much interest as implantable sensors, and current research includes drug delivery for diabetes, cancers, cardiac disease, and neurological disorders. The most widely used implantable drug delivery device today is the drug-eluting stent, used to keep coronary arteries open after an angioplasty procedure. The complex structure of a metal stent must allow the device to collapse to travel through the blood vessels, then deploy (i.e., increase in diameter) and lock at the desired location. To prevent tissue from growing over the stent and blocking the artery again, drugs are imbedded into the stent to prevent this. This requires placing the drug onto complex features of 50–150 μm width. A number of companies are using inkjet for this process (Shekalim and Shmulewitz, 2003; Tarcha *et al*., 2004). To illustrate this application, Figure 6.13 shows a cardiac stent with 100 μm wide structures which is a model stent with a double-helix configuration. In the top image, a fluorescently labeled drug analog is shown printed onto one helix. In the middle image, a second fluorescently labeled drug analog is printed onto the other helix. At the bottom, a composite of the two images is shown. This illustrates the ability of inkjet to print multiple drugs onto a single implantable device and in a complex pattern. In addition to multiple drugs, the linear concentration could be varied by placing more drug at the ends, where restenosis (reclosing of the artery) occurs.

6.7 Assembly and Packaging

Assembly of MEMS devices presents challenges that are different and, in general, greater than those encountered in standard IC packaging. Difficulties include the inherent

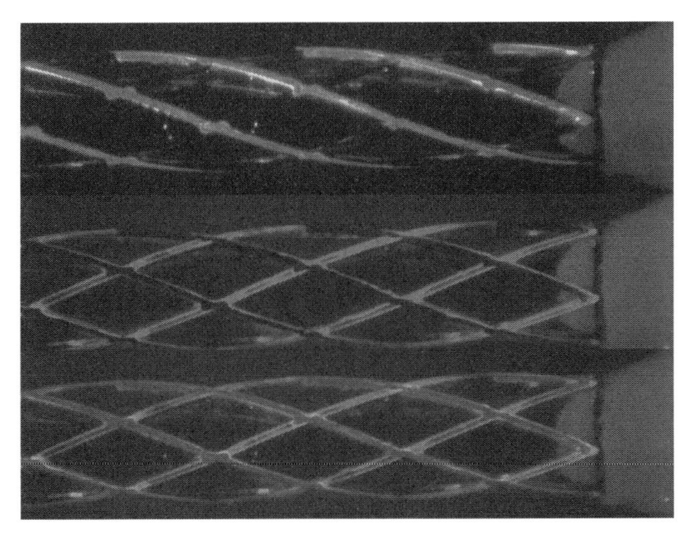

Figure 6.13 *Cardiac stent with 100 μm wide structures, model double-helix configuration. Top: Fluorescently labeled drug analog printed onto one helix. Middle: Second fluorescently labeled drug analog printed onto other helix. Bottom: Composite image (See plate section for coloured version).*

three-dimensionality of MEMS devices, fragile structures, hermetic sealing, and so on. An example of a 3D electrical interconnect is the assembly of the signal-conditioning IC to the flex circuit containing the read-write head for a hard disk drive. Figure 6.14a shows a top view of the IC and flex circuit, and Figure 6.14b the side view. The IC has 80 μm electrical interconnect pads on the edge that interface with the flex circuit which has 60 μm traces. Solder has been dispensed and partially reflowed into the corner formed by the IC and flex using Solder Jet® printing hardware, which uses piezoelectric drop-on-demand inkjet technology to dispense droplets of molten solder (Hayes and Wallace, 1998). Comparable 3D electrical interconnect requirements exist for other integrated devices. If a wafer with VSCELs, posts, and lenses similar to the one discussed in Section 6.5 is diced into segments with four VSCELs, it can be placed on its side next to conductive pads for a corner interconnect requirement comparable to that of the disk drive assembly discussed here, but with the much more difficult addition of the post and lens overhanging the electrical interconnect region. Figure 6.15a shows a MEMS gripper that was fabricated onto the substrate to position a VCSEL bar correctly relative to the electrical interconnect pads. Figure 6.15b shows the gripper having positioned the VCSEL bar, and Solder Jet® printing hardware having deposited a solder ball into a groove in the gripper. The groove is over the corner formed by the VSCEL bar and the substrate. For this assembly, the posts are approximately 100 μm tall, 100 μm wide, and on 225 μm centers. Note that solder balls are also deposited on the back side of the VCSEL bar into similar features to connect to the common for the four lasers. Figure 6.15c shows the final assembly with one of the lasers emitting at 850 nm, as seen by an infrared-sensitive camera (Nallani *et al*., 2006).

Another packaging application of Solder Jet® printing that is applicable to MEMS assembly is depositing the sealing material for hermetic packages. Figure 6.16 shows

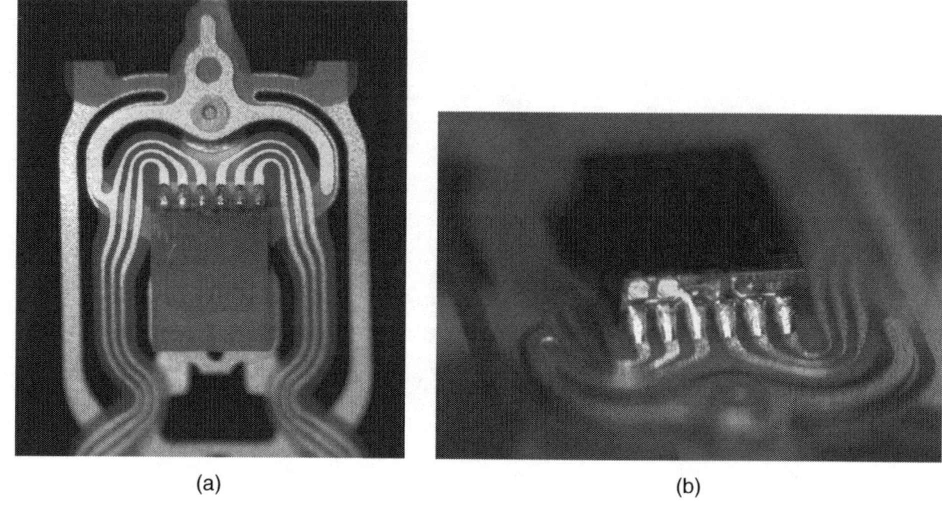

Figure 6.14 (a) Top view of signal-conditioning IC and flex circuit of disk drive read head assembly. (b) Side view showing 80 μm electrical interconnect pads on IC edge that interfaces with the flex circuit (60 μm traces). Solder has been dispensed and partially reflowed into the corner formed by the IC and flex using Solder Jet® printing hardware, which uses piezoelectric drop-on-demand inkjet technology to dispense droplets of molten solder.

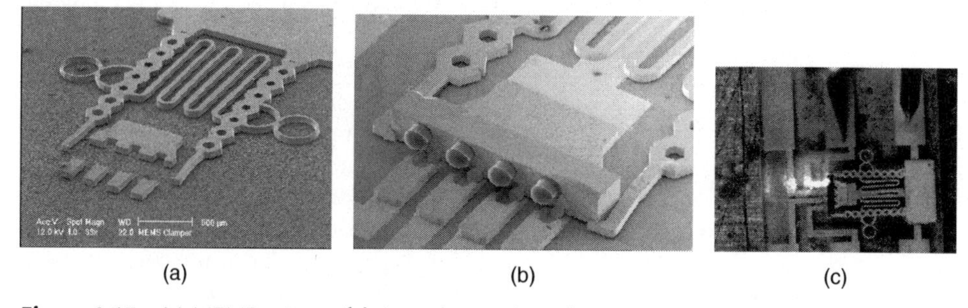

Figure 6.15 (a) MEMS gripper fabricated onto the substrate to position VCSEL bar correctly relative to electrical interconnect pads. (b) Shows gripper having positioned the VCSEL bar, and Solder Jet® printing hardware having deposited a solder ball into a groove in the gripper. The groove is over the corner formed by the VSCEL bar and the substrate. (c) Shows final assembly with one laser emitting at 850 nm, as seen by an infrared-sensitive camera.

a portion of the sealing ring of a gallium arsenide device that has had 80 μm solder bumps deposited on it in four rows. Hermetic sealing is generally used in packaging moisture-sensitive devices, particularly imaging devices. Conventionally, solder preforms are placed on the sealing ring since solder paste, and thus screen printing, is not allowed because outgassing of the flux during reflow would fill the volume to be protected by the seal. Use of printed solder for this function allows for much smaller sealing rings and thus much higher device density at the wafer level, resulting in lower device cost. For applications that do not require the quality of hermetic seal that solder can provide,

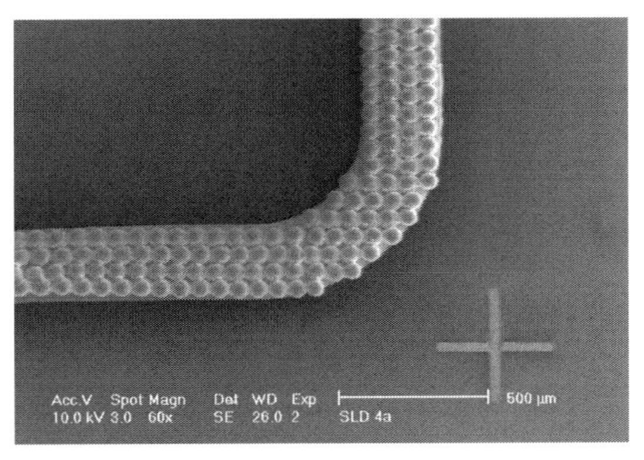

Figure 6.16 *Sealing ring of a gallium arsenide device that has had 80 µm solder bumps deposited on it in four rows.*

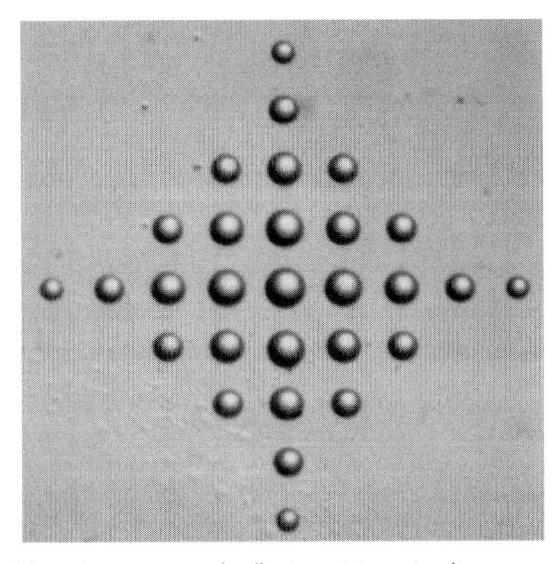

Figure 6.17 *Variable volume spots of adhesive, 80 µm in diameter and upward, printed by inkjet.*

adhesives can be printed with inkjet. The waveguides shown in Figure 6.10 are also examples of printed adhesive, since this material is an epoxy resin. Figure 6.17 shows variable volume spots of adhesive, 80 µm diameter and upward, printed in a complex pattern by inkjet.

Most of the recent developments in printed electronics, as discussed in more detail in Chapters 7, 9, and 11, have focused on low-cost, high-volume macroscale devices such as radiofrequency identification (RFID) tags (Gamota *et al*., 2004). However, as noted elsewhere in this volume, inkjet printing processes for conductors, dielectrics, resistors, capacitors, inductors, antennae, organic transistors, batteries, fuel cells, and other functional elements have been demonstrated and may have application to both MEMS device manufacturing and packaging.

6.8 Conclusions

The use of inkjet printing technology has been described for several MEMS manufacturing applications. The processes in the examples would be difficult, expensive, or impossible to accomplish using conventional, photolithographically based MEMS manufacturing methods. Use of inkjet methods has the potential to drastically increase the range of functions in MEMS devices because of the additive, noncontact, data-driven nature of inkjet printing technology. These same characteristics will also result in lower production cost and in less environmentally unfriendly waste.

Acknowledgements

The author would like to thank the following for their valuable contributions: Donald Hayes, Ting Chen, Mike Boldman, Virang Shah, David Silva, Bogdan Antohe, Royall Cox, Rick Hoenigman, Lee Weiss, Gerald Schultz, Peter Huettl, Pooja Talauliker, J.B. Lee, and Arun Nallani.

References

Bedair, S.S. and Fedder, G.K. (2008) Controlled pico-gram material placement on suspended structures using solution wicking. Solid-State Sensors, Actuators, and Microsystems Workshop Hilton Head Island, SC, pp. 272–275.

Bochenkov, V.E. and Sergeev, G.B. (2007) Nanomaterials for sensors. *Russian Chemical Reviews*, **76**, 1084–1093. doi: 10.1070/RC2007v076n11ABEH003735

Brand, O. (2005) CMOS-based resonant sensors. Proceedings of IEEE Conference on Sensors, pp. 129–132, doi: 10.1109/ICSENS.2005.1597653

Cox, W.R. and Chen, T. (2001) Micro-optics fabrication by ink-jet printing. *Optics & Photonics News*, **12** (6), 32–35.

Cox, W.R., Guan, C., Hayes, D.H. and Wallace, D.B. (2000) Microjet printing of micro-optical interconnects. *International Journal of Microcircuits and Electronic Packaging*, **23** (3), 346–351.

Day, B.K., Pomerleau, F., Burmeister, J. *et al.* (2006) Microelectrode array studies of basal and potassium-evoked release of l-glutamate in the anesthetized rat brain. *Journal of Neurochemistry*, **96** (6), 1626–1635.

Daw, R. and Finkelstein, J. (eds) (2006) Nature insight. *Lab on a Chip*, **442**, 7101.

Elliott, D.J. (1989) *Integrated Circuit Fabrication Technology*, 2nd edn, McGraw-Hill, New York.

Epstein, J.R. and Walt, D.R. (2003) Fluorescence-based fibre optic arrays: a universal platform for sensing. *Chemical Society Reviews*, **32**, 203–214.

Fedder, G.K., Barkand, D.T., Bedair, S.S. *et al.* (2008) Jetted nanoparticle chemical sensor circuits for respirator end-of-service-life detection. Abstracts of the 12th International Meeting on Chemical Sensors, Columbus, OH.

Ferguson, J.A., Healey, B.G., Bronk, K.S. *et al.* (1997) Simultaneous monitoring of pH, CO2 and O2 using an optical imaging fiber. *Analytica Chimica Acta*, **340**, 123–131.

Gamota, D., Brazis, P., Kalyanasundaram, K. and Zhang, J. (2004) *Printed Organic and Molecular Electronics*, Kluwer Academic Publishers, New York.

Grove, M., Hayes, D., Cox, R. *et al*. (1999) Color flat panel manufacturing using inkjet technology. Proceedings of Display Works '99, San Jose.

Hayes, D.J. and Chen, T. (2004) Next generation optoelectronic components enabled by direct write microprinting technology. *Proceedings of SPIE, Defense and Security Symposium (Orlando, FL)*, **5435** (83), 83–90.

Hayes, D.J., Cooley, P. and Wallace, D.B. (2004) Miniature chemical and biomedical sensors enabled by direct-write microdispensing technology. *Proceedings of SPIE, Defense and Security Symposium (Orlando, FL)*, **5416** (73), 73–83.

Hayes, D.J. and Wallace, D.B. (1998) Solder jet printing: wafer bumping and csp applications. *Chip Scale Review*, **2** (4), 75–80.

Jaeger, R.C. (2002) *Lithography: Introduction to Microelectronic Fabrication*, Prentice Hall, Upper Saddle River, NJ.

James, C.D., Okandan, M., Galambos, P.C. *et al*. (2006) Surface micromachined dielectrophoretic gates for the front-end device of a biodetection system. *ASME Journal of Fluids Engineering*, **128** (1), 14–19.

Kiguchi, H., Katagami, S., Yamada, Y. *et al*. (2001) Technical Digest of Asia Display/IDW01, 1745.

Kim, K. and Lee, J.B. (2007) High aspect ratio tapered hollow metallic microneedle arrays with microfluidic interconnector. *Microsystem Technologies*, **13** (3–4), 231–235.

Kishen, A., John, M.S., Lim, C.S. and Asundi, A. (2003) A fiber optic biosensor (fobs) to monitor mutants streptococci in human saliva. *Biosensors and Bioelectronics*, **18**, 1371–1378.

Kuehnel, W. and Sherman, S. (1994) A surface micromachined silicon accelerometer with on-chip detection circuitry. *Sensors and Actuators A*, **45** (1), 7–16.

Liu, Y. (2002) Heterogeneous integration of OE arrays with Si electronics and microoptics. *IEEE Transactions on Advanced Packaging*, **25**, 43–49.

Mlcak, D., Doppalapaudi, J., Chan, J. and Tuller, H.L. (2004) MEMS sensor arrays for explosives and toxic chemical detection. 3rd Microsensors Workshop, ASU Kerr Center, Scottsdale, AZ.

Nallani, A.K., Chen, T., Hayes, D.J. *et al*. (2006) A method for improved vcsel packaging using MEMS and inkjet technologies. *Journal of Lightwave Technology*, **24** (3), 1504–1512.

NASA (2012) Success Story: Endotronix Licenses Two NASA Patents to Develop More Accurate, Less Invasive Ways to Measure Cardiovascular Conditions in Human Body. https://technology.grc.nasa.gov/SS-Endotronix.shtm (accessed July 2012).

Oosterbroek, E. and van den Berg, A. (eds) (2003) *Lab-on-a-Chip: Miniaturized Systems for (Bio)Chemical Analysis and Synthesis*, 2nd edn, Elsevier Science, Amsterdam.

Park, H. and Park, K. (1996) Biocompatibility issues of implantable drug delivery systems. *Pharmaceutical Research*, **13** (12), 1770–1776.

Preejith, P.V., Lim, C.S., Kishen, A. *et al*. (2003) Total protein measurement using a fiber-optic evanescent wave-based biosensor. *Biotechnology Letters*, **25**, 105–110.

Sakai, Y., Sadaoka, M. and Matsuguchi, M. (1996) Humidity sensors based on polymer thin films. *Sensors & Actuators*, **B35**, 85–90.

Sampsell, J.B. (1993) An overview of Texas Instruments digital micromirror device (DMD) and its application to projection displays. Society for Information Display International Symposium Digest of Technical Papers, vol. 24, pp. 1012–1015.

Seto, S., Nakamura, H., Murata, M. and Morita, N. (2008) Thin film piezo inkjet printhead having matrix nozzle arrangement using MEMS technology. Proceedings of Pan-Pacific Imaging Conference '08, Imagine Society of Japan, pp. 52–55.

Shah, V.G. and Wallace, D.B. (2004) Low-cost solar cell fabrication by drop-on-demand inkjet printing. Proceedings of IMAPS 37th Annual International Symposium on Microelectronics, Long Beach, CA.

Shekalim, A. and Shmulewitz, A. (2003) Stent coating device. US Patent 6,645,547.

Shimoda, T., Kanbe, S., Kobayashi, H. *et al.* (1999) Multicolor pixel patterning of light-emitting polymer by inkjet printing. Technical Digest of SID99, p. 376.

Staples, M., Daniel, K., Cima, M.J. and Langer, R. (2006) Application of micro- and nano-electromechanical devices to drug delivery. *Pharmaceutical Research*, **23** (5), 847–863.

Tarcha, P.J., VerLee, D., Hui, H.W. *et al.* (2004) Drug loading of stents with inkjet technology. Proceedings of BioInterface '04.

Tatum, J.A. (2000) Packaging flexibility propels VCSELs beyond telecommunications. *Laser Focus World*, **36** (4), 131–136.

Tseng, A., Tang, W.C., Lee, Y.C. and Allen, J.J. (2000) NSF 2000 workshop on manufacturing of micro-electro-mechanical systems. *Journal of Materials Processing and Manufacturing Science*, **8** (4), 292–360.

Tuller, H.L. (2005) MEMS-based thin film and resonant chemical sensors, in *Electroceramic-Based MEMS: Fabrication Technology and Applications* (ed. N. Setter), Springer, New York, pp. 3–17.

7

Conductive Tracks and Passive Electronics

Jake Reder
Celdara Medical, LLC, USA

7.1 Introduction

Imagine a world without electronics. It is difficult. Few technologies in human history can rival the impact that electronics have had, or the degree to which they have been integrated into our daily lives. Digital fabrication has developed with the importance of electronics in clear view; within this focus, both active and passive electronics development has created a rich and fascinating body of technology.

The term passive electronics includes any element that is useful for the management of electrons but does not produce gain. A conductive track is the simplest example, managing electrons only by defining their path. To date, the vast majority of research, production, sales, and use of digitally printed passive electronics have been focused on conductive tracks. The rest of this chapter will reflect this and discuss conductive tracks as the model passive electronic system. Other passive devices will be detailed only in their specific subsections.

7.2 Vision

To date, two distinct visions for the development of digitally printable passive electronics have emerged. Industrial manufacturers are already applying the technology. Applications are currently quite limited, but many forward-thinking electronics manufacturers are

Inkjet Technology for Digital Fabrication, First Edition. Edited by Ian M. Hutchings and Graham D. Martin.

squarely focused on the development of digital, or at least additive, processes for their flexibility, cost saving, and environmental benefits. This is an attractive first market entry for the technology as customers are technologically sophisticated, expectations of the nascent technology are reasonable, and volumes could eventually become very large. Of course, with increasing volumes comes margin pressure, and over time the consumables and equipment that are cutting edge today will become commodities.

The second vision is further into the future, and certainly less likely, but it has the potential to create a mass-market appeal and a monumental shift in how society views manufacturing itself. As seen throughout this book, the continuous development of three-dimensional (3D) printing technology (with variants also known as additive fabrication, rapid prototyping, rapid manufacturing, and the like) has followed an aggressive experience curve with rapidly dropping prices, improving quality, and an array of newly enabled applications (Wohlers, 2007). These systems cover a wide array of fabrication technologies, but all can produce intricate objects in three dimensions. Many use plastics, but final products can be ceramic, metal, or elastomeric, depending on the printer technology. Today's 3D printers (as discussed in Chapter 14) are priced similarly to color copying machines, making them affordable to small businesses or even home enthusiasts. The logical extension of this technology is into electronics. If a user could create both structural and electronic components within a single printer, the possibilities would be endless. However, as the rest of this chapter will detail, the technical challenges to the realization of this vision are daunting.

7.3 Drivers

Despite the fascinating science and technology associated with printed electronics, there are very few examples of "new-to-world" applications that could not have been fabricated without digital printing. Of course, Henry Ford did not invent the automobile, either; he created a new process to dramatically reduce the cost of production and thereby created an enormous market. The key driver for adoption of digital, additive processes, at least for industrial uses, is cost. These production cost advantages appear in a variety of places.

7.3.1 Efficient Use of Raw Materials

As noted in Chapter 1, the formation of conductive tracks can be described as additive or subtractive, that is, the useful pattern is formed by either building it up (like painting) or removing everything that is not the pattern (like stone sculpting). Additive processes used for the production of conductive traces include silkscreen, offset, and gravure printing, and electroplating. Subtractive processes include photoengraving, printed circuit board (PCB) milling, and photolithographic processes. If a surface has 10% of its area covered in conductive traces, then 90% of that area would need to be removed in a subtractive process. The efficient use of raw materials is not a digital advantage but rather an additive advantage, and it is shared by other printing processes. Photolithography is taken as a subtractive process example for contrast.

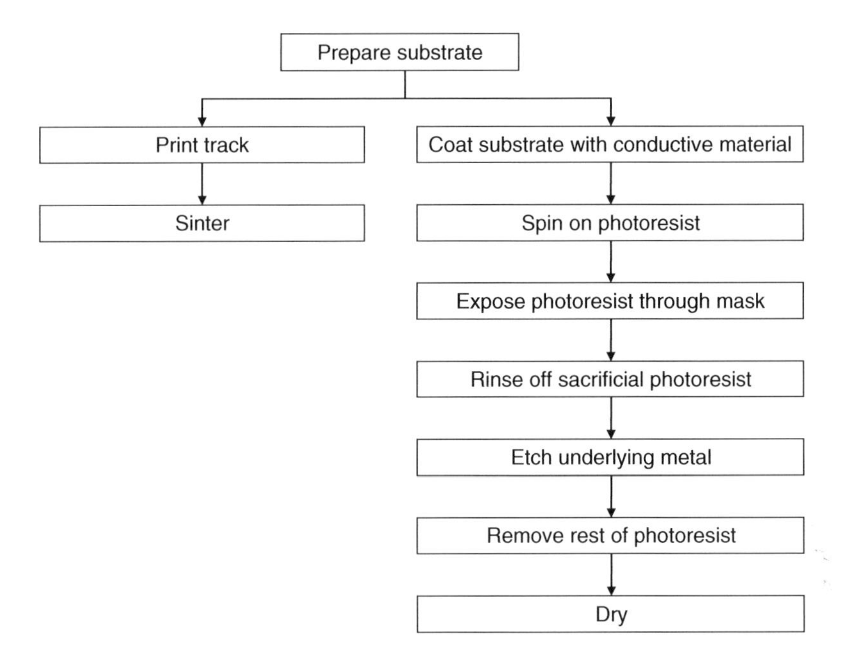

Figure 7.1 *Process steps involved in digital (left) and photolithographic (right) fabrication.*

As seen in Figure 7.1, in a typical photolithographic process for the creation of conductive tracks, the desired pattern is created in a (negative) mask, while a thin foil of conductive material on a substrate is coated with a photosensitive emulsion. This emulsion is selectively cross-linked by exposing it to light through the mask. The non-cross-linked emulsion is then washed away, and the metal underneath etched. Finally, the cross-linked photopolymer is then removed. This process provides remarkable resolution and fidelity. Masks, while expensive, can be used for millions of exposures. The issue, however, is that more than 90% of the metal used is removed by etching, effectively increasing the cost of this raw material by an order of magnitude. Recycle loops can be added, but they have their own associated costs in equipment, complexity, maintenance, and the like. The photopolymer emulsion is also a process consumable with considerable associated cost. PCB milling and other etching processes (as discussed in detail in Chapter 8) have the same "subtractive" downside. By putting metal only where you want it, only the solvent and additives in the ink are "wasted," and solvent can easily be recovered where it is economical to do so.

7.3.2 Short-Run and Single-Example Production

Inherent but not explicit in the foregoing discussion are the costs of making a photolithographic mask, or for that matter any of the many and varied process tools required in any production process. A comparative example is shown in Figure 7.1. Only by spreading these costs across many devices can the process become economical. The flexibility of digital printing allows a fabricator to produce short runs or even single examples of designs for about the same unit cost as long runs. This is especially advantageous for niche applications and design experimentation and optimization. Vis-à-vis other additive processes, for example silkscreen printing, digital printing's inherent flexibility can minimize retooling costs. Of course, if flexibility is not a priority, other production methods will result in lower overall costs. In most situations, there is a crossover point of a certain number of units produced (either overall or per campaign) beyond which the digital advantage is overtaken by economies of scale of nondigital additive processes.

7.3.3 Capital Equipment

The cost of a digital printer and the associated postprocessing equipment for conductive track fabrication can be significantly lower than photolithographic equipment with similar capabilities. Many other competitive technologies also require expensive vacuum processes. If a process requires expensive production equipment, then that equipment should at least be run near capacity to maximize profit potential by spreading depreciation costs over more units. To run near capacity with dedicated machines is extremely difficult in practice. The inherent flexibility of digital printing processes allows increased capacity utilization through the production of "noncore" products during what otherwise would be downtime.

7.4 Incumbent Technologies

In order to understand the value of a new technology, it must be evaluated in the context of competitive, and especially incumbent, technologies. The challenge with such a general descriptor as "passive elements" is that there are hundreds of incumbents, and any diligent evaluation must therefore depend on the specific application. This simple fact has been the undoing of many a business plan. That said, it is generally the case that the incumbent technologies have the advantage of existing relationships and contracts with customers, existing capital equipment (sometimes fully depreciated), realized economies of scale, and products that perform acceptably at an acceptable cost. The generalized disadvantages are waste (or the cost associated with recycle loops, especially for subtractive processes), a lack of flexibility, and an increase in the number of process steps required to achieve a similar function. The disruptive nature of digital fabrication pits this promising technology against many incumbents – with deep pockets. It will be interesting.

7.5 Conductive Tracks and Contacts

7.5.1 What Is Conductivity?

Conductivity is simply the ratio of current density to field. When a conductor is placed in an electric field, charge flows to minimize the potential difference. It is an inherent property of the material and measures how much charge flows per unit of potential:

$$\sigma = \frac{j}{E} \tag{7.1}$$

where σ is conductivity in siemens per meter, j is current density in amperes per square meter, and E is the electric field in volts per meter. As noted, conductivity is a material property which does not take the physical parameters of size or shape into account. Electrical conductance incorporates these physical dimensions, making it a good measure for the properties of a specific device or track.

$$G = \frac{\sigma A}{\ell} \tag{7.2}$$

where the conductance G is in siemens, A is the cross-sectional area of the conductive track in square meters, and l is the length of the conductive track in meters. Note that conductance is the inverse of resistance, that is, siemens are the inverse of ohms.

$$1S = 1\Omega^{-1} \tag{7.3}$$

Conductive tracks are valued for the amount of current they can carry through a small cross-section with low resistivity.

To simplify discussions with non-experts, conductive tracks are often spoken of in terms of their conductivity compared with the bulk material from which they are made, for example "30% of the conductivity of bulk silver." This is a simple and useful measure of the total effect of the different defects in the track, while also allowing for calculation of effective conductance given some basic geometric assumptions.

When digitally printing conductive tracks, there are myriad process parameters, all of which are regularly used to maximize conductance. Starting with a material with high conductivity, making the track dense in conductive material, maximizing the potential for sintering, and increasing the cross-sectional area though multiple passes of the print-head are all common techniques designed to maximize conductance.

7.5.2 Conductive Tracks in the Third Dimension

Only the most rudimentary of today's electronic devices are constructed in two dimensions, and yet, since digital printing equipment was originally designed for the graphics industry, there are serious challenges in building 3D structures. As mentioned in this chapter, there is a significant global effort to improve fully 3D capable printing technology (Wohlers, 2007), but in the interim, simple structures like vias and steps have been demonstrated with more traditional equipment, as shown in Figure 7.2 and elsewhere (Baron, Hofmann, and Schneider, 2006; Hayton *et al.*, 2006).

7.5.3 Contacts

Contacts form a special subset of conductive tracks in that they do not comprise the bulk of the conductive path, but do allow a fabricator to take advantage of some properties of digital processes. The two most common examples involve machine vision.

Radiofrequency identification (RFID) tags, discussed in more detail in Chapter 11, are simple devices comprising an antenna and an integrated circuit (IC or "chip"). As the poster child for disposable electronics, RFID manufacturers are under extreme cost pressures. Since each antenna is identical, a digital process is not advantageous, so they are stamped from foil or screen-printed. Of course, each chip is also identical. Variability is introduced when trying to attach the chip to the antenna. It may be advantageous to

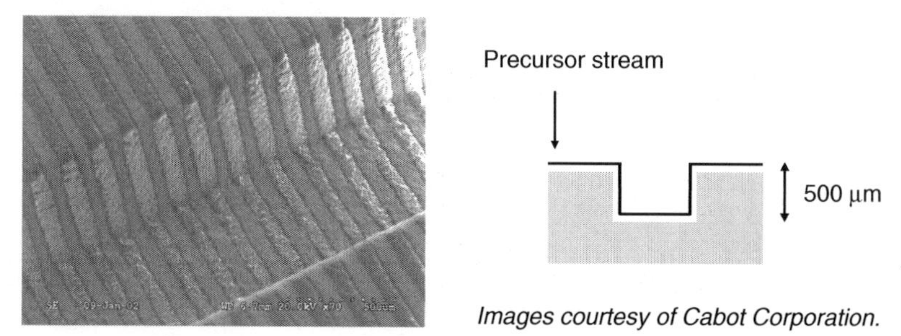

Figure 7.2 Scanning electron micrograph of noncontact printed silver nanoparticle-based ink lines (Reproduced with kind permission from Cabot Corporation Copyright (2012) Cabot Corporation).

make this conductive connection using digital printing guided by machine vision rather than expensive registration equipment, depending on the process requirements.

Somewhat similarly, defects in conductive tracks can also be identified using machine vision and corrected with digital printing. This process can be used for offline repairs, as part of quality control, or in a recycle loop.

Both of these applications are enabled by the spatial flexibility and process simplicity afforded by the additive digital process.

7.6 Raw Materials: Ink

The majority of conductive inks on the market and in development today contain inorganic particles, and this section will focus on this class of ink, developing and extending the general concepts introduced in Chapter 4. Conductive polymer-containing inks are briefly discussed at the end of this section.

It is useful to describe the outcomes of opposing process parameters in terms of "trade-offs," not just to clarify what is being gained and given up through the manipulation of known process parameters, but especially to focus innovators on ways to "solve the trade-off," for example by introducing a new process parameter. This paradigm will be used throughout the rest of this chapter.

7.6.1 Particles

Most inks used for conductive tracks contain solid particles for a few reasons. As discussed in Section 7.5.1, conductance is a function of the cross-sectional area of the track. While reduction of metal salts will yield metal (Kodas and Kunze, 2006), it is difficult to create a thick, or even a continuous, conductive path due to the low density of metal atoms in the as-printed track. The use of solid (typically metal) particles imparts the as-printed track with a high density of metal atoms. Of course, by depositing a conductive particle, no redox chemistry is necessarily required, thereby also removing a process step vis-à-vis a pure metal salt approach. Of course, the dispersion of particles in ink results in trade-offs which must be balanced or overcome, some of which are seen in Figure 7.3.

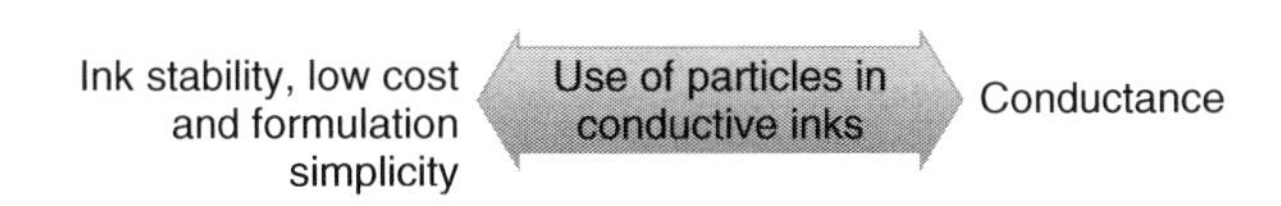

Figure 7.3 *Trade-offs in the use of particles in conductive inks.*

Particles can be defined by their composition, size, shape, surface, and associated dispersants. Each of these parameters must be optimized to maximize stability, printability, and ultimately, conductance.

7.6.1.1 Composition

The high conductivity of metals makes them an obvious choice. Their ability to sinter is a critical, if secondary, factor. Silver, copper, nickel, gold, and other easily sinterable metals are commonly used, with the properties listed in Table 7.1. There are few examples of alloys, presumably due to the technical challenges associated with creating homogeneous alloys at the nanoscale.

Metals are prone to oxidation, and high-surface-area metals even more so, which means special care must be taken in their production, handling, and processing. Deposition and postprocessing can be carried out in an inert atmosphere, though this adds cost and process complexity. After deposition, it is sometimes advantageous to coat the conductive track with an inert, less oxidizable, or sacrificial layer to prevent oxidation.

Conductive carbon particles are another interesting compositional class for the fabrication of conductive tracks. A variety of particles and concepts have been explored, including the printing of carbon nanotubes (Cho, Jung, and Hudson, 2007; Wei *et al*., 2007; Mustonen *et al*., 2008; Song *et al*., 2008), fullerenes (Mort, 1992), and modified versions of each. The inability of carbon to sinter at low temperatures makes combinations of metals and carbon-based particles attractive. Carbon nanotubes are challenged by the fact that their long linear dimension, while ideal for conduction, is also highly thixotropic, making them very difficult to print, and by the fact that they do not sinter. If an innovation can resolve these trade-offs (Figure 7.4), provide a printable ink, and reduce the energy barrier to trans-tube electron hopping, carbon nanotubes may find wide applicability in printed electronics.

Table 7.1 *Properties of metals commonly used for printing conductive tracks.*

Metal	Resistivity (Lange and Speight, 2005) (μΩ cm) @ 20 °C	Common oxide(s)	Energy of formation of the oxide (kJ/mol)	Melting point (°C)
Ag	1.587	Ag_2O	−11.2	962
Au	2.214	n/a[a]	>0	1064
Cu	1.678	Cu_2O, CuO	−127	1085
Ni	6.93	NiO	−92	1453
			(Erri and Varma, 2009)	

[a]Gold is the only metal which cannot be oxidized in air or oxygen, even at elevated temperatures.

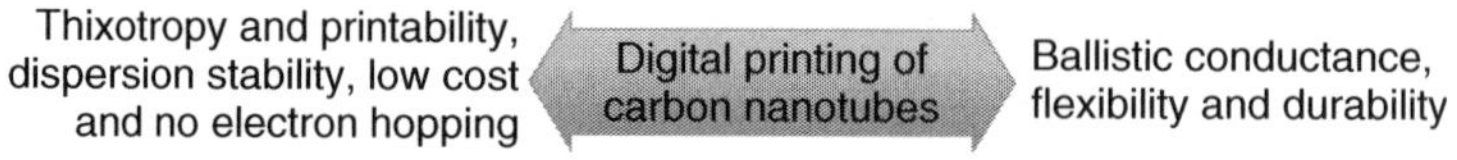

Thixotropy and printability, dispersion stability, low cost and no electron hopping

Digital printing of carbon nanotubes

Ballistic conductance, flexibility and durability

Figure 7.4 *Trade-offs in the use of carbon nanotubes in conductive inks.*

More complex particles have most often been employed to resolve the conductivity–price trade-off, or at least take advantage of resistive sintering.[1] Metal-coated carbon particles might provide for a metallic conduction path without the associated costs of pure metal dispersions (Probst, Grivei, and Fockedey, 2003; Callen and Walkhouse, 2007). Of course, it is also quite easy to develop a hybrid particle that costs much more to produce than a pure metal analog.

Some ceramics can also be electrically conductive. The majority of the work on printable ceramic conductors has focused on transparent systems such as indium tin oxide (ITO) and the like, for example, zinc oxides (Eguchi, Sasakura, and Yamaguchi, 2004; Imanishi and Nishimura, 2008).

7.6.1.2 Particle Shape

The nanotube trade-off illustrates the importance of a particle's shape to various parameters of the process and ultimate performance. Particles must be stable in dispersion, be jettable, and then readily contribute to the primary performance attribute: conduction. Spheres are the lowest energy shape requiring the minimum amount of dispersant, and also contribute the least to shear thickening and other non-Newtonian thixotropic effects. Of course, spheres are not ideally suited to move electrons over long (linear) distances. Additional trade-offs are seen in Figure 7.5.

Imagine a monodisperse collection of spherical particles, as printed on a surface, and remember that in order to maximize conductance, the highest possible density of particles in the track is needed. If the deposition is perfect, the resulting particles look like oranges stacked at the market. The problem is that this is only 74% dense (ignoring dispersant). In order to increase this packing density, particles can be made either nonspherical or polydisperse, as seen in Figure 7.6.

7.6.1.3 Particle Size

For a particular particle composition, dispersant, and medium, there is a maximum particle size that can be dispersed. Any particles beyond this size eventually settle and can cause many obvious problems, including clogged print-heads.

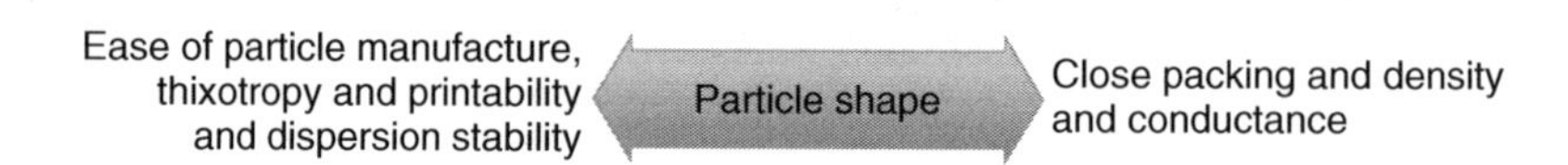

Figure 7.5 *Trade-offs in the use of nonspherical particles for conductive inks.*

[1] The process by which particles sinter as current is passed through the material. The heat required for sintering is caused by electrical resistance in the material and, as such, is optimally localized in the track.

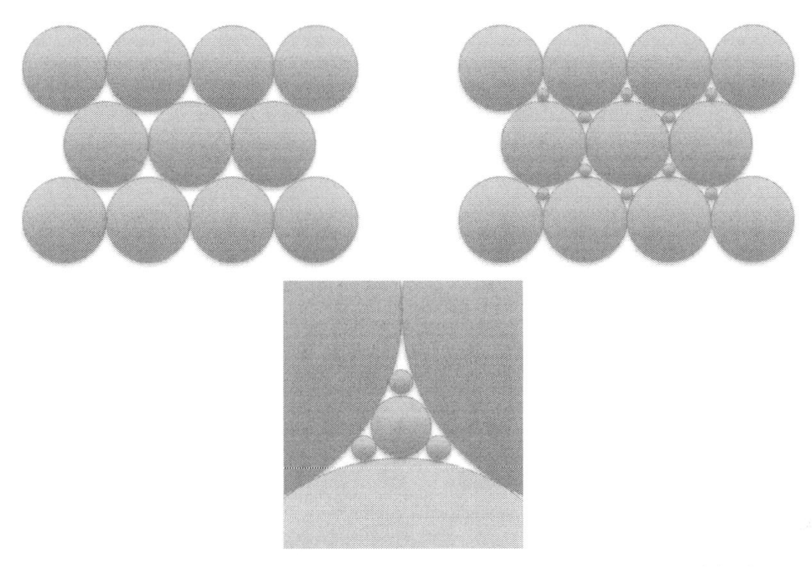

Figure 7.6 *Monodisperse close packing and ideal bimodal and trimodal close packing (in two dimensions).*

As particles get smaller, however, a few things happen. Their ratio of surface area to volume begins to rise dramatically. This concept is critical, as each new unit of surface area requires stabilization to avoid aggregation in solution. When particles get very small (around 10 nm), the volume of requisite dispersant can quickly overtake the volume of the particles themselves.

A particle's mass scales with its volume (note that there are sometimes apparent exceptions at very small particle sizes due to changes in crystal structure and/or the species present on the particle surface), and volume scales with the cube of radius while surface area scales with the square, as seen in Figures 7.7 and 7.8.

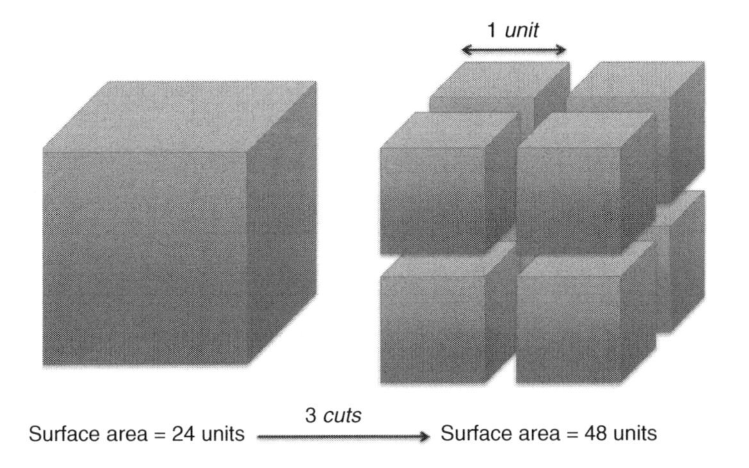

Figure 7.7 *Shrinking cubes reveal additional surface area.*

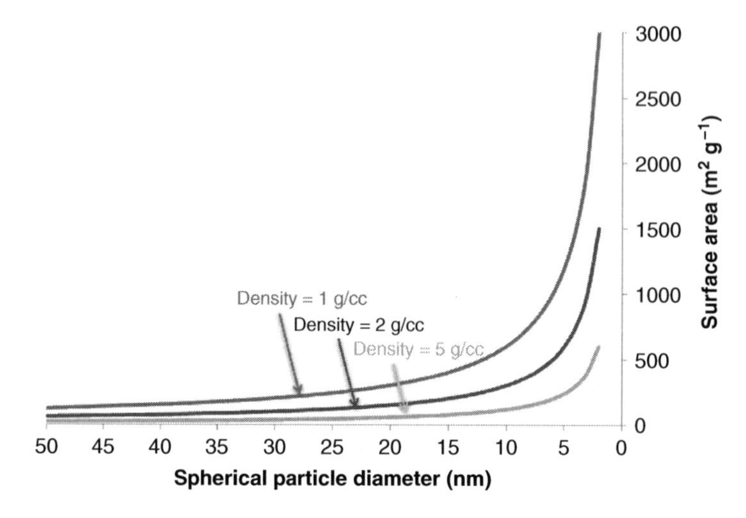

Figure 7.8 *At small diameters, the surface area of spheres grows geometrically.*

Perhaps not as obvious is that as particles get smaller, their surface energy also begins to rise. This is best understood with the perspective of a materials scientist, though simple geometry (curvature) is also a contributor.

Even in bulk materials, surfaces have higher energy than bulk phases. It is not uncommon for surface layers to change their crystal structure to minimize this energy. As particles shrink to smaller than 100 nm, the percentage of atoms at the surface rapidly begins to rise (see the discussion of surface area in this subsection), leaving more and more atoms with incomplete coordination shells as seen in Figure 7.9. This surface energy increase is responsible for the decrease in required sintering temperatures seen in many systems.

As particles get smaller, they need more dispersant to stabilize their new surface, resulting in a lower density of metal atoms per volume of ink, but they gain the advantage of lower sintering temperature. This implies that, at least for monodisperse systems, there is probably an optimum, and there are certainly trade-offs, as shown in Figure 7.10.

The use of particles in ink brings with it a variety of technical challenges, primarily in the stability of the dispersion, whether in the holding vessel, at the tip of the print-head nozzle, or on the substrate. What is most difficult is that the stability requirements change. A highly stable dispersion is desired in the holding vessel and at the tip of the print-head nozzle, but a highly unstable dispersion is desired once the ink hits the substrate, particularly in the case of porous substrates. Instability of the dispersion on the substrate can allow for a higher local density of particles by avoiding transport of particles into the pores or across the surface of the substrate. The best way to modify dispersion stability is through the use of dispersants.

7.6.2 Dispersants

Particles experience the van der Waals force in the same way as molecules. As particles get smaller, the strength of this attractive force increases relative to their mass, increasing

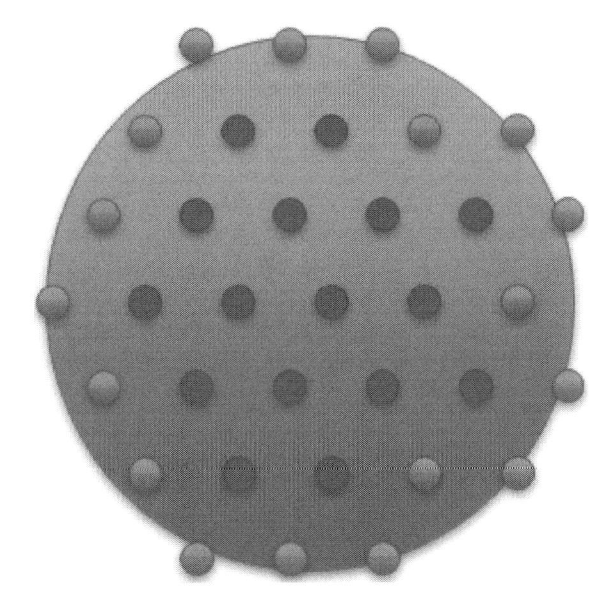

Figure 7.9 *Surface atoms with incomplete coordination spheres comprise a large proportion of the whole at small particle radii (large circle = particle, small dark circles = atoms with complete coordination spheres, small light circles = atoms with incomplete coordination spheres). Note that the effect is more pronounced in three dimensions.*

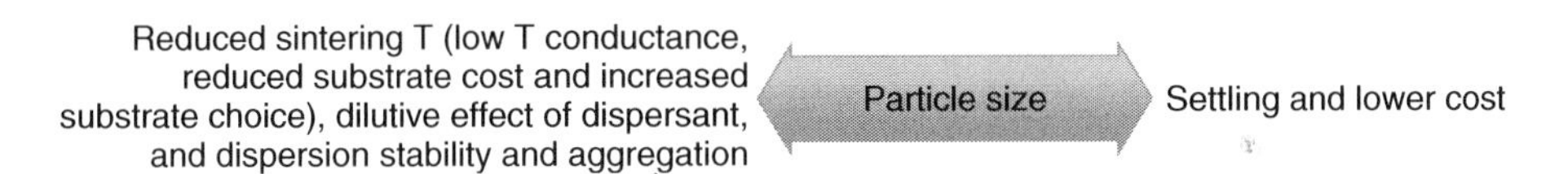

Figure 7.10 *Trade-offs in the use of small particles.*

the chance of flocculation. In order to maintain a stable dispersion, at least one repulsive force must be introduced. Dispersants include surfactants and other moieties capable of increasing dispersion stability. The available methods of stabilization are electrostatic, steric, and combinations of these (Morrison and Ross, 2002), as seen in Figure 7.11.

Because the inks used to print conductive tracks must be highly loaded and extremely stable, particles are typically sterically stabilized. The stabilizing moiety is ordinarily an organic species which may be one or more small molecules, oligomers, or polymers. This species may be bound to the particle through van der Waals interactions and hydrogen, ionic, or covalent bonds. Single or multipoint attachments are possible, with the latter further segmented into discrete chelate-like structures (Boils *et al.*, 2001; Uozumi and Yamamoto, 2005; Nakano, Yamaguchi, and Fukuhara, 2006; Jackson, 2008) or polyelectrolytes (Park *et al.*, 2004; Cheng *et al.*, 2005). In those cases where the concentration of free dispersant is not a function of the concentration of particles, that is, the agent is strongly bound to the particle surface, the dispersant may be considered part of the particle. In those cases where surface-bound molecules are in equilibrium with free molecules in the medium, the effects of changing environments on

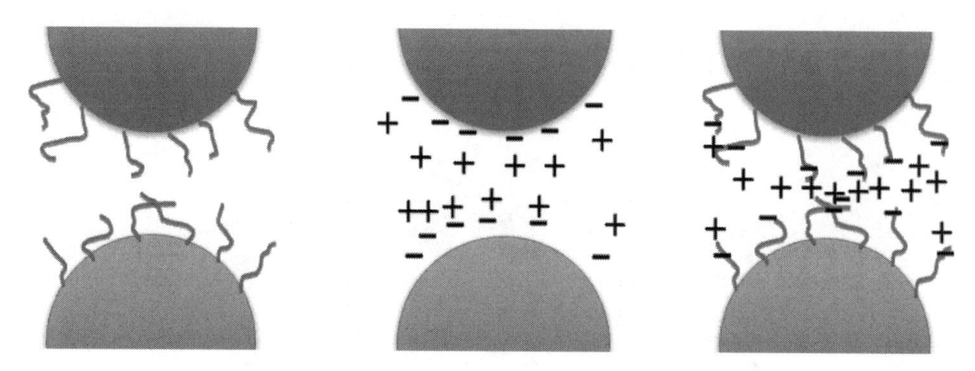

Figure 7.11 Steric, electrostatic, and electrosteric stabilization (L to R).

equilibrium concentrations, as well as competitive surface binding, must be considered, making formulation much more challenging.

7.6.3 Carriers (Liquid Media)

The liquid medium is the continuous phase of the ink, and therefore must stabilize the conductive ingredients (whether particles, polymers, or precursors) within it. Furthermore, it must be easily removed from the ink once deposited (by evaporation) without leaving undesirable residues which may physically block or simply reduce the density of the conductor in the conductive path, resulting in decreased conductance. Evaporation at the print-head nozzle, however, can lead to solids deposition and clogging. Most conductive inks are formulated using non-aqueous solvents.[2] Solvents may include alcohols, esters, amides, aromatics, and any other which satisfies the multiple requirements of the medium used in the specific application.

7.6.4 Other Additives

The art of ink formulation includes the use of humectants, co-solvents, wetting agents (surfactants), adhesion promoters, biocides, and chemicals to improve a particular property of the as-printed pattern, for example durability. Graphics inks, as discussed in Chapter 4, have developed sophisticated formulation technologies over centuries of development. The expectations of and demands on a modern graphics inkjet ink far surpass the state of the art in conductive inks, though as digital printing of conductive tracks grows in application, these expectations and demands will grow in tandem.

7.6.4.1 Humectants

Humectants are used in aqueous systems to prevent drying and clogging of the ink in the nozzle at the air–ink interface. Humectants are common additives in everything from inks to cosmetics to food products. A wide range of humectants for aqueous systems are available, all of which have one or more hydrophilic groups, for example hydroxyls,

[2] If the vision for small office and home office (SOHO) applications of printable conductive tracks is to be fulfilled, aqueous ink formulations will almost certainly be required.

amines, or carboxylic acids. These groups attract and retain water through hydrogen bonding. This in turn slows evaporation, drying, and solid precipitation and deposition at the liquid–air interface. Of course, in a complex system such as a conductive ink, there are always unintended consequences. Large amounts of humectant can increase ink viscosity, thereby reducing jettability, and the hydrogen-bonding capability of humectants can affect particle stabilization, surface tension, and adhesion to the substrate. In many co-solvent systems, one solvent contributes humectant benefits either as a primary function or as a secondary benefit to viscosity modification.

7.6.4.2 Surfactants

Surfactants (also called wetting agents) serve two major purposes in inkjet ink formulation. First, they are commonly used to stabilize the dispersion of particles in the medium, and as such comprise a significant subset of all dispersants. Even in systems where covalently bound dispersants are used, surfactants may be beneficial to dispersion stability. The other major role of surfactants is in the control of surface tension (at the liquid–air interface, as opposed to at the liquid–particle interface). Control of surface tension is crucial for a number of reasons. If the system's surface tension is too high, ink may not move through the nozzle, or printing may be sporadic as ink beads within the print-head. If surface tension is too low, the opposite problem occurs, and ink may leak out of the nozzle. Once the ink hits the substrate, as discussed in Chapter 5, surface tension becomes even more important. In order to create a continuous track, the deposited droplets must not bead. But in order to maximize the density of deposited particles, the droplets must not be significantly absorbed into the substrate, nor should they "feather" or spread laterally. Thus, there are three distinct optima to consider: dispersion stability, printability, and as-deposited wettability. In order to address each of these, arbitrarily complex "surfactant systems" can be employed, but as with any other additive, large amounts of surfactant can decrease the effective density of metal particles in the deposited track, stabilize the particles to prevent sintering, and decrease adhesion to the substrate (in fact, many are designed to do exactly this). Thankfully, many decades of formulation have led to an excellent literature on the experimental and theoretical uses of surfactants (Morrison and Ross, 2002; Rosen, 2004; Farn, 2006).

7.6.4.3 Adhesion Promoters

After creating an affordable, stable, jettable ink; depositing it in a dense, thin track on a substrate; and sintering the particles together to form a conductive track, someone – perhaps even someone you employ – will take a piece of adhesive tape, apply it over the surface of your beautiful track, and then try to tear it off. With all the dispersant required to stabilize the miniscule particles and all the surfactant needed to adjust the surface tension, it is surprising that the conductive track does not just curl up and lift off the substrate of its own accord (indeed sometimes it does). In order to prevent this, and to generate an acceptable level of adhesion to the substrate, adhesion promoters are used. Note that "adhesion" assumes both a defined track and substrate composition, making the chemistry completely application dependent. For graphic inks printed on paper, acrylates are often employed. To print on glass, alkoxysilanes may be used. For certain metal substrates, mercaptans may be useful. Whatever the specific situation, the addition of a significant amount of a chemical which is strongly attracted to certain elements in your ink (and possibly in your print-head) can be problematic for all the reasons discussed in this chapter, for example prevention of sintering, dispersion destabilization, and modification of surface tension.

7.6.4.4 Biocides

Because most conductive inks are formulated in solvent-based or at least co-solvent systems, biocides are not typically necessary. In aqueous systems, and where shelf life is a consideration, small amounts (0.01–0.5%) of biocides may be necessary to prevent mold and fungus growth. Even in such small amounts, care must be taken to avoid unintended effects, especially on surface tension.

7.7 Raw Materials: Conductive Polymers

As discussed in this chapter, the introduction of particles into an ink formulation brings with it significant technical challenges. The use of conductive polymers (Sirringhaus, Friend, and Kawase, 2001; Cui *et al*., 2005; Müller *et al*., 2005; Ngamna *et al*., 2007; Wallace, In Het Panhuis, and Innis, 2008) removes many of these complications, but usually at a cost of overall conductance. Some conductive advantage can be gained with standard polymer chemistry concepts, though usually with a cost, as seen in Figure 7.12. Conductivity can be significantly improved with copolymers, blends, or dopants, though this adds additional complexity, and even doped polymers still fall far short of the conductance of metallic tracks. Combinations of conductive polymers and particles are certainly possible, but as in many systems, the high formulation, shipping, storage, and processing demands of each component can leave the unprepared innovator with the "worst of both worlds." Some basic conductive polymers, which can be used alone or in combinations as described here, are shown in Figure 7.13.

Figure 7.12 *Trade-offs in one parameter of conductive polymer use in conductive inks.*

7.8 Raw Materials: Substrates

The choice of substrate is, of course, highly application dependent. It is related to ink selection through maximum processing temperature (and time), wetting and adhesion interactions, and the requirement (or lack thereof) for physical flexibility. Substrates include almost any solid material, so a comprehensive discussion is impractical. The most common substrates used in printable conductive applications are summarized in Table 7.2.

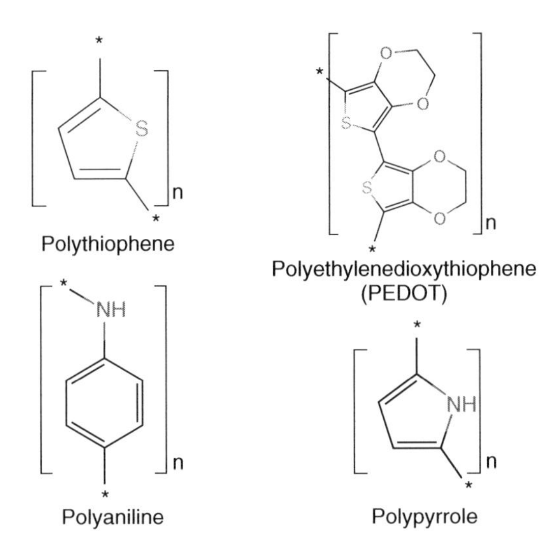

Figure 7.13 *Common conductive polymers.*

Table 7.2 *Typical substrate materials.*

Substrate	$T_{failure}$ (°C)	Flexible	Cost	Common uses	Notes
Paper	150[a]	Yes	+	RFID, greeting cards, and disposable applications	Porosity of uncoated paper can be a challenge.
Glass	>1000	When very thin	++	Displays	Well understood and commonly used
Metals	>1000	When thin	++	Display backpanels	Metal may be coated with insulating polymer or oxide layer.
Ceramics	>1000	No	+++	Multilayer ceramic capacitors, specialty	Uncommon
Polymers	100–400+	Usually	+ to +++	Printed circuit boards, flexible electronics, and RFID	There is a huge array of available polymers that are suitable for use as substrates.
Textiles	150[a]	Yes	+	Clothing, keyboards, and bags	Most "electronic textiles" actually have embedded polymer substrates.

[a]Decomposition of organic matter is a highly complex phenomenon. An excellent discussion is available in Borch *et al.* (2001).

7.9 Printing Processes

Because the process under discussion is digital, it is trivial to print almost any 2D pattern within the resolution constraints of the printing device. Some specialized patterning approaches have been employed; however, these are most often used to increase the density of conductive particles in the track. Usually layers are printed in multiple passes with a drying step, or at least drying time, in between. This allows for different inks to be used in different layers, thereby mitigating some of the trade-offs detailed here, but note that each new layer reduces the throughput of the capital equipment, thereby rapidly increasing costs.

7.10 Post Deposition Processing

7.10.1 Sintering

In order to attain a conductive track, individual particles that have been deposited in an appropriate pattern must then fuse together to form a continuous path through which electrons can flow, as shown in Figure 7.14. Sintering is a common metallurgical process which coalesces particles into a continuous whole by heating. The critical advantage of sintering is that it is performed below the melting point of the material, thereby accommodating less thermally stable substrates and reducing thermal stresses while keeping processing costs low.

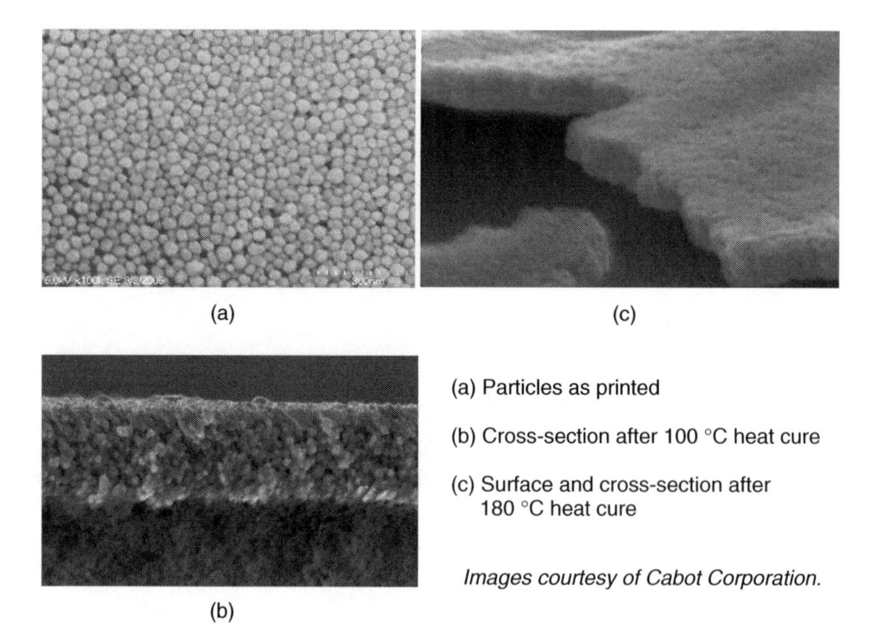

Figure 7.14 *Scanning electron micrograph of as-deposited and sintered conductive tracks (Reproduced with kind permission from Cabot Corporation Copyright (2012) Cabot Corporation).*

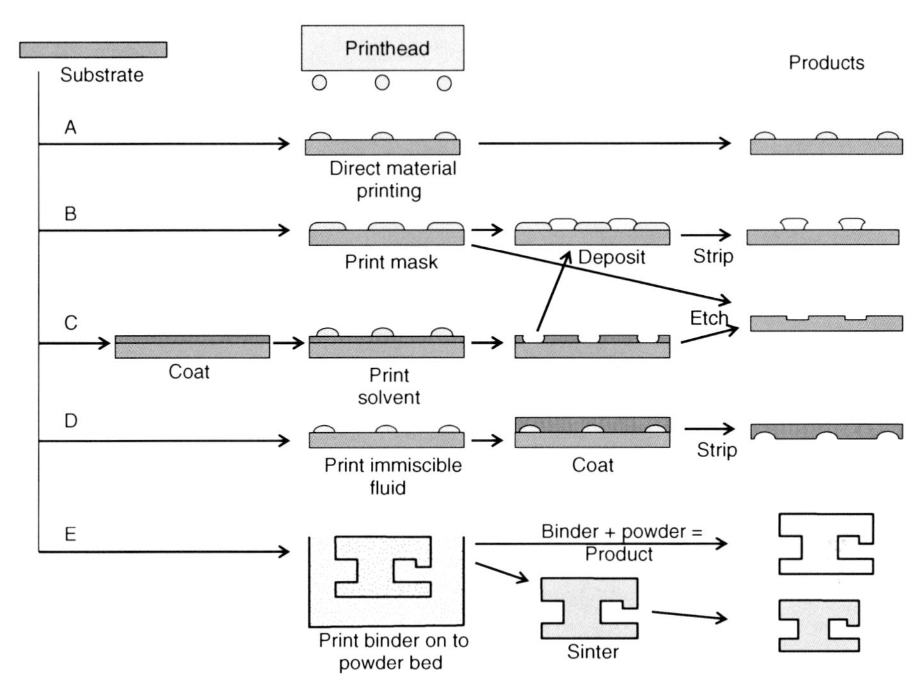

Figure 1.5 *Classification of process routes by which inkjet printing can be used to create structures: (A) direct material printing; (B) printing of a mask followed by material deposition or etching; (C) inkjet etching; (D) inverse inkjet printing and (E) printing onto a powder bed.*

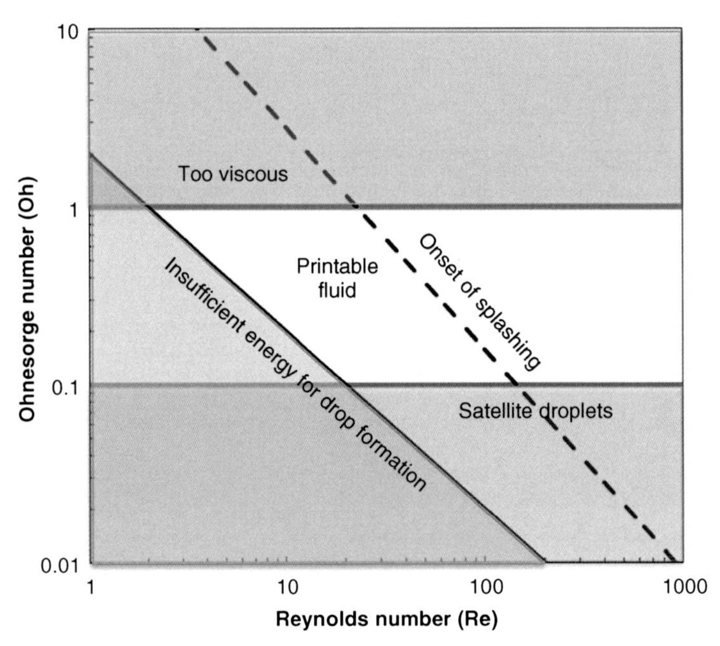

Figure 2.4 *Schematic diagram showing the operating regime for stable operation of drop-on-demand inkjet printing, in terms of the Ohnesorge and Reynolds numbers (Reproduced with permission from McKinley and Renardy (2011) Copyright (2011) American Institute of Physics).*

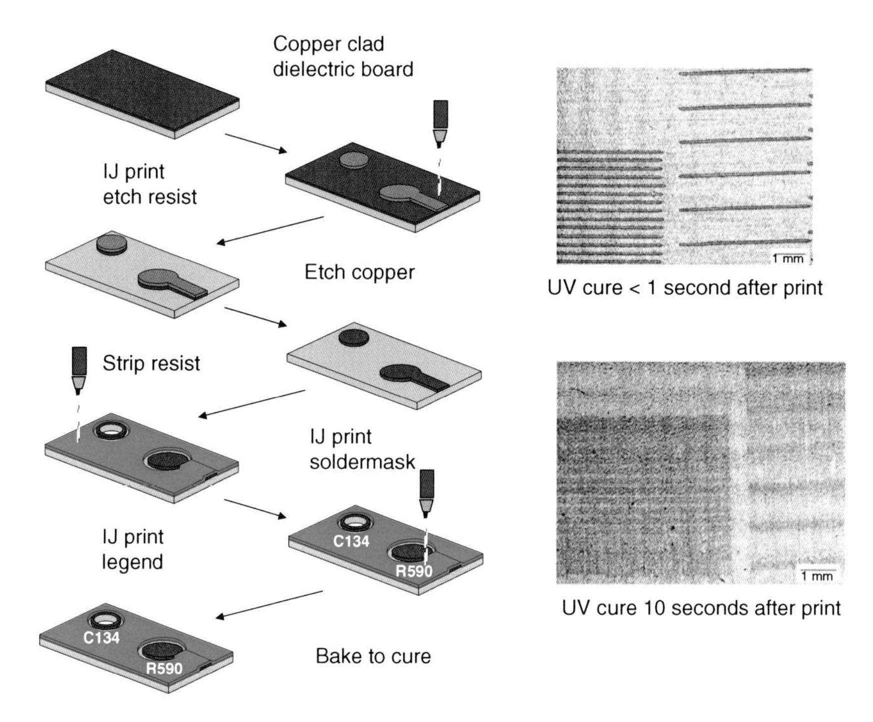

Figure 4.9 PCB production using IJ printing.

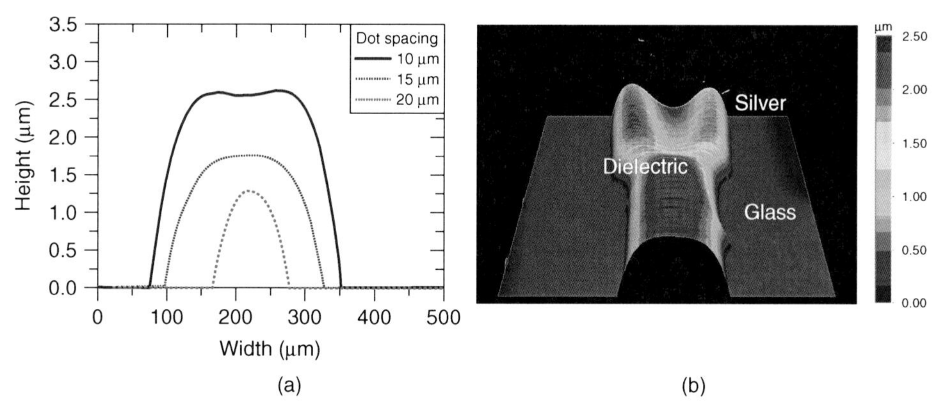

Figure 4.10 *(a) Line profile of SU-8 dielectric layer on glass after UV cure and (b) image of SU-8 dielectric traversing a glass–silver film interface (Sanchez-Romaguera, Madec and Yeates, 2008).*

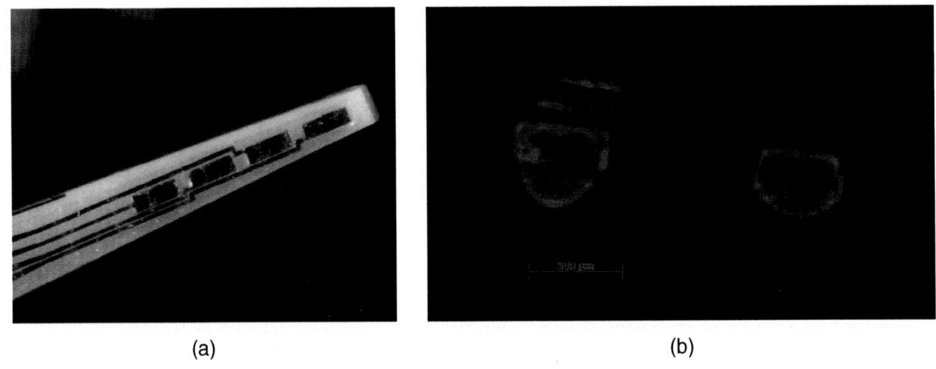

(a) (b)

Figure 6.12 *Implantable (brain) microelectrode with four 20 × 60 μm electrochemical measurement sites on a ceramic substrate. All four sites and the surrounding regions in (a) have been coated with a glutamate oxidase enzyme and overcoated with glutaraldyhyde, a fixative, both by inkjet printing. (b) Two of the four electrodes have been coated with a fluorescent dye to illustrate the capability of localized deposition control (Images courtesy of University of Kentucky Center for Microelectrode Technology).*

Figure 6.13 *Cardiac stent with 100 μm wide structures, model double-helix configuration. Top: Fluorescently labeled drug analog printed onto one helix. Middle: Second fluorescently labeled drug analog printed onto other helix. Bottom: Composite image.*

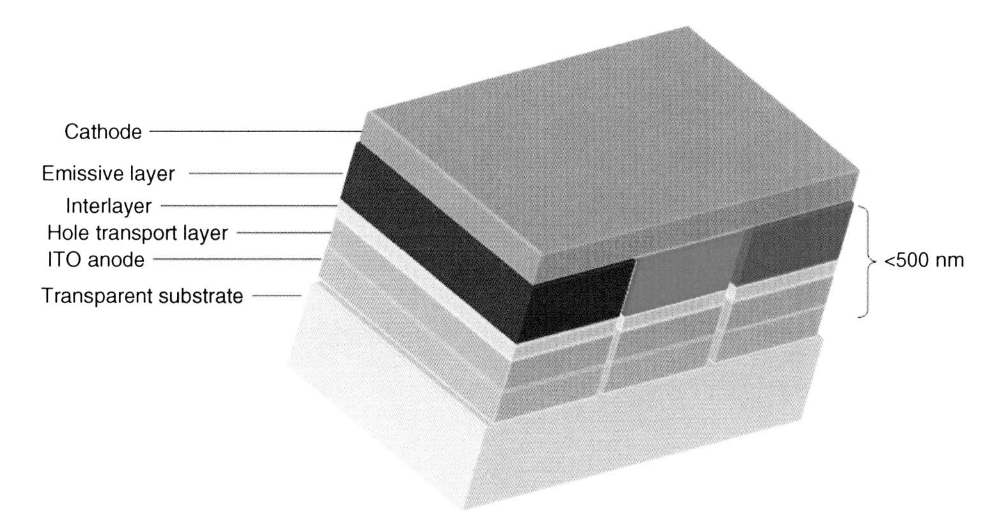

Figure 10.1 *The typical structure of a P-OLED cell.*

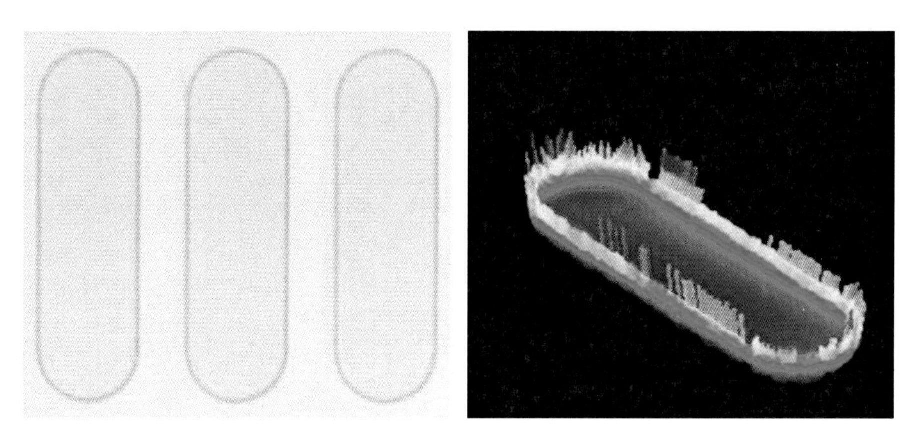

Figure 10.2 *Photomicrograph of printed PEDOT and white light interferometry representation of a PEDOT film profile in one of the wells – the uniformly coloured areas represent a thickness variation of* ± 2 *nm.*

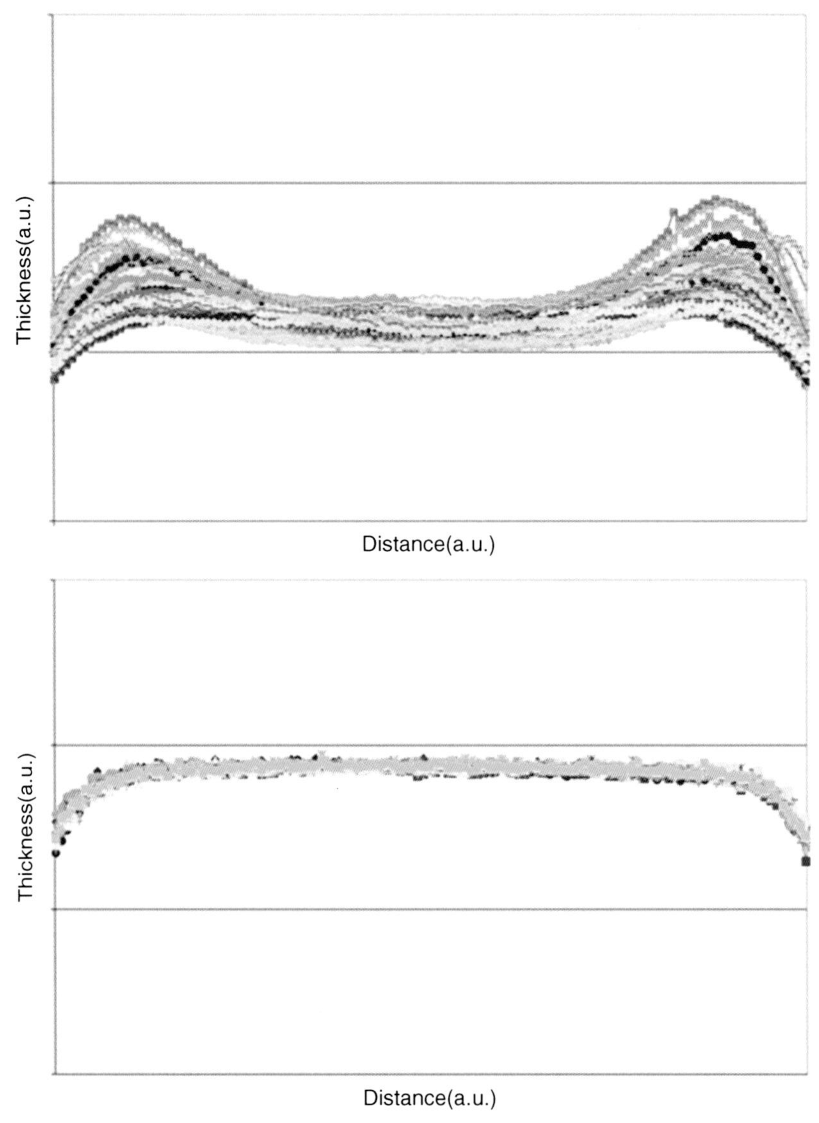

Figure 10.3 *The graphs show PEDOT film profiles for pixels on either side of a swathe join. The top graph shows profiles for the case where the swathe join is significant, whereas the lower graph shows profiles for a reformulated ink that was developed to remove the sensitivity to swathe edges.*

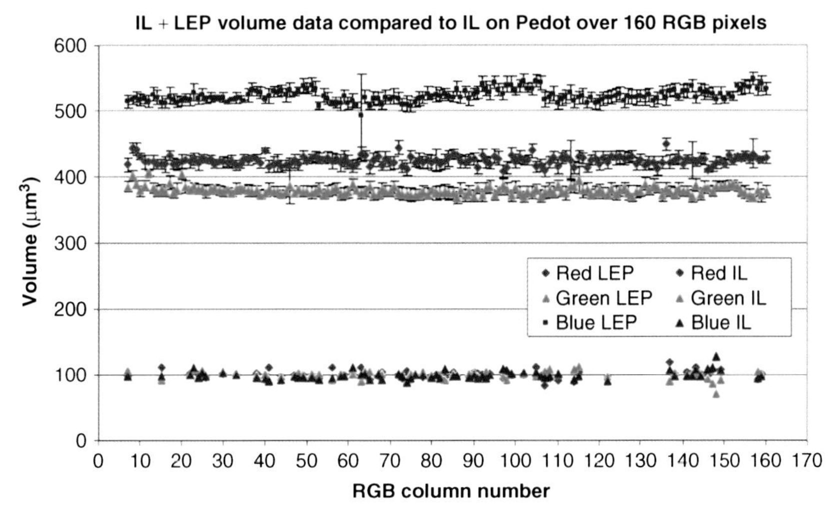

Figure 10.4 Dry pixel volumes of interlayer and of emissive layer + interlayer as measured over several printed swathes on a segment of a wide extended graphics array (WXGA) display test substrate. Consistent variation in the blue is clearly visible and could be correlated to the driven display.

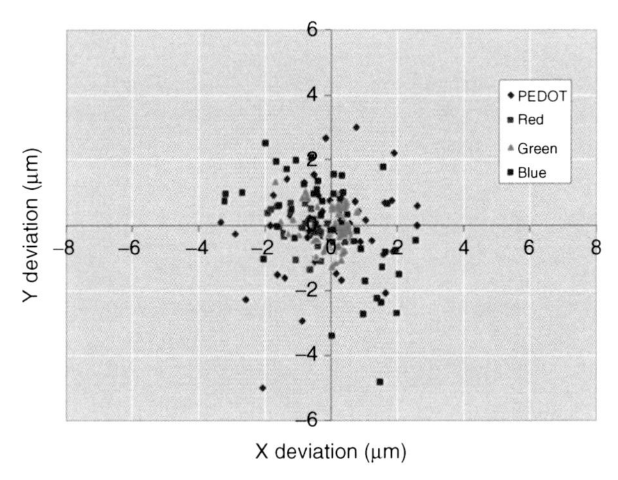

Figure 10.5 Drop deviation from a range of nozzles of PEDOT, red, green and blue emissive layer inks for a particular location on a substrate.

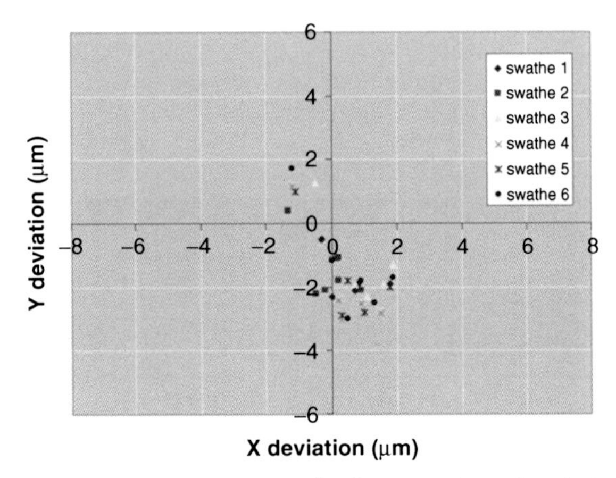

Figure 10.6 Drop deviation across a printed substrate where the deviation due to jetting deviation has been averaged out. In this case, six swathes of the head are required to cover the substrate and the drop positions are taken from six positions along the length of the swathe.

Figure 10.7 Images of drop formation taken at two different strobe delays showing how the tail coalesces to form a spherical drop.

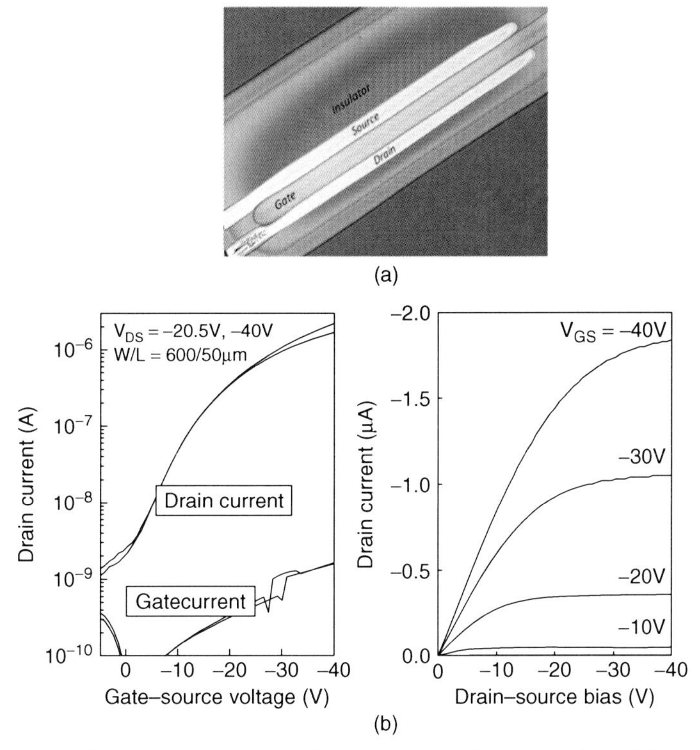

Figure 11.12 (a, b) Micrograph and characteristics of a printed self-aligned transistor (Reproduced with permission from Tseng (2009) Copyright (2009) IEEE).

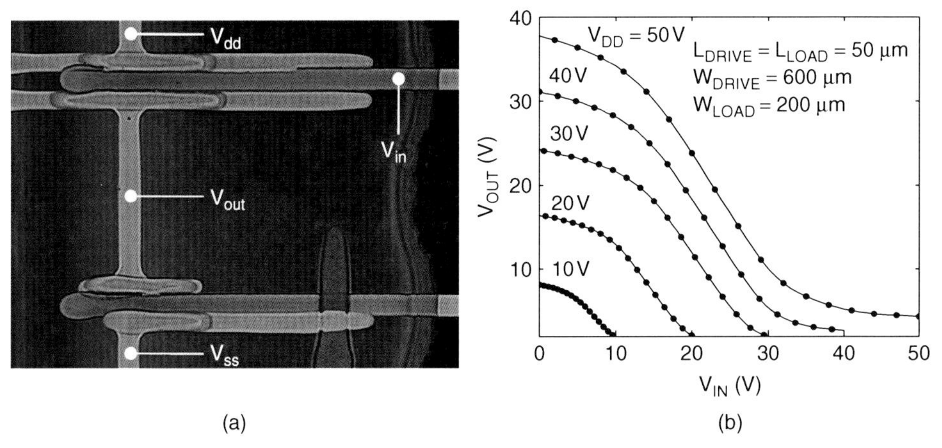

Figure 11.13 (a,b) Optical micrograph and transfer characteristics of self-aligned inverter. Switching speeds of several kilohertz were measured (Reproduced with permission from Tseng (2009) Copyright (2009) IEEE).

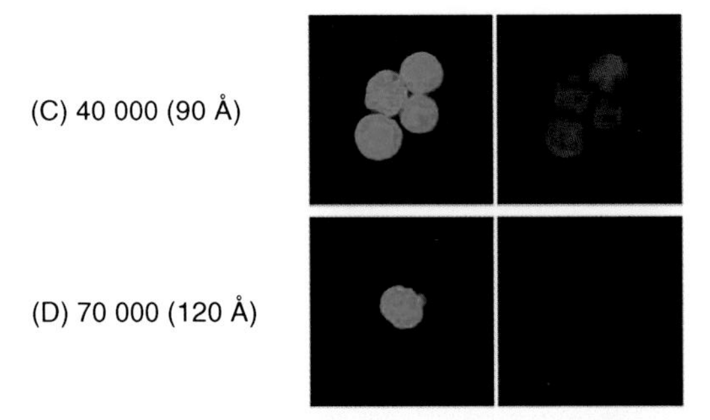

Figure 12.3 *Inkjet-printed cells labeled with high-molecular-weight fluorescent probe, showing that the small (40 kDa) probe penetrates pores in the cell membrane caused by the printing process but the larger probe does not (Reproduced with permission from Cui (2010) Copyright (2010) John Wiley & Sons Inc).*

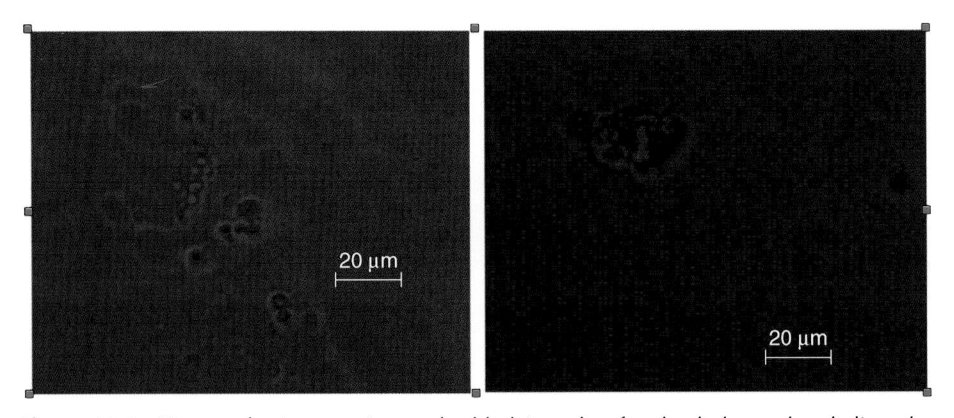

Figure 12.8 *Yeast colonies growing embedded in gels of polyethylene glycol diacrylate (PEGDA). Cured by blue light after co-extrusion of cells and gel (Mishra and Calvert, 2009). After 7 hours, cells started to grow in all three dimensions and formed a colony. Cells are entrapped and grown in a PEGDA thin-film hydrogel sample (20 × 10 × 9 mm). Left after 1 day; right after 7 days (Reproduced with permission from Mishra (2009) Copyright (2009) IEEE).*

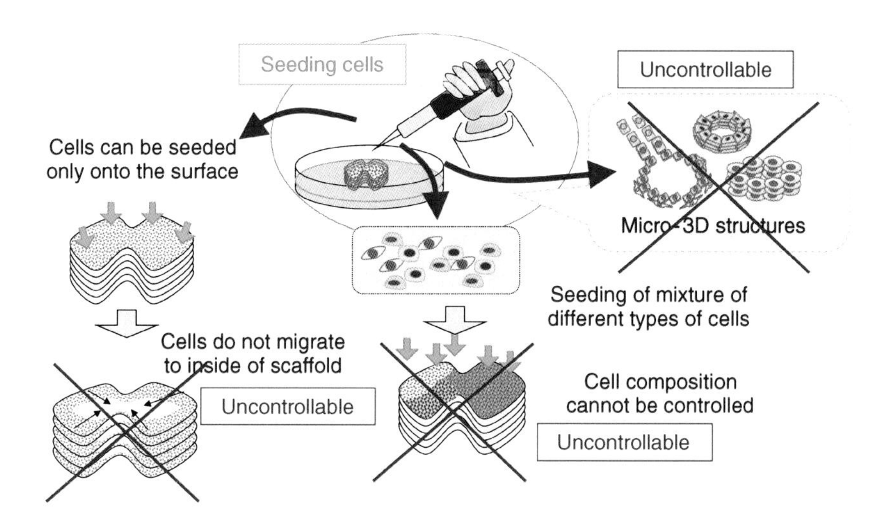

Figure 13.2 *Limitation of the approach based on subsequent cell seeding.*

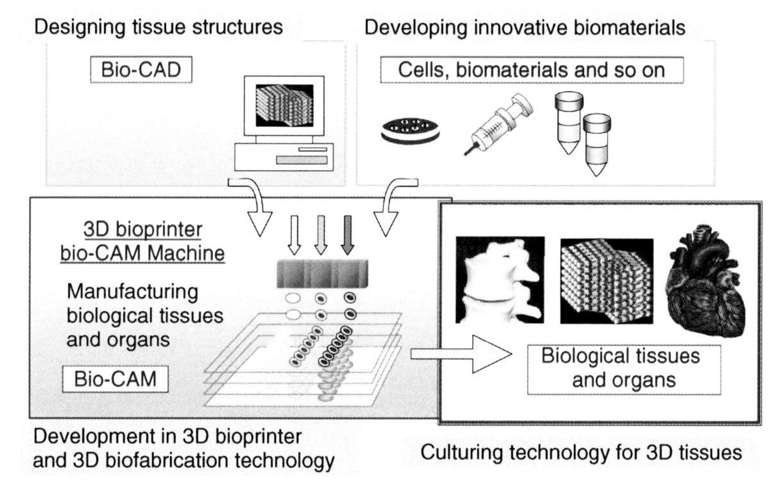

Figure 13.3 *Concept of inkjet-based direct 3D biofabrication. The original data is designed on a computer (bio-CAD), and based on the digital data, the 3D bioprinter fabricates 3D biological structures as a bio-CAM machine. The materials for 3D biofabrication are living cells and various biomaterials including natural and synthetic materials. Fabricated biological structures are cultured in succession to form functional tissues and organs.*

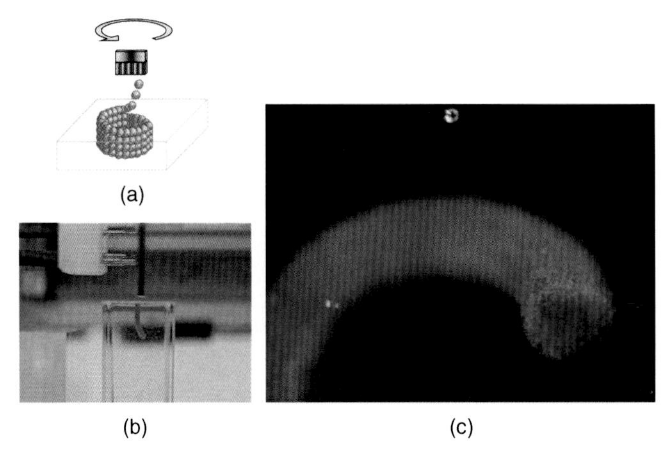

Figure 13.10 *Fabrication of 3D gel tube containing HeLa cells. (a,b) Schematic and photograph showing fabrication method; and (c) 3D gel tube containing HeLa cells. See http://www.youtube.com/watch?v=g2ZTWHsO8l0&feature=player_embedded.*

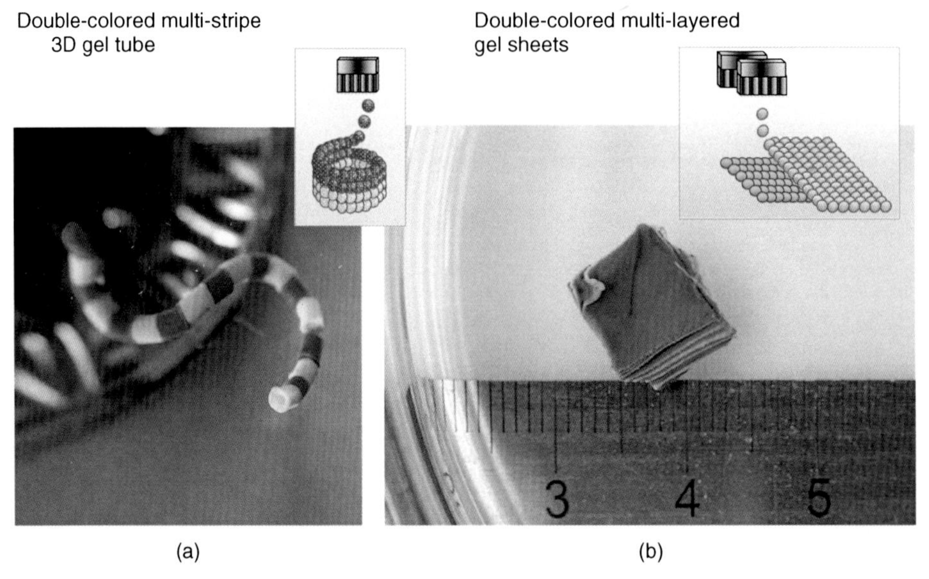

Figure 13.11 *Fabrication of multicolored 3D structures. (a) Two-colored multistripe 3D gel tube; and (b) two-colored multilayered gel sheets.*

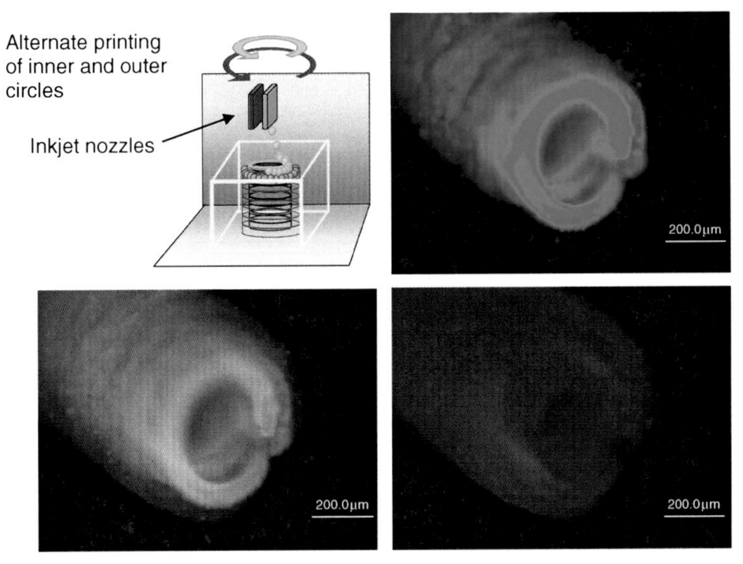

Figure 13.12 Fabrication of double-walled 3D gel tube with different fluorescent colored inks. (a) Top left: Schematic showing fabrication method. (b–d) Top right, bottom left and bottom right: Double-walled 3D gel tube with different fluorescent colored inks.

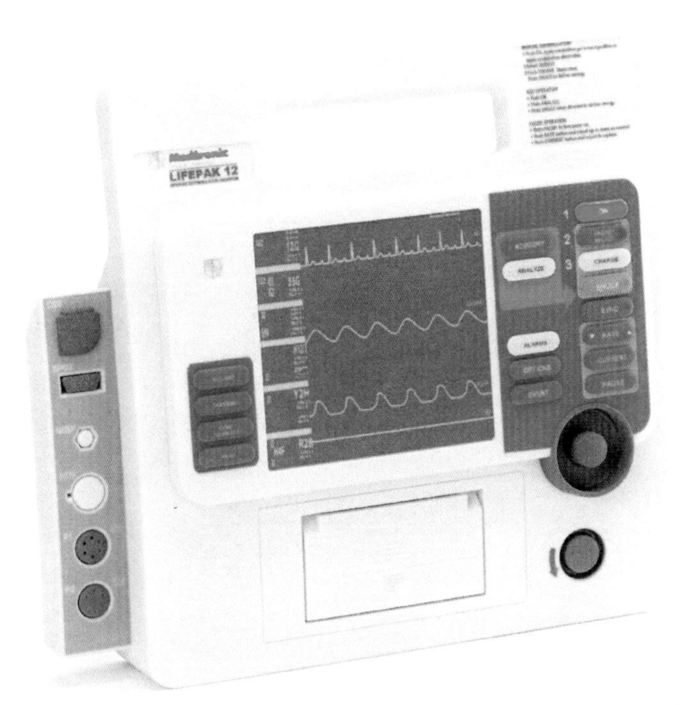

Figure 14.11 A colored model produced by a Spectrum Z510 (Reproduced from Z Corporation Copyright (2012) Z Corporation).

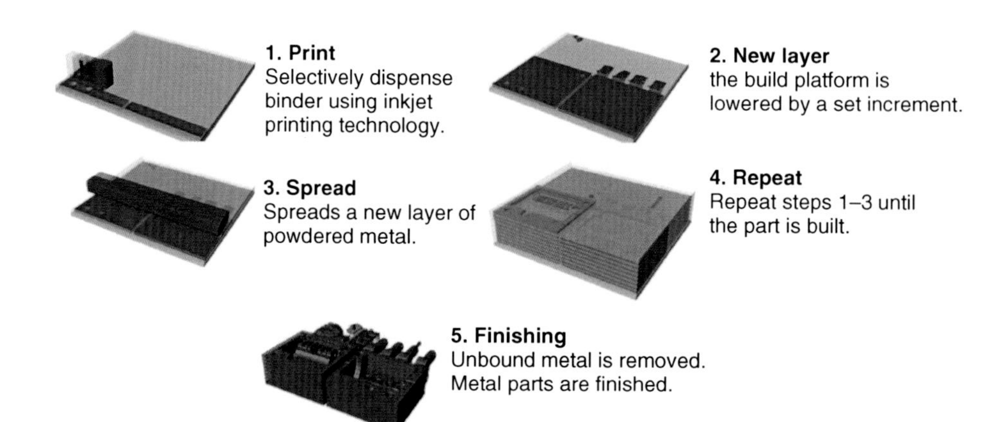

Figure 14.12 *A schematic of the ProMetal™ process (Reproduced from Prometal Inc. Copyright (2012) Prometal Inc).*

The degree of sintering, and therefore of conduction, is a function of temperature, particle size, particle composition (especially small amounts of certain additives, e.g., fluxing agents), particle surface species, and time. In a production environment, one can rarely afford to wait for thermodynamic equilibrium. When comparing inks, a plot of sintering temperature and time versus conductance is a good first step.

In order to overcome these trade-offs, unique sintering systems have been designed. Most rely on management of heat in time and/or space, for example through the use of lasers (Todorohara and Amako, 2006; Ko *et al*., 2007a; Khan *et al*., 2008a, b), high-intensity light (Kamyshny, Steinke, and Magdassi, 2011) or other flash-heating processes. These approaches certainly have advantages, but the required equipment and controls add cost and complexity to the system, and therefore will be useful in some, but not all, applications.

7.10.2 Protective Layers

A metal conductive track may be susceptible to oxidation due to its composition or the environment in its intended use. It may also be prone to mechanical damage. In these cases, it may be advantageous to print a second layer on top of the conductive track. This layer may be another metal, a metal alloy, or a polymer. It may contribute to the track's conductance, or it may simply provide an oxygen and/or moisture barrier, sacrificial material, or mechanical durability.

7.11 Resistors

Resistors are amongst the most basic of passive components, and are certainly the most numerous, both in the number of different designs and in total units shipped. Their name arises from their resistance to the flow of electric current. As current rises, the voltage drop across the resistor also rises as described by Ohm's law. The quality of a resistor is measured primarily by its tolerance (%), a measure of variability (i.e., precision), and its temperature coefficient (ppm/K), a measure of temperature stability. While the temperature coefficient is largely determined by the materials of construction (with the notable exception of biphasic phenomena, e.g., the positive temperature coefficient (PTC) switch-like effects detailed in Section 7.13.1), creating low-tolerance resistors by digital fabrication is especially challenging for a number of reasons.

First and foremost, dimensional (thickness, length, and width) variability in the printed component will define the maximum achievable resistive tolerance. Droplet scatter, substrate surface variability, and droplet size and velocity variability (due to, e.g., evaporation from the nozzle, particle settling, and temperature variations in the print-head during start-up) will all negatively impact dimensional variability. Furthermore, spatial compositional variability due to any of the above reasons or due to uneven drying parameters (e.g., the "coffee ring" effect discussed in Section 5.6.2), flocculation events, or other seemingly chaotic phenomena will also add to device variability. Very precise control of the ink, the substrate, the deposition, and the drying process is required to make high-quality digitally printed resistors.

A resistor ink may contain conductive or moderately conductive particles or conductive particle precursors (to control resistivity rather than to maximize conductance), and insulating particles, insulating particle precursors, or simply an insulating continuous phase. Multiple inks may also be advantageous for device fabrication, though a single ink can suffice. Many variants on these themes have been described (Kodas *et al.*, 2007).

7.12 Capacitors

Capacitors are another subset of passive components that are simple in concept but can be quite sophisticated in actual use. They have a wide array of applications in electronics, but invariably they oppose changes in voltage. Composed of two conductive elements separated by a dielectric, capacitors formed by digital fabrication require two different material compositions, that is, two different inks, though many variants have been demonstrated (Redinger *et al.*, 2004; Szczech *et al.*, 2004; Howarth, Edwards, and Vanheusden, 2006; Ko *et al.*, 2006a, 2006b, 2007; Chang, Lee, and Chang, 2007; Kowalski and Edwards, 2007; Seh and Min, 2007; Jeong, Kim, and Moon, 2008; De La Vega and Rottman, 2009).

Capacitance is proportional to the area of the conductive elements and the dielectric constant of the insulator, and is inversely proportional to the distance between the conductors. The optimal geometry for a high capacitance capacitor in two dimensions is therefore three interdigitated spirals – two conductive spirals separated by an exceedingly thin, high-dielectric-constant insulator layer. Of course, one electrical short ruins the device, so tolerances of the printer and print-head will limit the pitch. Many common capacitors are fabricated in "vertical" layers (i.e., in the third dimension), which may also be created with digital printing processes.

7.13 Other Passive Electronic Devices

With an appropriate printable materials set including conductors and high- and low-κ dielectrics, many passive devices can be created. Manufacturing flexibility is limited only by the limitations of the digital printing equipment. A few words are said about each in this section.

7.13.1 Fuses, Circuit Breakers, and Switches

Though both fuses and circuit breakers are designed to address the same issue, that is, circuit faults (usually overcurrent protection), they can lie at opposite ends of the passive electronics complexity spectrum. Fuses can be extremely simple, consisting of little more than a thin conductor which will burn, break, or otherwise cause an open circuit when current exceeds a threshold value. Circuit breakers, in contrast, almost always have moving parts, making fabrication by digital printing exceptionally challenging. Between these two extremes is a third device, called a resettable fuse. These devices are based on

PTC materials, usually a conductive particle (often carbon black) highly loaded within a PTC polymer. During regular operation, current flows as conductive paths can be found between the particles. As temperature increases (due to overcurrent or, e.g., as an active sensor), the polymer expands, separating the conductive particles within and opening the circuit. The polymer then cools and contracts, and the fuse is "reset" as conductive paths are again created. This is how the rear-window defogger or demister in a car is limited. There is no fundamental reason that these devices could not be digitally printed.

Switches are a simpler version of a circuit breaker in that they open and close a circuit, but without any current-sensing capacity. That said, they too require moving parts. As printable micro-electro-mechanical systems (MEMS) technology improves, these devices will become possible, and potentially even practical.

7.13.2 Inductors and Transformers

Inductors and transformers require a third dimension in their construction, and therefore digitally fabricated examples have so far been rare. The composition of the core should be straightforward, and conductor inks are commercially available, so no invention is required to create these devices, but rather, improvements in existing digital fabrication methods should suffice.

7.13.3 Batteries

Battery technology has enjoyed a renaissance in recent years, owing in large part both to the rapidly increasing capabilities of materials scientists, and to the public's interest in alternative engine technologies and portable electronics. Myriad materials sets have been designed, and some have even been printed (Tamminen, 1966; McTaggart, 1998; Parker, 2002; Langan *et al.*, 2003; Zucker, 2003; Southee *et al.*, 2007), though the advantage of a digital process (e.g., vis-à-vis screen printing) for batteries is not immediately obvious.

7.13.4 Passive Filters

Filters are used to remove unwanted signal from an electric current. There are many types, but the simplest are passive filters, which can be constructed from capacitors, resistors, and inductors. The comments in Sections 7.11, 7.12, and 7.13.2 on these individual component types apply.

7.13.5 Electrostatic Discharge (ESD)

Unwanted electrostatic charge can build up in dielectrics through a variety of processes, the simplest of which is friction. Most children have rubbed a balloon on their clothing and stuck it to a wall; ESD is concerned with unsticking that balloon. All that is required for ESD is a conductive path to ground. In applications where unwanted charge can cause significant problems, most materials are filled with conductive particles that provide a sufficient level of electrical conductivity. ESD elements are trivial to print, but the advantage over printing a simple ground or filling the substrate with a moderately conductive filler is unclear.

7.13.6 Thermal Management

Though not technically within the realm of passive electronics, thermal management is a critical aspect of electronic device design. The connection to digital printing is even stronger, in that many of the same materials sets used for electronically conductive tracks are also highly thermally conductive, and could be easily employed in this application. As devices are miniaturized and power densities increase (e.g., in light-emitting diodes), thermal management will become increasingly important. As volumes of digitally printed electronic conductors rise, prices will fall, and at some point a thermal management market may open.

7.14 Outlook

Despite initial hyperbole about the plethora of electronic devices that could be digitally printed, the industry has returned to a more conservative, pragmatic approach centered on conductors, eschewing one-off, unreasonably expensive laboratory demonstrations to focus on manufacturable products that meet the needs of real customers. This progression is neither good nor bad: it is simply the way that technologies grow from new to established. There is little doubt that digitally printed conductive tracks will grow into a tool in the component manufacturers' toolbox; the question is only how often it will be used. Strong economic cases have already been made for a number of lead applications, and more will be sure to follow. Similar economic cases for other passive components have been less evident, and make no mistake – the technical challenges are every bit as daunting as they are for conductive tracks. So for the foreseeable future, conductive tracks will dominate digital printing of passives. In the medium term, a collaborative effort between printer, print-head, ink, and particle manufacturers, or a vertical integration of the same, might produce useful systems with suites of electronic inks allowing full passive-printing capabilities for the industrial user. A single ink maker could, through R&D and/or strategic acquisitions, assemble a similar suite. Again, a compelling economic case, which clearly demonstrates the costs and benefits of the afforded production flexibility, has yet to be made (at least publicly). Finally, taking the long view, it is easy to imagine copier-like, desktop, or even portable printers that can print electronics for the SOHO user. But as with any new technology, it is the vision of how the technology can be used to benefit the customer that will determine whether it is successful. Digital printing has the potential to fundamentally change manufacturing, moving production from enormous, lean, full-scale plants into the hands of creative, highly variable, individual users. Similar shifts have taken place in information-based markets on the back of the internet, for example software, music and videos, and retailing, but it remains to be seen if individual consumers will pay for, or even want, the unbounded ability to print their own electronics.

References

Baron, D.T., Hofmann, H-P. and Schneider, R. (2006) Method of manufacturing an electronic circuit device through a direct write technique. Patent WO/2006/010639.

Boils, D.C., Mayo, J.D., Gagnon, Y. and Mackinnon, D.N. (2001) Ink composition and processes thereof. US Patent 6,210,473, Xerox Corporation, Stamford, CT

Callen, B.W. and Walkhouse, W.K. (2007) Enhanced performance conductive filler and conductive polymers made therefrom. US Patent 20070012900, Sulzer Metco Inc., Fort Saskatchewan, AB.

Chang, C-H., Lee, D. and Chang, Y-J. (2007) Solution deposition of inorganic materials and electronic devices made comprising the inorganic materials. US Patent 20070184576, Oregon State University, Eugene, OR.

Cheng, K., Yang, M-H., Chiu, W.W.W., Huang, C-Y., Chang, J., Ying, T-F. and Yang, Y. (2005) Ink-jet printing, self-assembled polyelectrolytes, and electroless plating: low cost fabrication of circuits on a flexible substrate at room temperature. *Macromolecular Rapid Communications*, **26**, 247–264.

Cho, G-J., Jung, M.H. and Hudson, J.L. (2007) Preparation of thin film transistors (TFTs) or radio frequency identification (RFID) tags or other printable electronics using ink-jet printer and carbon nanotube inks. Patent WO/2007/089322, Rice University, Houston, TX.

Cui, T., Liu, Y., Chen, B. *et al*. (2005) Printed polymeric passive RC filters and degradation characteristics. *Solid-State Electronics*, **49** (5), 853–859.

De La Vega, F.B.S. and Rottman, C. (2009) Ink jet printable compositions for preparing electronic devices and patterns. Patent application 20090053400, Cima Nano Tech Israel Ltd., Caesarea, Israel.

Eguchi, K., Sasakura, H. and Yamaguchi, Y. (2004) Patterned films of ITO nanoparticles fabricated by ink-jet method. AIChE Annual Meeting, Conference Proceedings, pp. 2449–2452.

Erri, P. and Varma, A. (2009) Diffusional effects in nickel oxide reduction kinetics. *Industrial and Engineering Chemistry Research*, **48** (1), 4–6.

Farn, R.J. (2006) *Chemistry and Technology of Surfactants*, Wiley-Blackwell Ltd, Oxford.

Hayton, C., Sirringhaus, H., Von Werne, T. and Norval, S. (2006) Electronic devices. Patent WO/2006/061589, Plastic Logic Limited, Cambridge.

Howarth, J.J., Edwards, C. and Vanheusden, K. (2006) Ink-jet printing of passive electricalcomponents. Patent WO/2006/076607, Cabot Corporation, Boston.

Imanishi, Y. and Nishimura, E. (2012) Zinc oxide thin film, transparent conductive film and display device using the same. Patent 8,137,594, Hitachi, Ltd., Tokyo.

Jackson, C. (2008) Additive for high optical density inkjet ink. Patent 7,351,278, E.I. du Pont de Nemours and Company, Wilmington, DE.

Jeong, S., Kim, D. and Moon, J. (2008) Ink-jet-printed organic–inorganic hybrid dielectrics for organic thin-film transistors. *Journal of Physical Chemistry C*, **112** (14), 5245–5249.

Tamminen, P.J. (1966) Printed battery and method of making the same. US Patent 3,230,115.

Kamyshny, A., Steinke, J. and Magdassi, S. (2011) Metal-based inkjet inks for printed electronics. *Open Applied Physics Journal*, **4**, 19–36.

Khan, A., Rasmussen, N., Marinov, V. and Swenson, O.F. (2008a) Laser sintering of direct write silver nano-ink conductors for microelectronic applications. Proceedings of SPIE – International Society for Optical Engineering, p. 6879.

Khan, A., Rasmussen, N., Marinov, V. and Swenson, O.F. (2008b) Laser sintering of direct write silver nano-ink conductors for microelectronic applications. *Photon Processing in Microelectronics and Photonics VII*, **6879**, 687910–687911.

Ko, S.H., Chung, J., Pan, H. *et al*. (2006a) Fabrication of multilayer passive electric components using inkjet printing and low temperature laser processing on polymer. Proceedings of SPIE – International Society for Optical Engineering, p. 6106.

Ko, S.H., Chung, J., Pan, H. *et al*. (2006b) Fabrication of multilayer passive electric components using inkjet printing and low temperature laser processing on polymer. *Photon Processing in Microelectronics and Photonics V*, **6106**, 127–135.

Ko, S.H., Chung, J., Pan, H. *et al*. (2007a) Fabrication of multilayer passive and active electric components on polymer using inkjet printing and low temperature laser processing. *Sensors and Actuators, A: Physical*, **134** (1), 161–168.

Ko, S.H., Pan, H., Grigoropoulos, C.P. *et al*. (2007b) All-inkjet-printed flexible electronics fabrication on a polymer substrate by low-temperature high-resolution selective laser sintering of metal nanoparticles. *Nanotechnology*, **18** (34), 345202.

Kodas, T.T., Edwards, C., Kunze, K. *et al*. (2007) Printed resistors and processes for forming same. Patent WO/2007/140480, Cabot Corporation, Boston.

Kodas, T.T. and Kunze, K. (2006) Printable conductive features and processes for making same. Patent 20060001726, Cabot Corporation, Boston.

Kowalski, M.H. and Edwards, C. (2007) Polyimide insulative layers in multi-layered printed electronic features. Patent WO/2007/140477, Cabot Corporation, Boston.

Langan, R.A., Schubert, M.A., Zhang, J. *et al*. (2003) Flexible thin printed battery with gelled electrolyte and method of manufacturing same. Patent WO/2003/069700, Eveready Battery Company, Inc., St. Louis, MO.

McTaggart, S.I. (1998) Laminated sheet product containing a printed battery. Patent WO/1998/022987.

Morrison, I.D. and Ross, S. (2002) *Colloidal Dispersions: Suspensions, Emulsions, and Foams*, Wiley-Interscience, Chichester.

Mort, J. (1992) Liquid ink compositions. US Patent 5,114,477, Xerox Corporation, Stamford, CT.

Müller, D.C., Reckefuss, N., Meerholz, K. *et al*. (2005) Electronic devices comprising an organic conductor and semiconductor as well as an intermediate buffer layer made of a crosslinked polymer. Patent WO/2005/024971, Merck, Darmstadt, Germany.

Mustonen, T., Mäklin, J., Kordás, K. *et al*. (2008) Controlled ohmic and nonlinear electrical transport in inkjet-printed single-wall carbon nanotube films. *Physical Review B: Condensed Matter and Materials Physics*, **77** (12), 125430.

Nakano, M., Yamaguchi, S. and Fukuhara, K. (2006) (Meth)acrylic acid copolymer, method for producing the same, and application thereof. Patent WO/2006/104255, Nippon Shokubai Co., Ltd., Osaka.

Ngamna, O., Morrin, A., Killard, A.J. *et al*. (2007) Inkjet printable polyaniline nanoformulations. *Langmuir*, **23** (16), 8569–8574.

Park, H-S., Seo, D-S., Choi, Y. *et al*. (2004) Synthesis of concentrated silver nano sol for ink-jet method. *Journal Korean Ceramic Society*, **41** (9), 670–676.

Parker, R. (2002) Printed display and battery. US Patent 6,369,793.

Probst, N., Grivei, E. and Fockedey, E. (2003) New ferromagnetic carbon based functional filler. *KGK-Kautschuk und Gummi Kunststoffe*, **56** (11), 595–599.

Redinger, D., Molesa, S., Yin, S. *et al*. (2004) An ink-jet-deposited passive component process for rfid. *IEEE Transactions on Electron Devices*, **51** (12), 1978–1983.

Rosen, M.J. (2004) *Surfactants and Interfacial Phenomena*, Wiley-Interscience, Chichester.

Seh, H. and Min, Y. (2007) Inkjet patterning for thin-film capacitor fabrication, thin-film capacitors fabricated thereby, and systems containing same. Patent WO/2007/115238, Intel Corporation, Santa Clara, CA.

Sirringhaus, H., Friend, R.H. and Kawase, T. (2001) Inkjet-fabricated integrated circuits. Patent WO/2001/046987, Plastic Logic Limited, Cambridge.

Song, J.W., Kim, J., Yoon, Y.H. *et al*. (2008) Inkjet printing of single-walled carbon nanotubes and electrical characterization of the line pattern. *Nanotechnology*, **19** (9), 095702/1–095702/6.

Southee, D., Hay, G.I., Evans, P.S.A. and Harrison, D.J. (2007) Lithographically printed voltaic cells – a feasibility study. *Circuit World*, **33** (1), 31–35.

Szczech, J.B., Megaridis, C.M., Zhang, J. and Gamota, D.R. (2004) Ink jet processing of metallic nanoparticle suspensions for electronic circuitry fabrication. *Microscale Thermophysical Engineering*, **8** (4), 327–339.

Todorohara, M. and Amako, A. (2006) Method for forming conductive film and electronic device. Patent CORPJP2006100381, Seiko Epson, Tokyo.

Uozumi, S. and Yamamoto, Y. (2005) Non-aqueous ink-jet ink composition containing a chelating agent. Patent 2005298585.

Wallace, G.G., In Het Panhuis, P.H.H. and Innis, P.C. (2008) Polymeric nanocomposites. Patent WO/2008/055311, University of Wollongong, Wollongong, NSW.

Wei, T., Ruan, J., Fan, Z. *et al*. (2007) Preparation of a carbon nanotube film by ink-jet printing. *Carbon*, **45** (13), 2712–2716.

Wohlers, T. (2007) Wohlers's Report 2007: State of the Industry Annual Worldwide Progress Report, Wohlers Associates, Fort Collins, CO.

Zucker, J. (2003) Printed battery. US Patent 20030219648, The InterTech Group, Inc., Elmhurst, IL.

8

Printed Circuit Board Fabrication

Neil Chilton
Printed Electronics Ltd., United Kingdom

8.1 Introduction

This chapter describes how inkjet methods can be applied to the multiple fabrication steps used in the manufacture of printed circuit boards (PCBs). We outline the current state of the art in PCB manufacturing using 'conventional' methods such as UV photolithography and laser direct imaging (LDI) and show where inkjet can augment or in some cases replace these processes.

8.2 What Is a PCB?

A PCB,[1] as illustrated in Figure 8.1, is at the heart of every electrical and electronic device; its primary function is to provide the physical connections between the various electronic components that populate the surface of the PCB and thus form a functional electronic product.

Although the PCB is commonly thought of as a straightforward, passive component, PCBs have become a key electronic component in their own right providing not only passive point-to-point connections but also designed in-board electronic characteristics such as controlled impedance, inductance, capacitance and resistance. In addition to providing the electronic interconnections, the PCB substrate also provides a mechanical platform and thermal connection to the electronic devices.

[1] The term printed wiring board (PWB) is also used – particularly in the United States.

Inkjet Technology for Digital Fabrication, First Edition. Edited by Ian M. Hutchings and Graham D. Martin.

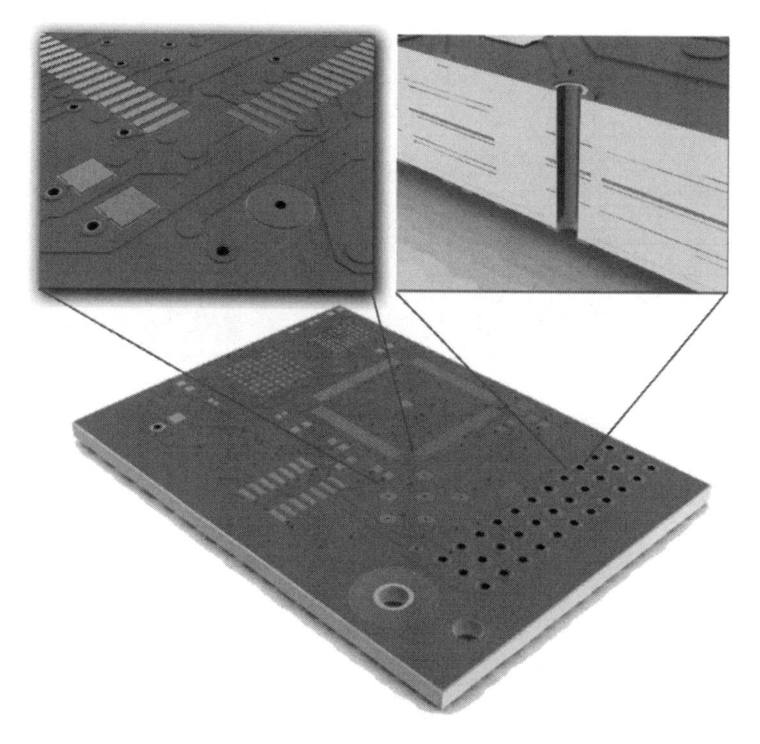

Figure 8.1 *Example of a multilayer PCB.*

The first rigid PCB was produced in the United Kingdom shortly after World War II by Dr Paul Eisler (1989). These PCBs were 'single-sided', in that they had a single copper layer which was patterned to provide the necessary interconnections. The single-sided PCB, whilst an important and still widely used product,[2] was superseded in the 1960s and 1970s by 'double-sided' or 'through-hole' PCBs. Double-sided PCBs have patterned copper structures on either side of an insulating glass–epoxy composite material, the two sides often being joined with a copper-plated drilled hole. In the 1980s, the first multilayer[3] PCBs were adopted to provide enhanced component density and function. Flexible printed circuit board (FPCB) technology using un-reinforced flexible dielectric layers such as polyimide has evolved parallel to that of rigid PCBs.

The PCB industry has evolved significantly[4] in its 60-year lifetime, and PCBs such as those used today in smartphones, tablets and laptop computers consist of many layers of high-density[5] circuitry coupled with advanced interlayer connections. The global PCB

[2] The single-sided PCB is used in very cost-sensitive applications and in applications where moisture absorption must be kept to a minimum (e.g. washing machines).

[3] A multilayer PCB is defined as one with multiple patterned copper layers within it. As a quick benchmark, most high-end mobile phones contain PCBs with six or more layers. Most laptop computers use a main PCB with some 10 layers or more. High-end telecommunications and semiconductor test products frequently use 30 or more layers.

[4] Although PCB track dimensions have reduced (from a few millimetres to $<100\,\mu m$), the reduction cannot be compared to the orders-of-magnitude reduction that has occurred in the same timeframe in silicon fabrication.

[5] Generally track widths of $100\,\mu m$ or smaller.

industry is by any measure large; in 2010 there were estimated to be well over 2000 manufacturing facilities globally with the top 100 manufacturers alone having combined sales of almost $45 billion.[6] Despite recent the economic slowdown, global PCB turnover in 2013 is forecast to reach $65 billion.[7] The fastest growing element of the PCB market is that of high-technology multilayer PCB production driving ever smaller, lighter and more powerful products. Having said that, even leading-edge PCB manufacturing relies almost entirely upon subtractive processes, something that has barely changed in the entire history of the product.

8.3 How Is a PCB Manufactured Conventionally?

Figure 8.2 shows a simplified schematic of the steps used to manufacture a conventional PCB. The PCB manufacturing process delivers *circuits* to its customers[8]; however, the entire manufacturing process is optimised for the manufacture of relatively standard-sized *panels* – these commonly have imperial (inch) sizes of 24″ × 20″, 12″ × 16″, 18″ × 24″, 21″ × 24″ or similar variants.[9] There is a commonality with the fixed-size 'Gen' variants used in the liquid crystal display (LCD) flat-panel industry. It is therefore the cost-effective and fast manufacture of *panels* that drives the PCB industry; similarly PCBs are manufactured using batch processes.

Figure 8.3 shows the relationship between the dimensions of the manufacturing panel and delivered circuit array.

8.4 Imaging

Imaging is the general term used in the industry to describe the process by which the circuit and other layers of the multilayer are defined. Imaging involves intensive multi-step processes and is a primary cost and resource driver in the manufacturing process. Even in the schematic of Figure 8.2, there are three separate imaging processes indicated.

Whilst screen printing of etch masks is still used for low-technology applications, it is not covered in detail here as the process is not a mainstream technique for advanced technology applications. In general, therefore, the imaging of the multiple layers in a PCB is done using UV photolithography. For a PCB with, for example, 14 layers of copper circuitry there will often be 18 imaging steps required: 14 for the copper layers, two for the top and bottom soldermask layers and a further two for the top and bottom component identification layers.[10]

[6] Nakahara Studio (N.T. Information Ltd.) published information on the top 100 manufacturers (2010 sales data).

[7] See, for example, Nakahara (2011).

[8] Customers here refers to the next step of the supply chain – the company that assembles the PCB with components.

[9] Panel sizes are cut from the maximum width of the woven E-glass cloth that is used to make FR4 laminate. These plants work on imperial gauge systems with commonly a 48″ width – hence two 24″ width PCB panels can be cut from the glass cloth.

[10] Strictly speaking, the imaging is usually done in pairs: the top and bottom sides of the material in one step.

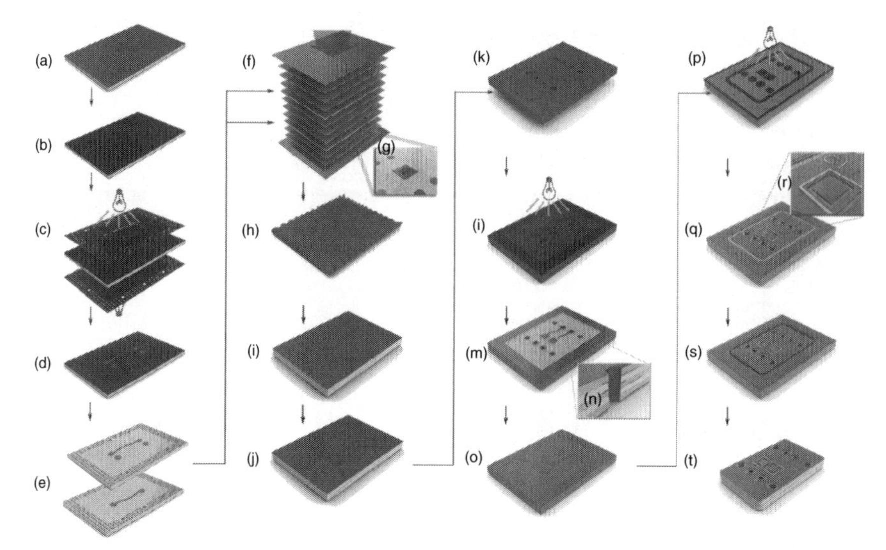

(a) The primary starting material for multilayer PCBs is copper-clad laminate *core*. This consists of two sheets of copper foil (generally of thickness 9–70 μm) separated by an insulating dielectric of woven glass and resin composite.
(b) Both surfaces of the core(s) to be used in the multilayer PCB are coated with a UV sensitive photo-imageable etch resist.
(c) The coated cores are then photolithographically imaged using either conventional UV lamps with a pair of large-format masks called a phototool, or using laser direct imaging on each side. The process of exposing the etch resist to UV light hardens the resist on the core.
(d) The imaged cores are then *developed* to remove the unexposed resist and then etched to form the desired circuit pattern. After etching, the hardened resist is removed in a strong alkali stripper.
(e) The etched cores are then scanned for defects using automated optical inspection (AOI). This process compares the etched traces to the CAD image. In the event of defects, the cores may either be repaired or rejected. It is important to note that a single multilayer PCB will contain many such cores – commonly 2–6, and perhaps as many as 40. Each of these cores undergoes the processes described here.
(f) The stack of cores that make up the PCB is interleaved with sheets of uncured glass–resin composite, termed *prepreg,* and capped the on top and bottom sides with a sheet of copper foil that will become the outer layers of the finished PCB.
(g) In order to align these multiple cores in the subsequent bonding or lamination process, an optically aligned post-etch-punch (PEP) is used to create four or more alignment slots.
(h) The stack is pressed under vacuum at a temperature of around 180°C (dependent on the resin system used) for perhaps 1–2 hours.
(i) After removing from the lamination press, the *now-multilayer* panels are trimmed of excess copper foil and resin, and readied for the subsequent drilling processes. An x-ray camera alignment system is used to view targets on the internal cores and drill new tooling holes in each panel.
(j) The panels are aligned to the tooling holes, and the required drill pattern is completed.
(k) After drilling, the panel is treated with a chemical dip process (*termed electroless copper*) to coat all exposed surface with a thin conductive layer of copper. Depending on the method employed by the PCB manufacturer, the entire surface area of the panel may then be electrolytically plated to its full and final copper thickness (termed *panel plating*), or, alternatively, the panel may undergo a process called *patter plating* where the final copper plating is applied selectively to exposed areas only.
(l) The panels are coated with UV-photosensitive resist and imaged on both sides using UV exposure or laser direct imaging.
(m) The panels are then stripped of resist and etched to form the final conductive pattern. Again, AOI is often used to eliminate defects and identify areas that may be repaired.
(n) The simplified cross-section shows the conductive 3D interconnection that is formed by the plated via (drilled hole).
(o) The electrically functional, but not yet complete, panel is coated on both sides with a liquid soldermask (an epoxy or acrylic-based polymer – often green in colour).
(p) Soldermask is another photolithographic process that requires imaging using phototools or, less frequently, LDI.
(q) Subsequent to the soldermask process, the panel is treated with a metal coating to ensure solderability of the finished PCB. Thereafter, the component markings or *legend* is applied by selective screen print, flood screen coating and photolithography or, more recently, inkjet printing of these often white or yellow legend markings.
(r) The close-up image here is shown to highlight one of the advantages of a digital legend printing process: it is, of course, essential that the legend markings are kept close to (but entirely clear of) the solderable pads. The camera alignment systems employed by the inkjet legend printers allow a greater level of accuracy than screen printing.
(s) The completed circuits are then mechanically routed to remove them from the manufacturing panel. In most real cases, the circuits are small in comparison to the panel and the panel will therefore contain many individual circuits.
(t) The circuits are then electrically tested and visually inspected.

Figure 8.2 *(a–t) A simplified schematic of the conventional manufacturing process for a PCB.*

Figure 8.3 *A PCB panel. The graphic shows a 'selling unit' (ready for component assembly) of four mobile phones.*

There are currently two predominant imaging techniques used in PCB manufacture: (i) photolithography using phototools and (ii) LDI.

8.4.1 Imaging Using Phototools

Phototools used for the imaging processes are essentially large sheets of (~700 mm × 600 mm) photographic film. Phototools themselves are imaged using (predominantly) cylindrical laser drum scanners. In a drum scanner, the unexposed film is wrapped around the drum and a galvanometer-driven laser scans the drum exposing the photosensitive film at very high speed as the drum rotates. Laser drum scanners have a very high native resolution – low-end systems operate at a few thousand ppi (pixels per inch) and today's state-of-the-art machines are capable of 50 800 ppi (e.g. 0.5 μm pixels).

After being exposed on the drum plotter, the phototool is then developed in a familiar photographic process. The resultant film – containing clear and dark areas – is then able to be used as a UV mask for the *PCB imaging processes*. The film is itself a multilayer structure consisting of a polyethylene terephthalate (PET) base with various interface layers and a silver halide in a gel. As such it is extremely prone to dimensional distortion caused by change in temperature, absorption of moisture and mechanical stress. PCB imaging processes using phototools are *analogue* printing methods that are susceptible to off-contact printing, registration errors, damage and other defects.

Manufacturing of a phototool is a machine- and labour-intensive process. The manufacture of a single phototool can cost in excess of $12 and most PCB facilities have a dedicated department specifically for the production of phototools. In the example in Section 8.4 where we need 18 tools for the one PCB design, that equates to a tooling outlay of around $200 per job.

Also, each phototool is unique to that PCB design and cannot be used for any other purpose. So, whether we use those phototools to make one or several hundred PCBs, the tooling-cost burden is the same. Moreover, a phototool has a finite life due to scratches, distortion and other defects, and with multiple-batch manufacturing it is often a serialisation requirement that each lot uses a new set of phototools that identifies the lot number.

8.4.2 Laser Direct Imaging

Within the last 5–10 years, LDI has become an efficient and effective replacement for phototools. LDI machines do away with the drum plotter and flat-scan the desired single-layer pattern directly onto a UV-sensitive etch resist on the PCB panel, thus negating the need for a phototool, increasing registration accuracy and eliminating some imaging defects. Using a digital method like LDI also enables serialisation to be easily performed.

The primary barrier to more widespread adoption of LDI is the cost of the machine – current estimates range up to $700 000 per machine. However, having said that, many PCB companies have proven that they can justify the expenditure against cost savings for reducing or eliminating phototool manufacture. As of 2012 industry sources indicate that close to 700 systems have been sold worldwide.[11] Whilst older LDI systems were criticised as being slower than manual printing using phototools, the fastest LDI systems in operation in 2012 can operate at an equivalent speed – around two panels per minute.

LDI, whilst a digital process, is not an additive process as it still relies upon etch resist being applied to the panel in a prior process. Also, the UV laser is a consumable item and needs replacing as it degrades over time.

It is perhaps clear from the description given here that the various imaging processes are amongst the highest cost and greatest overhead elements of the manufacturing process. It is in these areas that inkjet brings the greatest potential for process simplification and time and cost saving.

8.5 PCB Design Formats

Electronic design formats for PCB manufacture are essentially vector-based. Simply put, most features on a PCB design layer will consist of either (i) *a draw* – a start point, an end point and an element width, or (ii) *a flash* – a single shape defined at a given position.

The most common data formats used for PCB manufacturing are Gerber, DPF, ODB++ and to a lesser extent DXF, the former being the long-standing industry standard. The Gerber (vector-based) data format makes for a small data file but contains no intelligence about the design itself. ODB++ and its related standard IPC-2581 are a format that is better suited to complex designs that can also embed information about the intended interconnections within the design; this can be very useful for design for manufacture (DfM) and CAD-to-CAM (computer-aided design to computer-aided manufacturing) data exchange.

Being vector-based means that there is no inherent 'resolution' in the data and so the design could be scaled according to any desired output size. However, when an electronic design is being created within a CAD system it is usual for a snap-grid to be set that facilitates the easy placement of components and routing of tracks. With many components designed to a metric standard footprint, it is common to use a metric

[11] Industry information courtesy of Dr H. Nakahara (2012).

grid. With such a grid, tracks of exactly 100 µm width could be placed exactly 100 µm apart. Such a placement grid does, of course, affect the placement of devices that are defined in the imperial system – for example, a 20 mil (=20/1000 in.) pitch integrated circuit package.

With high-resolution plotter and LDI output, it is unlikely that the PCB manufacturer need be concerned about metric-to-imperial (or vice versa) conversion. However, as we address in this chapter, for low-resolution print systems such as inkjet this can become a significant concern.

The photo-plotters that we have described deliver a resolution as high as 50 800 ppi, which is equivalent to sub-micrometre pixel sizes. LDI generally operates at lower resolutions, but still provides in excess of 8000 ppi, which is equivalent to pixels smaller than 3 µm.

The PCB designs themselves are vector-formatted, and so a conversion from vector to raster format – or raster image processor (RIP) engine – is used in the data pipeline from a CAM station to a laser plotter or LDI system. Such RIP engines are highly efficient and cope very well with the conversion to a finite (but very small) pixel grid. Therefore, except perhaps for the production of ultra-high-density substrates,[12] it is unusual for PCB manufacturers to ever need to be concerned about line acuity and 'staircasing' of non-orthogonal lines, an area where inkjet practitioners will already be very familiar with the pitfalls of working at much lower resolutions.

Whilst the high resolution of phototools and LDI enables the imaging of traces as fine as 10 µm in width, the vast majority of mainstream PCB imaging capacity could be adequately covered by a system that is capable of reliably and accurately manufacturing tracks and gaps that are greater than 100 µm; it is in this region that inkjet is potentially able to offer a viable alternative. Any successful system for inkjet printing of PCBs therefore has to incorporate intelligent RIP engines that cope well with the lower resolution of the inkjet head together with multiple inkjet heads and/or scanning arrays to provide a reliable and fast system capable of printing such features.

8.6 Inkjet Applications in PCB Manufacturing

8.6.1 Introduction

The primary process and cost advantage of inkjet is that it is *both* a digital technique *and* an additive technique. Being digital negates the need for analogue phototools, allowing very fast set-up and allowing inline linear and non-linear image manipulation to significantly improve feature registration and product accuracy. Being additive means that materials are utilised only in the areas of the PCB where they are directly used for the process or design.

Conventional PCB manufacturing, in contrast, is mostly analogue and subtractive. For example, the patterning method for the conductive connections is that of photolithography of UV-sensitive etch resist on double-sided copper-clad laminate materials. This core is

[12] The manufacture of chip carriers – the small PCB that interfaces directly to the silicon die and has a ball grid array (BGA) or other interconnection structure on the back side – uses essentially the same manufacturing processes as have been described in this chapter. However, these devices have conductor dimensions of 15 µm or smaller.

then etched leaving only the track features that are needed by the design ('subtractive'). On average more than 65% of the starting copper surface area is etched away, with the copper going into chemical solution. In addition to being costly to recover, copper is also toxic to the environment so handling costs are substantial.

It is also important to consider that the etch resist (a photosensitive polymeric material) is also an entirely sacrificial material in that it must be removed from the PCB. Both the 'developed-off' and 'stripped-off' etch resist materials go into the waste stream. These processes are therefore inherently energy-intensive, producing significant waste streams that must be carefully managed. It has been calculated that in the United Kingdom, the cost of environmental compliance and water alone can account for 8% of PCB manufacturing costs (Ellis, 2001).

With inkjet printing we can, in principle, massively reduce the manufacturing cycle time, cost and waste.

In its fullest implementation, an inkjet-based PCB would have its conductive tracking pattern directly printed onto a base dielectric substrate. An insulation layer would then be over-printed using an inkjettable dielectric, leaving apertures for any interlayer connections (thus replacing the drilled and plated holes used in conventional manufacture). Subsequent tracking layers and dielectric layers would follow in an iterative build-up manner. As of the time of writing, this completely additive manufacturing method remains at the proof-of-concept stage and has certainly not yet made it from the laboratory to the PCB factory. PCB manufacturing is a conservative industry that is, for many very good reasons, firmly wedded to photolithography.

8.6.2 Legend Printing

There is one PCB manufacturing process – albeit one that is more graphic art than electronic – where inkjet is already in wide and fast-growing use: the printing of the *legend* or component identification layer as indicated in Figure 8.2q.

The legend layer has four main purposes:

1. to identify electrical connection points to the end user or for repair
2. as a cosmetic layer or for branding with a logo or similar: the legend has high contrast to the underlying materials
3. as a light reflecting material near to light emitting diodes (LEDs) and other light emitters
4. for serialisation – identifying the manufacturing lot.

Screen printing is the prevailing method used currently. Now, because the manufacturing process for the 'screen' itself also uses photolithography, that makes screen printing a high-cost process. Recent estimates for the cost of making a screen range up to $75 each. In addition, because screens are challenging to manufacture, they need skilled labour and often have a short working life. If a screen is used for serialising the lot number or date code, then the code pattern itself must be highly simplified. So human-readable and barcoded serialisation with screen printing is almost impossible.[13]

[13] Some (often smaller) facilities use photo-imageable legend resist which is applied by screen printing over the entire surface area. The required areas of the legend text are then imaged using a phototool, and the >95% of unexposed ink is then removed in a developer.

Within the last few years, a number of PCB equipment companies have successfully launched inkjet 'legend printers'. These machines have now reached widespread use, especially for PCB manufacturers making small-lot quick-turnaround production, where the cost and time for making the screen can be avoided by the use of digital inkjet techniques. In addition, the added benefit of improved registration and serialisation adds to the return-on-investment argument. Although legend printing is an important part of the PCB manufacturing process, it can still be regarded as a graphics application: the material deposited is not electronically functional, and acceptance of the printed layer is defined only by its legibility and visual quality.

Legend printers available today utilise either single-head or multiple-head arrays offering on average 300–720 dots per inch (dpi). Table 8.1 indicates the pixel pitch distance (which relates to the print-head nozzle pitch as explained in Chapter 2) for the stated dpi.

Of course, printed drop volume and drop formation on the substrate play a very important factor in forming a legible printed feature. Smaller drop volumes, particularly with heads that support multiple subdrops (often termed greyscale heads – see Chapter 2), will give finer printed detail. Even with relatively low resolutions coupled with somewhat large drops from binary print-heads, it is true to say that the achievable printed acuity is sufficient for the task.

Table 8.2 lists some information that is publicly available on current legend printers, and Figure 8.4 shows an example of a PCB with inkjet-printed legends.

Machine integrators will generally use industrial inkjet heads from the major inkjet original equipment manufacturers (OEMs), for example Fujifilm Dimatix, Konica

Table 8.1 *Comparison of dpi versus pixel size.*

dpi	300	360	600	720	1440
Pixel size (μm)	84.7	70.6	42.3	35.3	17.6

Table 8.2 *Comparison of inkjet legend printers available in 2012.*

Manufacturer[a] information	Microcraft[b] JetPrint range	Orbotech[c] Sprint	Printar[d] LGP and GreenJet
DPI	360–2400 configurable by application	Selectable, for example 720 dpi	375 and 750
Number of heads	Configurable – up to 6	Configurable	Up to 6
Drop volume (pl)	4/14/10 and 30 pl	Not stated	Not stated
Stated minimum feature (μm)	50 (for 4 pl head)	75	100
Minimum text height (mm)	0.3	0.5	0.5

[a]Information taken from manufacturer's web sites.

[b]http://www.microcraft.jp/en/lineup/index.html (accessed July 2012).

[c]Orbotech information as stated in http://www.pcb007.com/pages/columns.cgi?clmid=26&artid=22128 (accessed July 2012).

[d]Printer was acquired by Camtek in 2009. Specification available is at http://www.camtek.co.il/products/pcb/greenjet (accessed July 2012).

Figure 8.4 *PCB showing white inkjet-printed legend layer (Courtesy of Microcraft Inc.).*

Minolta, Kyocera, Seiko Epson, Toshiba, Xaar and the like. The machine integrator will almost always design their own ink delivery system that allows for optimal control of vacuum and ink temperature and viscosity.

One of the primary benefits that a digital process can bring is individual serialisation during manufacturing. This is already used to identify the manufacturing lot, but it can be used for a more rigorous identification of the part in manufacture, for example to allow the end user to trace the individual circuit back to the panel on which it was made. In this case 2D data matrix barcodes are used, as shown in Figure 8.5. Serialisation is not limited to readable markings on the outer layers of the PCB. In a digitally printed PCB, each of the inner layers could also be uniquely identified by serialisation.

In common with all industrial inkjet applications, the inkjet platform uses a tightly specified combination of inkjet head and ink. So the platform vendor will specify and control which inks may be used within their system. Most legend inks used in such systems are UV-curable, and the UV lamps and/or pinning units are included as core features to allow printing and curing to take place in the same machine.

It is important to note that the PCB industry, especially when driven by end users in safety-critical applications such as aerospace, automotive or military, can be extremely conservative and the materials used for manufacturing PCBs are under tight control of both the end customer and industry bodies such as Underwriters Laboratories (UL). Therefore one of the hurdles that early adopters of inkjet legend printers had to overcome was a reluctance to use a legend material that was *different* from the screen-printed materials in use for many years. This situation improved greatly once the major PCB industrial chemical suppliers started to provide inkjet formulations of their legend inks.

Figure 8.5 *Example of 2D barcodes printed with an inkjet legend printer (Reproduced with kind permission from Microcraft Inc. Copyright (2012) Microcraft Inc.).*

This growing acceptance of inkjet legend printing has been a vital step in moving towards digital inkjet methods for replacing photolithography for both soldermask and perhaps eventually for conductive track formation.

As of 2012, there are estimated to be more than 340 inkjet legend printers in operation.[14] When this is compared with the total surface area of legend-printed PCBs in Table 8.3, it can be seen that there is significant room for growth of machine systems should inkjet continue to prove a robust replacement for screen printing.

Table 8.3 *Estimate of the number of inkjet machines required to legend-print for 2010 global production volumes.*

Global PCB output (m^2)	200 000 000[a]	Per year
PCB panels per m^2	3.50	Per panel
Percentage with legend	75	–
Percentage with double sided print	60	–
Prints	630 000 000	Prints
At 60 prints per hour	10 500 000	Printing hours/year
Hours in a year	8760	–
Machine efficiency	75%	–
Machines required	1600	Units
At 6 heads per machine	9600	Inkjet heads

[a]Estimate by the author based on published information from NTI (Nakahara, 2011) and market value of $50 billion for 2009–2010.

[14] Estimate by author based on discussions with major platform manufacturers and industry sources.

8.6.3 Soldermask

Soldermask is the name used for the (often green) epoxy or acrylic material that gives PCBs their distinctive colour. The primary purpose of the soldermask layers on a PCB is to define (i.e. leave exposed) the solderable pads where components will be attached. At the same time, the areas of the PCB which are covered are insulated and protected from electrical misconnection and damage. It may appear that soldermask is therefore another 'graphic' layer. However, the requirements for extremely tight alignment accuracy, coupled with the fact that the deposited thickness of soldermask material is critical to maintain electrical isolation, make soldermask application by inkjet a significantly greater technical challenge than that of legend printing (Figures 8.6).

Whilst inkjet-applied soldermask is challenging, there are very real cost reduction and allied commercial arguments for its adoption. The conventional soldermask process is a subtractive photoimaging-based process that is labour- and energy-intensive. Soldermask material is a thick deposit (around 50 μm thick), and UV-imaging and post-curing require substantial energy. The threshold energy density for conventional soldermask is often up to $1\,\mathrm{J\,cm^{-2}}$. It is worth noting that LDI is also used for limited applications but this requires a specially photo-sensitised soldermask often with a $200\,\mathrm{mJ\,cm^{-2}}$ threshold. But even with an 'LDI soldermask resist' the energy requirement is high, and therefore LDI of soldermask is significantly slower than LDI imaging of photo-resist.

A few machine integrators[15] have offered direct inkjet soldermask printers, but the only company still active is Printar (now owned by Camtek, Migdal Ha'emek, Israel) with

Figure 8.6 Examples of soldermask clearances. Typically the designed distance between copper-pad and soldermask is 50 μm or less (Reproduced with kind permission from Microcraft Inc. Copyright (2012) Microcraft Inc.).

[15] Some early manufacturers have now left this market or ceased trading.

their GreenJet system. This system uses a bespoke soldermask ink. Inkjet soldermask printers use comparable platforms, inkjet print-heads and ink systems to those employed for legend printing, and so have similar print specifications. The manufacturer of the GreenJet system has stated at an industry forum that the process time can be reduced from 2 hours or more in a conventional system to around 2 minutes in a fully digital system.

Soldermask is by its nature applied to a heavily textured surface with areas of 'raised' copper tracks and holes in the PCB, and therefore directly applied inkjet may not be suitable in all cases. As discussed in Chapter 2, jettable inks generally need to have a low viscosity, one that is some orders of magnitude lower than conventionally applied soldermask materials. With a low-viscosity material, it is difficult to maintain coverage on raised copper areas, possibly leading to concerns over dielectric breakdown. In addition, for a growing number of PCB designs it is necessary to have soldermask that fills holes and planarises the surface. An inkjet-printed material cannot achieve this in the straightforward manner achievable with a conventional screen-printed soldermask. It is therefore the case that directly applied inkjet soldermask, whilst showing large potential process savings, does not currently have quite the same level of synergy with the PCB industry as inkjet legend enjoys. Moreover, the conservative nature of the PCB industry has made it rather challenging to gain universal customer approval for a bespoke soldermask ink.

However, there is still a case for the digital processing benefits that inkjet can bring to bear.

A method that is gaining critical attention at present is a hybrid soldermask inkjet imaging method. In this approach (Sutter, 2005) the soldermask is applied in a conventional manner: the PCB panel is pre-cleaned, then coated uniformly with the 'standard' solder mask and pre-dried. A vision system then captures fiducials on the panel and aligns the panel and printed image with greater accuracy than is often achieved by conventional means.

The inkjet printing machine then prints a UV-blocking material onto the panel, thereby defining the pattern required for the soldermask. The board is then exposed to UV radiation with the inkjetted layer now becoming a sacrificial mask. The panel is then *developed* to produce a finished board with soldermask openings or pads. A key advantage of this hybrid approach is that conventional soldermask coatings are used on the board, so there is no need to change or re-qualify a new soldermask material or to reformulate it to be jettable in an inkjet head. This method could be termed a inkjet-printed UV block system.

8.6.4 Etch Resist

8.6.4.1 Introduction

The majority of inkjet system manufacturers targeting the PCB industry are directing their efforts at converting PCB manufacturers from screen-printed legend to inkjet legend. However, as we outlined at the beginning of the chapter, the area where the greatest potential benefit can be achieved is by direct application of etch resist by inkjet.

Using an inkjet system for the patterning of etch resist is similar in principle to the methods explained in this chapter but with one key difference: here the deposited inkjet image is now responsible for directly defining the copper tracking pattern, the electronically functional part of the board.

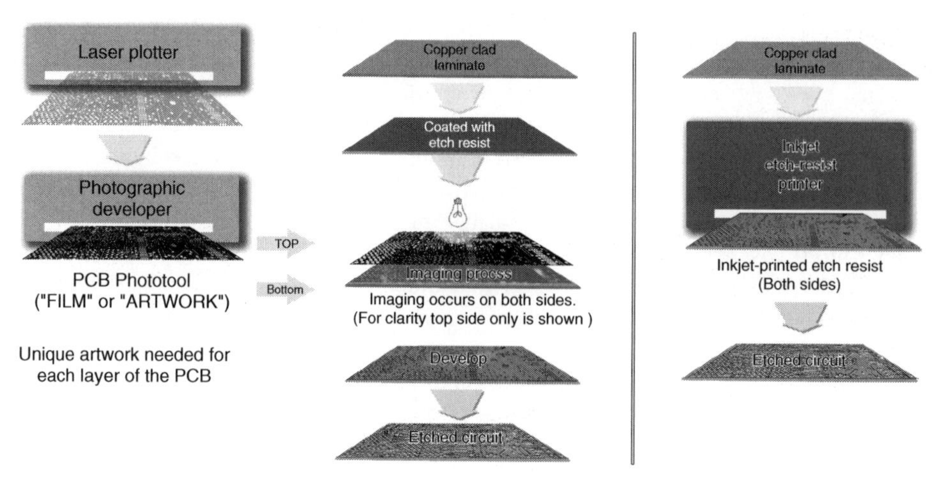

Figure 8.7 *Inkjet-based PCB etch process: comparing (a) the conventional manufacture of a double-sided core (five process steps) with that of (b) inkjet methods (two process steps).*

Figure 8.7 shows how inkjet application of etch resist can significantly reduce and simplify the process by eliminating phototools. In this case the etch resist is applied in only the exact pixel-by-pixel areas where it is required for the design so the quantity of etch resist deposited is a small fraction of the material that is conventionally applied in sheet form. In this example the entire copper etching (and recovery) processes remain unchanged. And likewise, the end product is identical in material and structure to a conventionally made PCB, so no new approval will be required by the customer.

Over the past few years the development of inkjet etch resist systems has proven to be significantly challenging, and a few 'false starts' have beset the adoption and reputation of inkjet in PCB manufacture. Of course, the primary difficulty with inkjet is that any defect in the printed image will manifest itself as a defect in the printed track: a potential open or short circuit. It has to be remembered that the inkjet process (heads, materials and platforms) had roots firmly in the graphic arts industry, and whilst a single missing or misplaced drop on a photographic image is unlikely to be seen, a *single* misplaced drop or misdirected nozzle in an electronic application will cause expensive scrap product. It is recognised in the inkjet industry that any inkjet nozzle will eventually fail, perhaps only momentarily due to air entrapment or a similar transient problem. It is this type of defect occurring for a short time that is most problematic. With this in mind, a number of inkjet head manufacturers have started to build 'failure prediction systems' into their heads. Such systems use redundant head arrays and monitor the electric or acoustic characteristic of each nozzle *whilst it is firing* in order to take the offending nozzle out of operation before it starts to misfire. Océ, an inkjet head manufacturer working with the University of Twente in the Netherlands (Jeurissen, 2009), has demonstrated a system which determines a potential nozzle failure many thousand drops before the nozzle would fail critically.

It is, of course, a very difficult proposition to guarantee that hundreds of billions of deposited drops of etch resist per hour will *all* be in the position and quality that are needed to ensure a good product. Because of this challenge, at the time of writing, there are few companies actively developing etch resist systems, and only two companies have launched commercial etch resist systems that have gained adoption in leading PCB facilities.

In the following list we state some of the key attributes of an etch resist, whether inkjet or conventionally applied. An etch resist:

- can define a pattern on a surface with suitable acuity;
- exhibits good adhesion to the surface and so protects the area beneath the resist during the chemical etch process yet is able to be removed completely from the surface by a chemical stripping operation;
- is robust enough to survive typical handling after printing and during the etching and stripping process;
- is chemically compatible with the process lines and waste streams;
- is cost-effective.

The PCB industry is used to dealing with dry film UV 'photoresists' which offer extremely high performance and are cost competitive. As indicated in Table 8.3, the global output of finished PCBs is around 200 million square metres per year. If one estimates the average layer count for a PCB to be six (and growing), then the total surface area of etched copper tracking and, therefore, etch resist is heading towards 1 billion square metres per year. The global infrastructure needed to support such a volume of material is therefore substantial.

A viable system for deposition of etch resist by inkjet has to be extremely capable to compete in this very cost-sensitive marketplace but, as is clear from the enormous surface area of the printed image, the potential market for systems, inks and heads is potentially very large. It is important to note also that whilst the PCB industry is only now starting to embrace inkjet etch resist as a possible method, the development path is not entirely un-trodden, as the solar industry has been using inkjet etch resist for some years now and the requirements of the two industries are similar in many ways. Table 8.4 outlines some of the main factors that define the needs of an inkjet etch resist system.

8.6.4.2 Print Resolution and Acuity

Whilst resolution is an important consideration for legend and soldermask, it is absolutely critical for the printing of etch resist. As noted above, the PCB industry is used to dealing with laser-defined images at dpi resolutions of many thousands. That is to be compared with current best-in-class inkjet processes that can be run at between 600 and 2400 dpi. Image quality even from a 'perfect' inkjet machine at such resolution will therefore show greater pixelation than the equivalent conventionally printed layer, as illustrated in Figure 8.8. However, because (i) the printed ink will exhibit spread on the surface and (ii) the chemical etching process will itself smooth some edges of features, the clear pixelation that is visible on a raw image is often not present in the final etched panel.

That having been said, inkjet is, and is likely to remain, a much lower resolution print system than the incumbent methods. Therefore some products will remain outside the capability of inkjet etch resist. Given current technology trends, it is likely that many tablet, laptop and smartphone PCBs and of course high-density chip interposer substrates will remain mostly, if not entirely, imaged by photolithography. In addition

Table 8.4 Requirements for an inkjet-based etch resist system.

	Performance requirement	Comments
Platform and print system		
Material handling	The system must be able to handle substrates from a few tens of micrometres to a few millimetres in thickness. The transport system will be scanning the panel relative to the inkjet heads at perhaps 1 m/s – the heads must be protected from striking the panel.	The thinnest PCB core material is thinner than 50 μm. Thicker materials, whilst more substantial, may exhibit bow and twist and may need to be flattened in order to print correctly.
Operation	The system must be able to print the required panels at suitable acuity, reliably and at high speed. It should be simple to operate.	The system will need to be robust and highly automated to complement other PCB process equipment.
Data pipeline	An inkjet RIP engine must be able to interface with existing CAM systems. Ideally the system should cope with dynamic data streams, for example barcode serialisation.	–
Registration	Image alignment to previously printed layers or to the back-side printed image is critical. A charge-coupled device (CCD)–based intelligent alignment system is required.	An ideal system will read fiducials on the already printed panel and dynamically scale the printed image with both isotropic and non-isotropic compensations.
Head maintenance	Minimising of head and ink 'maintenance' (e.g. purging, wiping, etc.) is critical as this activity, whilst essential, decreases process 'up time'.	Head maintenance should be automatic. The print-heads should ideally be protected from manual handling.
Ink delivery system	The ink needs to be easily replenished.	–
Repair	In the event of an inkjet head problem, the head should be replaceable.	On-site replacement is ideal – hence modular systems that have pre-aligned inkjet heads offer benefits.

Table 8.4 *(continued)*

	Performance requirement	Comments
Ink		
Etch resist 'ink' selection	Etch resist materials are selected by application and, for example, a different etch resist is used for inner and outer layer processing. In the PCB industry there are two predominant etch methods: (i) acid etch, for example cupric chloride or ferric chloride, and (ii) alkaline etch or ammoniacal etch. It is essential that the etch resist chosen can withstand the method in use. It is not an absolute requirement that the same material can withstand both extremes.	PCB manufacturers also have differing requirements of the ink in the stripping operation: an etch resist is chemically formulated to either break into small solid fragments that can be removed from solution by a cyclonic or mesh filter or be dissolvable in the strip solution.
Ink: UV cure or phase change	The liquid ink drops must be 'fixed' into position on the surface. The primary methods are (i) to use a UV curing method to 'pin' or cure the drops onto the surface, (ii) to use a phase-change (or wax-like) ink that sets on the surface on contact and (iii) to use a 'hybrid' ink that contains phase-change components together with monomers and UV photo-initiators hence being able to be hardened.	Current mainstream activity is focussed on hybrid inks as these offer the closest match in performance to the 'standard' process. Many studies have shown that phase-change inks are able to reduce printed drop diameter on a copper surface.[a]
UV curing	If a UV curing ink is used – is the ink fully cured *in situ* on the machine, or is a separate cure required?	This is a process and handling time consideration.
Ink adhesion to the surface and 'robustness'	It must protect the copper pattern beneath the resist during the chemical etch process whilst being able to be removed from the surface after etch by a stripping operation.	A careful balance of properties is required. It is essential that the ink can be removed from the Cu surface in the strip operation (generally a solution of NaOH or KOH).
Chemical compatibility	Management of waste streams for an inkjetted etch resist should be no more challenging than for a conventional etch resist.	–

(*continued overleaf*)

Table 8.4 *(continued)*

	Performance requirement	Comments
Substrate and handling		
Preparation of the substrate and substrate–ink interactions	The process should be compatible with conventionally used substrate preparation methods. Some UV-curing inks exhibit a high level of spread on a copper surface (due to the surface's high surface energy). In order to use these inks, it is necessary to coat the copper laminate with a specialised 'de-wetting agent' – this ensures that the ink drops maintain high acuity. Such coatings are commercially available.	In the PCB industry, the predominant preparations are mechanical scrubs (e.g. pumice and Al oxide) and chemical microetching. These processes deliver quite different surface topographies. It has been reported that brush-scrubbed surfaces can contains multiple capillary-like scratches which draw liquid UV-curable etch resist into the channels and may contribute to circuit defects.[b]

[a]Ink drop deposition and spreading in inkjet-based printed circuit board fabrication (Hsiao *et al.*, 2008).
[b]Jones, Chilton and Kasper (2008).

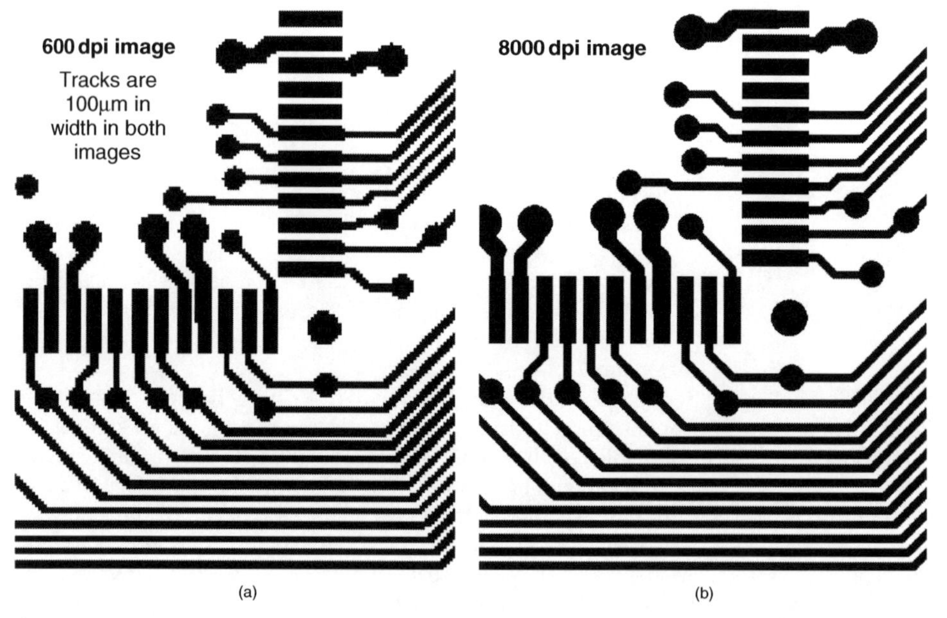

Figure 8.8 *Comparison of resolution between (a) a track image at an inkjet resolution of 600 dpi (42 μm pixel spacing) and (b) the same image with LDI resolution of 8000 dpi (0.3 μm pixel spacing).*

some high-frequency applications (antenna, waveguide and high-end radiofrequency (RF) PCBs, for example) will require photolithography to maintain the high resolution needed to minimise skin-effect deviations.[16]

The complexity of a PCB is primarily defined by its layer count and its track and gap. The track and gap (minimum pitch of printed conductors and gaps on the same layer) varies from around 200 μm for simple commercial and industrial applications to less than 75 μm for a high-end mobile phone. At a print resolution of, say, 1200 dpi, the native printed drop pitch will be 21 μm. Taking into account the spread of the drop as it hits the surface, this may take the effective pixel (drop) size to some 20–50 μm. So whilst printing 50 μm track and gap with such a system may be overly challenging, it is feasible to print tracks of say 100 μm with such a system.

In any application, it is the substrate–ink interaction that defines the printed drop size whilst the inkjet head defines the single-pass dpi. By using multiple heads, or by microstepping[17] the panel position with multiple passes under the head, the platform can of course deliver higher effective dpi. However, because inkjet is an additive method, each extra printed pass adds another layer of material which is not always beneficial, especially if the print-head is depositing relatively large drops. In some cases it may be necessary to consider the use of greyscale or drops per drop (dpd) print-heads as discussed in Section 2.6. Using a system that allows the ejection of variable volumes of ink can give better control over the size of the drop on the surface and thereby improve the printed resolution. In such cases the RIP engine will become more complex, but the benefits in acuity will be substantial.

Figure 8.9 shows the benefit of a dpd system in an electronics application: in this case, the gap between connectors is maintained by use of a dpd inkjet head.

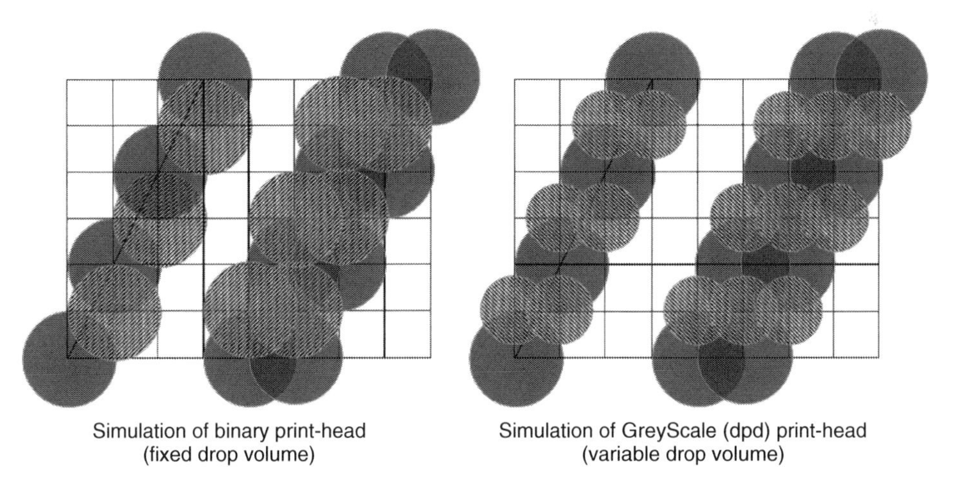

Figure 8.9 *Comparison of printing from binary and greyscale (dpd) heads at the same resolution.*

[16] At 1 GHz, the skin depth in copper is around 2 μm. At 10 Ghz, it is just 0.65 μm. So conductor surface 'meandering' has a significant effect on the path length at high frequencies.

[17] Microstepping refers to offsetting or rotating the print-head or panel to enhance effective print resolution.

8.6.4.3 Commercial Implementation of Inkjet Etch Resist

The foremost etch resist systems currently commercially available are from Microcraft and Mutracx. Microcraft have long experience in PCB legend systems. Mutracx (http://www.mutracx.nl/), a spin out from Océ, was started exclusively for inkjet etch resist systems.

State-of-the-art inkjet etch resist systems use a robust, high accuracy base platform similar to that used for PCB drilling or legend printing, and so offer a proven user interface and panel handling system. The drop-positional accuracy for such a system is of the order of 10 μm. For etch resist applications the inkjet heads, ink delivery system, panel curing and software must be carefully optimised for the application.

As has been covered elsewhere, the ink–surface interaction is crucial to good print quality and so a critical enabling factor for this application is the ink that is used. The two systems noted here have many differing characteristics but both share one common feature – they use inks that can be termed hot melt or phase-change. These inks are semi-solid pastes at room temperature but melt at an elevated temperature, achieving the viscosity required for jetting at a head temperature of between 60 and 95 °C dependent on the formulation. These inks jet as a liquid, but set almost instantaneously upon hitting the room-temperature PCB panel. In addition, it is also possible to add cross-linking to the hot-melt ink – thereby creating a hybrid UV curing phase-change ink. In such a case a cured ink can attain a good pencil hardness of 3H to 6H when cured. The key advatage of such phase-change inks for the PCB industry is that the copper foil substrate itself is not by any means defect free; and such a printed ink is much more forgiving than a UV curing liquid that can run into surface defects and scratches.

An inkjet etch resist, especially when coupled with the 'inkjet-printed UV block' soldermask process and legend print, can deliver very substantial saving over the existing time-consuming and costly analogue processes. This is especially so for a company newly entering the market: that is, one that does not have incumbent photolithographic equipment installed. In such a scenario it is becoming possible to imagine a single inkjet platform able to perform most of the imaging steps necessary for PCB manufacture.

It is not in any way realistic to expect that inkjet will take over from the already very capable and cost-effective photolithography-based methods in use globally.

In fact, if a manufacturer is making many tens of thousands of the same products daily there are few advantages to digital printing. However, it is clear to the author that inkjet does have a role to play, particularly in rapid-prototype and fast-turnaround PCB applications, although it remains to be seen how long the transition will take and what fraction of the market will adopt it.

8.7 Future Possibilities

This chapter's description of the benefits of inkjet etch resist has one caveat: the process described is, as it has been for more than 50 years, still entirely subtractive. We remove most of the copper that we start with by an etching process. In an ideal 'additive digital world' we would deposit both conductive tracks and all dielectric materials by inkjet or another digital print method. As described in Chapters 4, 7 and 11, the printing of nano-particle inks containing metals such as silver, copper and gold is now a common staple

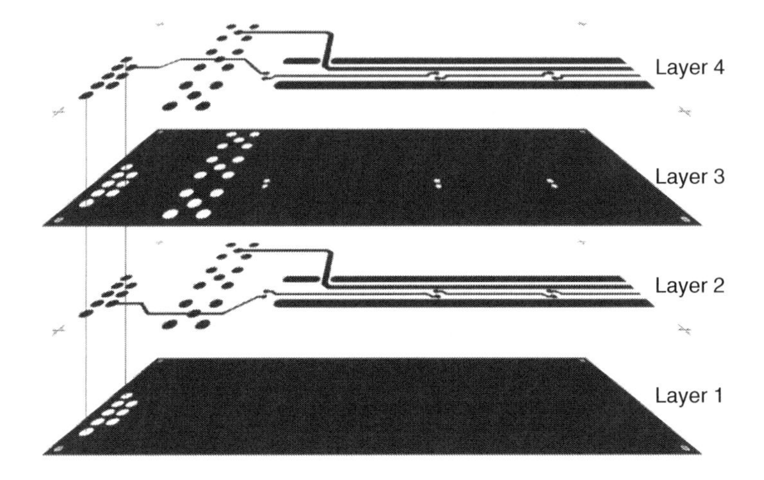

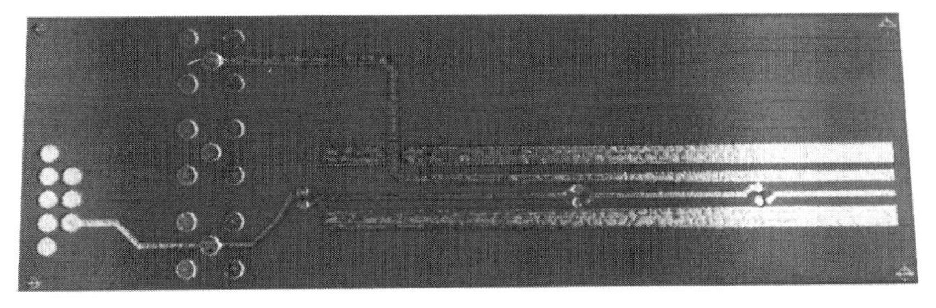

Figure 8.10 *An early multi-layer structure made entirely by inkjet. Layers 2 and 4 are nano-silver; layers 1 and 3 are inkjettable dielectric. The entire structure is less than 0.1 mm in thickness.*

of the inkjet electronics community. An example of a multilayer all-inkjet *PCB-like structure* is shown in Figure 8.10. The author is not alone in demonstrating applications where single- and multiple-layer circuit structures have been formed on paper, polyimide, FR4, plastic and other substrates whilst being cured *in situ* at low temperatures. All of this is interesting, 'benchmark changing', and potentially valuable, but the circuits fabricated in this way are not in the strictest sense PCBs as the (somewhat conservative) industry would recognise them. That primary benefit of inkjet of being able to print onto a flexible substrate adds little merit if the PCB in question needs to provide a 3.2 mm thick thermal substrate and robust mechanical mounting. The nano-metal conductive tracks deposited by inkjet are typically a fraction of a micrometre in thickness (less than 1% of the copper thickness of a standard PCB layer) and so cannot carry substantial current. The PCB industry also has a long-held mistrust of silver conductors as they are prone to electro-migration and dendritic growth.

Above all, whilst the price of nano-metallic inks will surely reduce as (i) demand increases, (ii) more suppliers bring out competing products and (iii) manufacturers are able to use lower cost metals, nano-metallic inks today remain extremely expensive. A number of groups have introduced nano-copper inks as a potential cost-effective inkjettable conductor material, but it remains a challenge to sinter these inks in a cost-effective manner that would be applicable to large-format manufacturing on the scale of a PCB operation.

In truth, it is impossible to imagine a world where the PCB as it is made today will be entirely replaced by the inkjet-printed alternative. Thankfully, however, that is not the end of the story. Firstly, there is growing acceptance of *PCB-like interconnect structures* formed entirely on flexible paper or plastic substrates being used in areas such as product packaging and brand protection. These are not PCBs in a strict sense but they perform an almost identical role, and these end users do not have entrenched views on acceptable manufacturing processes and are keen to be early adopters, something the PCB industry naturally shies away from. Moreover, there are substantial benefits that a hybrid approach to PCB manufacture utilising additive inkjet can bring.

As will be clear from the discussion here, the metal of choice for a PCB conductor remains copper. It is well known and understood by the community. Now, whilst it remains industrially challenging to inkjet copper in a reliable manner, there is a commercialised approach that uses inkjet to define the tracking pattern *and* provides to the end customer a copper track of more substantial thickness. That process was developed by Conductive Inkjet Technology Ltd (CIT, Cambridge, United Kingdom). CIT uses a two-step process to create copper circuitry.

The first step of the process uses inkjet deposition to print the circuit pattern with a catalytic ink. This ink is then cured, often on a flexible substrate. The substrate is then immersed in an electroless plating solution[18] to chemically deposit solid copper onto the desired pattern. The copper thickness can be deposited to a few micrometres in thickness and so can carry substantially more current than a directly printed nanometal. The CIT process can deliver resistivity of around 30 mΩ/square, which is about 2% of a conventional PCB track and which, whilst being substantially lower, for many applications would be acceptable provided that the circuit designer uses suitable track dimensions and compensates for the voltage drop. At present, the commercially available system runs at lower resolutions than the etch resist systems described here, but there is no fundamental reason why higher resolution print-head systems could not be used if the application required. The process has been made available[19] for single-sided conductors printed onto flexible substrates such as PET or Melinex™, and its major advantage is for reel-to-reel processing. The system allows very cost-effective conductor patterns to be formed in a digital process with no hard tooling and minimal setup costs. Whilst the technology is of course suited to radiofrequency identification (RFID) and similar structures, there is potential for the CIT system or variants of it also to be used in more mainstream PCB applications. Whilst strictly speaking not a PCB process, in that it would not fit within current PCB manufacturing processes, the CIT methodology is a very important step towards the fully additive digital PCB. The process additively builds circuits, with ink and metal being deposited only where required by the design, resulting in a significantly smaller waste stream and environmental footprint than for traditional circuit fabrication processes.

[18] Electroless copper plating was developed primarily for the PCB industry and so is well understood by the PCB community.
[19] See http://www.inkjetflex.com.

To attain the benefits of digital additive fabrication fully, it is likely that a rather fundamental rethink of the design ethos will be required. As has been described, a conventional subtractive PCB process is able to deliver an electronic substrate that performs multiple roles (electrical, thermal and mechanical). However, the true cost of manufacture and the environmental impact of the multiple waste streams are substantial.

In a multilayer PCB each conductive layer adds considerable cost to the manufacturing process and, of course, the layer is present across the whole surface area of the PCB, whether a component connection to that layer is required in one small area of the PCB or across the entire surface.[20]

To fully realise a PCB structure that uses additive metal and dielectric deposition, it is useful to consider a hybrid approach where some of the benefits of the traditional PCB manufacturing method are retained and augmented by digital printing. We might call this approach a *vari-layer PCB*. Perhaps one of the easiest structures to imagine would be one where we use a conventional two-layer PCB with its copper tracking able to carry all the necessary current for the design. On top of that structure we additively build up the requisite number of layers to provide interconnection to the complex arrangement of components. Such a structure may use just two layers in one area of the PCB, such as an area that needs high power connections; these connections would be delivered by vias from the conventionally made PCB substrate. However, in localised dense circuitry areas of the PCB we may need 10 or more additive layers built by sequentially digitally printing conductor and dielectric layers (in a similar manner to Figure 8.10). Such an approach could use a single low-cost conventional PCB as the substrate from which many different digitally defined circuit layouts could be spawned.

The addition of directly printed components (resistors, capacitors, inductors and perhaps actives) into the PCB during its manufacture is another area in which inkjet can play a role – the technology required for that is covered in Chapters 7, 9 and 11.

In the author's view it is clear that inkjet has a role in PCB applications but in order to utilise it fully we may have to throw out some of the established manufacturing methodologies. That will take both time and the development of a new design mind-set, but it will surely be worth the effort.

References

Eisler, P. (1989) *My Life with the Printed Circuit*, Lehigh University Press, Bethlehem, PA.

Ellis, B. (2001) The printed circuit board industry: an environmental best practice guide. *Circuit World*, **27** (2), 24.

Hsiao, W.K., Martin, G.D., Hoath, S.D. and Hutchings, I.M. (2008) Ink drop deposition and spreading in inkjet based printed circuit board fabrication. NIP24: International Conference on Digital Printing Technologies and Digital Fabrication 2008, Pittsburgh, PA, Society for Imaging Science and Technology, pp. 667-670.

Jeurissen, R. (2009) Bubbles in inkjet printheads: analytical and numerical models. PhD thesis, University of Twente.

[20] A PCB is manufactured with n layers (usually an even number), and the number of layers necessitates a number of sheet materials in the stack-up. It therefore contains these n layers over its entire layer surface area.

Jones, S., Chilton, N. and Kasper, A. (2008) Development of industrial inkjet processes. Joint paper presented at EIPC Conference, Rome, January 24.

Nakahara, H. (2011) The Asian persuasion. *Printed Circuit Design and Fab*, 31 August, http://pcdandf.com/cms/fabnews/8280-nti-100 (accessed July 2012).

Sutter, T. (2005) An overview of digital printing for advanced interconnect applications. *Circuit World*, **31** (3), 4–9.

9
Active Electronics

Madhusudan Singh,[1] *Hanna M. Haverinen,*[2] *Yuka Yoshioka*[3] *and Ghassan E. Jabbour*[*,4]
[1]*Department of Materials Science and Engineering, University of Texas at Dallas, USA*
[2]*Solar and Alternative Energy Engineering Research Center, King Abdullah University of Science and Technology (KAUST), Saudi Arabia*
[3]*Heraeus Materials Technology, USA*
[4]*Department of Electrical Engineering, Department of Materials Science and Engineering, Solar and Alternative Energy Engineering Research Center, King Abdullah University of Science and Technology (KAUST), Saudi Arabia*

9.1 Introduction

According to various industry estimates, including the organic electronics roadmap, the size of the printed electronics industry is expected to exceed billions of dollars in the coming decades. A recent IDTechEx report (Das and Harrop, 2011) states that the market for printed and thin-film electronics will reach US$44.25 billion in 2021. Inkjet printing has the potential to contribute significantly in this rapidly growing area. We have seen in Chapter 7 how electrical conductors and passive components can be fabricated by inkjet methods; here we shall review the application of inkjet printing to active electronic devices. In this field, the particular merits of inkjet include the fact that it provides a noncontact, maskless additive patterning process, which can be used to deposit and pattern a very wide range of materials (Lewis, 2006). With the high costs of materials used in traditional lithographic patterning processes, the reduced material waste and scalability to large-scale manufacturing are further attractive attributes. Several reviews consider the process and its application in device fabrication (Bao, Rogers,

* Corresponding author. ghassan.jabbour@kaust.edu.sa

Inkjet Technology for Digital Fabrication, First Edition. Edited by Ian M. Hutchings and Graham D. Martin.

and Katz, 1999; De Gans, Duineveld, and Schubert, 2004; Sun and Rogers, 2007; Tekin, Smith, and Schubert, 2008; Arias *et al*., 2010; Singh *et al*., 2010).

The performance of active devices which involve the movement of charge carriers, whether they are light-emitting diodes (LEDs), photovoltaic cells, transistors, or other devices, depends critically on the use of specialized materials with reproducible and well-characterized properties, structured over small and precisely controlled length scales. For these devices to be successfully manufactured by inkjet printing, stringent control of the process is needed. In particular, a good understanding is of all stages of drop formation, deposition, and the processes such as drying or curing needed to form the final product.

Although, as reviewed in Chapter 2, several methods are available for drop generation, current inkjet printers for material deposition in active device fabrication employ piezeoelectric drop-on-demand (DOD) technology, and we will here focus primarily on applications of such methods.

Computational fluid dynamics (CFD) techniques using dynamic modeling (Lunkad, Buwa, and Nigam, 2007), finite element methods (Sprittles and Shikhmurzaev, 2012), measurements of contact angles on different substrates, viscosity, surface tension estimation with sessile drops, and so on have been used to analyze the flow processes and drop–substrate interactions. Such a combination of approaches is typically sufficient in determining the suitability of an ink recipe, supplemented by trial and error (as a "black art"). However, direct measurements of fluid flow patterns, such as a rheofluorescent technique applied to polymers in a shear field ((Hill, Watson, and Dunstan, 2005), offer another handle on this problem. This method involves the investigation of the fluorescence from poly[2-methoxy-5-(2′-ethyl-hexyloxy)-1,4-phenylene vinylene)] (MEH–PPV) in a Couette flow situation between two cylinders with relative rotation to create a shear field. As a result of the applied shear, the fluorescence peak of the fluid shifts and this effect is used to monitor changes in polymer segment length. Since the segment length is usually directly related to molar mass and viscosity and the shearing experienced in the process of drop ejection is replicated, estimates of shear stress can possibly be used to provide a direct visualization of inkjet print flow stress through suitable sensor placements.

Direct imaging offers a more direct measurement of ink flow patterns (Dong, Carr, and Morris, 2006). A DOD system can be used with a pulsed laser and a flash photography setup to obtain sufficient time resolution for direct imaging measurements of the drops. The processes of ejection and stretching of fluid, necking and pinch-off of liquid ligaments from the nozzle, recoil of the freely falling liquid ligament, breakup of the ligament, formation and breakup of the primary drop and satellites, stretching of the liquid drop, and so on were visualized for several different fluids (Tsai *et al*., 2008). A more recent similar study seeks to explain disparities between the first ejected drop and subsequent drops through optical means (Famili, Palkar, and Baldy, 2011). Another study has examined bubble formation via infrared imaging of the acoustic field and its contribution to clogging of nozzles (Van Der Bos *et al*., 2011).

It may be noted that the methods discussed here employ a more elaborate measurement system than is practicable for production units. While fundamental characterization studies for various ink formulations can be carried out, a more phenomenological and direct measurement on the effects of changes in ink parameters can involve the printing of inorganic nanoparticles and subsequent imaging to map the spreading of the drop (Perelaer *et al.*, 2008). These methods can be used in conjunction with a more recent direct model involving an equivalent length and optical measurements of meniscus positions to obtain a full picture of the printing process and gain predictive control over its stability (Moon *et al.*, 2008).

What happens after drop ejection determines the deposited layer morphology and, thus, device performance. The momentum of the drop is dissipated in the deformation and motion of the drop after it has the substrate, as is also discussed in Chapter 5 (Pesach and Marmur, 1987). As the solvent evaporates, the drop dries, depositing the solute at its final location. The time-scale of the drying process depends on environmental conditions, viscosity and concentration of the ink (Perelaer *et al.*, 2008), molar mass of the solute (Perelaer *et al.*, 2009), and the chosen solvent, and it can determine the nature of packing of polymers, molecules, or even nanoparticles (Lim *et al.*, 2008).

The surface energy of the substrate can have an important influence on the final quality of the print. Hydrophobic surfaces reduce the size of the drop, while hydrophilic surfaces do the opposite as confirmed by contact angle measurements that provide a direct picture of the detailed shape of the drop on a given substrate. Traditionally, plasma treatments have been used to modify the hydrophobicity of the surface prior to device deposition (Oh and Lim, 2010). Since the hydrophilic nature of the substrates is closely related to the occurrence of $-OH$ groups at the surface, prolonged immersion of flexible substrates in low-molecular-weight alcohols is also often of assistance in this regard to produce a temporary shift in contact angles. The printing of conductive metal nanoparticles on untreated substrates (Van Osch *et al.*, 2008) compared with treated substrates provides a stark illustration of this fact, as illustrated in Figure 9.1.

The use of pre-patterned substrates at multiple levels of processing in functional device structures complicates the fluid flow situation further. Modeling of drop interactions (which should depend strongly on the ratio of the size of the drop to the minimum feature size, among other factors) has been previously carried out (Khatavkar *et al.*, 2005) to determine the importance of wettability of pre-existing barriers to free lateral flow of the ink. The size of the ejected drop is one of the factors in determining the final print size on the substrate. Jetting from a smaller nozzle, while expected to lead to a smaller drop size, is more likely to lead to clogging, while fabrication of smaller nozzles requires the use of lithographic techniques, thereby increasing cost. Processes that can produce drops smaller than the nozzle diameter have been developed (Goghari and Chandra, 2008). In most lab situations, careful design of actuating waveforms in piezoelectric DOD printers can lead to significant reductions in drop size through control of the length of the liquid column ejected (Gan *et al.*, 2009). Real-time monitoring of an inkjet-printed film has been reported recently, which opens up the possibility of real-time feedback and control of film quality (Im, Sengupta, and Whitten, 2010).

As minimum feature sizes decrease, as for display backplanes, tolerances on line non-uniformities tighten. It is thus important to control the interaction of the ink and substrate through the coffee stain effect referred to in Chapter 5 (simultaneous drying of the solvent, changes in viscosity, lateral transport of the solute, interaction with the substrate, etc.). Depending on the nature of the ink, two contrasting approaches can be undertaken toward this end: slowing down the process of drying using surfactants, allowing more of the initial spatial ripples to die out (Hanyak, Darhuber, and Ren, 2011),

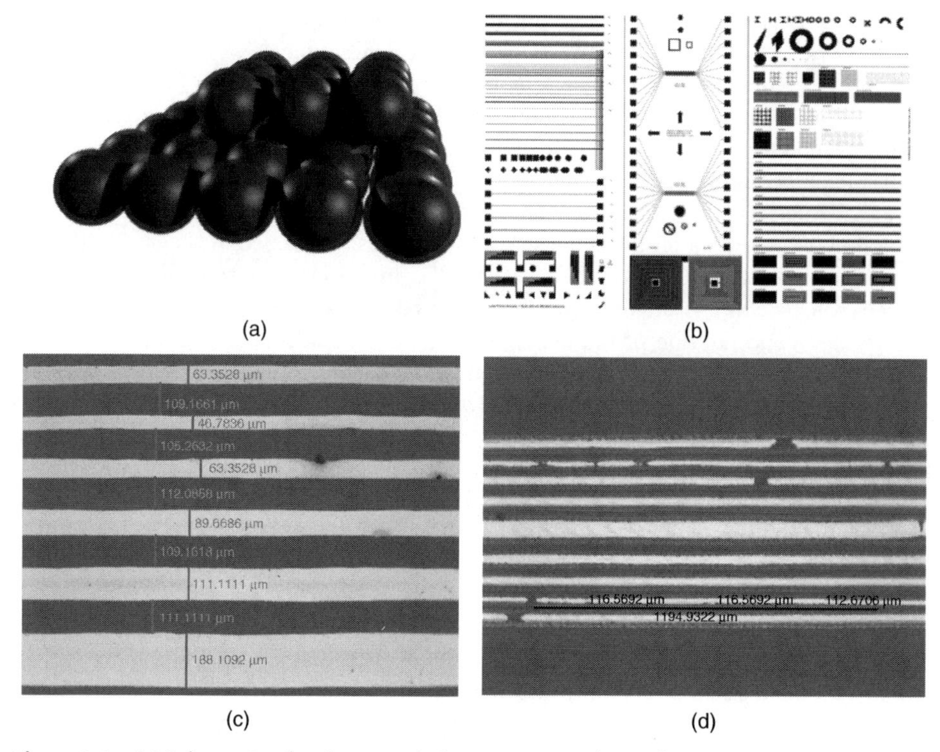

Figure 9.1 *(a) Schematic of resin-coated silver nanoparticles (Cabot Inc., Boston); (b) custom test pattern for the print process; (c) printed lines on untreated clean glass; and (d) printed lines on poly(methyl methacrylate) (PMMA)–coated glass (Previously unpublished).*

or using thermally gelating polymers to arrest the shape of the deposited ink patterns shortly after deposition (Van Den Berg *et al*., 2007a). The choice between these two methods would critically depend on the role of the surface tension in transport of the solute. A stronger contribution from surface tension would indicate the use of surfactants, as well as situations where a pre-patterned well (with well-defined confines) has to be filled with printed material (as in the example of organic light-emitting diode (OLED) display manufacture discussed in Chapter 10). Even higher integration densities are potentially achievable by using electrohydrodynamic jet printing (Park *et al*., 2007), pre-patterned surfaces (fabricated using e-beam or nano-imprint lithography) (Sirringhaus *et al*., 2000; Wang *et al*., 2004), contact printing (Kim *et al*., 2008), molecular jet printing (Chen *et al*., 2007), and so on. Interesting recent work involving electrohydrodynamic jet printing exploits the reduction of the coffee ring effect and seeks to reduce costs by using larger aluminum particles (Kang *et al*., 2011).

Many kinds of chemical species are accessible to inkjet printing. Providing that the inkjet cartridge is not destroyed by the ink, almost any species that is either soluble or can form a suspension stable over the time period of the print can be printed. A particular set of challenges arises when aqueous inks are required. The high surface tension of water at room temperature interferes with jet formation in most current piezoelectric DOD printers. One of the common methods for surmounting this problem involves the use of miscible surfactants, like low-molecular-weight alcohols, as additive species to lower the overall surface tension (Vazquez, Alvarez, and Navaza, 1995). As the miscibility of higher molecular weight alcohols declines, the use of higher boiling point alcohols in aqueous solutions is not possible. For printing nanoparticles, two approaches are in general possible: functionalization that renders them soluble in organic solvents, and the use of high-specific-gravity solvents that make it easier to form colloidal suspensions (Van Den Berg *et al*., 2007b; Perelaer *et al*., 2008).

Some of the issues involved in the inkjet printing process and its fundamentals for device fabrication discussed in this section. In the remainder of this chapter, we shall review some of the major applications of inkjet printing to active device fabrication.

9.2 Applications of Inkjet Printing to Active Devices

9.2.1 OLEDs

Combinatorial approaches to evaluation of electroluminescent materials (Sun and Jabbour, 2002; Yoshioka, Calvert, and Jabbour, 2005a; Yoshioka and Jabbour, 2006a; Lucas *et al*., 2007; Shinar, Shinar, and Zhou, 2007) were one of the early reasons for the use of inkjet printing for fabrication of OLEDs (Figure 9.2), besides the attractiveness of using a process with low material wastage for the often expensive lumophores which can be used in a direct-write process suitable for large-area manufacturing (Ren *et al*., 2011). For instance, inkjet printing has been used to study the effect of side chains and film thicknesses in poly(phenylene-ethynylene)–poly(phenylene-vinylene) (PPE–PPV) based π-conjugated polymers (Tekin *et al*., 2006), photoluminescence from inorganic CdTe nanoparticle–polymer composites (Tekin *et al*., 2007), and so on. The merit of these combinatorial approaches lies in:

- preservation of the precise common process steps for the entire experiment for all cases;
- reduction of waste of usually costly materials;
- no limit on the number of parallel experiments that can be run (the only limitations being the device geometry, resolution of the print process, and substrate size); and
- the possibility of quantitative evaluation of laterally non-uniform processes, such as thermal evaporation and so on.

9.2.1.1 Lighting

While the major application of printed OLEDs has been for displays, there is increasing use of solid-state lighting for energy conservation. For these applications, it is important to develop a white OLED that operates at high efficiency at high current levels and suffers from minimal environmental degradation or operational failure. Since several modes of failure of OLED devices are known (Fenter *et al*., 1997; Kervella, Armand, and Stephan, 2001; Ke *et al*., 2002), opportunities exist in the chemical design of suitable species, such as Ir-based phosphorescent macromolecules anchored in a polyhedral oligomeric silsesquioxane (Sellinger *et al*., 2005; Froehlich *et al*., 2007;

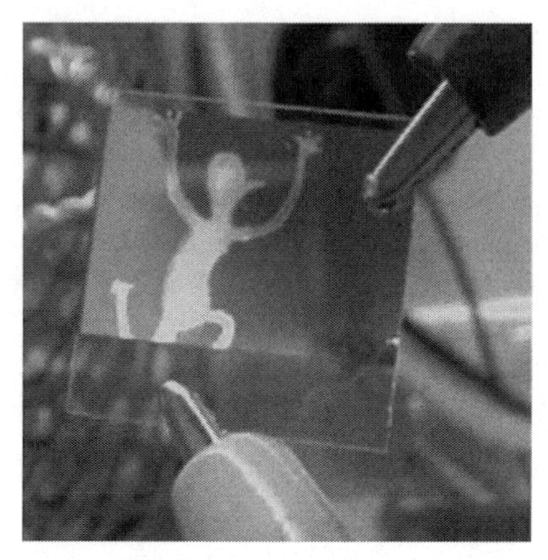

Figure 9.2 *OLED with an inkjet-printed PEDOT:PSS anode. In this work, an HP thermal inkjet cartridge was used (Reproduced with permission from Yoshioka and Jabbour (2006a) Copyright (2006) Elsevier Ltd.).*

Yang *et al*., 2009). Inkjet processes for deposition of such hybrid organic–inorganic materials have been developed, and luminances exceeding 10 kcd m^{-2} for rigid (Singh *et al*., 2009) (Figure 9.3) and 9.6 kcd m^{-2} for flexible substrates have been demonstrated.

Another possibility for countering the failure of OLEDs exists in avoiding the use of organic materials entirely and using inorganic quantum dots (QDs) as the light emitters. Several issues have been identified in this case (Wood and Bulović, 2010), especially the charging of QDs and the quenching of QD photoluminescence. Overall, the use of printing and solution-processing methods for solid-state lighting applications remains an intense area of research.

9.2.1.2 Displays

The three-component basis for most colors in the gamut of human vision, the maskless nature of inkjet printing, the relatively large size of component OLEDs all combine to make inkjet printing an ideal method of fabricating displays (Chapter 10). However, owing to the low mobility of charge carriers in organic materials, transport-dominated components of a display such as the backplane have been resistant to the use of organic materials. As a result, backplanes based on a-Si:H or polysilicon have traditionally been used. Recent commercial demonstration of OLED TVs and their active incorporation into cell phone displays have shown OLED-based displays to be at a fairly developed stage. Owing to the air and water sensitivity of OLEDs and their relatively poorer color purity materials (in addition to the lack of ideal organic blue emitter), the use of QDs in display applications, as an alternative candidate, is now an active area of work. Besides air stability, QD-LEDs offer unprecedented tunable pure colors, which can help to simplify display design.

Inkjet printing of functional LEDs using CdSe–ZnS QDs has been demonstrated (Figure 9.4) using small molecule organic materials as electron and hole transport layers (Haverinen, Myllyla, and Jabbour, 2009). Full red-green-blue (RGB), d.c.-driven

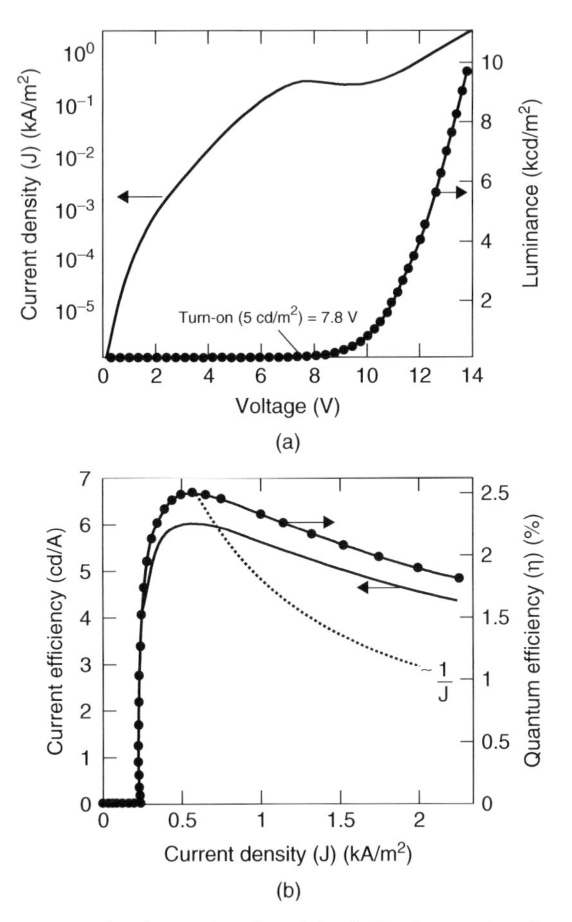

Figure 9.3 *Performance of inkjet-printed polyhedral oligomeric silsesquioxane (POSS) OLEDs (Reproduced with permission from Singh* et al. *(2009) Copyright (2009) Royal Society of Chemistry).*

displays were also demonstrated for the first time using different-sized QDs in quarter video graphics array (QVGA) format at a brightness of $100\,cd\,m^{-2}$ (Haverinen, Myllyla, and Jabbour, 2010). Other approaches to QD-based displays involve the use of polymer–QD blends that address issues of quenching and improve bulk charge transport (Wood *et al*., 2009), and transfer printing to obtain higher resolutions (Kim *et al*., 2011a). A limitation of these approaches lies in the choice of inorganic materials which potentially more elementally rare and costly (Cohen, 2007).

9.2.2 Other Displays

Owing to the high carrier mobility in carbon nanotubes (CNTs), the use of single-wall carbon nanotubes (SWCNTs) as printable electrodes for field emission displays with indium tin oxide (ITO) coated with a phosphor as the counter-electrode has been demonstrated using an electrostatic inkjet printing technique (Shigematsu *et al*., 2008).

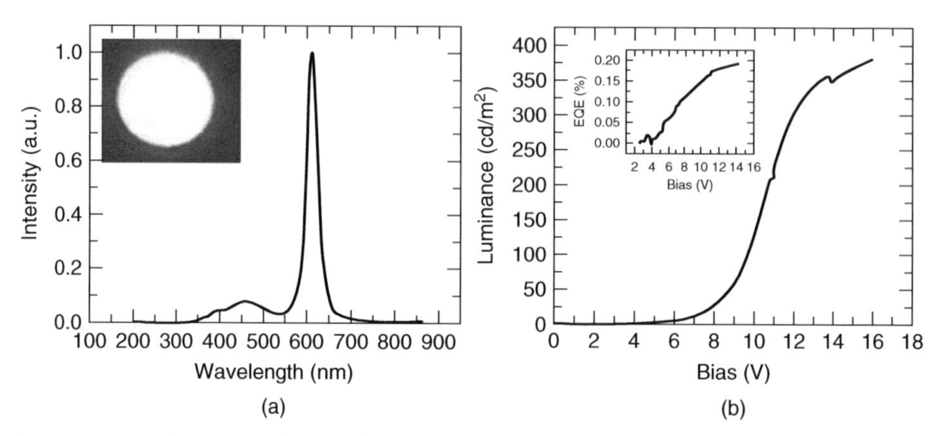

Figure 9.4 *Performance of printed QD-LEDs (Reproduced with permission from Haverinen* et al. *(2009) Copyright (2009) American Institute of Physics).*

Electrochromic displays involve change in color as a function of applied external bias. Printed aqueous multiwall carbon nanotube (MWCNT) layers have been used to exhibit large changes in CIE coordinates from yellow to green and to blue over a voltage change of 1 V (Small *et al.*, 2007). E-paper or electrophoretic displays involving inkjet printing have been demonstrated utilizing organic thin-film transistors (OTFTs) (Kawase *et al.*, 2003; Kawashima *et al.*, 2009). The low switching speeds achievable are consistent with this application.

9.2.3 Energy Storage Using Batteries and Supercapacitors

Electronics applications memory devices and autonomous sensors require the availability of on-chip power or charge sources when the system is disconnected from the grid. While batteries are electrochemical in nature and charge slowly when connected to a power source, supercapacitors charge as rapidly as the external circuit permits. For flexible or paper (Tobjörk and Österbacka, 2011) electronics applications, solution-processed, low-temperature inkjet printing methods are likely to be ideal (Nishide and Oyaizu, 2008). In 2007, an important advance was made in developing a supercapacitor–battery hybrid based on cellulose paper (Pushparaj *et al.*, 2007). While the authors did not employ a printing approach, most of the elements of their process are capable of being modified to use printing methods to create multifunctional devices with on-chip storage. In the same year, another group proposed a MnO_2–$ZnCl_2$–SWCNT-based device architecture to produce supercapacitors specifically for inkjet-printed structures (Kiebele and Gruner, 2007). More recently, reports of a screen-printed Zn–C–polymer anode with a poly(3,4-ethylenedioxythiophene)-poly(styrenesulfonate) (PEDOT:PSS) cathode and a LiCl electrolyte have shown that simple paper-based approaches can be successful in producing working storage devices (Chen *et al.*, 2010). Very recently, cloth-based inkjet-printed SWCNT–RuO_2-based supercapacitors with a reasonable performance (power density $\sim$96 kW kg^{-1} and energy density $\sim$18.8 Wh kg^{-1}) have been demonstrated.

Printing methods are especially attractive for fabric-based devices since many fabrics can be color printed at the same time as they are enhanced with such multifunctional devices for future human-integrated computing applications.

9.2.4 Photovoltaics

According to publicly available sources of data, including the IEA 2010 report (IEA, 2010), the earth receives nearly 5000 times the energy it needs per year (~13 TW) from the sun (89 PW at the surface of the earth). Even though the rate of increase of global energy consumption is twice that of the world population, the potential of solar energy alone as a vast, largely untapped, clean source of energy provides a major impetus to solar cell research.

However, the share of solar energy in the world's energy mix is very small due to several factors:

- solar cell materials for the most efficient devices are expensive (e.g., III–V multijunction concentrators, crystalline Si, etc.);
- most current commercial solar cells require costly fabrication processes; and
- cost-effective materials (like polymers, small molecule organics, a-Si:H, etc.) have poor device efficiencies.

In this situation, it makes sense to consider (among other possibilities) reducing the cost of the overall fabrication process for low-cost materials such as polymer–fullerene blends (Deibel and Dyakonov, 2010; Marin *et al.*, 2005), though the promise of efficient printed QD solar cells based on inorganic materials exists (Jabbour and Doderer, 2011), including solution-processed approaches (Tang and Sargent, 2011) for harvesting the infrared region of the solar spectrum (Sargent, 2009). Early work in printed organic solar cells or light-harvesting structures was prompted by such considerations (Shaheen *et al.*, 2001; Kärkkäinen *et al.*, 2002; Shaheen, Ginley, and Jabbour, 2005; Yoshioka *et al.*, 2005b). Combinatorial approaches made possible by inkjet printing to optimize organic solar cell design have been and continue to be employed (Teichler *et al.*, 2011).

Inkjet printing of PEDOT:PSS layers to create conductive polymer layers (Yoshioka *et al.*, 2005b; Yoshioka and Jabbour, 2006a, 2006b) has involved printing of the material and a study of the effect of oxidation of PEDOT:PSS through peroxide or hypochlorite exposure. Such conductive layers have been used to fabricate polymer organic solar cells with printed PEDOT:PSS electrodes (Eom *et al.*, 2009). An alternate approach has involved the use of inkjet-printed composite Ag–ITO electrodes for lower sheet resistance (Jeong, Kim, and Kim, 2011).

Bulk-heterojunction solar cells based on polymer–fullerene blends have shown the highest efficiencies of any organic solar cells (Chen *et al.*, 2009). Konarka Technologies (Lowell, MA, United States) have developed an inkjet printing process for depositing these materials on various substrates and shown efficiencies in the range of 2.9–3.5% (Hoth *et al.*, 2007, 2009). A similar efficiency level (3.71%) was achieved by a group using an inkjet printing approach to deposit a PEDOT:PSS and poly(3-hexylthiophene)–phenyl-C61-butyric acid methyl ester (P3HT:PCBM) active layer (Eom *et al.*, 2010). The practical applicability of printing to produce entire large-area solar cells has been underscored by a recent demonstration (Krebs *et al*., 2009). Other related methods have involved the use of a screen print-deposited etching paste to pattern ITO, followed by gravure printing of polymer–fullerene blend solar cell modules as shown in Figure 9.5 (Kopola *et al*., 2011).

Other reports involving roll-to-roll coating of organic solar cells based on organic and organic–inorganic hybrid materials show the promise of scaling laboratory research into industrial production of flexible organic solar cells (Blankenburg *et al*., 2009; Krebs, Senkovskyy, and Kiriy, 2010).

The synthesis of chalcogenide materials like copper indium gallium selenide (CIGS) and copper zinc tin sulfide (CZTS) requires the use of high-temperature and toxic processes (Guo, Hillhouse, and Agrawal, 2009; Chory *et al*., 2010; Kameyama *et al*.,

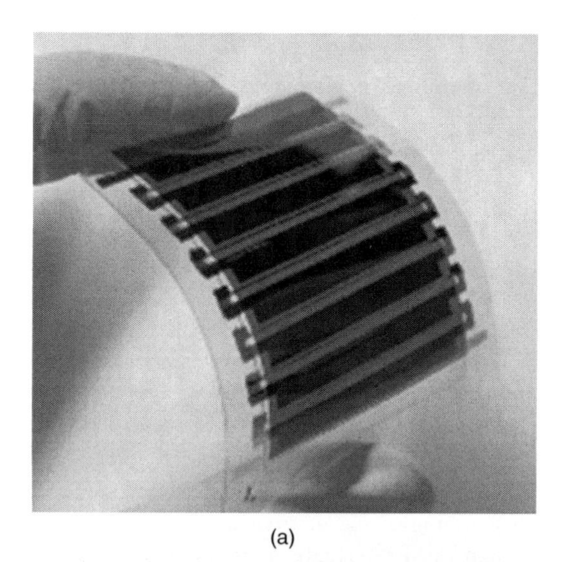

(a)

(b)

Figure 9.5 *Gravure-printed bulk-heterojunction organic solar cells (Reproduced with permission from Kopola* et al. *(2011) Copyright (2011) Elsevier).*

2010), while they can show fairly good performance in solar cell devices (Schock and Noufi, 2000; Contreras *et al*., 2005). Nanosolar has developed a printing process for deposition of CIGS-based inks for rapid industrial production of CIGS-based solar cells (Morton, 2006). Commercial interest in this technology also comes from entities like HelioVolt (Austin, TX, United States, ISET (International Solar Electric Technology, Chatsworth, CA, United States), and so on (Kapur, Fisher, and Roe, 2001; Morton, 2006). In the meantime, Innovalight (Sunnyvale, CA, United States) has developed a Si-based ink for manufacturing Si-based solar cells using Czochralski–Si substrates (Antoniadis, 2009).

9.2.5 Sensors

Sensors play a vital role in homeland security, industrial process monitoring and safety, home protection, transport, and many other fields. Numerous classes of sensors include photodetectors, gas sensors, pH sensors, stress sensors, and so on.

While it is fashionable sometimes to refer to low-efficiency solar cell as a "photodetector," it is essential for photodetectors to have a well-defined response to incident radiant intensity and in the wavelength range of interest. Photodetectors have been demonstrated that are composed of inkjet-printed HgTe nanoparticles with a detectivity of $D^* = 3.9 \times 10^{10}$ cm $Hz^{1/2}$ W^{-1} at 1.4 μm wavelength, with a linear response to applied bias, and that are usable up to a wavelength of 3 μm (Böeberl *et al*., 2007).

Ammonia sensors based on monitoring of electrical current have been demonstrated using inkjet-printed dodecylbenzene sulfonate acid (DBSA)–doped polyaniline (PANI) nanoparticles on a screen-printed carbon electrode, with a detection limit of ~2.58 μM (Crowley *et al*., 2008). Another group reported the use of a method for producing an aqueous dispersion of PANI–PSS nanoparticles for use in an inkjet-printed ammonia sensor (Figure 9.6) with sensitivity of 10 ppb (Jang, Ha, and Cho, 2007).

A composite of coiled-conformation biopolymers and SWCNTs printed onto a polyethylene terephthalate (PET) substrate has been shown to operate as a humidity sensor dependent on mechanical swelling of the biopolymer matrix (Panhuis *et al*., 2007) or possibly the enhancement of electrolytic conductivity. A more novel recent approach involves the use of passive ultrahigh frequency (UHF) radiofrequency identification (RFID) tags that respond to the humidity-dependent permittivity of the flexible Kapton substrate (Virtanen *et al*., 2011). Micro-electro-mechanical systems (MEMS)–based sensors have been discussed already in Chapter 6. While MEMS-based sensors involving inkjet deposition of self-assembled monolayers (SAMs) sensitive to ionic concentrations and pHs in liquids have previously been demonstrated (O'Toole *et al*., 2009), a somewhat less complicated solution for gas sensing has recently been developed involving a pH indicator dye, a phase transfer salt, and polymer ethyl cellulose. This chemical formulation was printed directly into the lens surface of an LED. Absorbance measurements were used to extract peak heights (in μs) from a paired emitter detector diode (PEDD) setup (Figure 9.7).

Recently, an all-printed temperature and pressure sensor exploiting the piezoelectric and pyroelectric properties of the ferroelectric block co-polymer polyvinylidene fluoride trifluoroethylene (PVDF–TrFE) has been developed (Zirkl *et al*., 2011). It employs a ferroelectric capacitor as the sensor, an OTFT as an amplifier, and an electrochromic

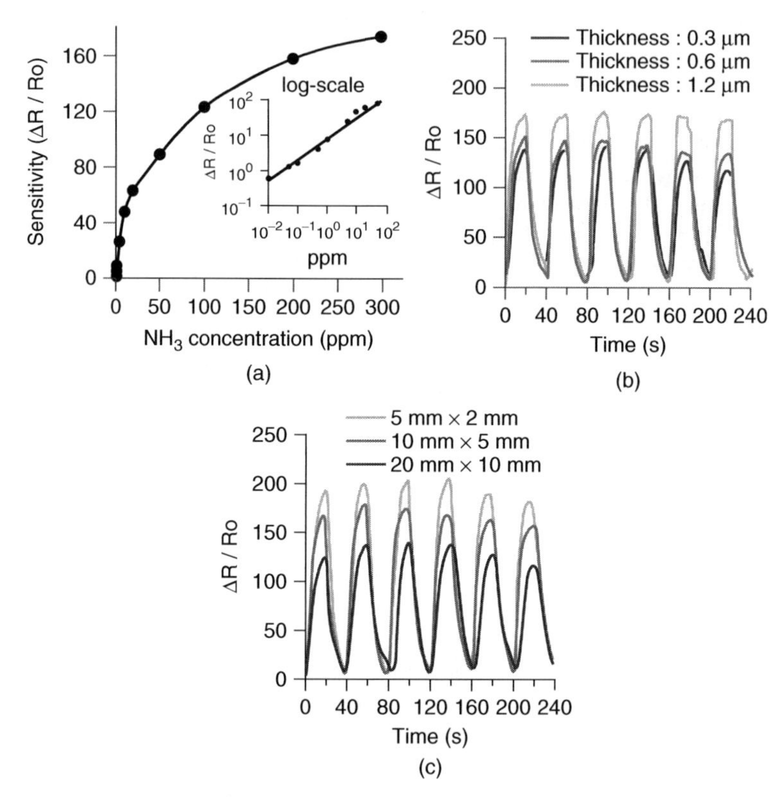

Figure 9.6 *(a) Sensitivity of PANI–PSS films as a function of NH_3 concentration: (b,c) increasing thickness and decreasing sensing area have positive impacts on sensor sensitivity (Reproduced with permission from Jang, Ha, and Cho (2007) Copyright (2007) Wiley-VCH).*

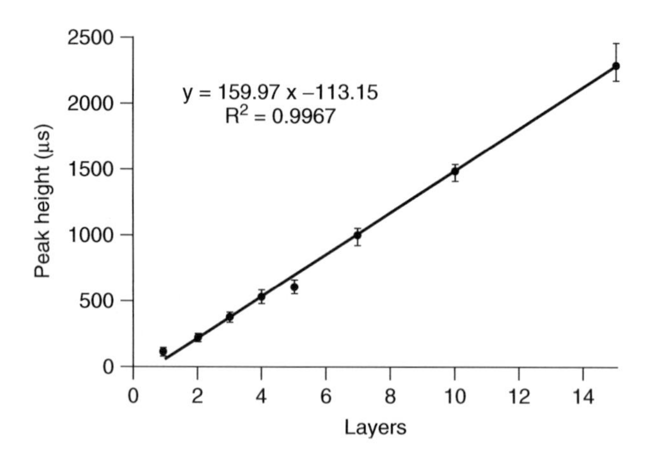

Figure 9.7 *Peak heights from a PEDD setup versus the number of chemical sensing layers (O'Toole* et al.*, 2009) (Reproduced with permission from O'Toole* et al. *(2009) Copyright (2009) Elsevier Ltd.).*

display. The same research group had earlier reported an all-printed infrared sensor for detecting human body heat (Zirkl *et al*., 2010).

9.2.6 Transistors, Logic, and Memory

All-organic circuits provide the advantage of being compatible with flexible substrates and being processible at low temperatures. The disadvantages are that the charge carrier mobilities in organic materials (such as rubrene and 6,13-bis(triisopropylsilylethynyl) pentacene (TIPS–pentacene)) are fairly low (typically $\sim 1\,cm^2\,V^{-1}\,s^{-1}$, though higher reported values exist), and there are not many good n-type organics – with some exceptions (Jones *et al*., 2004). The primary building block for organic circuits is the OTFT, a four-layer device with two layers of electrode materials (source–drain and the gate) and one layer each of a dielectric and the organic semiconductor. A printed OTFT thus necessarily requires a layer-by-layer additive approach (Yoshioka *et al*., 2002; Yoshioka and Jabbour, 2006a). This raises solvent orthogonality issues: subsequently printed layers must not normally dissolve previous layers. While it is possible to "assemble" a multifunctional device on a flexible substrate through the use of discrete components (Kim *et al*., 2009), it lacks the economies possible through adopting an all-printed approach (Chung *et al*., 2011).

Problems unique to flexible substrates involve local nonplanarity during fabrication and adhesion and strain issues for reliability during flexion. Additionally, the switching speed of a circuit depends upon the mobility and channel length-to-width (L/W) ratio of the transistor. Given the multimicrometer resolution of the inkjet printing process, only low-frequency circuits are currently feasible without any elaborate circuit-level design.

The reduction of channel length and contact resistance to overcome the speed limitations has been the focus of several reports over the past decade (Sirringhaus *et al*., 2000; Noh *et al*., 2007a; Herlogsson *et al*., 2008; Cheng *et al*., 2009). Chosen SAM species the work function changes in morphology, tunneling probabilities, and so on; and injection of both electrons and holes can be improved. The species used in the relevant study (Cheng *et al*., 2009) are all capable of being inkjet-printed. In combination with nano-imprint lithography (Wang *et al*., 2006) and self-aligned methods (Sele *et al*., 2005), the modification of the surface energy landscape (Wang *et al*., 2004) has been used to reduce channel lengths. Combination of a self-aligned gate with channel lengths between 60 nm and 4 μm with a high-mobility organic semiconductor poly(2,5-bis(3-dodecylthiophene-2-yl)thieno[3,2-b]thiophene) (pBTTT) has been used to obtain respectable mobility values of $0.1–0.2\,cm^2\,V^{-1}\,s^{-1}$, resulting in inverters with frequencies of $\sim$1.6 MHz (Noh *et al*., 2007b).

As mentioned orthogonality of solvents is an overriding concern in layer-by-layer assembly of all-printed OTFTs. Given this tight constraint, all-printed transistors have been demonstrated using pentacene precursors, poly-4-vinylphenol (PVP) (Molesa *et al*., 2004; Chung *et al*., 2011), and polypyrrole and PVP-K60 (Liu, Varahramyan, and Cui, 2005). Active semiconductors used include poly[5, 5′-bis(3-dodecyl-2-thienyl)-2, 2′-bithiophene] (PQT–12) (Arias *et al*., 2004), F8T2 (Paul *et al*., 2003), air-stable P3HT derivatives (Ko *et al*., 2007), numerous reports with TIPS–pentacene (Kim *et al*., 2007;

Chung *et al.*, 2011), other pentacene precursors and oligothiophenes, CNTs (Beecher *et al.*, 2007), polysilicon (Shimoda *et al.*, 2006), zinc oxide nanowires (Noh *et al.*, 2007b). Use of solvent mixtures has been shown to promote the growth of large self-organized TIPS–pentacene crystals (Lim *et al.*, 2006, 2008) that improve transport properties. Electrode materials include PEDOT:PSS, and Ag–Au–Cu nanoparticle inks (Hong and Wagner, 2000; Huang *et al.*, 2003; Wu *et al.*, 2005a; Gamerith *et al.*, 2007; Ko *et al.*, 2007). Simultaneous printing of encapsulation layers has also been demonstrated (Arias, Endicott, and Street, 2006). In a recent study, a combination of Monte Carlo simulations, inkjet printing, and measurements was carried out to study scaling of conductance of CNT-based thin-film transistors (Li *et al.*, 2011).

Registration of computer-generated features requires accurate relative placement of subsequent layers. For instance, if the gate is not placed accurately in the channel (since the gate is not self-aligned as in the standard complementary metal–oxide semiconductor (CMOS) process (Jaeger and Gaensslen, 1988)), the gate–drain and gate–source parasitic capacitance can be high and the frequency response suboptimal upon scaling (Wu, Singh, and Singh, 2005b). The disadvantages inherent in the large dimensions of inkjet-printable structures can be avoided by organic electrochemical transistors (Basiricò *et al.*, 2011).

Top contact architectures, which exploit the picoliter and lower level control available with inkjet printing, have also been demonstrated (Noguchi *et al.*, 2008). The use of smaller sized nanoparticles reduces the sintering temperature (Moon *et al.*, 2005), thereby reducing the damage to the underlying active organic layer. Alternately, infrared (Denneulin *et al.*, 2011a) or laser (Ko *et al.*, 2007) sintering can be used to further alleviate the damage. Further, comparison of various other methods of depositing TIPS–pentacene in all-printed OTFTs demonstrates that inkjet-printed film quality compares very favorably with other techniques of deposition – see Figure 9.8 (Chung *et al.*, 2011).

In conventional CMOS scaling, increased capacitance per unit area is seen as a method to maintain control over the channel even as dimensions shrink. This often involves the use of high-k dielectrics or ferroelectrics (Moon *et al.*, 2005; Noguchi *et al.*, 2008) or very thin dielectric layers in OTFTs (Singh, Wu, and Singh, 2003; Wu, Singh, and Singh, 2003). A barium titanate–nickel perovskite layer was reported to be inkjet-printed using

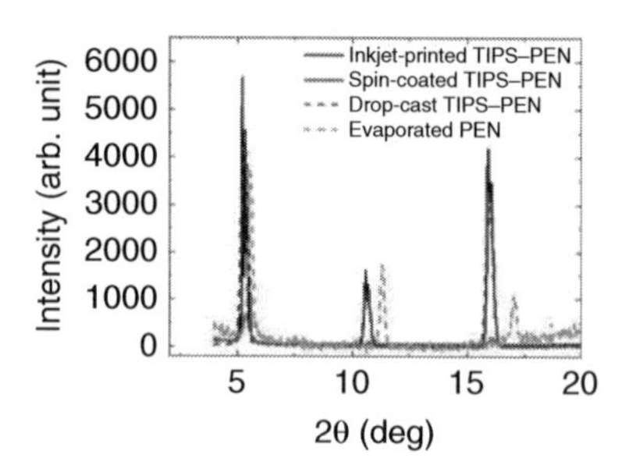

Figure 9.8 *X-ray diffraction patterns for different methods of depositing TIPS–pentacene (with evaporated pentacene as control) in OTFTs (Reproduced with permission from Chung* et al. *(2011) Copyright (2011) Japan Society of Applied Physics).*

an ethanol–isopropanol blend (Tseng, Lin, and Wang, 2006), which may be noted to be a poor solvent for TIPS–pentacene, thereby showing its applicability for use in the most common of all current OTFT materials.

In addition to the energy storage applications discussed in Section 9.2.3, multifunctional flexible devices also need reliable memory storage devices. There are two major possibilities in this regard: the use of ferroelectric block co-polymers such as polyvinylidene–TrFE (PVDF–TrFE) (Noh and Sirringhaus, 2009), or the use of magnetic nanoparticles to print magnetic data devices. Screen-printed lead zirconate titanate (PZT)–PVDF–TrFE composites have been analyzed to show a minimum in pyroelectric response at 40% PZT loading, which is attributed to the opposite sign of the response of the constituent pure phases (Tseng, Lin, and Wang, 2006). Though the authors used 0–3 connectivity for their layers, it is conceivable that with the use of appropriate surface treatments, it might be possible to use a 2–2 connectivity scheme as well, with a very different response expected in the coupling between the two phases. The use of inkjet printing has also been demonstrated for magnetic data storage applications with synthesized magnetic nanoparticles of iron oxide ((Fe_2O_3, γ-Fe_2O_3) and Fe_3O_4) in a stabilizing dextran or polysyrene coating (Voit *et al.*, 2003). Superconducting quantum interference device (SQUID) magnetometry measurements were used to reveal a coercivity of 15 Oe at room temperature for dextran-coated nanoparticles, and similar values obtained for polystyrene-coated particles dispersed in dipropylene glycol.

9.2.7 Contacts and Conductors

Perhaps in no other application is the maskless direct-write nature of inkjet printing such a perfect fit than in the printing of high-aspect-ratio metal conductors composed of costly metallic nanoparticles and their subsequent sintering to form interconnects. This topic was introduced in Chapter 7.

High-density integration, as in backplanes, requires tight tolerances on printed conductors to minimize wasted space in the layout and design and extra-resistive drops along long lines. Inkjet printing of silver nanoparticles studied at different drying temperatures reveals that the morphology of printed lines changes from the common hill-like to ring formation (Li *et al.*, 2007). More rapid removal of the solvent and sintering of the nanoparticles are likely to leave a signature of the underlying coffee ring effect, a contention further supported by the strong dependence on the drop size. While there are several reports on the printing of silver nanoparticles (Meier *et al.*, 2009; Van Osch *et al.*, 2008), inkjet-printed contacts and conductive lines consisting of Ag–Cu nanoparticles (Gamerith *et al.*, 2007), Cu (Kim *et al.*, 2009), and Au (Huang *et al.*, 2003; Zhao *et al.*, 2007) have also been developed. Nonmetallic species, such as PANI (Ngamna *et al.*, 2007), CNTs (Kordás *et al.*, 2006; Mustonen *et al.*, 2007; Song *et al.*, 2008), and graphene (Shin, Hong, and Jang, 2011a) (see Figure 9.9), offer alternate materials for printing conductive layers. A process employing a graphene oxide ink inkjet printed onto a PET substrate and subsequently reduced to form patterned graphene was recently developed (Shin, Hong, and Jang, 2011b). Further, SWCNT–PEDOT:PSS composite inks have been shown to limit the coffee ring effect observed in aqueous SWCNT inkjet-printed layers (Denneulin *et al.*, 2011b).

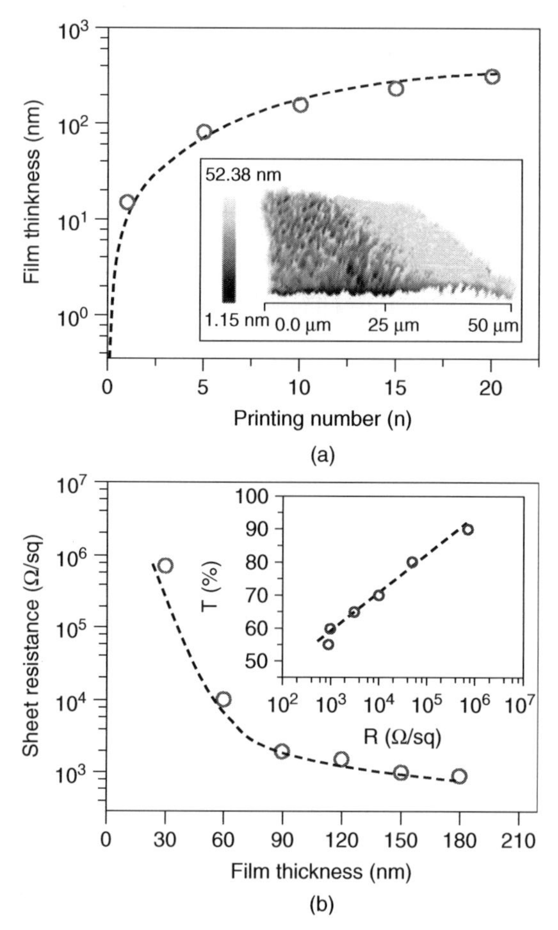

Figure 9.9 *Film thickness and sheet resistance as functions of number of inkjet-printed graphene layers (Reproduced with permission from Shin* et al. *(2011a) Copyright (2011) Royal Society of Chemsitry).*

Owing to the ease of using inkjet printing at room temperature, it is suitable for depositing temperature-unstable conductive compounds such as TCNQ derivatives $(BO_9(C_{14}\text{-TCNQ})_4$ (where BO is bis ethylenedioxy tetrathiafulvalene and C_{14}-TCNQ is tetradecyltetracyano–quinodimethane) (Hiraoka *et al*., 2006). Recently, new methods for synthesizing (Shankar *et al*., 2011), annealing (Kim *et al*., 2011b), and environmental protection of inkjet-printed silver conductors have been developed (Halonen *et al*., 2011). It has also been found that inkjet-printed silver lines in microchannels exhibit lower resistance than lines printed on unpatterned substrates (Loeffelmann *et al*., 2011). In the past, more exotic applications of printing conductive materials in three-dimensional (3D) structures, such as MEMS, have been demonstrated (Fuller, Wilhelm, and Jacobson, 2002); see also Chapter 6.

9.2.8 In Situ Synthesis and Patterning

In perhaps one of the first examples of such a process, *in situ* synthesis of CdS and PbS has been demonstrated using an inkjet printing technique (Teranishi *et al*., 2002). The authors used a sodium sulfide precursor in conjunction with lead nitrate and cadmium chloride to form patterns of PbS and CdS. This work can be extended by combining it with circuit layout to form n-type (CdS) and p-type (PbS) semiconductor regions for rapid manufacture of chalcogenide circuitry. This idea develops chemical bath deposition–based aqueous chemistry (Salas-Villasenor *et al*., 2010) whereby thiourea is used as a sulfur precursor, while chalcogenide salts (such as cadmium chloride, zinc acetate, etc., for n-type; and lead acetate, copper chloride, etc., for p-type) are used for forming *in situ* semiconductor layers.

While inkjet printing is a maskless direct-write process, it is also usable as a patterning process, which involves printing the solvent on an unpatterned layer (De Gans, Hoeppener, and Schubert, 2007; Xia and Friend, 2007), or using a combination of a protective layer and SAM-based surface modification to carry out patterning, in a form of digital lithography (Chabinyc *et al*., 2002; Paul *et al*., 2003; De Gans *et al*., 2006; Wong *et al*., 2006; Daniel *et al*., 2007).

Additive manufacturing is another area in which inkjet printing has recently been used, as discussed further in Chapter 14. Inkjet printing of caprolactam (a nylon 6 precursor) was studied to develop a manufacturing method for nylon components (Fathi, Dickens, and Hague, 2012).

9.2.9 Biological Applications

We have discussed in this chapter the efficacy of combinatorial approaches in material discovery for OLEDs and solar cells. Such methods are also useful to carry out complex biological studies by printing species of interest onto substrates or into prefabricated vials (Goldmann and Gonzalez, 2000). This ability has been used to study multistrain bacterial patterns, with viabilities for *Escherichia coli* exceeding 98.5% (Merrin, Leibler, and Chuang, 2007). In the present time of heightened biological threats and through the rise in drug-resistant infectious diseases, the ability to rapidly test the efficacy of new vaccines or antibiotics over thousands of different strains of pathogens (Yatsushiro *et al*., 2011) is vitally important because of the accelerated time constraints imposed by the rapid movement of people and goods.

The room-temperature, low-damage, and solution-processible nature of inkjet printing (Calvert, 2007) makes it an ideal tool for handling species like bacteria, viruses, vesicles, proteins (Demarche *et al*., 2011), living cells in general, and so on in the lab (Derby, 2008), or in industrial prototyping applications such as pharmaceutical research (Yu *et al*., 2008), especially with picoliter (or better) level of control over dispensing volume (Hauschild *et al*., 2005; Jayasinghe, Qureshi, and Eagles, 2006). Given that inkjet printing can mimic biological growth (Yoshioka, Jabbour, and Calvert, 2002) (as in biofilms), inkjet printing has been used to study the dynamics of *in vivo* dendrimer-mediated silica condensation reactions (Deravi *et al*., 2008), using fluorescence-labeled biomimetic inks, to acquire an understanding of biomineralization processes at cell walls.

Biological components such as peptide nanotubes and nanospheres can be jetted with inkjet printing (Adler-Abramovich and Gazit, 2008), removing one of the roadblocks toward their use in hybrid organic-biological devices of the future, since no conventional lithographic technologies can be used cheaply and in solution phase to deposit controlled arrays of such species. Further, the picoliter-level control over fluid volume provided by inkjet printing allows controlled encapsulation of single cells in sensitive drop microfluidic studies on large cell populations (Edd *et al.*, 2008).

Aggregation and sedimentation of cells remain a problem in inkjet printing of such species. To alleviate this, surfactants and gentle stirring have both been used to lengthen the time available for optimal deposition of living cells without any significant loss of cell viability (Parsa *et al.*, 2010). This underlines the unique applicability of inkjet printing methods for depositing biomaterials in tissue-engineering applications, which is further illustrated in Chapters 12 and 13.

9.3 Future Outlook

Deposition processes today still follow the fundamental processes in microfabrication: deposition of a material, followed by patterning that results in increased energy and material consumption and wastage. An interesting recent review examines various issues and alternative fabrication methods (Liddle and Gallatin, 2011). Solution-based printing methods, such as inkjet printing, offer a method to turn that paradigm on its head and are thus successfully making inroads into fields previously considered the preserve of conventional lithography and planar processing.

The locally anisotropic nature of printing also gives rise to certain new challenges that relate to the uniformity of deposited films. A better understanding of the flow processes and their precise relation to the process steps and ink chemistry is being slowly acquired empirically and also theoretically which is of predictive value. The downward scaling in device dimensions and the consequent reduction in drop size are bringing about a revolution in the kind of printing methods used: screen, gravure, inkjet, microcontact, electrohydrodynamic jet, and so on. The predicted parallel revolution in the shrinking of devices to the scale of single semiconductor nanoparticles participating in photoelectrochemical photocurrent switching effect (PEPS) is well suited to take advantage of printing methods for future devices (Gawęda *et al.*, 2011).

The scope of inkjet printing is also undergoing a subtle transformation. From being a method for industrial and prototype development, it is now aggressively being retargeted to the lab space as an ink verification and development method, while scaled-up industrial processes such as roll-to-roll (R2R) printing become more accurate in conjunction with techniques like imprint lithography, in cases where high throughput is essential. In the coming years, as printing methods take over ever-increasing fractions of fabrication workload, the role of inkjet printing will continually change in response to the changing application space and economics.

References

Adler-Abramovich, L. and Gazit, E. (2008) Controlled patterning of peptide nanotubes and nanospheres using inkjet printing technology. *Journal of Peptide Science*, **14** (2), 217–223.

Antoniadis, H. (2009) Silicon ink high efficiency solar cells. Proceedings of the 2009 34th IEEE Photovoltaic Specialists Conference (PVSC 2009), pp. 000650–000654.

Arias, A.C., Endicott, F. and Street, R.A. (2006) Surface-induced self-encapsulation of polymer thin-film transistors. *Advanced Materials*, **18** (21), 2900–2904.

Arias, A.C., Mackenzie, J.D., McCulloch, I. *et al*. (2010) Materials and applications for large area electronics: solution-based approaches. *Chemical Reviews*, **110** (1), 3–24.

Arias, A.C., Ready, S.E., Lujan, R. *et al*. (2004) All jet-printed polymer thin-film transistor active-matrix backplanes. *Applied Physics Letters*, **85** (15), 3304–3306.

Bao, Z.N., Rogers, J.A. and Katz, H.E. (1999) Printable organic and polymeric semiconducting materials and devices. *Journal of Materials Chemistry*, **9** (9), 1895–1904.

Basiricò, L., Cosseddu, P., Fraboni, B. and Bonfiglio, A. (2011) Inkjet printing of transparent, flexible, organic transistors. *Thin Solid Films*, **520** (4), 1291–1294.

Beecher, P., Servati, P., Rozhin, A. *et al*. (2007) Ink-jet printing of carbon nanotube thin film transistors. *Journal of Applied Physics*, **102** (4), 043710.

Blankenburg, L., Schultheis, K., Schache, H. *et al*. (2009) Reel-to-reel wet coating as an efficient up-scaling technique for the production of bulk-heterojunction polymer solar cells. *Solar Energy Materials and Solar Cells*, **93** (4), 476–483.

Böeberl, M., Kovalenko, M.V., Gamerith, S. *et al*. (2007) Inkjet-printed nanocrystal photodetectors operating up to 3 μm wavelengths. *Advanced Materials*, **19** (21), 3574–3578.

Calvert, P. (2007) Printing cells. *Science*, **318** (5848), 208–209.

Chabinyc, M.L., Wong, W.S., Salleo, A. *et al*. (2002) Organic polymeric thin-film transistors fabricated by selective dewetting. *Applied Physics Letters*, **81** (22), 4260–4262.

Chen, P., Chen, H., Qiu, J. and Zhou, C. (2010) Inkjet printing of single-walled carbon nanotube/RuO(2) nanowire supercapacitors on cloth fabrics and flexible substrates. *Nano Research*, **3** (8), 594–603.

Chen, H.Y., Hou, J., Zhang, S. *et al*. (2009) Polymer solar cells with enhanced open-circuit voltage and efficiency. *Nature Photonics*, **3** (11), 649–653.

Chen, J., Leblanc, V., Kang, S.H. *et al*. (2007) High definition digital fabrication of active organic devices by molecular jet printing. *Advanced Functional Materials*, **17** (15), 2722–2727.

Cheng, X., Noh, Y.Y., Wang, J. *et al*. (2009) Controlling electron and hole charge injection in ambipolar organic field-effect transistors by self-assembled monolayers. *Advanced Functional Materials*, **19** (15), 2407–2415.

Chory, C., Zutz, F., Witt, F. *et al*. (2010) Synthesis and characterization of Cu(2)ZnSnS(4). *Physica Status Solidi C: Current Topics in Solid State Physics*, **7** (6), 1486–1488.

Chung, S., Jang, J., Cho, J. *et al*. (2011) All-inkjet-printed organic thin-film transistors with silver gate, source/drain electrodes. *Japanese Journal of Applied Physics*, **50** (3), 03cb05.

Cohen, D. (2007) The last place on earth. *New Scientist*, **194** (2608), 34–47.

Contreras, M.A., Ramanathan, K., Abushama, J. *et al*. (2005) Diode characteristics in state-of-the-art ZnO/CdS/Cu(In(1-x)Gax)Se-2 solar cells. *Progress in Photovoltaics*, **13** (3), 209–216.

Crowley, K., O'Malley, E., Morrin, A. *et al*. (2008) An aqueous ammonia sensor based on an inkjet-printed polyaniline nanoparticle-modified electrode. *Analyst*, **133** (3), 391–399.

Daniel, J., Arias, A.C., Wong, W. *et al*. (2007) Jet-printed active-matrix backplanes and electrophoretic displays. *Japanese Journal of Applied Physics Part 1: Regular Papers Brief Communications and Review Papers*, **46** (3B), 1363–1369.

Das, R. and Harrop, P. (2011) *Printed, Organic and Flexible Electronics Forecasts, Players and Opportunities 2011–2021*, IDTechEx, Cambridge.

De Gans, B.J., Duineveld, P.C. and Schubert, U.S. (2004) Inkjet printing of polymers: state of the art and future developments. *Advanced Materials*, **16** (3), 203–213.

De Gans, B.J., Hoeppener, S. and Schubert, U.S. (2006) Polymer-relief microstructures by inkjet etching. *Advanced Materials*, **18** (7), 910–914.

De Gans, B.J., Hoeppener, S. and Schubert, U.S. (2007) Polymer relief microstructures by inkjet etching. *Journal of Materials Chemistry*, **17** (29), 3045–3050.

Deibel, C. and Dyakonov, V. (2010) Polymer-fullerene bulk heterojunction solar cells. *Reports on Progress in Physics*, **73** (9), 096401.

Demarche, S., Sugihara, K., Zambelli, T. *et al*. (2011) Techniques for recording reconstituted ion channels. *Analyst*, **136** (6), 1077–1089.

Denneulin, A., Blayo, A., Neuman, C. and Bras, J. (2011a) Infra-red assisted sintering of inkjet printed silver tracks on paper substrates. *Journal of Nanoparticle Research*, **13** (9), 3815–3823.

Denneulin, A., Bras, J., Carcone, F. *et al*. (2011b) Impact of ink formulation on carbon nanotube network organization within inkjet printed conductive films. *Carbon*, **49** (8), 2603–2614.

Deravi, L.F., Sumerel, J.L., Sewell, S.L. and Wright, D.W. (2008) Piezoelectric inkjet printing of biomimetic inks for reactive surfaces. *Small*, **4** (12), 2127–2130.

Derby, B. (2008) Bioprinting: inkjet printing proteins and hybrid cell-containing materials and structures. *Journal of Materials Chemistry*, **18** (47), 5717–5721.

Dong, H., Carr, W.W. and Morris, J.F. (2006) An experimental study of drop-on-demand drop formation. *Physics of Fluids*, **18** (7), 072102.

Edd, J.F., Di Carlo, D., Humphry, K.J. *et al*. (2008) Controlled encapsulation of single-cells into monodisperse picolitre drops. *Lab on a Chip*, **8** (8), 1262–1264.

Eom, S.H., Park, H., Mujawar, S.H. *et al*. (2010) High efficiency polymer solar cells via sequential inkjet-printing of PEDOT:PSS and P3HT:PCBM inks with additives. *Organic Electronics*, **11** (9), 1516–1522.

Eom, S.H., Senthilarasu, S., Uthirakumar, P. *et al*. (2009) Polymer solar cells based on inkjet-printed PEDOT:PSS layer. *Organic Electronics*, **10** (3), 536–542.

Famili, A., Palkar, S.A. and Baldy, W.J.Jr., (2011) First drop dissimilarity in drop-on-demand inkjet devices. *Physics of Fluids*, **23** (1), 012109.

Fathi, S., Dickens, P. and Hague, R. (2012) Jetting stability of molten caprolactam in an additive inkjet manufacturing process. *International Journal of Advanced Manufacturing Technology*, **59** (1–4), 201–212.

Fenter, P., Schreiber, F., Bulović, V. and Forrest, S.R. (1997) Thermally induced failure mechanisms of organic light emitting device structures probed by X-ray specular reflectivity. *Chemical Physics Letters*, **277** (5–6), 521–526.

Froehlich, J.D., Young, R., Nakamura, T. *et al.* (2007) Synthesis of multi-functional POSS emitters for OLED applications. *Chemistry of Materials*, **19** (20), 4991–4997.

Fuller, S.B., Wilhelm, E.J. and Jacobson, J.M. (2002) Ink-jet printed nanoparticle microelectromechanical systems. *Journal of Microelectromechanical Systems*, **11** (1), 54–60.

Gamerith, S., Klug, A., Scheiber, H. *et al.* (2007) Direct ink-jet printing of Ag-Cu nanoparticle and Ag-precursor based electrodes for OFET applications. *Advanced Functional Materials*, **17** (16), 3111–3118.

Gan, H.Y., Shan, X., Eriksson, T. *et al.* (2009) Reduction of droplet volume by controlling actuating waveforms in inkjet printing for micro-pattern formation. *Journal of Micromechanics and Microengineering*, **19** (5), 055010.

Gawęda, S., Kowalik, R., Kwolek, P. *et al.* (2011) Nanoscale digital devices based on the photoelectrochemical photocurrent switching effect: preparation, properties and applications. *Israel Journal of Chemistry*, **51** (1), 36–55.

Goghari, A.A. and Chandra, S. (2008) Producing droplets smaller than the nozzle diameter by using a pneumatic drop-on-demand droplet generator. *Experiments in Fluids*, **44** (1), 105–114.

Goldmann, T. and Gonzalez, J.S. (2000) DNA-printing: utilization of a standard inkjet printer for the transfer of nucleic acids to solid supports. *Journal of Biochemical and Biophysical Methods*, **42** (3), 105–110.

Guo, Q., Hillhouse, H.W. and Agrawal, R. (2009) Synthesis of Cu(2)ZnSnS(4) nanocrystal ink and its use for solar cells. *Journal of the American Chemical Society*, **131** (33), 11672–11673.

Halonen, E., Pynttari, V., Lilja, J. *et al.* (2011) Environmental protection of inkjet-printed Ag conductors. *Microelectronic Engineering*, **88** (9), 2970–2976.

Hanyak, M., Darhuber, A.A. and Ren, M. (2011) Surfactant-induced delay of leveling of inkjet-printed patterns. *Journal of Applied Physics*, **109** (7), 074905.

Hauschild, S., Lipprandt, U., Rumplecker, A. *et al.* (2005) Direct preparation and loading of lipid and polymer vesicles using inkjets. *Small*, **1** (12), 1177–1180.

Haverinen, H.M., Myllyla, R.A. and Jabbour, G.E. (2009) Inkjet printing of light emitting quantum dots. *Applied Physics Letters*, **94** (7), 073108.

Haverinen, H.M., Myllyla, R.A. and Jabbour, G.E. (2010) Inkjet printed RGB quantum dot-hybrid led. *Journal of Display Technology*, **6** (3), 87–89.

Herlogsson, L., Noh, Y.Y., Zhao, N. *et al.* (2008) Downscaling of organic field-effect transistors with a polyelectrolyte gate insulator. *Advanced Materials*, **20** (24), 4708–4713.

Hill, E.K., Watson, R.L. and Dunstan, D.E. (2005) Rheofluorescence technique for the study of dilute MEH-PPV solutions in couette flow. *Journal of Fluorescence*, **15** (3), 255–266.

Hiraoka, M., Hasegawa, T., Abe, Y. *et al*. (2006) Ink-jet printing of organic metal electrodes using charge-transfer compounds. *Applied Physics Letters*, **89** (17), 173504.

Hong, C.M. and Wagner, S. (2000) Inkjet printed copper source/drain metallization for amorphous silicon thin-film transistors. *IEEE Electron Device Letters*, **21** (8), 384–386.

Hoth, C.N., Choulis, S.A., Schilinsky, P. and Brabec, C.J. (2007) High photovoltaic performance of inkjet printed polymer: fullerene blends. *Advanced Materials*, **19** (22), 3973–3978.

Hoth, C.N., Choulis, S.A., Schilinsky, P. and Brabec, C.J. (2009) On the effect of poly(3-hexylthiophene) regioregularity on inkjet printed organic solar cells. *Journal of Materials Chemistry*, **19** (30), 5398–5404.

Huang, D., Liao, F., Molesa, S. *et al*. (2003) Plastic-compatible low resistance printable gold nanoparticle conductors for flexible electronics. *Journal of the Electrochemical Society*, **150** (7), G412–G417.

IEA (2010) *Key World Energy Statistics 2010*, OECD Publishing, Paris.

Im, J., Sengupta, S.K. and Whitten, J.E. (2010) Photometer for monitoring the thickness of inkjet printed films for organic electronic and sensor applications. *Review of Scientific Instruments*, **81** (3), 034103.

Jabbour, G.E. and Doderer, D. (2011) The best of both worlds. *Nature Photonics*, **4** (9), 604–605.

Jaeger, R.C. and Gaensslen, F.H. (1988) Low temperature semiconductor electronics. InterSociety Conference on Thermal Phenomena in the Fabrication and Operation of Electronic Components: I-THERM'88 (Cat. No.88CH2590-8).

Jang, J., Ha, J. and Cho, J. (2007) Fabrication of water-dispersible polyaniline-poly(4-styrenesulfonate) nanoparticles for inkjet-printed chemical-sensor applications. *Advanced Materials*, **19** (13), 1772–1775.

Jayasinghe, S.N., Qureshi, A.N. and Eagles, P.A.M. (2006) Electrohydrodynamic jet processing: an advanced electric-field-driven jetting phenomenon for processing living cells. *Small*, **2** (2), 216–219.

Jeong, J.A., Kim, J. and Kim, H.K. (2011) Ag grid/ITO hybrid transparent electrodes prepared by inkjet printing. *Solar Energy Materials and Solar Cells*, **95** (7), 1974–1978.

Jones, B.A., Ahrens, M.J., Yoon, M.H. *et al*. (2004) High-mobility air-stable n-type semiconductors with processing versatility: dicyanoperylene-3,4: 9,10-bis(dicarboximides). *Angewandte Chemie-International Edition*, **43** (46), 6363–6366.

Kameyama, T., Osaki, T., Okazaki, K.I. *et al*. (2010) Preparation and photoelectrochemical properties of densely immobilized Cu(2)ZnSnS(4) nanoparticle films. *Journal of Materials Chemistry*, **20** (25), 5319–5324.

Kang, D.K., Lee, M.W., Kim, H.Y. *et al*. (2011) Electrohydrodynamic pulsed-inkjet characteristics of various inks containing aluminum particles. *Journal of Aerosol Science*, **42** (10), 621–630.

Kapur, V.K., Fisher, M. and Roe, R. (2001) Nanoparticle oxides precursor inks for thin film copper indium gallium selenide (CIGS) solar cells. II–VI Compound Semiconductor Photovoltaic Materials. Symposium (Materials Research Society Symposium Proceedings Vol. 668).

Kärkkäinen, A.H.O., Tamkin, J.M., Rogers, J.D. *et al*. (2002) Direct photolithographic deforming of organomodified siloxane films for micro-optics fabrication. *Applied Optics*, **41** (19), 3988–3998.

Kawase, T., Shimoda, T., Newsome, C. *et al*. (2003) Inkjet printing of polymer thin film transistors. *Thin Solid Films*, **438**, 279–287.

Kawashima, N., Kobayashi, N., Yoneya, N. *et al*. (2009) A high resolution flexible electrophoretic display driven by OTFTs with inkjet-printed organic semiconductor. 2009 Sid International Symposium Digest of Technical Papers, vol. 40, pp. 25–27.

Ke, L., Chua, S.J., Zhang, K. and Yakovlev, N. (2002) Degradation and failure of organic light-emitting devices. *Applied Physics Letters*, **80** (12), 2195–2197.

Kervella, Y., Armand, M. and Stephan, O. (2001) Organic light-emitting electrochemical cells based on polyfluorene – investigation of the failure modes. *Journal of the Electrochemical Society*, **148** (11), H155–H160.

Khatavkar, V.V., Anderson, P.D., Duineveld, P.C. and Meijer, H.H.E. (2005) Diffuse interface modeling of droplet impact on a pre-patterned solid surface. *Macromolecular Rapid Communications*, **26** (4), 298–303.

Kiebele, A. and Gruner, G. (2007) Carbon nanotube based battery architecture. *Applied Physics Letters*, **91** (14), 144104.

Kim, L., Anikeeva, P.O., Coe-Sullivan, S.A. *et al*. (2008) Contact printing of quantum dot light-emitting devices. *Nano Letters*, **8** (12), 4513–4517.

Kim, T.H., Cho, K.S., Lee, E.K. *et al*. (2011a) Full-colour quantum dot displays fabricated by transfer printing. *Nature Photonics*, **5** (3), 176–182.

Kim, N.R., Lee, J.H., Yi, S.M. and Joo, Y.C. (2011b) Highly conductive Ag nanoparticulate films induced by movable rapid thermal annealing applicable to roll-to-roll processing. *Journal of the Electrochemical Society*, **158** (8), K165.

Kim, Y.H., Han, S.M., Lee, W. *et al*. (2007) Organic thin-film transistors using suspended source/drain electrode structure. *Applied Physics Letters*, **91** (4), 042113.

Kim, H.S., Kang, J.S., Park, J.S. *et al*. (2009) Inkjet printed electronics for multifunctional composite structure. *Composites Science and Technology*, **69** (7–8), 1256–1264.

Ko, S.H., Pan, H., Grigoropoulos, C.P. *et al*. (2007) All-inkjet-printed flexible electronics fabrication on a polymer substrate by low-temperature high-resolution selective laser sintering of metal nanoparticles. *Nanotechnology*, **18** (34), 345202.

Kopola, P., Aernouts, T., Sliz, R. *et al*. (2011) Gravure printed flexible organic photovoltaic modules. *Solar Energy Materials and Solar Cells*, **95** (5), 1344–1347.

Kordás, K., Mustonen, T., Toth, G. *et al*. (2006) Inkjet printing of electrically conductive patterns of carbon nanotubes. *Small*, **2** (8–9), 1021–1025.

Krebs, F.C., Jorgensen, M., Norrman, K. *et al*. (2009) A complete process for production of flexible large area polymer solar cells entirely using screen printing-first public demonstration. *Solar Energy Materials and Solar Cells*, **93** (4), 422–441.

Krebs, F.C., Senkovskyy, V. and Kiriy, A. (2010) Preorganization of nanostructured inks for roll-to-roll-coated polymer solar cells. *IEEE Journal of Selected Topics in Quantum Electronics*, **16** (6), 1821–1826.

Lewis, J.A. (2006) Direct ink writing of 3D functional materials. *Advanced Functional Materials*, **16** (17), 2193–2204.

Li, Y., Fu, C. and Xu, J. (2007) Topography of thin film formed by drying silver nanoparticle dispersion droplets. *Japanese Journal of Applied Physics Part 1: Regular Papers Brief Communications and Review Papers*, **46** (10A), 6807–6810.

Li, J., Unander, T., Cabezas, A.L. *et al*. (2011) Ink-jet printed thin-film transistors with carbon nanotube channels shaped in long strips. *Journal of Applied Physics*, **109** (8), 084915.

Liddle, J.A. and Gallatin, G.M. (2011) Lithography, metrology and nanomanufacturing. *Nanoscale*, **3** (7), 2679–2688.

Lim, J.A., Cho, J.H., Park, Y.D. *et al*. (2006) Solvent effect of inkjet printed source/drain electrodes on electrical properties of polymer thin-film transistors. *Applied Physics Letters*, **88** (8), 082102.

Lim, J.A., Lee, W.H., Lee, H.S. *et al*. (2008) Self-organization of ink-jet-printed triisopropylsilylethynyl pentacene via evaporation-induced flows in a drying droplet. *Advanced Functional Materials*, **18** (2), 229–234.

Liu, Y., Varahramyan, K. and Cui, T.H. (2005) Low-voltage all-polymer field-effect transistor fabricated using an inkjet printing technique. *Macromolecular Rapid Communications*, **26** (24), 1955–1959.

Loeffelmann, U., Korvink, J.G., Hendriks, C.E. *et al*. (2011) Ink jet printed silver lines formed in microchannels exhibit lower resistance than their unstructured counterparts. *Journal of Imaging Science and Technology*, **55** (4), 040302.

Lucas, L.A., Delongchamp, D.M., Vogel, B.M. *et al*. (2007) Combinatorial screening of the effect of temperature on the microstructure and mobility of a high performance polythiophene semiconductor. *Applied Physics Letters*, **90** (1), 012112.

Lunkad, S.F., Buwa, V.V. and Nigam, K.D.P. (2007) Numerical simulations of drop impact and spreading on horizontal and inclined surfaces. *Chemical Engineering Science*, **62** (24), 7214–7224.

Marin, V., Holder, E., Wienk, M.M. *et al*. (2005) Ink-jet printing of electron donor/acceptor blends: towards bulk heterojunction solar cells. *Macromolecular Rapid Communications*, **26** (4), 319–324.

Meier, H., Loeffelmann, U., Mager, D. *et al*. (2009) Inkjet printed, conductive, 25 mu m wide silver tracks on unstructured polyimide. *Physica Status Solidi A: Applications and Materials Science*, **206** (7), 1626–1630.

Merrin, J., Leibler, S. and Chuang, J.S. (2007) Printing multistrain bacterial patterns with a piezoelectric inkjet printer. *PLoS ONE*, **2** (7), e663.

Molesa, S.E., Volkman, S.K., Redinger, D.R. *et al*. (2004) A high-performance all-inkjetted organic transistor technology. IEEE International Electron Devices Meeting 2004, Technical Digest, pp. 1072–1074.

Moon, K.S., Choi, J.H., Choi, D.J. *et al*. (2008) A new method for analyzing the refill process and fabrication of a piezoelectric inkjet printing head for lcd color filter manufacturing. *Journal of Micromechanics and Microengineering*, **18** (12), 125011.

Moon, K.S., Dong, H., Maric, R. *et al*. (2005) Thermal behavior of silver nanoparticles for low-temperature interconnect applications. *Journal of Electronic Materials*, **34** (2), 168–175.

Morton, O. (2006) Solar energy: a new day dawning? Silicon valley sunrise. *Nature*, **443** (7107), 19–22.

Mustonen, T., Kordás, K., Saukko, S. *et al*. (2007) Inkjet printing of transparent and conductive patterns of single-walled carbon nanotubes and PEDOT-PSS composites. *Physica Status Solidi B: Basic Solid State Physics*, **244** (11), 4336–4340.

Ngamna, O., Morrin, A., Killard, A.J. *et al*. (2007) Inkjet printable polyaniline nanoformulations. *Langmuir*, **23** (16), 8569–8574.

Nishide, H. and Oyaizu, K. (2008) Materials science – toward flexible batteries. *Science*, **319** (5864), 737–738.

Noguchi, Y., Sekitani, T., Yokota, T. and Someya, T. (2008) Direct inkjet printing of silver electrodes on organic semiconductors for thin-film transistors with top contact geometry. *Applied Physics Letters*, **93** (4), 043303.

Noh, Y.Y., Zhao, N., Caironi, M. and Sirringhaus, H. (2007a) Downscaling of self-aligned, all-printed polymer thin-film transistors. *Nature Nanotechnology*, **2** (12), 784–789.

Noh, Y.Y., Cheng, X., Sirringhaus, H. *et al*. (2007b) Ink-jet printed ZnO nanowire field effect transistors. *Applied Physics Letters*, **91** (4), 043109.

Noh, Y.Y. and Sirringhaus, H. (2009) Ultra-thin polymer gate dielectrics for top-gate polymer field-effect transistors. *Organic Electronics*, **10** (1), 174–180.

Oh, J.H. and Lim, S.Y. (2010) Precise size control of inkjet-printed droplets on a flexible polymer substrate using plasma surface treatment. *Journal of Micromechanics and Microengineering*, **20** (1), 015030.

O'Toole, M., Shepherd, R., Wallace, G.G. and Diamond, D. (2009) Inkjet printed led based pH chemical sensor for gas sensing. *Analytica Chimica Acta*, **652** (1–2), 308–314.

Panhuis, M.I.H., Heurtematte, A., Small, W.R. and Paunov, V.N. (2007) Inkjet printed water sensitive transparent films from natural gum-carbon nanotube composites. *Soft Matter*, **3** (7), 840–843.

Park, J.U., Hardy, M., Kang, S.J. *et al*. (2007) High-resolution electrohydrodynamic jet printing. *Nature Materials*, **6** (10), 782–789.

Parsa, S., Gupta, M., Loizeau, F. and Cheung, K.C. (2010) Effects of surfactant and gentle agitation on inkjet dispensing of living cells. *Biofabrication*, **2** (2), 025003.

Paul, K.E., Wong, W.S., Ready, S.E. and Street, R.A. (2003) Additive jet printing of polymer thin-film transistors. *Applied Physics Letters*, **83** (10), 2070–2072.

Perelaer, J., Smith, P.J., Hendriks, C.E. *et al*. (2008) The preferential deposition of silica micro-particles at the boundary of inkjet printed droplets. *Soft Matter*, **4** (5), 1072–1078.

Perelaer, J., Smith, P.J., Van Den Bosch, E. *et al*. (2009) The spreading of inkjet-printed droplets with varying polymer molar mass on a dry solid substrate. *Macromolecular Chemistry and Physics*, **210** (6), 495–502.

Pesach, D. and Marmur, A. (1987) Marangoni effects in the spreading of liquid-mixtures on a solid. *Langmuir*, **3** (4), 519–524.

Pushparaj, V.L., Shaijumon, M.M., Kumar, A. *et al*. (2007) Flexible energy storage devices based on nanocomposite paper. *Proceedings of the National Academy of Sciences of the United States of America*, **104** (34), 13574–13577.

Ren, M., Gorter, H., Michels, J. and Andriessen, R. (2011) Ink jet technology for large area organic light-emitting diode and organic photovoltaic applications. *Journal of Imaging Science and Technology*, **55** (4), 40301.

Salas-Villasenor, A.L., Mejia, I., Hovarth, J. *et al*. (2010) Impact of gate dielectric in carrier mobility in low temperature chalcogenide thin film transistors for flexible electronics. *Electrochemical and Solid State Letters*, **13** (9), H313–H316.

Sargent, E.H. (2009) Infrared photovoltaics made by solution processing. *Nature Photonics*, **3** (6), 325–331.

Schock, H.W. and Noufi, R. (2000) CIGS-based solar cells for the next millennium. *Progress in Photovoltaics*, **8** (1), 151–160.

Sele, C.W., Von Werne, T., Friend, R.H. and Sirringhaus, H. (2005) Lithography-free, self-aligned inkjet printing with sub-hundred-nanometer resolution. *Advanced Materials*, **17** (8), 997–1001.

Sellinger, A., Tamaki, R., Laine, R.M. *et al*. (2005) Heck coupling of haloaromatics with octavinylsilsesquioxane: solution processable nanocomposites for application in electroluminescent devices. *Chemical Communications*, (29), 3700–3702.

Shaheen, S.E., Ginley, D.S. and Jabbour, G.E. (2005) Organic-based photovoltaics: toward low-cost power generation. *MRS Bulletin*, **30** (1), 10–19.

Shaheen, S.E., Radspinner, R., Peyghambarian, N. and Jabbour, G.E. (2001) Fabrication of bulk heterojunction plastic solar cells by screen printing. *Applied Physics Letters*, **79** (18), 2996–2998.

Shankar, R., Groven, L., Amert, A. *et al*. (2011) Non-aqueous synthesis of silver nanoparticles using tin acetate as a reducing agent for the conductive ink formulation in printed electronics. *Journal of Materials Chemistry*, **21** (29), 10871–10877.

Shigematsu, S., Ishida, Y., Nakashima, N. and Asano, T. (2008) Electrostatic inkjet printing of carbon nanotube for cold cathode application. *Japanese Journal of Applied Physics*, **47** (6), 5109–5112.

Shimoda, T., Matsuki, Y., Furusawa, M. *et al*. (2006) Solution-processed silicon films and transistors. *Nature*, **440** (7085), 783–786.

Shin, K.Y., Hong, J.Y. and Jang, J. (2011a) Flexible and transparent graphene films as acoustic actuator electrodes using inkjet printing. *Chemical Communications*, **47** (30), 8527–8529.

Shin, K.Y., Hong, J.Y. and Jang, J. (2011b) Micropatterning of graphene sheets by inkjet printing and its wideband dipole-antenna application. *Advanced Materials*, **23** (18), 2113–2118.

Shinar, J., Shinar, R. and Zhou, Z. (2007) Combinatorial fabrication and screening of organic light-emitting device arrays. *Applied Surface Science*, **254** (3), 749–756.

Singh, M., Chae, H.S., Froehlich, J.D. *et al*. (2009) Electroluminescence from printed stellate polyhedral oligomeric silsesquioxanes. *Soft Matter*, **5** (16), 3002–3005.

Singh, M., Haverinen, H.M., Dhagat, P. and Jabbour, G.E. (2010) Inkjet printing-process and its applications. *Advanced Materials*, **22** (6), 673–685.

Singh, M., Wu, Y.R. and Singh, J. (2003) Examination of LiNbO3/nitride heterostructures. *Solid-State Electronics*, **47** (12), 2155–2159.

Sirringhaus, H., Kawase, T., Friend, R.H. *et al*. (2000) High-resolution inkjet printing of all-polymer transistor circuits. *Science*, **290** (5499), 2123–2126.

Small, W.R., Masdarolomoor, F., Wallace, G.G. and Panhuis, M. in het, (2007) Inkjet deposition and characterization of transparent conducting electroactive polyaniline composite films with a high carbon nanotube loading fraction. *Journal of Materials Chemistry*, **17** (41), 4359–4361.

Song, J.W., Kim, J., Yoon, Y.H. *et al*. (2008) Inkjet printing of single-walled carbon nanotubes and electrical characterization of the line pattern. *Nanotechnology*, **19** (9), 095702.

Sprittles, J.E. and Shikhmurzaev, Y.D. (2012) Finite element framework for describing dynamic wetting phenomena. *International Journal for Numerical Methods in Fluids*, **68** (10), 1257–1298.

Sun, T.X. and Jabbour, G.E. (2002) Combinatorial screening and optimization of luminescent materials and organic light-emitting devices. *MRS Bulletin*, **27** (4), 309–315.

Sun, Y. and Rogers, J.A. (2007) Inorganic semiconductors for flexible electronics. *Advanced Materials*, **19** (15), 1897–1916.

Tang, J. and Sargent, E.H. (2011) Infrared colloidal quantum dots for photovoltaics: fundamentals and recent progress. *Advanced Materials*, **23** (1), 12–29.

Teichler, A., Eckardt, R., Hoeppener, S. *et al*. (2011) Combinatorial screening of polymer:fullerene blends for organic solar cells by inkjet printing. *Advanced Energy Materials*, **1** (1), 105–114.

Tekin, E., Smith, P.J., Hoeppener, S. *et al*. (2007) Inkjet printing of luminescent cdte nanocrystal-polymer composites. *Advanced Functional Materials*, **17** (1), 23–28.

Tekin, E., Smith, P.J., and Schubert, U.S. (2008) Inkjet printing as a deposition and patterning tool for polymers and inorganic particles. *Soft Matter*, **4** (4), 703–713.

Tekin, E., Wijlaars, H., Holder, E. *et al*. (2006) Film thickness dependency of the emission colors of PPE-PPVs in inkjet printed libraries. *Journal of Materials Chemistry*, **16** (44), 4294–4298.

Teranishi, R., Fujiwara, T., Watanabe, T. and Yoshimura, M. (2002) Direct fabrication of patterned PbS and CdS on organic sheets at ambient temperature by on-site reaction using inkjet printer. *Solid State Ionics*, **151** (1–4), 97–103.

Tobjörk, D. and Österbacka, R. (2011) Paper electronics. *Advanced Materials*, **23** (17), 1935–1961.

Tsai, M.H., Hwang, W.S., Chou, H.H. and Hsieh, P.H. (2008) Effects of pulse voltage on inkjet printing of a silver nanopowder suspension. *Nanotechnology*, **19** (33), 335304.

Tseng, W.J., Lin, S.Y. and Wang, S.R. (2006) Particulate dispersion and freeform fabrication of BaTiO3 thick films via direct inkjet printing. *Journal of Electroceramics*, **16** (4), 537–540.

Van Den Berg, A.M.J., De Laat, A.W.M., Smith, P.J. *et al*. (2007a) Geometric control of inkjet printed features using a gelating polymer. *Journal of Materials Chemistry*, **17** (7), 677–683.

Van Den Berg, A.M.J., Smith, P.J., Perelaer, J. *et al*. (2007b) Inkjet printing of polyurethane colloidal suspensions. *Soft Matter*, **3** (2), 238–243.

Van Der Bos, A., Segers, T., Jeurissen, R. *et al*. (2011) Infrared imaging and acoustic sizing of a bubble inside a micro-electro-mechanical system piezo ink channel. *Journal of Applied Physics*, **110** (3), 034503.

Van Osch, T.H.J., Perelaer, J., De Laat, A.W.M. and Schubert, U.S. (2008) Inkjet printing of narrow conductive tracks on untreated polymeric substrates. *Advanced Materials*, **20** (2), 343–345.

Vazquez, G., Alvarez, E. and Navaza, J.M. (1995) Surface-tension of alcohol plus water from 20-degrees-c to 50-degrees-c. *Journal of Chemical and Engineering Data*, **40** (3), 611–614.

Virtanen, J., Ukkonen, L., Bjorninen, T. *et al*. (2011) Inkjet-printed humidity sensor for passive UHF RFID systems. *IEEE Transactions on Instrumentation and Measurement*, **60** (8), 2768–2777.

Voit, W., Zapka, W., Belova, L. and Rao, K.V. (2003) Application of inkjet technology for the deposition of magnetic nanoparticles to form micron-scale structures. *IEE Proceedings-Science Measurement and Technology*, **150** (5), 252–256.

Wang, J.Z., Gu, J., Zenhausem, F. and Sirringhaus, H. (2006) Low-cost fabrication of submicron all polymer field effect transistors. *Applied Physics Letters*, **88** (13), 133502.

Wang, J.Z., Zheng, Z.H., Li, H.W. *et al*. (2004) Dewetting of conducting polymer inkjet droplets on patterned surfaces. *Nature Materials*, **3** (3), 171–176.

Wong, W.S., Chow, E.M., Lujan, R. *et al*. (2006) Fine-feature patterning of self-aligned polymeric thin-film transistors fabricated by digital lithography and electroplating. *Applied Physics Letters*, **89** (14), 142118.

Wood, V. and Bulović, V. (2010) Colloidal quantum dot light-emitting devices. *Nano Reviews*, **1**, 5202.

Wood, V., Panzer, M.J., Chen, J. *et al*. (2009) Inkjet-printed quantum dot-polymer composites for full-color AC-driven displays. *Advanced Materials*, **21** (21), 2151–2155.

Wu, Y.L., Li, Y.N., Ong, B.S. *et al*. (2005a) High-performance organic thin-film transistors with solution-printed gold contacts. *Advanced Materials*, **17** (2), 184–187.

Wu, Y.R., Singh, M. and Singh, J. (2005b) Sources of transconductance collapse in III–V nitrides – consequences of velocity-field relations and source/gate design. *IEEE Transactions on Electron Devices*, **52** (6), 1048–1054.

Wu, Y.R., Singh, M. and Singh, J. (2003) Gate leakage suppression and contact engineering in nitride heterostructures. *Journal of Applied Physics*, **94** (9), 5826–5831.

Xia, Y. and Friend, R.H. (2007) Nonlithographic patterning through inkjet printing via holes. *Applied Physics Letters*, **90** (25), 253513.

Yang, X., Froehlich, J.D., Chae, H.S. *et al*. (2009) Efficient light-emitting devices based on phosphorescent polyhedral oligomeric silsesquioxane materials. *Advanced Functional Materials*, **19** (16), 2623–2629.

Yatsushiro, S., Akamine, R., Yamamura, S. *et al*. (2011) Quantitative analysis of serum procollagen type I C-terminal propeptide by immunoassay on microchip. *PLoS ONE*, **6** (4), e18807.

Yoshioka, Y., Calvert, P.D. and Jabbour, G.E. (2005a) Simple modification of sheet resistivity of conducting polymeric anodes via combinatorial ink-jet printing techniques. *Macromolecular Rapid Communications*, **26** (4), 238–246.

Yoshioka, Y., Williams, E., Wang, Q. *et al*. (2005b) Progress in printed organic electronics and hybrid photovoltaics. 2005 Conference on Lasers & Electro-Optics.

Yoshioka, Y. and Jabbour, G.E. (2006a) Desktop inkjet printer as a tool to print conducting polymers. *Synthetic Metals*, **156** (11–13), 779–783.

Yoshioka, Y. and Jabbour, G.E. (2006b) Inkjet printing of oxidants for patterning of nanometer-thick conducting polymer electrodes. *Advanced Materials*, **18** (10), 1307–1312.

Yoshioka, Y., Jabbour, G. and Calvert, P. (2002) Multilayer inkjet printing of materials. *Nanoscale Optics and Applications*, **4809**, 164–169.

Yu, D.G., Zhu, L.M., Branford-White, C.J. and Yang, X.L. (2008) Three-dimensional printing in pharmaceutics: promises and problems. *Journal of Pharmaceutical Sciences*, **97** (9), 3666–3690.

Zhao, N., Chiesa, M., Sirringhaus, H. *et al*. (2007) Self-aligned inkjet printing of highly conducting gold electrodes with submicron resolution. *Journal of Applied Physics*, **101** (6), 064513.

Zirkl, M., Sawatdee, A., Helbig, U. *et al*. (2011) An all-printed ferroelectric active matrix sensor network based on only five functional materials forming a touchless control interface. *Advanced Materials*, **23** (18), 2069–2074.

Zirkl, M., Scheipl, G., Stadlober, B. *et al*. (2010) Fully printed, flexible, large area organic optothermal sensors for human-machine-interfaces. *Eurosensors Xxiv Conference*, **5**, 725–729.

10

Flat Panel Organic Light-Emitting Diode (OLED) Displays: A Case Study

Julian Carter,[1] *Mark Crankshaw*[2] *and Sungjune Jung*[3]
[1]*Technology Consultant, United Kingdom*
[2]*R&D – Engineering, Xaar PLC, United Kingdom*
[3]*Department of Physics, University of Cambridge, United Kingdom*

10.1 Introduction

The period from 1995 to 2010 saw the emergence and then dominance of flat panel display technology over incumbent cathode ray tube technology. Laptop computers created the demand for what were then expensive displays, but flat panel displays have become ubiquitous not only because of their form factor and weight but also due to a significant effort to reduce the cost of production through a decrease in materials utilisation and capital equipment and by the economies that arise through processing larger substrates.

Flat panel displays, whether plasma, liquid crystal displays (LCDs) or organic light-emitting diode (OLED), are based on thin film technology, and as part of the cost reduction process manufacturers have been adopting thin film deposition technologies that utilise materials efficiently and are scalable to large substrate sizes. In addition, it has been essential that these technologies have a reasonable capital cost and can achieve high throughput. Inkjet printing has been developed for display applications precisely because it holds promise in decreasing the cost of manufacture of certain thin films used in displays. In particular, inkjet printing has been demonstrated for the alignment layer (Hiruma, *et al.*, 2010; Hwang and Chien, 2006), colour filters in LCD

Inkjet Technology for Digital Fabrication, First Edition. Edited by Ian M. Hutchings and Graham D. Martin.

displays (Moon *et al.*, 2008; Shin and Smith, 2008; Kim *et al.*, 2002), liquid crystal (LC) material in cholesteric colour displays (Chen *et al.*, 2005) and carrier transport, injection and emission layers in OLEDs (de Gans, Duineveld and Schubert, 2004; Tekin, Smith and Schubert, 2008; Singh *et al.*, 2010). However there have been significant challenges to make inkjet printing a viable technology for these applications. These include:

- formulating materials so that they are capable of being inkjet printed;
- achieving the specified film properties – for example film thickness, uniformity and electro-optical properties;
- developing inkjet equipment for manufacturing with high reliability and throughput;
- developing suitable metrology tools to keep the process within specification.

What has become apparent over the last 10 years is that the promise of inkjet as a manufacturing technology for displays as demonstrated by many different groups is one thing, but actually developing a manufacturing process capable of meeting the requirements for high yield and throughput is quite another, representing a significantly greater capability.

In this chapter, the focus will be on the use of inkjet printing to deposit polymer organic light-emitting diode (P-OLED) materials not only because this is the area with which the authors are most familiar but also because it is the most challenging application. The use of inkjet printing for P-OLED has required significant development of printing equipment, inkjet heads, ink formulation and the overall process.

10.2 Development of Inkjet Printing for OLED Displays

Plastic electronics evolved from fundamental work conducted in the field of molecular electronics in the 1960s and 1970s. Electrical activity in plastic materials was first observed in 1977 when H. Shirakawa, A.G. MacDiarmid and A.J. Heeger, who were later awarded the Nobel Prize for Chemistry, discovered that polymers could be made electrically conductive after some modifications (Chiang *et al.*, 1977; Shirakawa *et al.*, 1977). They found that when silvery films of the semiconducting polymer polyacetylene were exposed to chlorine, bromine or iodine, uptake of halogen occurs, and the conductivity increased dramatically by more than seven orders of magnitude.

In 1989, a research group led by Richard Friend at the Cavendish Laboratory of the University of Cambridge developed light-emitting diodes (LEDs) which were made from the conducting polymers (Burroughes *et al.*, 1990). Jeremy Burroughes, a research student at that time, had been researching the applications of conducting polymers when he noticed one sample was emitting a slight glow. 'At first I thought it was the reflection of the computer screen', Burroughs later commented; 'I turned it off to see if the glow remained, and it did. I was absolutely amazed by this as it was something that was supposed to be impossible' (Seldon, Probert and Minshall, 2003). Since the startling discovery of organic electro-luminescence from polymers in the Cavendish Laboratory, P-OLEDs have received significant attention as the basis for the most promising next-generation flat panel displays (Friend *et al.*, 2002; Haskal *et al.*, 2009; Forrest, 2008). Compared to conventional displays such as LED or LCD, the self-luminous display does not require a backlight, so it can be thinner and lighter, consumes less energy and can offer higher brightness and contrast. Inkjet printing technology has been demonstrated to be well-suited to deposition of light-emitting polymer (LEP) solutions.

OLED display substrates are typically patterned with an acrylate or polyimide photoresist in the form of banks to define pixel wells into which the organic material is to be printed. Prior to printing, the substrate is preconditioned in order to maximise wetting within the well and give a non-wetting (low surface energy) surface to the photoresist bank in order to confine the drying inks to within the pixel. In an ideal situation, ink completely fills the pixel well, such that the ink pins at a point dependent on only the volume of ink, rather than other factors, most notably the precise position of ink impact. Three organic layers are placed between the anode and cathode: a hole transport layer (HL), an inter (or primer) layer (IL) and an emission layer (EL), as shown in Figure 10.1. The conducting anode is transparent and has a large work function. Indium tin oxide (ITO) is commonly used, but conducting polymers such as PEDOT:PSS (i.e. poly(3,4-ethylenedioxythiophene) (PEDOT)–poly(styrenesulphonate)(PSS)), carbon-nanotubes or graphenes have been studied for use as transparent electrodes due to the scarcity of indium and the brittleness of conductive oxides (Fehse *et al.*, 2004; Hu, Hecht and Gruener, 2008; Yim, Park and Park, 2010; Novoselov, 2011). Charge carriers are injected into the polymer layers from the electrodes: 'electrons' from the cathode and 'holes' from the anode. Those transported electrons and holes are captured through electrostatic interaction, and the radiative recombination of electron and hole generates light within a thin emission layer which is less than 100 nm thick. A thin organic IL is inserted to improve the efficiency and lifetime of the device. The wavelength of the emitted light depends on the band gap of the light-emitting polymer used in the device. P-OLEDs can produce light with a very wide range of wavelengths including both visible and invisible light by modifying the precise structure of the light-emitting polymer. The key feature of P-OLED technology is that the conjugated polymers can be processed

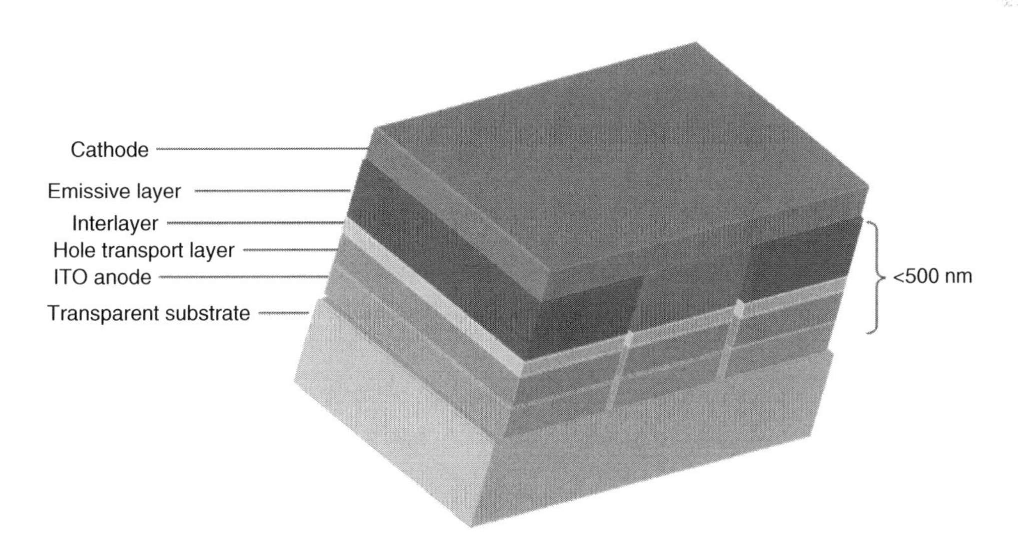

Figure 10.1 *The typical structure of a P-OLED cell (See plate section for coloured version).*

as solutions, allowing printing techniques to be used to directly pattern red-green-blue (RGB) materials on a substrate.

In 1998, Hebner *et al*. first used inkjet printing to directly deposit patterned luminescent doped-polymers (polyvinylcarbazol (PVK)) films, with dyes of coumarin 6 (C6) and coumarin 47 (C47), and fabricated P-OLEDS from inkjet-deposited doped polymers (Hebner *et al.*, 2006). In the same year, Bharathan and Yang also demonstrated a P-OLED device made with a conventional Epson desktop printer by printing an aqueous solution of semiconducting PEDOT (Bharathan and Yang, 1998). They also successfully fabricated dual colour light-emitting pixels by printing a 2 wt% aqueous solution of orange-emitting poly[5-methoxy-(2-propanoxy-sulfonide)-1,4-phenylene vinylene] (MPS-PPV) on a spin-coated film of blue-emitting poly[2,5-bis[2-(N,N,N-triethyl-ammonium)ethoxy]1,4-phenylene] (PPP-Net3+) (Yang *et al.*, 2000).

A significant step to fabricate an active matrix RGB multicolour panel was made by Cambridge University's Cavendish Laboratory in collaboration with Seiko Epson Corp. Kobayashi *et al*. developed a system for depositing the light-emitting polymer poly(para-phenylene vinylene) (PPV) for a green or a red emitter with an Epson inkjet print-head followed by spin coating of poly(di-octyl fluorene) (F8) to form an electron-transferring layer or a blue emitter (Kobayashi *et al.*, 2000). Using this system, they successfully patterned EL on a thin-film transistor (TFT) substrate and displayed a RGB multicolour image. In 2002, Duineveld and colleagues reported on the inkjet fabrication of a true full-colour 80 ppi active and passive matrix display (Duineveld *et al.*, 2007). Much effort had been made in increasing pixel resolution and improving uniformity, longevity and manufacturability. The last decade has seen an increasing activity in the development of inkjet printing for depositing electronic materials. In particular, there have been demonstrations of inkjet printing of both HL and EL of OLED devices by more than a dozen display manufacturers and research institutes (Funamoto *et al*., 2006; Grove *et al*., 2002; Letendre, 2004; Carter *et al*., 2005; Lee *et al*., 2005; Gohda *et al*., 2003; Iino and Miyashita, 2009; Dijksman *et al*., 2007; Frazatti and De Andrade, 1999; Sonoyama *et al*., 2008; Haverinen, Myllyla and Jabbour, 1998; Seki *et al*., 2009; Suzuki *et al*., 2009; Villani *et al*., 2009; Wood *et al*., 2009; Ren *et al*., 2011).

More recently, Singh *et al*. demonstrated bright inkjet-printed OLEDs based on Ir-based phosphorescent macromolecules anchored on a polyhedral oligomeric silsesquioxane (POSS) molecular scaffolding used as a phosphorescent dye in a polymer inkjet containing a hole-transporting polymer, poly(9-vinylcarbazole), and an electron-transporting polymer, 2-4-biphenylyl-5-4-tertbutyl-phenyl-1,3,4-oxadiazole (PBD) (Singh *et al*., 2010). A peak luminance of more than 6000 cd m^{-2}, a low turn-on voltage (6.8 V for 5 cd m^{-2}) and a relatively high quantum efficiency of 1.4% were achieved in their research. Through improvement in dye chemistry and print morphology, the authors were able to achieve a peak luminance of 10 000 cd m^{-2}. A simple and scalable printing method has also been demonstrated to achieve patterned pixels for flexible, full-colour, large-area, AC-driven displays operating at video brightness (Wood *et al.*, 2009). They showed that a quantum dot–polymer composite could be inkjet-printed with stable ink solutions, and that it contributed to efficient and robust device architecture. They also reported that the inkjet printing technique was well-suited for integration with metal oxide dielectric layers, which could enable improved optical and electrical performance.

10.3 Inkjet Requirements for OLED Applications

10.3.1 Introduction

There are significant differences between the application of inkjet printing for graphic arts and P-OLED displays. For printing P-OLED displays, it is necessary to achieve a high specification of both drop placement and drop volume. A single misplaced or outsized drop might produce an imperceptible defect in a graphics print, but the same error in a display might cause a pixel to short-circuit and fail completely, or emit at a different brightness, and hence result in a much more visible defect. Averaging techniques used to mask failing nozzles are not always possible to implement, or are not always effective. In addition, the cost of the substrates is orders of magnitude higher than that of paper and card, so that the cost of scrap is significant and minimising scrap is crucial to the viability of the process.

Once the ink lands on the substrate, the processes that occur are also different. The display substrate is not porous and significant development work has been undertaken to transform the printed droplets into films that achieve the electro-optical performance required for the display. The wet droplets must dry by the removal of the solvents used to carry the active material without the benefit of absorption into the substrate; the solute initially distributed uniformly in the hemispherical wet drop has to dry to a flat film. Also, the solvents and other additives in the ink must evaporate to avoid contaminating the organic semiconductor and affecting the performance of the resulting device. The development of inkjet printing for P-OLED displays has therefore included developments of the inkjet printer, the inkjet head, ink formulation, surface processes and printing and drying methods.

10.3.2 Display Geometry

OLED displays are arrays of pixels made up of (in most cases) red, green and blue sub-pixels, with the geometry shown in Figure 10.2. The number of pixels and their size are determined by the application. For high-definition televisions (HDTVs), the pixel array is 1080×1920 and typical sizes range from 37 to 65 in. (940–1651 mm) on the diagonal, corresponding to sub-pixel pitches of 142 and 250 μm respectively. For hand-held devices such as smartphones, the pixel arrays are up to a wide video graphics array (WVGA) or 480×800 pixels with sizes ranging from 2.9 to 3.8 in. (74–97 mm) diagonal corresponding to sub-pixel pitches of 26 and 35 μm respectively. To prevent overspill of ink from one pixel to the neighbouring one, it is necessary to create areas between the pixels of low surface energy: typically photo-patterned resins. As a result, the active area is somewhat smaller than the total area, but the aperture ratio (the ratio of active to total pixel area) is arranged to be as high as possible.

10.3.3 Containment and Solid Content

Our first consideration is whether we can apply enough material in our pixel. For a target dry film thickness, given solid loading of the ink and well area, the volume of ink is completely defined. What is not immediately apparent is that the applicability of inkjet printing then is determined both by whether this amount of ink can be contained by the well for small pixels and by whether this volume of ink can 'wet out' the well for large

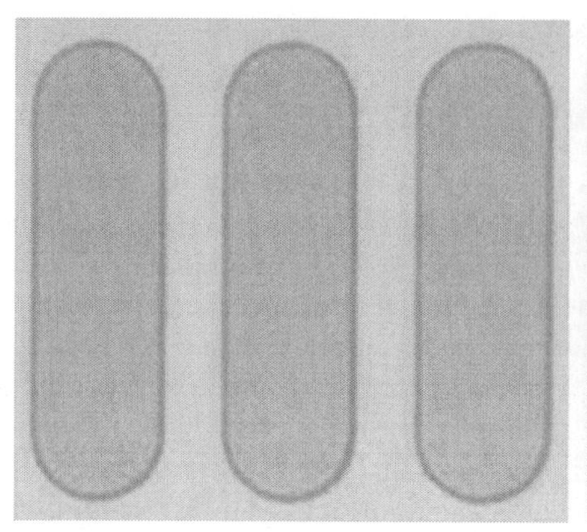
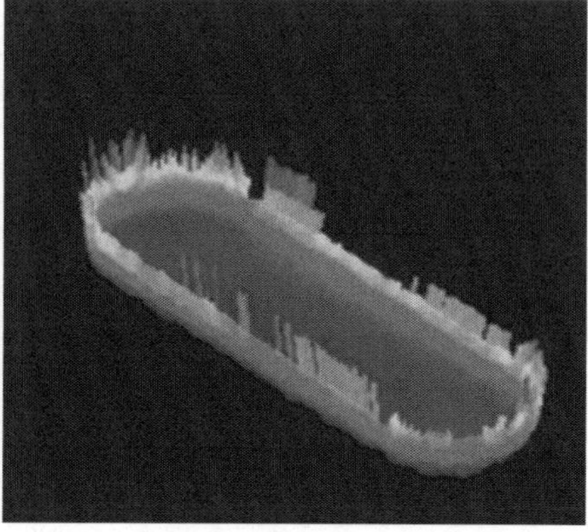

Figure 10.2 *Photomicrograph of printed PEDOT and white light interferometry representation of a PEDOT film profile in one of the wells – the uniformly coloured areas represent a thickness variation of ±2 nm (See plate section for coloured version).*

pixels. To demonstrate this point, we consider the inkjet printing of two sizes of pixels: a small pixel 30 μm across that might be found in a smartphone display and the other 180 μm across that might be found in a large TV display. Data for these examples is given in Table 10.1. We consider in the first instance that the solid content of the ink is 1% by weight and assume densities of 1000 kg m^{-3} for both solvent and solid, with a target film thickness of 70 nm. If we define the capacity of the well as the amount of ink that is contained within the well area at a contact angle defined by the bank and ink surface energies (which, for this example, is assumed to be 70°), then we can see that the maximum containable volume is smaller than that required to achieve a 70 nm film: that is, in this case it will not be possible to print such a small pixel with an ink of this solid content. At the other extreme, we consider the case of the larger pixel. In this case, we define a 'just-wetted' volume to be the volume of ink where the meniscus just touches

Table 10.1 *Data relating to examples of printing into two different well sizes, as discussed in Section 10.3.3.*

Parameter	Smartphone display: 30 μm sub-pixel pitch	Large TV display: 250 μm sub-pixel pitch
Sub-pixel pitch (μm)	30	250
Bank width (μm)	5	20
Maximum aperture ratio	0.74	0.83
Required wet volume (μm^3)	13 936	1 095 866
Maximum containable volume (μm^3)	11 664	8 381 157
Just wetted volume (μm^3)	1782	1 387 903
Drop volume (pl)	3	30
Drop volume (μm^3)	3000	30 000
Drop diameter (μm)	18	39
Number of drops to fill	4.6	36.5

the well edge but the contact angle is defined by the ink and well surface energies and in this case is assumed to be 15°. We can see that the amount of ink required to create a 70 nm film is not enough to wet out the well, that is, to print this well will require an ink with a lower solid content so that the volume dispensed is sufficient to fill the well.

Table 10.1 illustrates the fact that, for a well dimension L, the ink volume required to attain a certain fixed depth scales with L^2, whereas the ink contained in the well scales as L^3. It follows that ideally, with everything else being equal, the solid content should scale as $1/L$. However, scaling the solid content is usually not possible since it has a direct impact on the quality of the jetting process and the drying process. The maximum solid loading is also limited by the solubility of the polymer, and changing this property involves making changes at the molecular level which will alter the electro-optical properties of the film.

10.4 Ink Formulation and Process Control

General aspects of ink formulation have been discussed in Chapter 4. Here we concentrate on aspects related to the OLED display application. The ink formulation is composed of a polymer material dissolved or dispersed in a solvent, or more usually a combination of solvents. The formulation activity has to take into account all of the processes by which the solid material in the ink is transformed from material in an ink container to become an active element in a display. The first requirement, in common with any inkjet ink, is that the ink should be stable. This is necessary not only to ensure that problems do not occur within the ink system or print-head, but also to ensure uniformity of the material on the substrate. Secondly, the solution should be jettable, meaning that the ink has suitable rheology to form well-defined and repeatable drops at a suitable frequency with minimal satellite formation and acceptable drop deviation. To make the jetting process more robust, it is also necessary to ensure that the ink does not dry too quickly and therefore dry in the nozzles, otherwise the nozzles will block. In common with inkjet for other applications, the jetting characteristics are determined by viscosity, surface tension and the variation of these with shear rate, and in turn these are influenced by the structure and molecular weight of the polymer, the solid content and the choice of solvents. For polymers used in P-OLED displays, the solid content is typically between 0.2 and 2.5% w/v. Solvent choice dictates the ink-drying rate, nozzle plate wetting, rheology, substrate-wetting characteristics, profile of the dry ink in the pixel and shelf life of the ink formulation.

The ink formulation has to have suitable properties so that it can fully wet the pixel but stay contained within it and not overspill the separating bank structure (typically photoresist) into the adjacent pixel. After correctly filling the pixel, it must then dry to form a flat film. The balance between wetting and containment is also adjusted through the careful control of the surface energies of the pixel and bank surfaces. Such surface energy control is achieved through selection of the material used for the bank structures, substrate-processing methods and surface treatments before printing such as plasma, UV-ozone or solution treatment.

In general terms, the behaviour of drying drops of ink is explained by the coffee-ring effect first modelled by Deegan *et al*. (1997, 1997) and also discussed in Chapter 5.

For the simplest case of circular pixels, the wet ink forms a section of a sphere, where the angle made by the drop surface with the substrate is the contact angle. When pinning occurs (which it invariably does for the inks and surfaces used in P-OLED display manufacturing), the drying drop maintains its diameter and solute is carried to the edges of the drop, since this is where solvent evaporation is fastest, forming a ring of material at the outer edges of the pixel. The amount of material carried to the edge depends on a number of factors, in particular how long the process of material transfer can occur before the drying drop gels, and the uniformity of the drying environment. If the polymer is very soluble and the drying time long, then all of the solute can be carried to the edge of the well. By selecting suitable solvents, solid content and drying environment, it is possible to get films that are uniform across the pixel, as shown in Figure 10.2. In this particular example, a high-boiling-point solvent is used to minimise the ambient evaporation rate and successfully separate the drying process from the printing process. Better control of the resulting profile can then be obtained by adjustment of the drying process parameters, for example temperature. The uniform colour of the white light micrograph indicates the high level of uniformity obtained within the pixel. White light interferometry is used to obtain volume and profile information, and the resulting three-dimensional representation is also shown in Figure 10.2. The blue region represents an area, greater than 80% of the pixel, within which the film uniformity is controlled to within 2 nm. When optimising the well-filling process, it is more usual to take sections of the pixel thickness profile such as along the major and minor directions to capture the behaviour.

The same drying phenomena can also lead to non-uniformities at swathe edges (i.e. the edge of the band of pixels printed by a single print-head pass). A formulation that leads to a flat film at the centre of a swathe can give a non-uniform film profile for pixels around a swathe join due to non-uniform drying conditions at the swathe edge in the time before the adjacent swathe is printed. Figure 10.3 shows the effect of a swathe edge on the profiles of a PEDOT ink printed into wells created by patterning photoresist. By reformulating the PEDOT so that it dries on a substantially longer timescale than the printing process, the change in film profile can be avoided. The same is also true for the EL ink. Figure 10.4 shows volume data collected for a fraction of a display printed in RGB and including an IL to improve device efficiency. The data is derived by subtracting integrated volume results for IL and for IL + EL from those for PEDOT in the same column. Note that the automated measurement of the thin interlayer film, only ~10 nm thick, can be difficult so some of the data is missing, and the combined IL + EL volume has to be used to infer variation in either layer. Swathe effects are characterised by their consistent periodic nature across the panel, and most noticeable is the almost linear 4% change in blue volume over each swathe. Also noticeable are the consistent outliers visible in the red pixel data, but these effects were not significant enough to cause a visual artefact in the display.

During the development of polymers for P-OLED devices, the standard method of assessing performance is by spin coating thin films, usually from mixed xylene or toluene solvents. Formulating an ink so that it has suitable jetting and film-forming properties requires the use of different solvents, and these have the potential to impact device performance. It follows that during the formulation process, a significant activity is to develop formulations where the device performance is minimally affected not only by

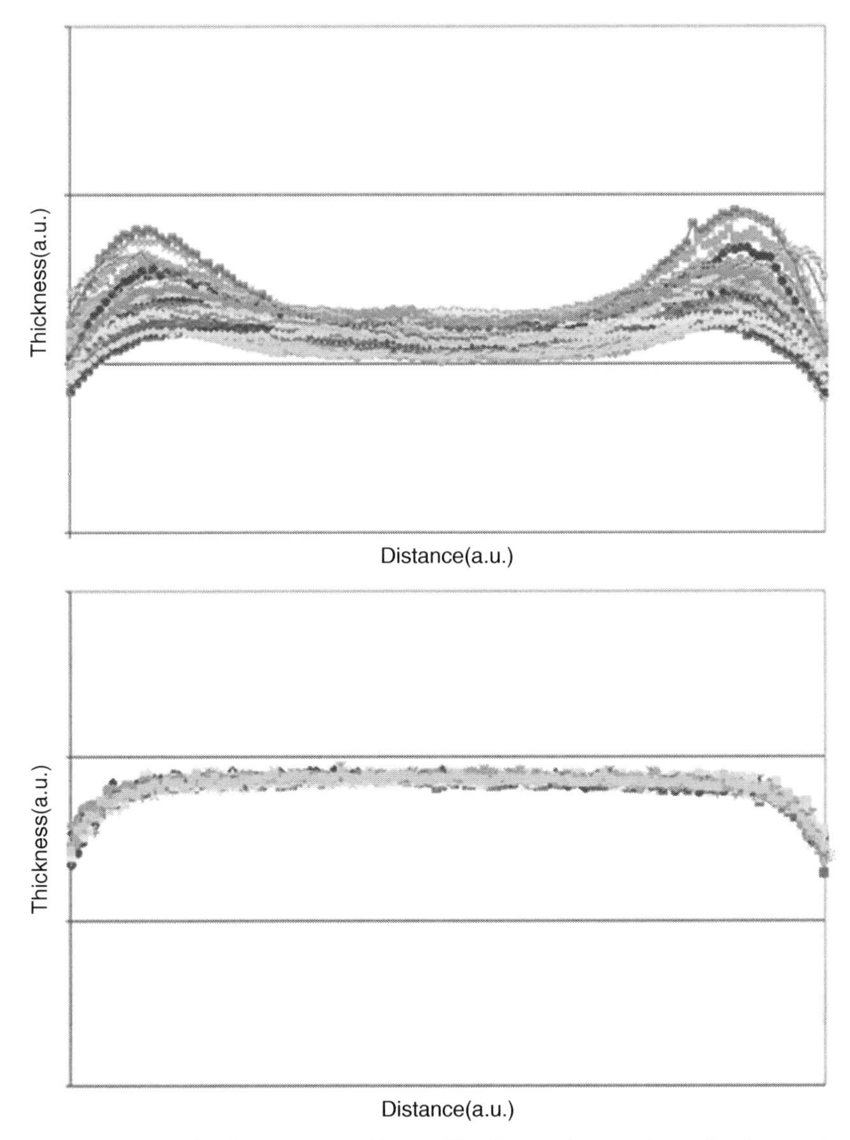

Figure 10.3 *The graphs show PEDOT film profiles for pixels on either side of a swathe join. The top graph shows profiles for the case where the swathe join is significant, whereas the lower graph shows profiles for a reformulated ink that was developed to remove the sensitivity to swathe edges (See plate section for coloured version).*

the formulation but also by the inkjet deposition process itself. One way to make this assessment is to make test devices by blanket printing or flood printing the polymers using the inkjet printer, and then drying the inks using a method analogous to that used for pixel printing. Apart from benchmarking ink performance, another advantage of inkjet printing for fabricating test cells is that it allows for combinatorial experiments, such as thickness optimisations to be carried out on a single substrate.

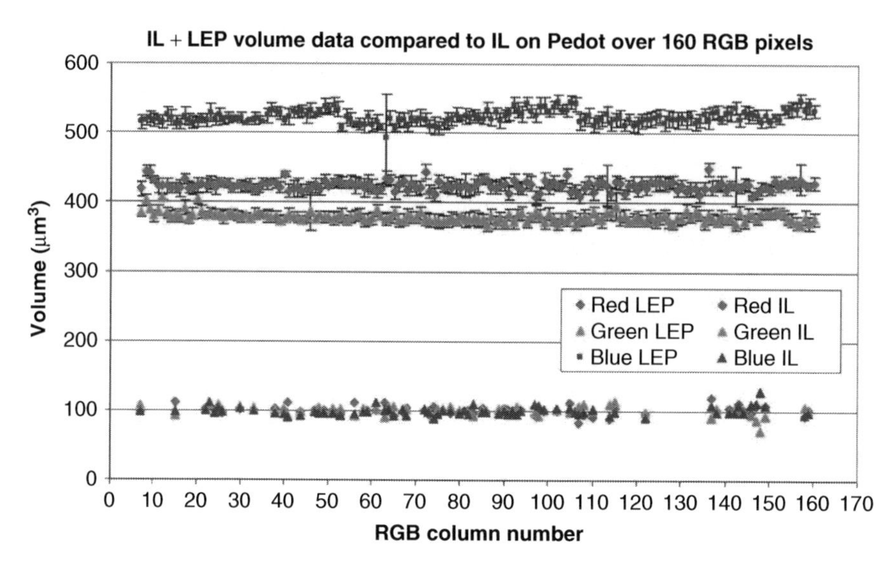

Figure 10.4 *Dry pixel volumes of interlayer and of emissive layer+ interlayer as measured over several printed swathes on a segment of a wide extended graphics array (WXGA) display test substrate. Consistent variation in the blue is clearly visible and could be correlated to the driven display (See plate section for coloured version).*

10.5 Print Defects and Control

The key requirements for the printer or print-head system are that some specified targets are met for drop position, drop volume, print reliability and throughput. The specification for drop position is determined by the geometry of the display to be printed, the drop size dispensed by the head and the surface energies of ink and surfaces. As an example, displays with a colour pixel pitch of between 100 and 150 ppi (corresponding to sub-pixel dimensions of $\sim$85–55 μm) printed with a 10 pl head (corresponding to a drop diameter of $\sim$25 μm) on suitable treated surfaces require drop-landing accuracies of better than $\pm$10 μm. For example, a 14 in. (356 mm) substrate would require between 10 and 20 million drops to be dispensed and, because a single misplaced drop could cause a pixel fault, each one of the drops has to land within 10 μm of where it was intended. The main components of drop-landing error are mechanical placement of the inkjet head over the pixel (stage accuracy) and deviation of the ink drop as it is ejected from the head (jetting deviation). The required mechanical placement accuracy can be achieved using high specification stages, for example air-bearing stages with suitable encoding. In addition, certain fixed stage bow and linearity corrections can be applied to the printed image by the printer system software. The jetting deviations depend to a large degree on the design and manufacture of the head, as well as the ink formulation. Print-heads are available that are specifically designed for printed electronics applications, or suitable for them, with drop deviations of less than 10 mrad. For a typical head–substrate distance, this would give rise to drop position errors of less than $\pm$5 μm. With the best heads and optimisation, deviation across the whole head within $\pm$2 mrad has been achieved, corresponding to the level of control required to print at 200 ppi.

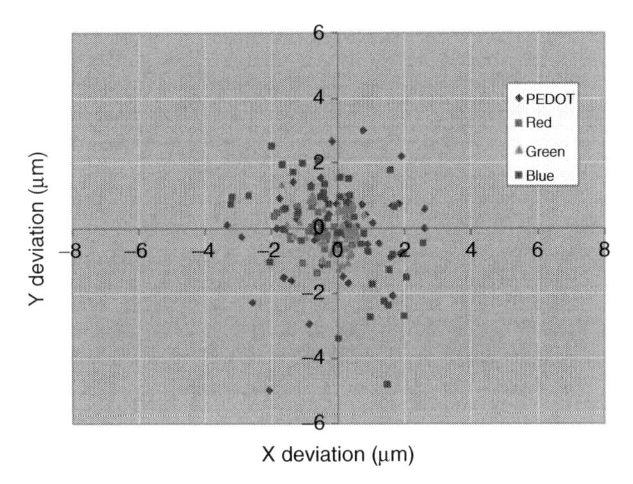

Figure 10.5 *Drop deviation from a range of nozzles of PEDOT, red, green and blue emissive layer inks for a particular location on a substrate (See plate section for coloured version).*

Figure 10.5 shows the typical landed drop deviation that is attributable to jetting deviation as measured with a calibrated vision measurement system (Quick Vision Pro 404, Mitutoyo Corp.). Each point represents the average error in the position of several drops from a particular nozzle. In Figure 10.5 ~30 nozzles are shown per ink, and in all cases the total deviation is less than 5 μm from the intended position. Note also the differences in jetting deviation between inks, indicating that the ink also has an impact on this parameter. The other feature to note from Figure 10.5 is that for a well-aligned head, the jetting deviation across a sufficiently large number of nozzles is random. This allows the mechanical deviation to be measured by averaging the drop deviation at any position on a printed substrate across a sufficiently large number of nozzles, since the jetting deviation averages to zero. Figure 10.6 shows the drop deviation due to mechanical errors obtained in this way. The deviation from the ideal position is measured at regular intervals across a 14 in. array of displays arranged in six columns and thus printed in six passes of the print-head (swathes). Measurements were made using the same calibrated metrology system used for nozzle deviation measurements that was described in this chapter.

The principal reason for a specification for drop volume is because, in common with the discussion here on film uniformity, the efficiency of the conversion of electrical energy into light is dependent to some extent on the thickness of the emissive layer. In inkjet printing, different pixels are printed with different nozzles, and differences in volume from nozzle to nozzle will give rise to different thickness of the emissive layer, resulting in a different brightness from adjacent pixels. The degree to which nozzle-to-nozzle volume variation will give rise to perceptible variations in display luminance depends on a number of factors, including the colour of the light emitted, but typically volume control to better than 2% can be required. This is typically beyond the inherent volume variation between nozzles, and so to achieve this level of volume control a drive-per-nozzle (DPN) system where the waveform to the inkjet head is adjustable on a nozzle-by-nozzle basis is employed. Hence, as an example, if a nozzle is producing too much ink, then the drive waveform applied to the nozzle can be adjusted individually to

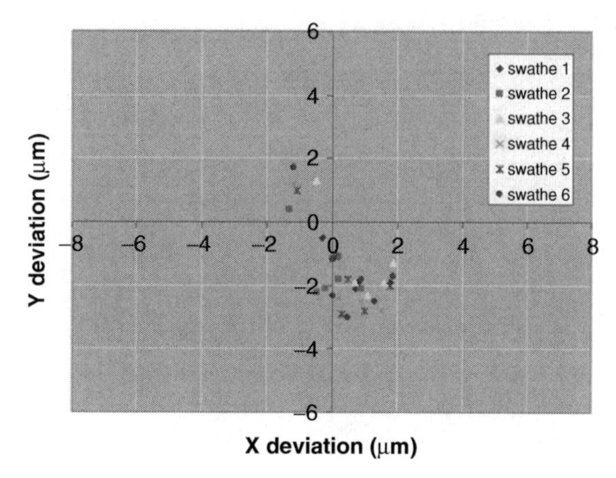

Figure 10.6 Drop deviation across a printed substrate where the deviation due to jetting deviation has been averaged out. In this case, six swathes of the head are required to cover the substrate and the drop positions are taken from six positions along the length of the swathe (See plate section for coloured version).

compensate. To determine whether the nozzle is in specification requires a system that is capable of measuring volume to significantly better than the ±2% specification. One method to measure volume is by using a drop visualisation system to inspect drops as they are ejected from the nozzle.

Figure 10.7 shows two sequential images taken with the visualisation system on a Litrex printer (Litrex 140P, Litrex Corp.). As the drop emerges from the nozzle at the top of the image, the tail detaches from the nozzle and at some distance below the nozzle plate forms a spherical droplet. Measurements of the area of the spherical drop have been used to estimate drop volume, and DPN electronics has corrected for variation. This measurement is, however, dependent on the formation of spherical droplets and for some inks, particularly when jetting at high frequency, this does not happen. The

Figure 10.7 Images of drop formation taken at two different strobe delays showing how the tail coalesces to form a spherical drop (See plate section for coloured version).

method is also limited by the accuracy and repeatability of the image capture and analysis system. It is particularly sensitive to where the edge of the drop is determined to be (itself sensitive to lighting conditions and thresholding in the image analysis) as image pixels on the edge of the two-dimensional circular image of the droplet represent much more of the droplet volume than those more certain pixels located nearer the centre of the projection of the droplet. An alternative approach is to use a relationship between the velocity of a drop and its volume and to control the velocity of the drop. Measuring velocity is more straightforward than volume since velocity can be measured even when the drop is non-circular, and it is much less sensitive to the imaging and analysis inaccuracies noted in this chapter. Of course, it is still important to confirm the velocity–volume relationship, but this can be done 'off-line' either by measuring the mass of a large number of drops with a sensitive balance or by measuring the volume of the dried drop on a suitable surface. However, the velocity–volume relationship may itself be subject to variation between individual nozzles and this has to be considered; indeed, the reason for any given nozzle's volume inaccuracy may also change the velocity–volume relationship for that nozzle (e.g. nozzle diameter). Alternatively, printed drop volumes can be measured directly, for example by using a white light interferometer-based optical profiler (e.g. a NewView 5000, Zygo Corp.) to perform measurements on arrays of dried ink drops. However, this method is significantly slower than measuring printed drops in flight. But by using sufficient sample sizes for each nozzle, by carefully controlling the measurement condition and by minimising noise from the laboratory environment, it is possible to obtain repeatable nozzle-to-nozzle measurements at the accuracy required for DPN adjustment to better than 2% variation.

Printing reliability is a function of both the ink formulation and the printing system. One key parameter to ensure reliable jetting is the pressure of the ink in the head, and its effect on control of the meniscus in the nozzle. If the head moves, it is possible to induce pressure fluctuations in the head that lead to nozzle plate flooding or to air ingestion, both of which could cause jetting to go out of specification. This kind of failure becomes more acute as the print frequency is increased. To overcome pressure fluctuations, it has been necessary to actively control the pressure using a mini-reservoir attached to the head itself. Using this system, it has been possible to print emissive layer inks at frequencies of >10 kHz. Figure 10.8 shows a comparison of displays produced by printing with and without active ink pressure control: the left-hand image is without control, and the right-hand image with control. Without pressure control, it is clear that jets from some nozzles may deviate and in some cases stop printing altogether.

10.6 Conclusions and Outlook

Although inkjet printing is a mature technology for the graphic arts, a significant development in printers, print-heads and ink formulation has been required to meet the requirements of P-OLED display manufacture. Inks that meet the requirements for reliable high-speed jetting and at the same time give acceptable film-forming and device properties have been formulated. These inks allied with DPN technology allow the manufacture of P-OLED displays using single pass printing. In addition, the commercial availability of inkjet printing equipment applicable to glass sizes up to gen7 (Albertalli, 2005) and

Figure 10.8 *The two images show the photoluminescence from printed displays where the print direction is vertically down the page. The left-hand display was printed without pressure control of the meniscus. The right-hand display was printed with pressure control at 15 kHz.*

the prospect of metrology tools to match make the scalable manufacture of P-OLED for TV-sized displays an achievable target within the next few years.

Acknowledgements

We would like to thank all members of the Technology Development Centre of Cambridge Display Technology and our colleagues at Litrex and Spectra/Dimatix for providing technical support in carrying out the work described in this chapter.

References

Albertalli, D. (2005) Gen 7 FPD inkjet equipment – development status. *SID International Symposium Digest of Technical Papers*, **36** (1), 1200.

Bharathan, J. and Yang, Y. (1998) Polymer electroluminescent devices processed by inkjet printing: I. Polymer light-emitting logo. *Applied Physics Letters*, **72**, 2660–2662.

Burroughes, J.H., Bradley, D.D.C., Brown, A.R. *et al*. (1990) Light-emitting-diodes based on conjugated polymers. *Nature*, **347**, 539–541.

Carter, J., Lyon, P., Creighton, C. *et al*. (2005) Developing a scalable and adaptable ink jet printing process for OLED displays. *SID International Symposium Digest of Technical Papers*, **36** (1), 523.

Chen, S.F., Huang, C.H., Lu, J.P. *et al*. (2005) IER film and inkjet printing method for full-color transflective cholesteric LCD. *Journal of Display Technology*, **1** (2), 225–229.

Chiang, C.K., Fincher, C.R., Park, Y.W. *et al*. (1977) Electrical-conductivity in doped polyacetylene. *Physical Review Letters*, **39**, 1098–1101.

Deegan, R.D., Bakajin, O., Dupont, T.F. *et al*. (1997) Capillary flow as the cause of ring stains from dried liquid drops. *Nature*, **389**, 827–829.

Deegan, R.D., Bakajin, O., Dupont, T.F. *et al*. (2000) Contact line deposits in an evaporating drop. *Physical Review E*, **62** (1, Pt B), 756–765.

de Gans, B.J., Duineveld, P.C. and Schubert, U.S. (2004) Inkjet printing of polymers: state of the art and future developments. *Advanced Materials*, **16** (3), 203–213.

Dijksman, J.F., Duineveld, P.C., Hack, M.J.J. *et al*. (2007) Precision ink jet printing of polymer light emitting displays. *Journal of Materials Chemistry*, **17** (6), 511–522.

Duineveld, P.C., De Kok, M.A., Buechel, M. *et al*. (2002) Ink-jet printing of polymer light-emitting devices. *Organic Light-Emitting Materials and Devices V*, **4467**, 59–67.

Fehse, K., Walzer, K., Leo, K. *et al*. (2007) Highly conductive polymer anodes as replacements for inorganic materials in high-efficiency organic light-emitting diodes. *Advanced Materials*, **19**, 441–444.

Forrest, S.R. (2004) The path to ubiquitous and low-cost organic electronic appliances on plastic. *Nature*, **428**, 911–918.

Frazatti, A. and De Andrade, A.M. (2008) A contribution to the development of organic semiconductor displays by the inkjet technique. *ECS Transactions*, **1**, 607–616.

Friend, R.H., Gymer, R.W., Holmes, A.B. *et al*. (1999) Electroluminescence in conjugated polymers. *Nature*, **397**, 121–128.

Funamoto, T., Matsueda, Y., Yokoyama, O. *et al*. (2002) 27.5: Late news paper: a 130-ppi, full-color polymer OLED display fabricated using an ink-jet process. *SID International Symposium Digest of Technical Papers*, **33** (1), 899–901.

Gohda, T., Kobayashi, Y., Okano, K. *et al*. (2006) A 3.6-in. 202-ppi full-color AMPLED display fabricated by ink-jet method. *SID International Symposium Digest of Technical Papers*, **37**, 1767.

Grove, M., Hayes, D., Wallace, D. and Shah, V. (2003) 39.2: variable ink-jet printing of polymer OLED display panels using array printheads. *SID International Symposium Digest of Technical Papers*, **34** (1), 1182–1185.

Haskal, E.I., Buchel, M., Duineveld, P.C. *et al*. (2002) Passive-matrix polymer light-emitting displays. *MRS Bulletin*, **27**, 864–869.

Haverinen, H.M., Myllyla, R.A. and Jabbour, G.E. (2009) Inkjet printing of light emitting quantum dots. *Applied Physics Letters*, **94** (7), 073108.

Hebner, T.R., Wu, C.C., Marcy, D. *et al*. (1998) Ink-jet printing of doped polymers for organic light emitting devices. *Applied Physics Letters*, **72** (5), 519–521.

Hiruma, K., Suzuki, K., Kasuga, O. *et al*. (2006) Ink jet fabrication of alignment layers on high-temperature polysilicon liquid crystal panels. *SID International Symposium Digest of Technical Papers*, **37**, 1583–1586.

Hu, L., Hecht, D.S. and Gruener, G. (2010) Carbon nanotube thin films: fabrication, properties and applications. *Chemical Reviews*, **110** (10), 5790–5844.

Hwang, J-Y. and Chien, L-C. (2008) Alignment of liquid crystal with inkjet printed polyimide for flexible liquid crystal displays. *SID International Symposium Digest of Technical Papers*, **39** (1), 1801.

Iino, S. and Miyashita, S. (2006) Invited paper: printable OLEDs promise for future TV market. *SID International Symposium Digest of Technical Papers*, **37** (1), 1463.

Kim, Y.D., Kim, J.P., Kwon, O.S. and Cho, I.H. (2009) The synthesis and application of thermally stable dyes for ink-jet printed LCD color filters. *Dyes and Pigments*, **81** (1), 45–52.

Kobayashi, H., Kanbe, S., Seki, S. *et al*. (2000) A novel RGB multicolor light-emitting polymer display. *Synthetic Metals*, **111**, 125–128.

Lee, D., Chung, J., Rhee, J. *et al*. (2005) Ink jet printed full color polymer LED displays. *SID International Symposium Digest of Technical Papers*, **36** (1), 527.

Letendre, W. (2004) 44.1: challenges in jetting OLED fluids in the manufacturing of FPD using piezoelectric micro-pumps. *SID International Symposium Digest of Technical Papers*, **35** (1), 1273–1275.

Moon, K.S., Choi, J.H., Choi, D-J. *et al*. (2008) A new method for analyzing the refill process and fabrication of a piezoelectric inkjet printing head for LCD color filter manufacturing. *Journal of Micromechanics and Microengineering*, **18** (12), 125011.

Novoselov, K.S. (2011) Nobel lecture: graphene: materials in the flatland. *Reviews of Modern Physics*, **83** (3), 837–849.

Ren, M., Gorter, H., Michels, J. and Andriessen, R. (2011) Ink jet technology for large area organic light-emitting diode and organic photovoltaic applications. *Journal of Imaging Science and Technology*, **55** (4), 040301.

Seki, S., Uchida, M., Sonoyama, T. *et al*. (2009) Current status of printing OLEDs. *SID International Symposium Digest of Technical Papers*, **40** (1), 593.

Seldon, S., Probert, D. and Minshall, T. (2003) Case Study: Cambridge Display Technology Ltd, University of Cambridge Centre for Technology Management.

Shin, D-Y. and Smith, P.J. (2008) Theoretical investigation of the influence of nozzle diameter variation on the fabrication of thin film transistor liquid crystal display color filters. *Journal of Applied Physics*, **103** (11), 114905.

Shirakawa, H., Louis, E.J., MacDiarmid, A.G. *et al*. (1977) Synthesis of electrically conducting organic polymers – halogen derivatives of polyacetylene, (CH)X. *Journal of the Chemical Society – Chemical Communications*, **16**, 578–580.

Singh, M., Haverinen, H.M., Dhagat, P. and Jabbour, G.E. (2010) Inkjet printing-process and its applications. *Advanced Materials*, **22** (6), 673–685.

Sonoyama, T., Ito, M., Seki, S. *et al*. (2008) Ink-jet-printable phosphorescent organic light-emitting-diode devices. *Journal of the Society for Information Display*, **16** (12), 1229–1236.

Suzuki, M., Fukagawa, H., Nakajima, Y. *et al*. (2009) A 5.8-in. phosphorescent color AMOLED display fabricated by ink-jet printing on plastic substrate. *Journal of the Society for Information Display*, **17** (12), 1037–1042.

Tekin, E., Smith, P.J. and Schubert, U.S. (2008) Inkjet printing as a deposition and patterning tool for polymers and inorganic particles. *Soft Matter*, **4**, 703–713.

Villani, F., Vacca, P., Nenna, G. *et al*. (2009) Inkjet printed polymer layer on flexible substrate for OLED applications. *Journal of Physical Chemistry C*, **113** (30), 13398–13402.

Wood, V., Panzer, M.J., Chen, J. *et al*. (2009) Inkjet-printed quantum dot-polymer composites for full-color AC-driven displays. *Advanced Materials*, **21** (21), 2151–2155.
Yang, Y., Chang, S.C., Bharathan, J. and Liu, J. (2000) Organic/polymeric electroluminescent devices processed by hybrid ink-jet printing. *Journal of Materials Science – Materials in Electronics*, **11** (2), 89–96.
Yim, Y., Park, J. and Park, B. (2010) Solution-processed flexible ITO-free organic light-emitting diodes using patterned polymeric anodes. *Journal of Display Technology*, **6** (7), 252–256.

11

Radiofrequency Identification (RFID) Manufacturing: A Case Study

Vivek Subramanian
Department of Electrical Engineering and Computer Sciences, University of California, USA

11.1 Introduction

As discussed in Chapter 9, inkjet printing of active electronics has received substantial attention in recent years. There have been several intriguing demonstrations of printed transistors and printed diodes, and various classes of printed passive components including conductors, capacitors, antennae, interconnects, and so on. The tremendous successes over the last decade in the realization of printed active electronics have resulted in increased emphasis on the development of novel electronic systems to exploit printed electronics. In general, the system focus of printed electronics has been aimed at displays, sensors, and radiofrequency identification (RFID) tags. In this chapter, we review the motivation, technology development, and state of the art of printed RFID with a particular focus on the applications and implications of inkjet printing on the same. We begin by considering the state of the art of conventional RFID, and then evaluate the advantages and disadvantages of printing in this regard. Next, based on the achievements of various classes of active electronic devices to date, we discuss the device, circuit, and system implementation of printed RFID with a particular focus on inkjet's implications on these. Finally, we review the outlook for printed RFID as a potential application of inkjet printing and discuss the remaining challenges and concerns.

Inkjet Technology for Digital Fabrication, First Edition. Edited by Ian M. Hutchings and Graham D. Martin.

11.2 Conventional RFID Technology

11.2.1 Introduction

RFID tags have been widely deployed in numerous applications ranging from applications in animal husbandry to library systems all the way up to sophisticated inventory control and factory automation. Some of the RFID tags that have been deployed contain batteries and are therefore called active tags, while others operate entirely based on power provided by the reader and are thus called passive tags. Similarly, RFID tags have been deployed that operate across a range of different frequencies. Therefore, in this overview, we begin by first examining the various classes of RFID tags that have already been developed and deployed. Next, we examine in more detail the specific classes of RFID that are candidates to exploit printed electronics. In particular, we look at the state of the art of the technology and economics of RFID in this regard.

11.2.2 RFID Standards and Classifications

In general, RFID tags may be classified based on two possible partitioning schemes:

1. the method by which they obtain power to operate;
2. the specific frequencies at which they communicate with the reader.

In the first scheme, tags are categorized based on whether they incorporate a battery (active tags) or depend on the reader to provide them with power (passive tags). Active tags, due to their higher costs and extended ranges (several meters or more), are currently using in inventory management, high-value asset tracking, and security access applications. Since passive tags do not contain built-in power sources, they depend on power that is supplied to them by the reader through electromagnetic coupling. The reader broadcasts large amounts of power (often isotropically), a small percentage of which is coupled into the antenna on the tag. The power transfer is often very inefficient. The maximum power that the reader can output is often constrained by licensing regulations and/or by system limitations imposed by the reader antenna size and power supply. Thus, given the poor coupling, the range of passive tags is usually limited; common ranges are a few centimeters to a few meters.

In the second classification scheme, RFID tags are classified based on the frequency at which the reader broadcasts information and power to the tag. In general, the frequency bands that are already used for RFID around the world are (i) a band below 135 kHz, (ii) a band at 13.56 MHz, (iii) a band at 900 MHz and (iv) a band at 2.4 GHz.

11.2.2.1 135 kHz RFID

135 kHz RFID is commonly called low-frequency (LF) RFID. Tags operating below 135 kHz are widely used in animal husbandry applications. This is because at this frequency, communication is remarkably robust despite the presence of liquids such as those present within living bodies. Tags operating below 135 kHz operate in the near field of the reader, thus interacting primarily with the reader field's magnetic component. Both the tag and reader antennae are inductors. When the tag is within a usable operating range of the reader, the two inductors are coupled. Since the inductors required

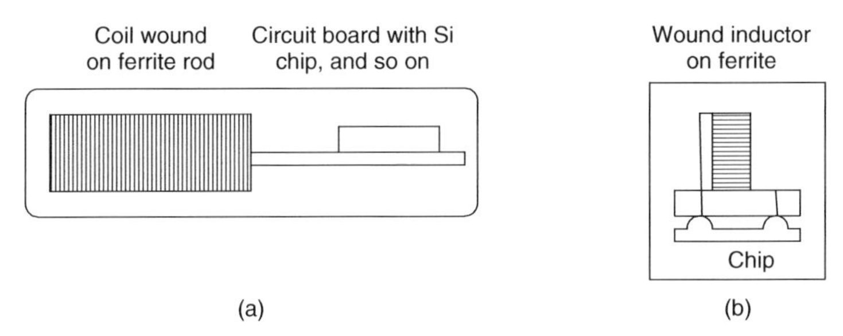

Figure 11.1 *(a,b) Prototypical 135 kHz RFID tag. The antenna is typically formed using a wound spiral inductor.*

to produce resonant circuits at such LFs are large, these inductors are typically wound coil inductors rather than planar spiral inductors (Finkenzeller, 2003). Hence, the fabrication costs associated with LF RFID are generally high, and their use is primarily limited to applications where LF is specifically advantageous, such as the applications discussed here. A schematic of a prototypical 135 kHz RFID tag structure is shown in Figure 11.1.

11.2.2.2 13.56 MHz RFID

13.56 MHz RFID is widely used in library systems and ticketing applications. Just as for 135 kHz RFID, 13.56 MHz RFID also uses near-field coupling and inductive antennae. 13.56 MHz RFID is commonly referred to as high-frequency (HF) RFID. Due to the use of a higher carrier frequency, unlike LF RFID, in HF RFID it is possible to implement the tag antenna as a planar spiral coil. These may be fabricated at substantially lower costs than the wound coil inductors used in LF RFID; as a result, 13.56 MHz RFID has received substantial attention as a candidate for low-cost RFID applications (Scharfeld, 2001). 13.56 MHz RFID tags are made by numerous manufacturers worldwide. Figure 11.2 shows a schematic of a typical 13.56 MHz RFID tag.

11.2.2.3 900 MHz RFID and 2.4 GHz RFID

RFID tags operating at 900 MHz and 2.4 GHz are called ultra-high-frequency (UHF) and microwave tags respectively. The near-field to far-field transition goes inversely with carrier frequency, such that these tags typically operate in the far-field region of the reader's

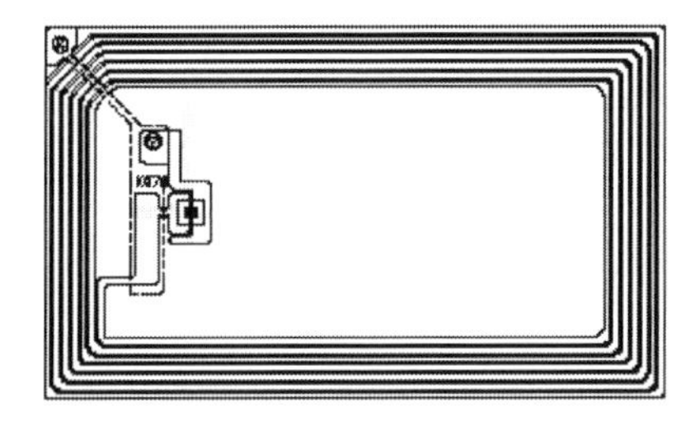

Figure 11.2 *Prototypical 13.56 MHz RFID tag. Note the planar spiral inductor geometry.*

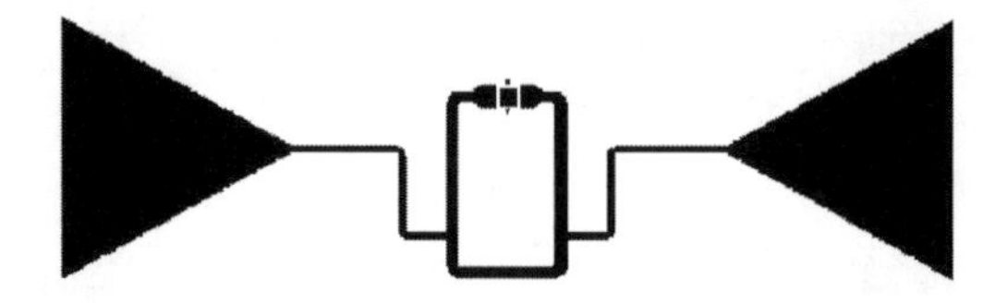

Figure 11.3 *Prototypical UHF RFID tag. Notice the planar dipole antenna configuration.*

electromagnetic field. Antennae are therefore implemented in a dipole configuration, as shown in Figure 11.3, rather than in the inductive configurations discussed in Section 11.2.2.2 (Glidden, Bockorick, and Cooper, 2004). This is almost always formed in a planar configuration. These systems are able to achieve relatively long-range operation and high data rates.

11.2.2.4 Summary of RFID Categories

Based on the categories discussed in Sections 11.2.2.1 through 11.2.2.3, we see in particular that HF and UHF RFID tags are commonly implemented in planar configurations. As a result, over the last decade, significant efforts have been devoted to reducing the cost of these tags, thus enabling their deployment in a range of low-cost applications. Also, for the same reasons, these particular classes of tags have received attention as candidate tags for the application of printing. Indeed, as we shall discuss in Section 11.4, the antennae are already printed in some cases. Prior to discussing the specifics of the tags, it is worthwhile to review the general architecture of HF and UHF RFID tags as implemented using silicon technology.

11.2.3 RFID Using Silicon

RFID technology to this point has had mixed success. While RFID has been tremendously successful in high-value asset-tracking applications, it has had only moderate success in the low-cost applications that offer the largest potential markets. This has been limited primarily by the current cost of RFID, which is still a little too high for the target markets (Homs and Metcalfe, 2004). This, in fact, is the primary motivation for the use of printing in low-cost RFID. Therefore, to facilitate a better understanding of the economics associated with printed RFID, an overview of conventional RFID based on silicon is provided.

The cost of silicon chips rises greatly with increase in chip size, due to a reduction in process yield and also due to a high cost per unit area associated with silicon. As a result, silicon chip size reduction is a dominant goal in silicon-based RFID. This has important consequences for architectural partitioning. The size of the antenna is independent of the process technology used and, in fact, can be very large, depending on the operating frequency and desired range. Therefore, in silicon-based RFID tags, the antenna is moved off-chip. In other words, everything but the antenna is fabricated on the silicon chip, which is then attached to a separate antenna, typically fabricated on a plastic inlay or strap. While this partitioning strategy reduces the cost of the silicon chip itself, it adds additional components into the cost equations associated with the overall tag. The overall cost equation for an RFID tag therefore includes three main components: (i) the cost of silicon, (ii) the cost of attach, and (iii) the cost of the antenna.

The cost per transistor of silicon technology has reduced in a sustained manner over 40 years, as a direct consequence of Moore's law. This reduction in silicon cost has been one of the enabling factors that have driven the cost reduction of silicon RFID. The area of silicon required for a typical silicon RFID device can be populated with the appropriate transistor-based circuitry such that the silicon's final cost is well below several US cents. The amount of silicon required for a typical RFID tag is exceedingly small – RFID chips smaller than 400 μm on a side have been demonstrated in 0.25 μm technology. Therefore, the actual silicon chip cost for RFID can be extremely low.

The second consequence of the small chip size associated with silicon RFID is related to attachment. The RFID chips do not contain an on-board antenna system. As discussed in this chapter, this antenna (e.g., a dipole for UHF tags, or an inductor and capacitor for HF tags) is mounted off-chip. It is therefore necessary to "attach" the silicon RFID chip to the external antenna (which is typically fabricated on a strip of plastic or paper) and establish appropriate electrical connections between the chip and the antenna. In normal circumstances, this involves two connections per tag as shown in Figure 11.4.

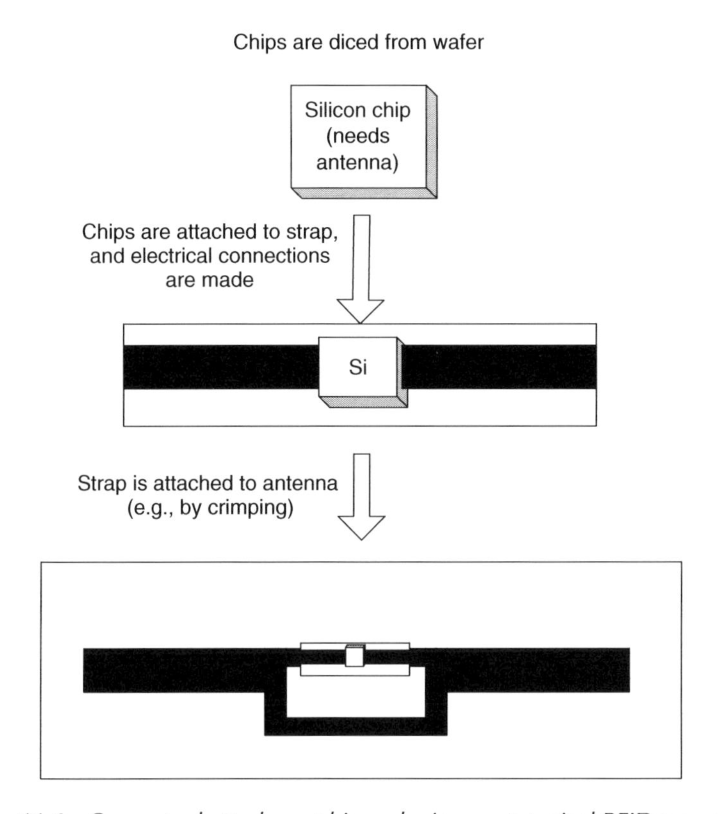

Figure 11.4 *Conceptual attachment hierarchy in a prototypical RFID tag process.*

The cost of attachment becomes significant when the cost of the chip is reduced to the level of typical RFID tags. In conventional pick-and-place approaches (in which a chip is placed using a robotic handler), the cost of attachment can be several cents. This raises the cost of the total tag, which, as was discussed in this section, limits the applications of RFID to a degree. As a consequence, several new techniques for attachment are under development, though none of these have been successfully commercially deployed to date. As will be discussed in Sections 11.3 through 11.5, this has opened up interest in printed electronics as a means of realizing low-cost RFID tags.

11.3 Applications of Printing to RFID

In general, printing has received attention for RFID fabrication in two main areas. Firstly, in the near term, printing has already been deployed for the realization of planar RFID antennae. Secondly, for future applications, printing is also being considered for the fabrication of the entire tags, including both the antenna and the active circuitry. Inkjet is one of several printing techniques that have been considered for these applications, and in Sections 11.4 and 11.5 we review these efforts and discuss their implications.

11.4 Printed Antenna Structures for RFID

11.4.1 The Case for Printed Antennae

Antenna technology has generally leveraged heavily off the technology used in the conventional process for fabricating printed circuit boards, which is discussed in Chapter 8. In the manufacture of typical printed circuit boards, wiring patterns are created lithographically on insulating substrates coated with copper. These are then etched using a copper etching solution (typically ferric chloride). This process is easily portable to the production of RFID antennae, and indeed the vast majority of RFID tags fabricated to date have made use of etched copper or aluminum antenna structures. Etched copper is attractive since it has very low resistivity, which ensures that the series resistance associated with antenna structures is minimized. This is important since the series resistance of the antenna directly impacts the efficiency of power harvesting by the tag from the reader, which reduces the tag's effective operating range. Unfortunately, etched copper is a comparatively expensive process. First, the lithography followed by the etch process typically requires at least five steps (deposit resist, expose, develop, etch, and strip), increasing its overall process cost. More importantly, the etching process produces substantial waste materials (ferrous and copper salts) which are expensive to abate and dispose.

Prior to discussing the specific applications of printing to RFID antenna manufacturing, we begin by reviewing the requirements for antenna materials. Again, we will focus primarily on HF and UHF RFID systems.

As discussed in Section 11.2.2.2, HF tags typically work by inductive coupling in the near-field region. As a consequence, the "antenna" consists of an inductor (and typically a capacitor as well). The inductor and capacitor together form a resonant circuit, as shown in Figure 11.5.

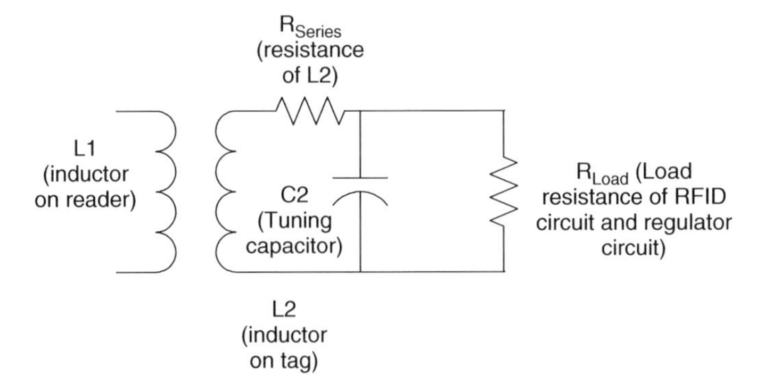

Figure 11.5 *Equivalent circuit for an HF RFID antenna stage.*

The two resistors are particularly important here. R_{series} is the series resistance associated with the inductor, while R_{load} is the equivalent resistance of the RFID circuitry. R_{series} purely contributes to loss of power. A typical measure of the efficiency of a resonant circuit is the quality factor, or Q. To maximize efficiency of power coupling, it is desirable to maximize Q; this is achieved by minimizing the series resistance. Typical Q ranges for HF tags are 1–10, depending on the antenna configuration, load resistance, and intended range of use. For typical HF tags, therefore, minimization of series resistance is extremely important.

Similarly, for UHF RFID tags, series resistance is also important, though for less important and perhaps less obvious reasons. The series resistance for UHF antennae is less important than for HF antennae for various reasons. Firstly, since the length of metal conductor in a typical dipole antenna is substantially shorter than in a spiral inductor, the overall resistance associated with UHF antennae is generally less of a concern. Secondly, the key to maximizing efficiency in these configurations is typically good matching. Therefore, provided the matching network is designed to account for the impedance of the antenna correctly, the range of the tag is typically less sensitive to the series resistance of the antenna; while series resistance does play a role in efficiency, it is not as dominant an effect as in HF tags.

11.4.2 Printed RFID Antenna Technology

Several well-known printing techniques have been demonstrated for the production of printed antennae, including screen, flexo, gravure, offset, and inkjet. Of these technologies, screen and flexo have been extensively demonstrated and are in commercial production. Inkjet has been used in conjunction with other techniques as well. Here, we will not review the more conventional techniques but will instead focus on the application of inkjet printing to antenna production.

Most printed antennae used in commercial RFID applications are fabricated using high-speed printing techniques as discussed in this chapter. Inkjet has generally been used much less for antenna fabrication. Unfortunately, inkjet has several drawbacks for the fabrication of antennae. Firstly, since inkjet uses a drop-by-drop approach using low mass-loaded inks, it typically requires several passes to produce relatively thick

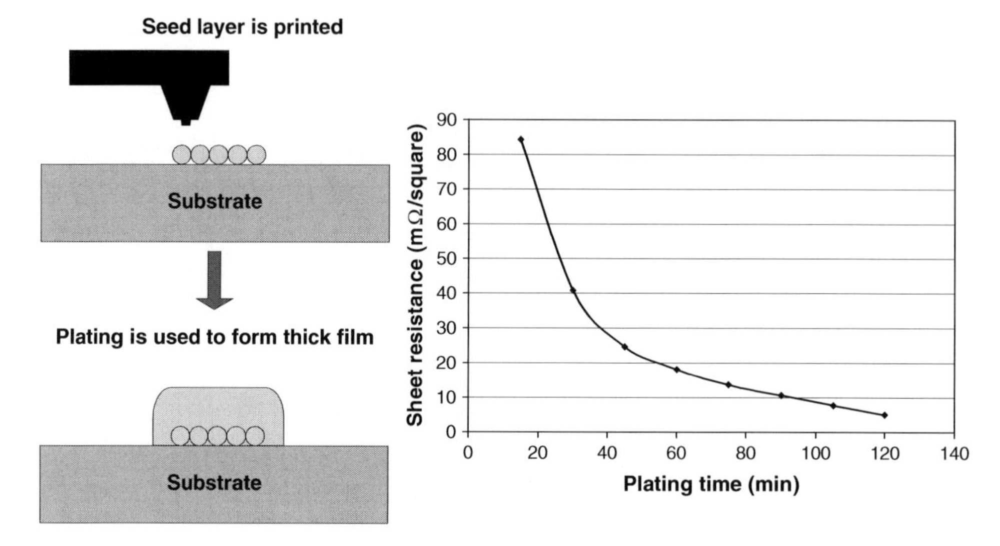

Figure 11.6 *Schematic antenna production process using inkjet printing followed by plating.*

films. Most inkjet-printed lines are less than 1 μm thick. This in turn reduces process throughput. Secondly, since inkjet tends to be a relatively slow technique compared to the high-speed techniques, it is necessary to use large arrays of heads to produce high-throughput inkjet systems; the manufacturability and process stability of such large arrays of print-heads are still uncertain for electronics applications, though page-wide heads are already in use in the graphic arts industry.

Inkjet does retain some advantages, however. Since it is able to deposit material with no binders in the ink (inkjet uses much lower viscosity inks than the aforementioned high-speed printing techniques, and therefore appropriate viscosities can be achieved without the need for additional binders), resulting films tend to be close to purely metallic in nature. Therefore, in recent years, inkjet has also been used to pattern thin seed layers of metal, which are subsequently plated to produce thicker features (as also discussed in Chapter 8). This process, illustrated in Figure 11.6, allows the realization of etched-copper-type resistivity in an additive process. While there are disposal costs associated with the plating solutions, these are generally expected to be lower than those associated with a typical etched copper process. Therefore, in recent years, there has been a resurgence in interest in inkjet for antenna applications.

11.4.3 Summary of Status and Outlook for Printed Antennae

Printed antennae currently represent a success story for the emergence of printed electronics. There is a fairly clear value proposition in support of printed antenna, at least for UHF applications. Current printed antenna technology generally does not meet the performance requirements for HF applications, at least for nonproximity applications; for near-contact applications, degraded antenna Q is potentially acceptable, so there may be a market for existing printed antenna technology in such near-contact applications as product authenticity verification and the like. In comparison to etched copper technology, printed antenna technology delivers a reduction in cost; with further deployment of the technology coupled with the increasing abatement cost associated with etched copper technology, this value proposition is expected to improve even further.

11.5 Printed RFID Tags

11.5.1 Introduction

In recent years, a tremendous amount of research has focused on the development of printed RFID tags. Conceptually, the idea is to print circuitry at the same time as and/or in a similar manner to the process used to print antennae. The key driver for this is, of course, cost. Therefore, to begin, it is worthwhile to perform a more detailed analysis of the cost imperatives for printed RFID.

As discussed in Section 11.2.3, the costs associated with silicon RFID do not scale rapidly due to the partitioning between the cost of the chip, the attachment process, and the antenna. The goal of printed RFID is to eliminate the cost of attachment in particular and, more generally, to reduce the overall cost of the entire RFID tag. The conceptual basis for the reduction in overall cost based on printing is the following.

From a cost perspective, printing is a process in which directed deposition of inks allows for the fabrication of arbitrary patterns without the need for the conventional subtractive approach used in conventional electronics (i.e., blanket deposition and lithography, followed by etching, stripping, and cleaning). However, existing high-speed printing techniques do not offer the resolution of state-of-the-art large-area lithography techniques. This trade-off between process throughput, capital expenditure, and resolution results in an important conclusion regarding the cost of printed electronics: printed electronics offers a large cost-per-unit-area advantage over conventional techniques, but, on the other hand, due to its limited resolution it will not offer an advantage in terms of cost per function or cost per transistor. Therefore, printed electronics is only economically attractive for applications that are area constrained, such as displays (for obvious reasons) and RFID tags (where the size of the antenna and passive components dominates the overall tag size).

Particularly with respect to RFID, the economic basis for printing is worthy of further consideration. In modern RFID tags, the cost of silicon has dropped rapidly to the point at which the costs of the antenna and the attachment process (with or without use of a strap) is now a significant fraction of the overall tag cost. Printing allows for the realization of tags printed with integrated antennae, or, at the very least, allows for the realization of circuitry printed directly on straps (note that strap attachment, given its lower resolution requirements, is much cheaper than chip attachment). Therefore, in an unusual way, printed electronics is not actually competing with the cost of silicon; rather, it is competing with the cost of attachment. This is likely to be beneficial for printed electronics, since the costs of attachment have not been reducing nearly as steadily as the cost of silicon, and therefore there is likely to be room at the bottom for printed electronics.

11.5.2 Topology and Architecture of Printed RFID

To begin our analysis of printed RFID, it is worthwhile to note that there are clear performance trade-offs when printed circuits are deployed to replace conventional silicon microelectronics. Firstly, the performance of printed semiconductors is much worse than the performance of silicon. Secondly, the line widths and layer-to-layer registration achievable by printing are substantially worse than those achievable in silicon technology. Based on these constraints, it is possible to examine possible circuit architectures for printed RFID tags and then discuss the implications of printed device performance on their viability.

As a first constraint, let us recall that all passive tags require that power be derived from the incident carrier frequency broadcast by the reader. For low-cost applications, this will likely be in either the HF or UHF bands, as discussed in Section 11.2.2.4. Given the performance limitations of printed materials, it is highly unlikely that harvesting UHF frequencies will be possible in the near future. As a result, the circuit analysis herein will focus exclusively on HF tags.

As discussed in Section 11.2.2.2, HF tags operate at 13.56 MHz. Power is inductively coupled from the reader to the tag. Figure 11.7 shows the block diagram architecture for a typical HF RFID tag. For the purposes of simplifying analysis, the tag has been broken down into several blocks, which will be individually discussed throughout the rest of this subsection.

11.5.2.1 Antenna Stage

As discussed in Section 11.2.2.2, the antenna in an HF tag operates through inductive power coupling. Conceptually, the reader antenna coil acts as a primary coil in a

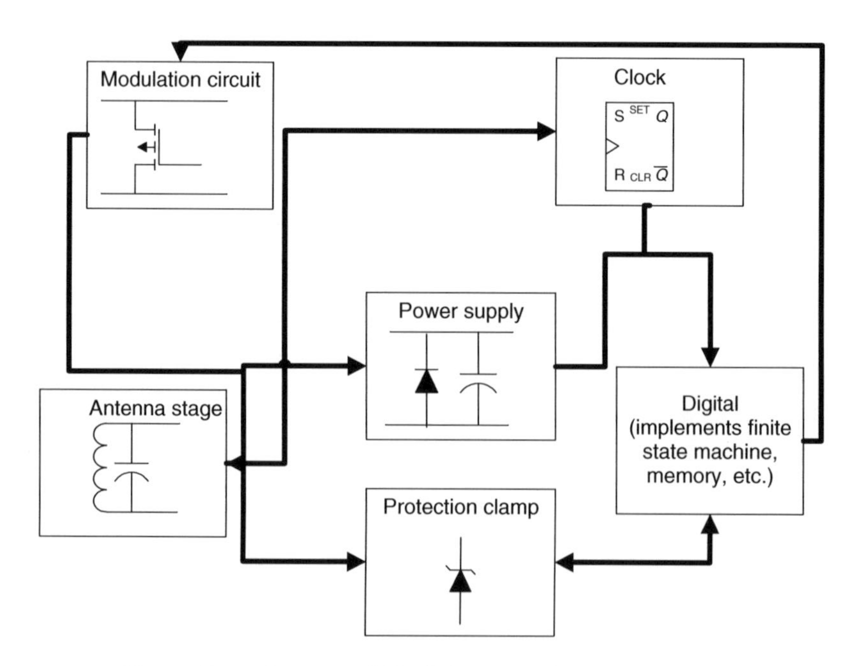

Figure 11.7 *Conceptual block diagram of a typical RFID tag.*

transformer, with the coil in the tag acting as a secondary. Coupling occurs through air; if the tag is within the near-field region of the reader antenna, magnetic coupling between the two coils occurs, and some voltage is "harvested" by the tag coil. To increase the voltage generated by the harvesting process, the antenna stage on the tag is typically a resonant circuit with the antenna inductor connected in parallel with a tuning capacitor. The specific values of the inductance and capacitance are chosen so as to cause resonance at 13.56 MHz. The consequence of this is that the voltage seen at the terminals of the tuned circuit is "Q boosted," where Q is the loaded quality factor of the antenna stage (Finkenzeller, 2003). As noted in Section 11.4.1, typical unloaded Q values in conventional RFID circuits range from 1 to 10, depending on the desired range. To achieve high Q, low series resistance of the inductor metallization is desired. As a result, substantial effort in recent years has been devoted to the development of low-resistance printed metals (Redinger, Farshchi, and Subramanian, 2004; Subramanian and Frechet, 2005). This antenna series resistance is extremely important, since the series resistance of the antenna acts as a loss mechanism, falling in series in the LC circuit, and thus reducing the Q of the circuit. This in turn reduces the power harvested and made available to the RFID circuit.

11.5.2.2 Rectifier, Power Supply, and Clamp

The voltage from the resonant circuit is applied to a rectification circuit. Rectification is performed using diodes, or diode-connected transistors. The output of the rectifier is connected into a filter capacitor, which smoothes out any ripple. This power is then used to drive the digital sections of the circuit. For printed RFID applications, rectification using both diodes and diode-connected transistors has been demonstrated. Diodes typically have the advantage of being faster than transistors; however, very few examples of high-performance fully printed diodes exist, and this is an area of intensive research today. Using nonprinted processes, however, several demonstrations of rectifiers running at 13.56 MHz have been made (Steudel *et al*. 2005).

Diode-connected transistors have also been demonstrated for use in rectifiers. Since the process steps used to fabricate these are similar to those used to fabricate the transistors used in the digital section (described in Section 11.5.2.3), the overall process flow is likely to be advantageous over diodes. However, as will be discussed in this chapter, realizing printed transistors that switch at 13.56 MHz is extremely difficult. However, by using evaporated materials coupled with lithography and sharing the gate overlap capacitance (discussed in Section 11.5.3.2) as part of the filter capacitance, there have been some initial demonstrations of transistor-based rectification at 13.56 MHz (Rotzoll *et al*., 2006).

11.5.2.3 Digital Section and Modulation Stage

The main function of the digital section, which includes hundreds to thousands of transistors, is to generate a unique ID as a bit string signal. In its simplest form, therefore, the digital circuit will include a memory, decoding circuit, and counter (Cantatore, Geuns, and Gelinck, 2006). Further complexity may be required if the circuit implements an anti-collision scheme.

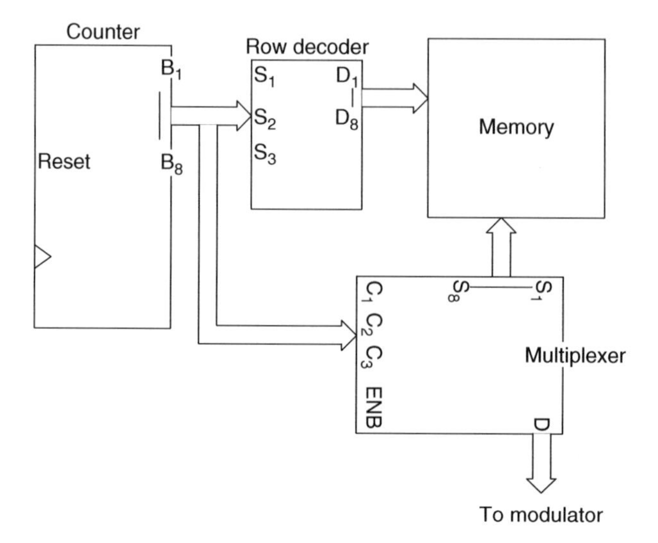

Figure 11.8 *Schematic block diagram of a typical RF barcode digital section.*

When the tag is energized by a reader, the rectifier charges up the smoothing capacitor. When the voltage on this capacitor reaches a sufficient value, the RFID circuitry begins to operate. In the simplest RF barcode applications, the tag circuitry will then simply broadcast a unique ID repetitively. This may be achieved by using a sequentially read memory connected to the output stage. This is shown in Figure 11.8.

More sophisticated RFID tags may require anti-collision, that is, they require a means by which numerous tags may exist and be energized within the reader field at the same time and yet enable the reader to identify each of the individual tags. There are numerous anti-collision protocols currently in existence, ranging from simple tags-talk-first (TTF) protocols to more sophisticated protocols as embodied in the electronic product code (Sarma, Brock, and Engels, 2001). Anti-collision protocol implementations will almost certainly require more complex circuits, which will make them difficult to implement in a fully printed process. As a result, at least in the near term, it is unlikely that printed RFID tags will make use of anything more complex than TTF if they implement anti-collision at all.

The output of the digital stage is fed to the modulation stage. This stage typically loads the antenna resonant circuit based on the output of the digital stage, causing the current in the antenna to change. This change in antenna current can be sensed by the reader and converted into the corresponding data stream. The precise circuitry of the modulation stage depends on the specific encoding methodology; however, it typically consists of at most a few transistors.

An important issue to consider is the operating frequency of the digital circuit. While the carrier signal is at 13.56 MHz, the clock for the digital circuit is substantially slower than this; typical HF tags have clock rates of a few hundred kilohertz. Based on current standards, HF tags generate the clock signal by dividing the carrier frequency, as shown in Figure 11.9. This has an interesting consequence: the only parts of the entire RFID

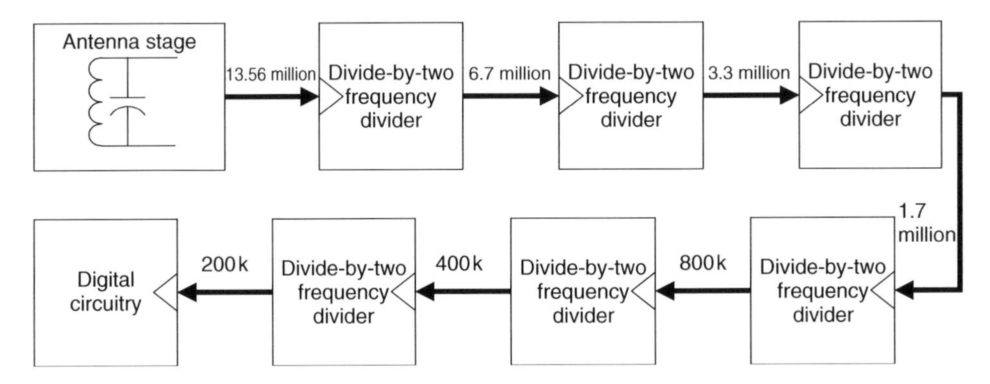

Figure 11.9 *Block diagram of a synchronous RFID tag.*

tag that run at the carrier frequency are the rectification stage (which has to rectify at 13.56 MHz) and the first stage of the divider. This is problematic if the performance of the circuitry is such that it is impossible to implement a 13.56 MHz divider. For printed transistors, this is almost certainly going to be the case, as will be discussed in this chapter.

Given the difficulties in dividing 13.56 MHz signals using low-performance printed circuitries, alternative strategies have been proposed and, indeed, have been used in some silicon-based RFID tags as well. For example, the clock signal could be generated on the tag using an oscillator running at the desired clock frequency, as shown in Figure 11.10. This is typically called asynchronous communication, since the tag clock is not synchronized to the carrier signal. This is somewhat problematic, since simple oscillators such as ring oscillators have an oscillation frequency that depends on the applied voltage. Since the applied voltage available on a tag depends on the coupling to the tag, this results in a clock frequency that varies depending on variations in tag manufacturing and power coupling. This will cause reading difficulties for the reader. This problem may be solved by using forgiving communication protocols and a more sophisticated reader to ensure that some variation in data rates is allowed.

The final portion of the digital circuitry to be considered is the memory, which must be nonvolatile. For some RFID architectures, it will be necessary to provide electrically erasable programmable read-only memory (EEPROM) or at least programmable read-only memory (PROM), since some programmability in the field will be required. For others, however, read-only memory (ROM) architecture will suffice. For example, in the simplest RF bar code implementations, the unique ID could be factory programmed (e.g. using an inkjet-printed metal mask). This is advantageous since it could likely be realized using existing technology and would eliminate the need for on-tag programming circuitry.

11.5.3 Devices for Printed RFID

Having discussed the architecture of a prototypical RFID tag, it is now possible to evaluate the state of the art in active and passive devices for printed RFID. We begin by reviewing some of the materials that have been used in printing such devices, and then

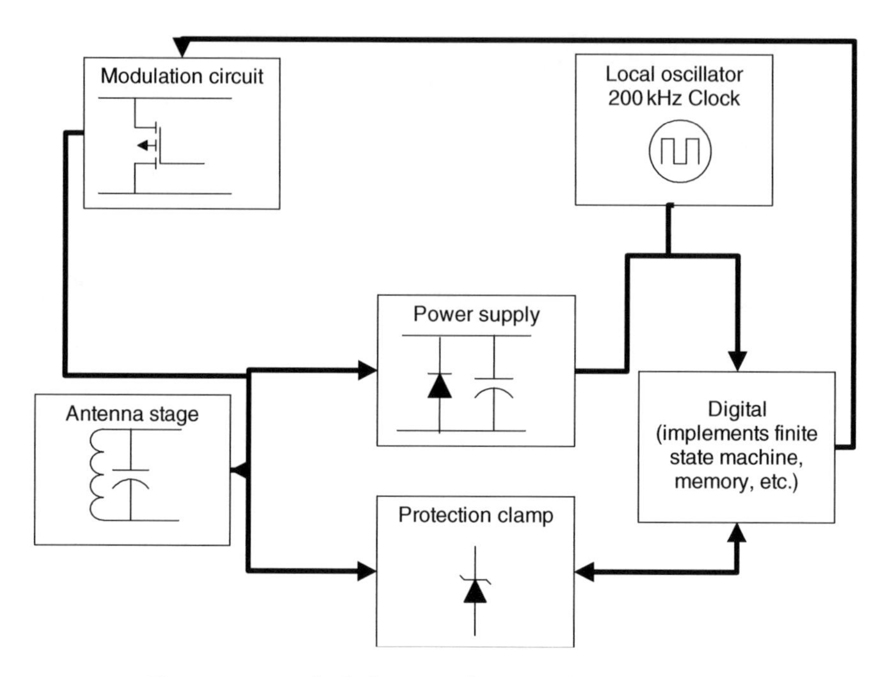

Figure 11.10 *Block diagram of an asynchronous RFID tag.*

review achievements in devices targeted at printed RFID. Note that details of printed active devices have been provided in Chapter 9.

11.5.3.1 Electronic Materials for Printed RFID

To print electronic devices and circuits, a range of printable semiconductor, conductor, and dielectric inks are required. For example, inorganic structures have been realized using nanoparticle precursors, and organic structures have been realized using soluble organic inks. In addition, significant work on printed RFID from companies such as Kovio has focused on the use of printed silicon precursors, while work on printed RFID from Sunchon National University has made use of solutions of printed carbon nanotubes.

Nanoparticles are attractive for printed electronics, since it is possible to realize soluble and therefore printable forms of a range of materials using nanoparticles; by appropriate design of the nanoparticles, it is possible to produce inks that may be printed and subsequently sintered to produce very high-quality thin films, as discussed in Chapter 7 (Huang, Liao and Molesa, 2003). This latter behavior is due to the fact that nanoparticles can be designed to show dramatically depressed melting points relative to their bulk counterparts, enabling low-temperature, plastic-compatible annealing of nanoparticle films to realize high-quality thin films with good grain structure and resulting high electrical performance. Indeed, using appropriately synthesized metallic nanoparticles, it has been possible to realize very high-quality conductors, with conductivity approaching as high as 70% of bulk conductivity. Using these in conjunction with polymer dielectrics, a range of printed passive components including inductors, capacitors, multilevel interconnects, and so on have been realized, as illustrated in Figure 11.11 (Scharfeld, 2001).

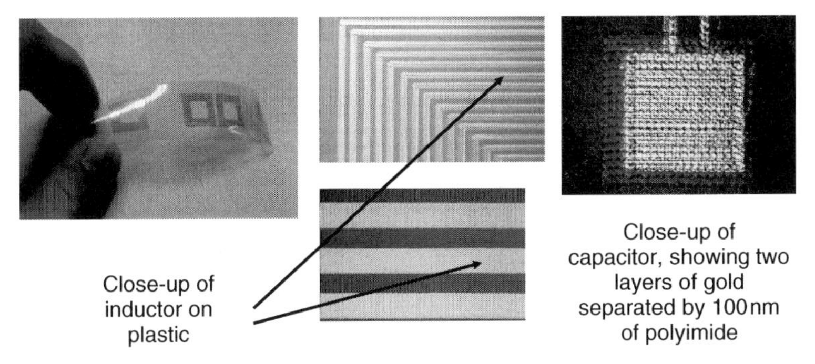

Figure 11.11 *Printed passive components formed using nanoparticles and polymers on low-cost plastic substrates (Reproduced with permission from Subramanian (2005) Copyright (2005) IEEE).*

11.5.3.2 Inkjet-Printed Transistors for RFID

These same materials can be utilized to realize printed transistors. Prior to examining specific achievements in printed transistor research, it is worthwhile to first lay out performance bounds based on what is likely to be achievable in the near future using inkjet-printed transistor technology. In general, inkjet-printed transistors will be subject to the following performance limitations:

- **Low carrier mobility**: Typical field-effect mobilities reported in the literature for inkjet-printed materials are in the range of 10^{-2} to $1\,\text{cm}^2\,\text{V}^{-1}\,\text{s}^{-1}$, which is approximately three orders of magnitude lower than field effect mobilities achievable in silicon. On the other hand, reports for inkjet-printed silicon transistors by Kovio suggest that much higher performance is achievable (mobility $>100\,\text{cm}^2\,\text{V}^{-1}\,\text{s}^{-1}$), albeit on steel foil rather than on plastic.
- **Long gate length**: Commercially viable inkjet printing techniques are currently incapable of producing line widths of $<10\,\mu\text{m}$ at high speed. They are also incapable of realizing line spaces much smaller than that, based on the bleed-out, spreading, and placement inaccuracy associated with high-speed printing. Consequently, printed transistors with gate lengths much shorter than $10\,\mu\text{m}$ will be difficult to realize in a manufacturing-worthy process. It should be noted, however, that several novel printing techniques are in development that may ease this constraint to a degree.
- **Large overlap capacitance**: In high-speed printing processes, achieving a layer-to-layer registration of better than $10\,\mu\text{m}$ is extremely difficult, and is generally not considered manufacturable using commercially viable technologies today. Typical printed transistors have printed sources, drain patterns, and a separately printed gate pattern. The alignment of these two layers to each other is achieved using the layer-to-layer registration capabilities of the printers in question. Since there is relatively poor control of the layer-to-layer alignment between these layers, it is critical that there be enough overlap between the layers to ensure that *all* devices have at least some overlap even in the worst-case misalignment, since device performance is dramatically degraded in the underlap condition, particularly using low-mobility printed materials. This results in a large overlap in typical printed devices, causing an

increase in gate-to-source and gate-to-drain overlap capacitance. This in turn reduces circuit speed.

- **High operating voltage**: Most printed materials are amorphous or polycrystalline at best. As a result, they typically show I–V characteristics that fit best to multiple-trap-and-release or variable-range-hopping models. In either case, they show a field-activated mobility behavior; printed field-effect transistors (FETs) typically deliver only the stated mobilities of $0.1-1\,\text{cm}^2\,\text{V}^{-1}\,\text{s}^{-1}$, at relatively high fields.
- **Thick gate dielectric requirements**: Typical gate dielectric thicknesses in most printed processes will be several tens to hundreds of nanometers. This is required to avoid concerns with pinholes and the like in typical high-speed printing processes. The net consequence of these facts is that typical printed transistors will have operating voltages in the range of 10 V; indeed, most current printed transistor demonstrations have operating voltages >20 V. Relating this back to RFID, either substantial Q boosting will be required or the tag range will be limited to a few centimeters.

Over the years, several groups have realized printed transistors (Sirringhaus *et al*., 2000; Molesa *et al*., 2004; Tseng and Subramanian, 2009a, b). In fully inkjet-printed transistors, mobilities as high as $0.2\,\text{cm}^2\,\text{V}^{-1}\,\text{s}^{-1}$, have been reported. By exploiting printed silicon, Kovio has reported transistors with mobility $>100\,\text{cm}^2\,\text{V}^{-1}\,\text{s}^{-1}$, albeit on stainless steel foils. Switching speed in these devices is often limited by the large overlap capacitance that results from the poor layer-to-layer registration inherent in most printing techniques. Using a fluid roll-off technique, the physics of drying droplets has been exploited to achieve self-alignment between the gate and source drain, thus lowering overlap capacitance and improving switching speed, as shown in Figure 11.12 (Tseng and Subramanian, 2009a, b).

11.5.3.3 Circuit Implementation Issues for Printed RFID

In addition to the direct device implications discussed in this chapter, there are also implications of printed device technology in general on circuit architecture. These include the general immaturity of printable n-type metal–oxide semiconductors (NMOS) relative to p-type metal–oxide semiconductors (PMOS), as well as issues related to device stability.

- **Logic family**: Most printed materials demonstrated to date are PMOS. Recently, as discussed in this section, there have been some demonstrations of printable NMOS materials as well, though the performance and stability of these materials still lag behind those of PMOS. To achieve low-power operation, the availability of complementary metal–oxide semiconductors (CMOS) will be highly desirable; otherwise, the operating range of tags will be degraded by the high power consumption requirements of the digital sections of the tag.
- **Environmental stability**: Perhaps one of the biggest concerns with inkjet-printed transistor materials (the vast majority of which are organic) to date relates to their poor environmental stability. Most organic semiconductor materials are prone to degradation on exposure to oxygen and/or moisture. This will necessitate the development of robust encapsulation processes.

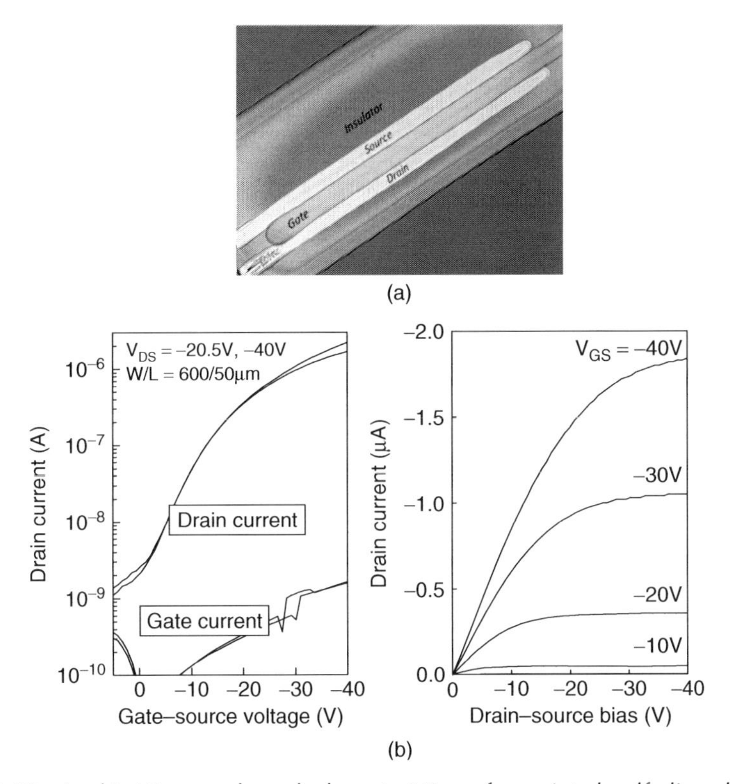

Figure 11.12 *(a, b) Micrograph and characteristics of a printed self-aligned transistor (Reproduced with permission from Tseng (2009) Copyright (2009) IEEE) (See plate section for coloured version).*

- **Bias stability**: Additionally, and perhaps of a greater concern for RFID applications, most organic devices to date show substantial bias stress effect, where their threshold voltage shifts during use. The mechanisms for this are currently being debated; however, the consequence is that organic devices show a history-dependent performance, which is problematic from a circuit design perspective for obvious reasons.
- **Poor diode performance**: As discussed in Section 11.5.2.2, typical printed diodes reported to date operate in the space-charge limited conduction regime, resulting in very high diode series resistance. Additionally, due to the large numbers of defects in the junction region, most printed diodes have a very poor ideality factor. These two phenomena together result in very poor rectification efficiency. To overcome this problem, it is necessary to either improve diode mobility or reduce diode layer thickness. In recent years, there have been several demonstrations of organic diodes operating at 13.56 MHz; however, they typically deliver very poor efficiency at these frequencies, limiting their use to extremely short ranges.

Based on these points, it is possible to generally summarize the circuit implications of printed organic electronics. Since rectification efficiency will be low and CMOS may not be realizable, it is likely that initial deployments of organic thin-film transistors in RFID will likely be in short-range, near-proximity applications in which adequate power will be available to operate highly inefficient tags. Additionally, communication protocols and circuit architectures will likely have to be conservative to account for the instability of organic transistors.

11.5.3.4 Printed RFID Demonstrators

Having reviewed various printed electronic components, we now examine their use in RFID tags and related circuit demonstrators. In general, demonstrators can be classified as follows: demonstrators using materials that are analogous to printable materials but use conventional fabrication techniques, and demonstrators actually realized by printing. Within the former category, several research groups have reported organic RFID tags fabricated using lithography, and have shown impressive levels of integration and performance (Cantatore, Geuns, and Gelinck, 2006; Rotzoll *et al*., 2006). However, since these demonstrators did not make use of printing, they will not be reviewed here. Rather, we will focus on what is actually realizable using printing technology.

In the arena of fully printed circuits, demonstrations in the kilohertz range have been achieved for simple building blocks using inverters and the like by exploiting techniques such as the self-aligned inkjet printing discussed here and illustrated in Figure 11.13. This has realized circuits with switching speeds approaching 5 kHz (Tseng and Subramanian, 2009b).

It is worth noting that a hybrid process combining printing for some layers while using lithographic techniques for others has also been employed by some researchers, including Kovio. Kovio, for example, has demonstrated synchronous RFID tags using this approach in conjunction with their high-performance transistors on steel.

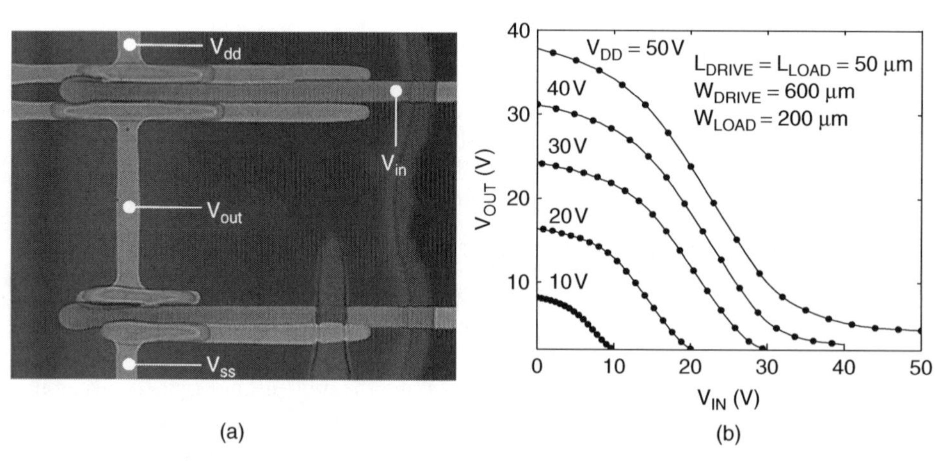

Figure 11.13 (a,b) Optical micrograph and transfer characteristics of self-aligned inverter. Switching speeds of several kilohertz were measured (Reproduced with permission from Tseng (2009) Copyright (2009) IEEE) (See plate section for coloured version).

11.6 Conclusions

In this chapter, we have reviewed trends in inkjet-printed RFID, and have defined the current state of the art. Inkjet is likely to have some good applications in RFID, both in antenna development and in the realization of fully printed RFID tags. For the realization of inkjet-printed antenna structures, inkjet has been used in conjunction with plating to realize high-quality antenna structures for both HF and UHF applications. For the realization of fully printed RFID circuits, inkjet has shown tremendous progress in recent years in the demonstration of high-quality printed transistors, other passive components, and circuit demonstrators.

Printed electronics inherently involves trade-offs in terms of performance to achieve the costs that are potentially realizable using printing. Based on the current performance of inkjet-printed devices, it is likely that the applications of inkjet-printed RFID tags will be limited to short-range, relatively simple, low-performance RF barcodes. This is certainly true at least for the foreseeable future.

The rapid progress in inkjet technology and in printed device technology in general offers hope, however, for the realization of new classes of printed devices offering higher performance and functionality. This in turn is expected to lead to the ultimate realization of fully printed RFID tags; inkjet will almost certainly play a role in these applications.

References

Cantatore, E., Geuns, T. and Gelinck, G. (2006) A 13.56-MHz RFID system based on organic transponders. Solid-State Circuits Conference, p. 15.2.

Finkenzeller, K. (2003) *RFID Handbook: Fundamentals and Applications in Contactless Smart Cards and Identification*, John Wiley & Sons, Inc., Hoboken, NJ.

Glidden, R., Bockorick, C. and Cooper, S. (2004) Design of ultra-low-cost UHF RFID tags for supply chain applications. *Communications Magazine*, **42**, 140.

Homs, C. and Metcalfe, D. (2004) *Exposing the Myth of the 5-Cent RFID Tag: Why RFID Tags Will Remain Costly This Decade*, Forrester Research, Cambridge, MA.

Huang, D., Liao, F., Molesa, S. *et al*. (2003) Plastic-compatible low resistance printable gold nanoparticle conductors for flexible electronics. *Journal of the Electrochemical Society*, **150**, 412.

Molesa, S.E., Volkman, S., Redinger, R. *et al*. (2004) A high-performance all-inkjetted organic transistor technology. IEEE International Electron Device Meeting Technical Digest, p. 1072.

Redinger, D., Farshchi, R. and Subramanian, V. (2004) An all-printed passive component technology for low-cost RFID. *IEEE Transactions on Electron Devices*, **51**, 1978.

Rotzoll, R., Mohapatra, S., Olariu, V. and Wenz, R. (2006) Radio frequency rectifiers based on organic thin-film transistors. *Applied Physics Letters*, **88**, 123502.

Sarma, S., Brock, D. and Engels, D. (2001) Radio frequency identification and the electronic product code. *IEEE Micro*, **21**, 50.

Scharfeld, T. (2001) An analysis of the fundamental constraints on low cost passive radio-frequency identification system design. MS thesis, Massachusetts Institute of Technology.

Sirringhaus, H., Kawase, T., Friend, R. and Shimoda, T. (2000) High-resolution inkjet printing of all-polymer transistor circuits. *Science*, **290**, 2123–2126.

Steudel, S., Myny, K., Arkhipov, V. and Deibel, C. (2005) 50 MHz rectifier based on an organic diode. *Nature Materials*, **4**, 597.

Subramanian, V. and Frechet, J. (2005) Progress toward development of all-printed RFID tags: materials, processes, and devices. *Proceedings of the IEEE*, **93**, 1330–1338.

Tseng, H. and Subramanian, V. (2009a) All inkjet printed self-aligned transistors and circuits applications. International Electron Devices Meeting (IEDM).

Tseng, H. and Subramanian, V. (2009b) All inkjet-printed, fully self-aligned transistors for low-cost circuit applications. IEEE Device Research Conference.

12

Biopolymers and Cells

Paul Calvert[1] *and Thomas Boland*[2]
[1]*College of Engineering, University of Massachusetts Dartmouth, USA*
[2]*Department of Biomedical Engineering, University of Texas at El Paso, USA*

12.1 Introduction

Bioprinting, tissue engineering, and regenerative medicine are linked aspiring technologies with the goal of building tissues in the laboratory. One ultimate application is implantable organs, such as a heart or liver. Bioreactors are a second area of application. 3D scaffolds that contain embedded cells and have a flow of nutrients could be used to generate biomolecules either when implanted or as an alternative to industrial fermentation methods. An implantable artificial pancreas containing embedded human or animal cells is an example. While simple 2D sensors can be made by conventional deposition methods, we could also envisage a need for 3D biosensor structures where greater sensitivity would come from increased surface area or from patterned arrays.

"Cells" covers a wide range from bacteria to yeast to human cells, all with a wide variety of needs. All need a supply of oxygen and nutrients, which may become limited by diffusion in 3D structures. Many types of bacteria and animal cells can build their own extracellular matrix, but the bioprinting process can provide at least an initial matrix. Most cell types will continually multiply or slowly die, so artificial tissues will need to accommodate cell growth or be constructed for a definite life cycle. Many types of animal cells are anchor dependent and will die if they do not attach to a suitable solid substrate. Once attached, they will divide until they achieve confluence, a single layer, and then stop. Some tumor cells are an exception to one or both of these constraints.

Inkjet Technology for Digital Fabrication, First Edition. Edited by Ian M. Hutchings and Graham D. Martin.

Human cells respond to cytokines, signaling proteins, that control their behavior in tissue and control cooperation between different cell types in tissue. They can also be controlled by drugs, such as dexamethasone.

Thus a bioprinter will need to deliver both cells and matrix into a 3D patterned structure. Any printed matrix may need to prevent the cells from escaping, prevent infection by other cell types, provide appropriate anchoring sites, and allow motion of large proteins. The matrix should be some combination of fibers, gel, and porosity.

Most complex structures in the world of hard engineering are built by combinations of casting, machining, and assembly, while biology builds by layerwise self-assembly. The closest synthetic approach to the multi-material layerwise growth seen in biology is in integrated circuit manufacturing where multilayer devices are formed by vapor phase deposition and photolithography. Inkjet printing and related techniques are the most promising approach to building similar structures with soft, low-temperature materials. We can also draw inspiration from rapid prototyping processes such as stereolithography and extrusion freeform fabrication, which build faster and allow easier "ink" formulation but are more constrained in combining materials by being contact processes (Calvert and Crockett, 1997). There are also many other methods for localized deposition that can be regarded as intermediate between inkjet printing and freeform fabrication techniques.

The nozzle orifice sizes of inkjet printers vary from <10 to $40\,\mu m$, as shown in Figure 12.1. The size of mammalian cells spans a wide range from a few micrometers in diameter for hematopoietic stem cells to hundreds of micrometers for skeletal muscle cells (axons of some neurons can grow up to 1 m in length or more). However, for most applications, the orifices of inkjet devices can rapidly deliver reagents over an area in the same way as a sheet of cells. Given a suitable gel acceptor layer, it would be possible to deliver a sequence of catalysts and reagents to build a layer of material as cells do. In addition to using inkjet printing as a potential method for constructing such a scaffold, it is a potential method for depositing the cells on this scaffold especially if patterns of cells or of different cell types are required. Thus a single printing system with enough "colors" ought to be capable of forming materials with the various structures

Figure 12.1 (a) Inkjet printer nozzles from a standard color cartridge; and (b) from a low-resolution cartridge (HP 6602).

and properties associated with tissues. Other printing processes could also be used, but noncontact, digital printing has many advantages.

12.2 Printers for Biopolymers and Cells

12.2.1 Printer Types

As discussed in Chapter 2, familiar home and office inkjet printers depend on two technologies to fire the droplets. Thermal printing, typified by the products of HP, Olivetti, and Canon, depends on pulse heating of a small resistor to create a vapor bubble that ejects a drop of ink from the nozzle. A typical heating pulse lasts a few microseconds, and drops are fired at a rate of at least 1 kHz. Piezoelectric printing technology, typified by Epson, depends on a pressure pulse generated by a bending cantilever or membrane. A combination of short forward and reverse pulses causes the drop to be ejected. These desktop printers are difficult to modify for laboratory use because the droplet ejection system cannot readily be decoupled from the page feed and head motion systems. However, in their research the authors have both opted for home-built systems dependent on consumer HP cartridges.

Many industrial processes depend on commercial inkjet printers that are purpose built. A number of manufacturers supply high-performance print-heads that are typically built into large production-line systems by integrators who draw on suppliers of motion systems, vision systems, fluid delivery, and control software. Where commercial equipment can be designed for use with a single ink type under controlled conditions, laboratory use often involves experimental inks and varying conditions. In due course as bioprinting develops, one may expect specialized bioprinter integration companies to emerge.

12.2.2 Piezoelectric Print-Heads

The principles of piezoelectric drop-on-demand print-heads have been discussed in Chapter 2. They dominate the market for high-speed industrial printing, and can also be used in laboratory applications (Tekin, Smith, and Schubert, 2008). Most piezoelectric inkjet printers have a fairly narrow window of ink surface tension and viscosity for optimum performance. Optimization of the pulse shape can adjust for fluid properties and allows for reduction of the formation of satellites (small extra drops from break-up of the tail), and for ensuring uniform drop speeds and sizes from adjacent nozzles for precise positioning and color control (Tsai and Hwang, 2008). In addition to rheological properties, the possibility of chemical attack on the head by the ink must be considered. Water-based inks are often a concern because the moisture and salt may eventually short out the high voltages needed to drive the piezoelectric elements. Particle-containing inks may abrade the nozzle and cause the drop size or firing direction to change.

Companies manufacturing commercial print-heads include Fujifilm-Dimatix, Xaar, Seiko Epson, Trident, and others. Such heads typically have hundreds or thousands of nozzles, cost hundreds or thousands of dollars, and are intended for production printing of one ink type with a tightly controlled formulation. The ink may be supplied by the company in cartridges or may be added to refillable reservoirs. Frequently they are sold as part of a complete system built for the customer by 'integrators' and containing cameras to monitor the drops and sample, transport and alignment systems, and sophisticated software to convert images into nozzle-firing sequences. Consumer piezoelectric printers have either a permanent print-head to which replaceable reservoirs can be attached, or a print-head integrated with the ink supply cartridge.

While it would seem sensible that a laboratory printing project should start from a simple consumer printer and then progress toward a commercial system, there are quite different constraints on a laboratory printer. The chief drivers for commercial printer development are speed, resolution (which is determined by droplet size), the precision of drop positioning, and the elimination of smaller satellite drops often produced in the firing process. These may also be important in some laboratory applications but are often not crucial. Laboratory printing can impose limitations that rarely occur commercially. Laboratory printing processes may involve solvents, strong acids, or bases that should not harm the printing system. The system may need to be sterilized. Inks may contain particles or polymers that aggregate over time so that print-heads may need to be disposable or readily cleanable. It may be desirable to print two inks in quick succession, as with a two-color printer. Inks may have high viscosity or low surface tension interfering with normal droplet formation.

A few companies do offer piezoelectric printer systems that can readily adapt to laboratory use. Microdrop (Norderstedt, Germany) and MicroFab (Plano, CA, United States) are examples. Dimatix (Fujifilm Dimatix, Inc., Santa Clara, CA, United States) and Unijet (Sungnam-si, Korea) have recently developed printers with cheap, replaceable piezoelectric print-heads that are designed primarily for laboratory use. The Unijet system is able to use a variety of commercial piezoelectric print-heads. These systems can combine printing, motion, and vision systems designed to optimize processes in the laboratory.

The Dimatix printer from Fujifilm Dimatix is a complete tabletop laboratory printing system with exchangeable multi-nozzle print-heads and excellent control of the printing environment. This printer has been very successful with developers of printed electronics. Users report great success when printing a consistent ink type for the study of substrate effects, patterns, or other downstream variables. Users seem to have less success with aqueous inks containing polymers where there is less freedom to optimize the ink for the printer. There are specifications for workable ranges of surface tension and viscosity, but a much more extensive set of parameters probably needs to be specified if there is to be a reasonable chance of knowing beforehand whether an ink will print.

To a great extent, the needs of bioprinting and printed electronics are at opposite poles in terms of resolution and ink properties. One place where there may be a common need is in printing speed. Many bioprinting projects envisage implantable bulk tissues being produced at short notice, while printed electronics is concerned with achieving high production speeds.

MicroFab makes full systems and stand-alone print-heads based on a glass tube squeezed by a piezoelectric cylinder. As a single-nozzle system, this is slower than multi-nozzle printers but the glass allows a wide range of liquids to be printed. The single-glass nozzle also allows many printing problems to be easily seen and diagnosed. Microdrop Technologies offers a somewhat similar single-nozzle printer that has been used successfully by Schubert and coworkers (Delaney, Smith, and Schubert, 2009).

12.2.3 Thermal Inkjet Print-Heads

For the authors, home-built systems based on HP pens have proved an excellent combination of robustness and inexpensive replacement. However, they offer only a narrow window of nozzle voltages and fluid properties in which jetting can be achieved. Therefore, ink formulations must fall within a relatively narrow range of viscosities and surface tensions. In addition, they have limited tolerance to solvents and strong acids. Customizing cartridges of thermal printers for biological inks is currently being investigated by manufacturers like HP and Olivetti; thus, in the future one may be able to use more viscous fluids and avoid the relative large amount of "dead volume" that is found in current pen designs.

In a thermal inkjet printer (also known as bubble-jet), a small heater (about 50 μm in diameter) causes a bubble of vapor to form in the ink. The expanding bubble pushes a drop of ink through the nozzle in a few microseconds. The ink then refills the cavity, and the process is repeated at a frequency of 1 kHz or more. The bubble formation process has been studied and modeled extensively for the various geometries of heater and nozzle position (Asai, Hara, and Endo, 1987; Chen, Chen, and Chang, 1997; Chen *et al*., 1998; Ruiz, 2007). These studies confirm that the fluid temperature does briefly reach close to the critical point (above 300 °C for water) but returns to close to room temperature within 10 μs.

Thermal inkjet heads are produced for consumer printers and can be cleaned for laboratory use. HP heads are identified by the level of technology embedded in them. TIJ1 and TIJ2.5 depend on a simple current pulse delivered to each nozzle through a pair of leads with one input corresponding to each nozzle and a return for each group of nozzles. Newer heads have transistorized switching of the nozzles, which would require a rather more sophisticated approach to address the nozzles directly on a head without a printer. In addition to the familiar consumer printers, print-heads are made for industrial applications such as labeling cans and boxes and printing bank checks. These low-resolution systems use large drop sizes, which may be more suitable for printing cells. Commercial systems for driving these heads are available from various suppliers, including Imtech (Corvallis, OR, United States) and MSSC (Collinsville, IL, United States).

The only thermal system designed for laboratory use is the HP Thermal Inkjet Pico-Fluidic System (TIPS), and this is available only by special agreement with HP (Fittschen and Havrilla, 2010). At a later stage of development, the speed and precision of commercial piezoelectric would become more important. Trident inkjet heads offer another compromise solution by being known as rugged and repairable piezoelectric systems. Olivetti has recently produced a thermal printing system for laboratory use, the Biojet (Tirella *et al*., 2011)

12.2.4 Comparison of Thermal and Piezoelectric Inkjet for Biopolymer Printing

As a technology, thermal inkjet is inferior to piezoelectric printing in that the maximum speeds are lower, there is less ability to control the pulse, and the head lifetime tends to be shorter. New piezoelectric technology allows drop size to be adjusted while printing by changing pulse shapes. This permits rapid printing of blocks of color combined with high definition of detailed regions and helps to overcome the contradiction between the desires for smaller drops and faster printing. For a laboratory printer, thermal printing may be seen as having advantages which outweigh the speed and control of the piezoelectric heads. Thermal print-heads can be mass-produced and are therefore inexpensive and disposable, which makes ink development considerably less expensive. Most piezoelectric print-heads can be cleaned ultrasonically, but irreversible damage is still common.

Existing thermal inkjet heads are mostly not very solvent resistant and have a limited resistance to acids and bases, but bioprinting solutions are mostly aqueous and neutral. While piezoelectric inkjet heads can be solvent and acid resistant, many commercial heads run at high voltages and can be vulnerable to slow degradation in water.

Thermal inkjet seems to be much more robust than piezoelectric inkjet in terms of the range of inks that can be printed, which is a valuable attribute for research systems. The ranges of viscosities and surface tensions printable by each technology are not well documented since in many cases the printer makers are also the ink makers. In general, viscosities up to 20 mPa s and surface tensions around 30–40 $\mathrm{mN\,m^{-1}}$ are printable. The authors' impression is that thermal printing is more forgiving of polymer solutions where there may be non-Newtonian effects, but there is no systematic experimental evidence for this. Thermal droplet ejection is a rugged process that is relatively insensitive to the ink properties, as long as some component of the ink can be vaporized. Rheological properties are less important to determine drop formation and jettability than they are for piezo-based print-heads. Thus even high-molecular-weight polymers at higher concentrations may be jettable. There may still be some limitation for biopolymers due to re-filling the ink chamber in the appropriate time to allow for subsequent nozzle firings. Ink formulation is discussed in Section 12.3.

In the long term, industrial printing of diagnostic sensors or other disposable biological systems would require the fast, reproducible printing that both thermal and piezoelectric systems can offer. On the other hand, it may be that systems for use during surgery would be better delivered from sterile disposable "pens" containing the biopharmaceutical or cellular ink, which may favor thermal systems.

12.2.5 Other Droplet Printers

Labcyte (Sunnyvale, CA, United States) makes the Deerac system which dispenses relatively large drops, in the microliter to nanoliter range, using a solenoid system (a 1 μl drop has a diameter of 1.3 mm). While the resolution is poor, printing on such a large scale is very simple and reliable. The same company also offers an acoustic jet printer that delivers much finer drops. The Deerac system can also print multiple inks via a suck-wash-and-spit method. The bigger drops, with a low surface-to-volume ratio, mean that there is much less concern about residual contamination on the nozzle surface.

Optomec (Albuquerque, NM, United States) produces a printing system based on aspiration of fluid droplets into an air stream that is then focused onto a substrate. Biopolymers have been printed, but there is not yet much information on any degradation to cells or biopolymers. Since aerosols are so important in disease transmission, there is no reason to expect serious damage. The high throughput of this system could give it a role in pilot production processes.

There are many other bioprinting methods under development with varying degrees of ability to print wet materials in patterns. MapleDW allows deposition of cell and matrix combinations by laser ablation from a prepared ribbon (Harris *et al.*, 2008). Electrospraying allows cells to be deposited but in a broad area without a simple patterning method (Hall *et al.*, 2008; Ng *et al.*, 2011).

12.2.6 Rapid Prototyping and Inkjet Printing

Since the development of stereolithography in 1986, a large number of methods have been developed to build 3D structures from a CAD file (Calvert and Crockett, 1997). Many of these techniques can be used to make 3D scaffolds for tissue engineering, either by direct deposition of the scaffold material (Zein *et al.*, 2002) or by forming a negative which is removed after the scaffold material is molded around it (Liu *et al.*, 2007). Stereolithography has also been used to directly form hydrogel scaffolds with embedded cells (Dhariwala, Hunt, and Boland, 2004; Arcaute, Mann, and Wicker, 2006).

A number of these rapid prototyping methods do involve inkjet printing of one component. Binder can be printed into a powder bed (Dimitrov, Schreve, and De Beer, 2006; Lozo *et al.*, 2008) or low-molecular-weight polymers and waxes can be printed directly, but they do not offer the materials versatility needed for biological structures.

There is a family of extrusion methods where a bead of material is pushed out through a fine nozzle that is applicable to a wide range of materials, including biomaterials, and is most comparable with inkjet printing in terms of applications. A fine stream of paste can be deposited and used to build structures. An ideal material has the rheology of toothpaste, with a yield point that lets it hold shape until some chemical change causes permanent gelation or solidification. Nozzle sizes down to about 100 μm can be used. Extrusion freeform fabrication has been applied to ceramics (Morissette *et al.*, 2000), polymers (Zein *et al.*, 2002), hydrogels (Calvert and Liu, 1998), composites (Calvert, Lin, and Martin, 1997; Peng, Lin, and Calvert, 1999; Liu *et al.*, 2002), and metals (Sercombe, Schaffer, and Calvert, 1999). Extrusion deposition of this type can be carried out with a simple three-axis computer-controlled stage and a pressure- or motor-driven syringe system. Nordson EFD (Providence, RI, United States) makes such deposition systems for the electronics industry.

Extrusion freeform fabrication is undoubtedly easier than inkjet printing, but inkjet printing offers some functions that extrusion cannot. The two could be regarded as complementary. In terms of resolution, they are similar. Line widths and features generated by inkjets are about 10–100 μm in the xy plane. A typical freeform extrusion generates a line width of 100 μm, although both can give much finer lines in special cases (Sirringhaus *et al.*, 2000; Gratson, Xu, and Lewis, 2004). In both cases, the discontinuity between lines or drops makes it difficult to form pore-free films. For extrusion methods, the vertical (z-axis) resolution is also about 100 μm, whereas an inkjet drop dried on a flat surface will form a pancake about 100 nm thick. Sun and coworkers have modeled the flow of encapsulated cells through various extrusion needles and compared the results to experimental studies. They concluded that the high pressures observed in needles smaller than 250 μm will result in high morbidity and cell death (Yan, Nair, and Sun, 2010). On the face of it, this contradicts observations that cells can be inkjetted through 80 μm nozzles, but there may be a complex response to time, pressure, and shear.

Extrusion methods can deliver materials of high viscosity, such as concentrated polymer solutions or particle suspensions, but there must be a rapid solidification process to hold the 3D shape after printing. Toothpaste is an ideal material, combining high viscosity with a yield point to prevent slumping. Final curing processes can occur slowly or after the part is built. Inkjet printing is limited to much lower viscosities and solids loadings but the drops dry rapidly, allowing structures to be built. Inkjet is also more suited to adding into porous or gel substrates. Inkjet also allows multiple inks to mix or react chemically, while thicker extruded layers remain distinct.

For bioprinting, several issues arise that do not occur with inkjet printing. Firstly, the stream is coarse relative to the typical diffusion length for these materials. With inkjet printing, mixing between drops of two different inks can lead to *in situ* reactions to form the desired final gel, but this is not possible with extrusion deposition. Also this is a contact method so each deposition cycle may disturb the previous layer. This is especially true with gels that may swell or shrink during deposition and so change height. It is possible to write wide but very thin lines of fluid on a hard flat substrate, but this normally does not work as thicker layers are built up. Height instabilities occur if the nozzle is not held accurately at one bead diameter above the current surface. Various laser-mapping strategies have been tried but do not seem to have solved this problem so far. Inkjet printing has the advantage of a large air gap.

Thus, inkjet printing and extrusion freeforming are both applicable to building soft, wet biomaterial structures and are complementary, with inkjet depositing reactive materials or small amounts of reagents while the bulk building is done by extrusion.

12.3 Ink Formulation

12.3.1 Introduction

A typical cellular or biopolymer ink will be an aqueous solution containing salts and other small solutes, coiled-chain polymers, globular proteins, cells, and particles. The solution may be very non-Newtonian. Aggregation, thermal denaturation, and other changes may occur during storage or printing. Printer makers and current commercial users have little experience of such problems, so it must be up to the researcher to tune the ink to the printer.

Consumer printer manufacturers design for compatibility with inks that they formulate for a specific system. In a laboratory, it would be preferable to have a well-defined window of ink parameters that define whether or not it can be printed. Generally this is not available and testing is necessary with any new group of liquids.

Manufacturers will normally specify whether or not their heads are compatible with acids, bases, and solvents. Piezoelectric heads tend to work well with normal solvents but may be vulnerable to aggressive solvents attacking adhesives in the system. The high voltages present in piezoelectric heads can lead to them being shorted by aqueous salt solutions or degraded by water penetration. The cartridges of thermal printers are often attacked by solvents. Poorer solvents such as alcohols or alkanes may print, while good solvents like toluene or acetone destroy the cartridge. In many cases, a head will work well for a few days before succumbing to attack and this can be satisfactory for laboratory use.

12.3.2 Printed Resolution

Typical droplet volumes from inkjet printers are from 1 to 100 pl, corresponding to diameters between 12 and 60 μm (see Chapter 2). On impact, drops will typically spread to become domes that are two or three times the initial drop diameter, and then dry down to pancakes with this same diameter and a thickness dependent on the solids content. Clearly, higher energy surfaces will result in higher domes of smaller diameter but spreading of isolated drops on low-energy surfaces will be limited by evaporation and surface roughness. Higher evaporation rates at the edge of the drops often result in "coffee rings" as discussed in Chapter 5, where the discs of dried ink have a raised rim (Derby, 2010). In printing yeast, we have also found that cells tend to migrate to the center of the disc during drying, as shown in Figure 12.2, driven by surface tension effects.

12.3.3 Major Parameters: Viscosity and Surface Tension

Viscosity is a major source of energy loss as the drop is expelled from the nozzle, so it would be expected that the speed and accuracy of the expelled drop will decrease as viscosity increases. Printing will also fail at high viscosities and high drop rates if the ink cannot flow into the chamber rapidly enough to refill it between firings. Modeling of the ink ejection process promises to clarify the relationship between viscosity and jettability (Asai, Hara, and Endo, 1987; Zeng *et al*., 2009). In general, inkjet printers will function over a viscosity range from 1 to 20 mPa s.

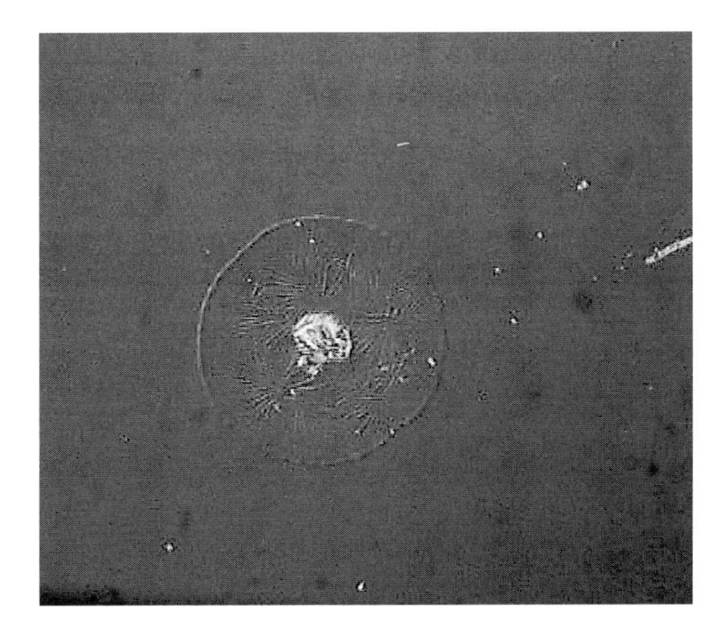

Figure 12.2 *Inkjet-printed drops on surface with centered yeast. Disk diameter is 150 μm.*

The viscosity of linear polymer solutions is very sensitive to concentration and molecular weight. For short chains, the relationships are roughly linear but there is a transition to more complex behavior when the open coils of polymer are long enough and concentrated enough to become entangled. Above this critical molecular weight, viscosity increases with molecular weight to the 3.4 power and concentration to the 5 power. The entangled regime may also give quite different behavior under the high shear rates and extensional flow found in an inkjet printer, when compared to low shear viscosity. Hence, in this region of concentrations above about 5% and molecular weights above about 10 000 Da, printability may vary greatly from batch to batch due to minor changes in molecular weight. Van Krevelen gives a concise summary of the complex topic of polymer solution viscosities (van Krevelen, 1976).

Recent work has addressed the role of viscoelastic effects in inkjet printing (Alamry *et al.*, 2010; Hoath, Martin, and Hutchings, 2010a, b). At low molecular weights, polymer molecules have a rapid relaxation time and so would be expected to show little shear-rate dependence of viscosity or extensional viscosity. At high molecular weights, viscoelastic damping is expected to prevent extension. In the intermediate range, chain extension and chain breakage are expected. Viscoelastic effects also affect drop and ligament formation.

These studies have mostly been carried out on solutions of synthetic polymers in organic solvents, for which samples with a wide range of well-characterized molecular weights are available. It is not immediately obvious how these results would translate to aqueous solutions of neutral or charged polymers where relaxational behavior might be quite different.

Biopolymer inks are non-Newtonian and will change the jetting behavior in piezoelectric systems. The effect of molecular weight and concentration of linear polymer inks have been studied by several authors. While 10–50 ppm concentrations pf 300 kDa polymers will prevent satellite droplet formation (Shore and Harrison, 2005), higher concentrations are often not printable. Hoath, Hutchings, and Martin have found that linear polymers are printable with piezoelectric nozzles if $c \times M < 100$, where c is the concentration of the polymer in wt.% and M is the molecular weight in kilodalton (Hoath *et al.*, 2009). Thus, for most biomedical linear polymers such as polysaccharides, only dilute solutions are expected to be printable. Most soluble proteins have a globular conformation in solution and so will have a much smaller effect on viscosity and little viscoelasticity.

Surface tension affects the tendency of the ink to wick out of the nozzle and form a film on the base plate that prevents drop expulsion. This depends on the contact angle between the ink and the base plate but can often be controlled by applying a small negative pressure to the print-head or adjusting the ink reservoir to be slightly below the nozzles. In addition, surface tension must affect the tendency of the ink column to break and form a drop in piezoelectric printing or to form a bubble in thermal printing. The presence of surfactants is clearly important in controlling surface tension and drop formation but the details are unclear, so their role in inkjet inks is not well understood (Liao *et al.*, 2004). Further, there have been few studies of these processes relevant to multicomponent inks. Typical printer manufacturers specify surface energies in the range of 30–60 mJ m^{-2}. Small amounts of surfactant can normally be added to achieve these values.

12.3.4 Drying

A concern for all inks is that they may dry in the nozzles. This can be partly avoided by maintaining environmental humidity and by regular ejection of a stream of drops between printing cycles. Many commercial inks contain humectants, such as ethylene glycol or glycerol, which cause the ink to dry only to an oil that can readily rehydrate.

While commercial inks will often be salt-free, biopolymer solutions and cell suspensions will normally be printed from a solution of 0.13 M sodium chloride plus other salts and buffer. The salt is likely to crystallize in the nozzle if the ink is allowed to dry. Care will need to be taken to maintain a humid environment during printing and to occasionally wash the nozzle plate during long pauses.

12.3.5 Corrosion

Salts in the ink may also corrode any metal lines that are in contact with the ink and may short connections through ionic conductivity. Chloride, in particular, tends to be aggressive in corroding many metals, including gold, under oxidizing conditions. It is possible to apply passivating treatments to commercial inkjet heads and some systems may have no corrodible materials exposed to the ink, but the technology is not well established as salts are not common in commercial inks.

12.3.6 Nanoparticle Inks

There is much current interest in printing nanoparticle pigment inks because pigments are more stable against fading than are dyes. Inkjet print-head nozzles are 20–80 μm in diameter, corresponding to 4–256 pl minimum drop volumes. In principle, particles smaller than the nozzle can be printed as long as they do not aggregate in suspension. In practice, the particles must also not settle during printing. The settling rate, dependent on solution viscosity, particle density, and particle diameter, can be estimated from Stokes law. Most dilute suspensions of inorganic particles below 0.5 μm diameter can be printed. Much larger biological particles or cells can be printed because their density is close to that of water, but occasional gentle stirring may be needed to prevent settling and aggregation.

As particles become smaller, the stabilizing surface layers, needed to prevent the formation of loose agglomerates in suspension, become more significant in volume and so the maximum stable concentration becomes lower. Once agglomeration occurs, suspensions will often develop a yield point and show a rapid increase in viscosity. As a result, it becomes challenging to formulate printable nanoparticle suspensions at greater than 1% by volume.

12.3.7 Biopolymer Inks

Conventional polymers are not often printed because the solution viscosity rises rapidly with molecular weight, limiting inks to dilute solutions. The common approach to printing a polymer is to use a "UV ink" where monomer and catalyst are printed and then polymerized by UV irradiation (Hancock and Lin, 2004). Dilute solutions of low-molecular-weight polymers can easily be printed if the viscosity is low enough. It is also possible to print dilute polymer emulsions, diluted emulsion paints for instance, as long as the droplet size is smaller than the nozzle size.

Biopolymers should be considered in two categories, globular and coiled. Coiled soluble biopolymers include the soluble polysaccharides such as alginates, chitosan, and various gums; soluble polypeptides such as polylysine and polyglutamate; and some forms of DNA. Given that the printer can tolerate aqueous ionic solutions and is subject to the constraints of viscosity and surface tension, these polymers can be printed readily. Depending on the printing environment, it may be useful to add 10–20% of a humectant such as glycerol or ethylene glycol to prevent drying to a hard cake in the nozzle. Most soluble biopolymers are partly ionized at the amine and carboxylate groups. The extent of chain expansion, and so the solution viscosity, will depend on the pH and the ionic strength of the solution.

12.3.7.1 Coiled Biopolymers

Alginates have long been used as matrix materials for tissue engineering (Rowley, Madlambayan, and Mooney, 1999). They are normally produced from seaweed for a variety of thickening applications in food and pharmaceuticals. They can also be produced in medical grades by bacterial fermentation. These medical grades are free of bacterial endotoxins, which will cause illness in humans and will kill cells in culture. The attraction of alginates is that they are rapidly cross-linked by solutions of calcium ions or other polyvalent cations. A drawback is that gelation is reversed if the calcium level falls below a level which depends on the chain structure but is in the region of 10 mM (Jorgensen *et al*., 2007).

Alginate gels can be inkjet-printed by printing drops of calcium chloride solution into alginate solution or vice versa. Nakamura and coworkers have described printing a suspension of cells in sodium alginate solution into a viscous solution of calcium chloride and polyvinyl alcohol (Nishiyama *et al*., 2009). Due to viscosity constraints, the alginate concentration is limited to about 1%. The viscous target solution allows the drops to gel and form solid structures, which slowly sink. This work is described in detail in Chapter 13. Xu *et al*. have described a system for printing 3D cellular constructs by printing calcium chloride and cells, or sodium alginate and cells, into a solution of alginate or calcium salt. As with stereolithography, a platform is submerged below the surface of the liquid and drops as each new layer of gel is added (Xu *et al*., 2008a, b).

Other polysaccharides can be printed. Gellan and xanthan have been inkjet printed as dispersants and matrices for carbon nanotubes (Panhuis *et al*., 2007). Structural proteins are insoluble, but dilute metastable solutions of silk and collagen can be prepared by acid dissolution and dialysis. These precipitate on aging or irreversibly solidify on drying and can be printed (Roth *et al*., 2004), but do tend to precipitate and block the nozzles so that the printing system will run only for a limited time before the head must be ultrasonically cleaned (Limem *et al*., 2006).

In principle, inkjet printing is an excellent route to preparing biopolymer hydrogel matrices for tissue engineering, drug delivery, and other applications, given that self-assembly, pH shift, enzymatic modification, or other routes can be used to convert soluble polymers into gels. In practice, most potential biopolymer matrix materials are high in molecular weight, which makes the solutions too viscous at concentrations above 1%. In addition, most commercial materials are intended for industrial use and are poorly characterized and quite variable between batches.

It has often been suggested that polymers, proteins, and cells in inks may be damaged by shear as the drop is ejected from a nozzle. An inkjet drop is expelled from the nozzle at a speed of about $10\,\text{m}\,\text{s}^{-1}$. For a nozzle of $20\,\mu\text{m}$ diameter, application of the Poiseuille equation gives shear rates of around $10^5\,\text{s}^{-1}$. In the case of solutions of very high-molecular-weight polymers (over 10^6 Da), very high strain rates in extensional (elongational) flow are known to break the chain in the middle where the extensional force is greatest (Ferguson, Hudson, and Warren, 1987; Odell, Muller, and Keller, 1990). As a result, narrow high-molecular-weight polymers show the appearance of a second peak at half the original molecular weight. While a dissolved polymer coil can be quite large, globular proteins of similar molecular weight tend to be tightly folded and so less subject to extension. Polymers are known to be oriented in shear flows, but there is little evidence for chain breakage in the absence of turbulence, which may also introduce extension (Buchholz and Wilson, 1986; Elbing *et al*., 2009). Because of their high molecular weight and possibly because of their chain structure with bulky monomer units, biopolymers may be especially susceptible to extensional flow effects (Chan *et al*., 2009).

12.3.7.2 Globular Proteins

Most nonstructural proteins have a globular structure. The primary sequence of amino acids contains regions that form strongly hydrogen-bonded secondary structures, including alpha helices and beta sheets. The whole molecule folds into a tertiary structure that usually has a hydrophobic core and a hydrophilic, predominantly anionic outer region. In many proteins, such as hemoglobin, multiple chains assemble into a cluster, the quaternary structure. The hydrophobic-hydrophilic folded structure is vulnerable to denaturation, especially in the presence of hydrophobic surfaces including plastics, air interfaces, and some metals. As a result proteins would be expected to denature on the surfaces in an inkjet printer cartridge, at the air interface of the ejected drop, and at the surfaces of the delivered drop. Likewise, they can be denatured by organic solvents and detergents. The unfolded, denatured structure of an enzyme is usually not active, and the process is normally irreversible because the initial folding occurs as the protein is made and may not be an equilibrium form.

The main families of globular proteins include enzymes, such as lysozyme, which catalyze chemical conversions; transport proteins, such as hemoglobin and serum albumin, which carry molecules that would otherwise have limited solubility in water; and signaling proteins, such as bone morphogenic protein, that modify cell behavior. A few transport proteins and enzymes are available in gram quantities; most are available only in milligram amounts or less but can still be extremely active. Thus the main interests in printing globular proteins are in immobilizing them on surfaces for sensors and catalysts or for controlled delivery of protein drugs.

As outlined here, globular proteins have a compact structure and so do not greatly increase the viscosity of an inkjet ink. Hence, the main concern relates to whether the printing process actually damages the protein. The major concerns include thermal damage by the heat pulse in a thermal inkjet printer, denaturation due to heat buildup in the print-head, and denaturation due to shear during droplet ejection. In addition, for dilute protein solutions, there is the potential for adsorption of protein to surfaces in the print-head with consequent loss of significant amounts of activity.

Experimental studies show no unfolding of cytochrome C (Jaspe and Hagen, 2006) or damage to a monoclonal antibody (Bee *et al*., 2009). However, insulin, which is a relatively small protein, does show aggregation effects even under quite low shear rates (Bekard and Dunstan, 2009). It is suggested that aggregation is triggered by unfolding of the protein under shear.

In addition to shear, piezoelectric printers set up a fast pressure pulse to eject a drop. Nishioka *et al*. suggest that this pulse causes denaturation of enzymes (peroxidase) which increases as the compression rate increases (Nishioka, Markey, and Holloway, 2004). The effect is reduced by added sugar, such as trehalose. It is perhaps surprising that denaturation can occur within the very short time of a pressure pulse, and it is possible that some other effect related to drop velocity is the immediate cause of degradation.

Denaturation is a kinetic process and so will not necessarily occur during the actual ejection process and the short (100 μs) journey of the drop to the substrate. It may occur on the surfaces of the print-head over a period of seconds, and it is believed that addition of a "sacrificial" protein such as serum albumin can saturate the surfaces and so limit denaturation of the enzyme or other protein of interest (Nishioka, Markey, and Holloway, 2004). In normal laboratory processes, using plastic labware for instance, such small volumes of liquid are not usually exposed to surfaces so there is little precedent for such concerns. Cell surfaces are very hydrophilic, and so cell damage might not be expected unless the cell-containing drops are allowed to reside on a dry substrate for a period, but cells would anyway be expected to die through dehydration under such circumstances. Delehanty and Ligler studied the loss of protein from nonspecific adsorption on the printer tubing and minimized this loss by the addition of a sacrificial protein (Delehanty and Ligler, 2003).

In thermal printers, the heating pulse raises the question of whether any damage is caused to biopolymers during the process. For small molecules it is possible that thermal degradation occurs, while very high-molecular-weight polymers and cells may also be damaged by shear processes. This has been studied in the context of the possible use of inkjet printing for delivery of drugs in aerosol form. Proteins (insulin) were found to be unaffected by passage through a thermal inkjet print-head (Goodall *et al*., 2002). A small-molecule drug (prednisolone) was not degraded but did recrystallize from the solvent in a different polymorph after drying (Melendez *et al*., 2008). In our experience of printing serum albumin with a thermal print-head, printing eventually stops after a time, which decreases at higher drop rates. This may reflect denaturation due to slow warming of the whole print-head.

"Gene chips" with arrays of DNA spots attached to a silicon or glass surface are typically made by photolithographic methods with individual A, T, C, or G bases added to the surface-bound growing chains in the desired sequence for each spot. An alternative approach suitable for smaller numbers of test spots is to attach complete chains to the surface by deposition using split-nib pens that withdraw samples from a large array of reservoirs.

The first approach requires a small set of reagents, including the four bases, and so lends itself to inkjet printing. Several groups have worked on developing such machines but have not yet been wholly successful (Lausted *et al*., 2004; Lozo *et al*., 2008). The second approach could be done by a suck-and-spit system that could draw from a large number of reservoirs, but the need to clean the nozzles between each print step would be problematic.

Beyond this application, there is not a clear large application for printed DNA. DNA is available as very high-molecular-weight polymers ($10^6 - 10^7$ Da), and there is interest in whether these very long chains will be damaged by shear during printing. Okamoto and coworkers were concerned with, but found no evidence of, damage caused by shear stress when printing DNA (Okamoto, Suzuki, and Yamamoto, 2000).

12.4 Printing Cells

12.4.1 Cell-Directing Patterns

In order to build structured tissues, we can deposit combinations of cell-adhesive and cell-repulsive polymers in patterns and then deposit cells over the patterns. In one of the first examples, collagen mixtures were printed on dried agarose-coated glass coverslips. Smooth muscle cells adhered to the patterns in less than 1 day, and dorsal root ganglion neurons were seen to attach to the patterns and extend neurites on the patterns (Roth *et al*., 2004). Other variations of this approach included collagen printed onto glass coverslips coated with polyethylene glycol (Sanjana and Fuller, 2004). After 8 days in culture, the neurons patterned on the adhesive regions.

A variation of this approach is the spotting of cytokines on hydrogels. Cooper *et al*. have printed the bone growth factor BMP−2 and inhibitory factors Noggin and transforming growth factor beta (TGF-β) onto the commercially available hydrogel Dema-Matrix. When printed alone, the growth factor caused bone to form in broad diffused patterns; however, when combined with the inhibitory factors, bone grew along the patterned semicircles (Cooper *et al*., 2010). This technology seems promising for directed growth of tissues.

12.4.2 Cell-Containing Inks

Most of the early cellular inks were a single cell type suspended in saline solution (Xu *et al*., 2005) at concentrations of typically 1–10 million cells/ml. Initially hypertonic solutions were used to reduce the cell volume; however, its use has largely been abandoned to date (Boland *et al*., 2006). Many types of ink contain additional nutrients such as glucose or amino acids; some contain serum-free media. Electromechanical printing systems will accept up to 10% or more serum to be added; however, thermal inkjet systems will stop ejecting drops when serum proteins are added, presumably due to the proteins adsorbing to the heating elements and thereby "poisoning" the mechanism.

More recently, cellular ink includes other functional macromolecules, such as proteins or DNA. One study took advantage of the transient pores that develop in membranes of thermally printed cells to deliver DNA vectors into porcine aortic endothelial cells and measured the effective transfer by green fluorescent protein expression in the cells (Xu *et al*., 2009b). Cui and Boland added thrombin and calcium ions to human capillary endothelial cells to build fibrin networks *in situ* (Cui and Boland, 2009). Much of the progress in building functional tissues will depend on the judicial choice of ions, macromolecules, and cells that make up the ink of a delivery device; thus it is thought that the area of biological ink development will see future growth.

Cell damage due to shear during drop ejection in inkjet printing is clearly a concern. Cell diameters are very dependent on cell type but are typically 20 μm or less. This corresponds to the size of a 1 pl nozzle, and nozzle diameters can thus range from smaller than a cell diameter up to the width of several cell diameters. Hence the degree of interaction between the cell and the fluid flow in the nozzle can be very variable. In most cases larger nozzles are used for cell printing.

Shear damage to cells is an issue in many medical devices. For instance, Lee *et al.* report on shear damage to red blood cells in a microchannel system (Lee *et al.*, 2009). Therefore the design of print-heads for cell printers must be undertaken to minimize small channels and chambers. Future designs should ensure that the nozzle orifice has the smallest constriction in the system.

12.4.3 Effects of Piezoelectric and Thermal Print-Heads on Cells

Living cells have been printed successfully with a number of systems, including two electro-mechanically driven systems, the electrostatically actuated print-head manufactured by Seiko Epson (Tokyo, Japan) (Nakamura *et al.*, 2005), and a commercially available system using a piezoelectrically driven Microjet MJ-AB-01 print-head (MicroFab, Plano, TX, United States) (Saunders, Gough, and Derby, 2008). These last studies used bovine endothelial cells and human fibroblast cells and showed 97% viability after ejection.

Cell survival has been studied after extrusion through a nozzle (Nair *et al.*, 2009), or electrostatic extraction through a nozzle, commonly referred to as electrospraying (Mongkoldhumrongkul, Flanagan, and Jayasinghe, 2009). However, these systems are limited by the resolution that can be achieved because of the cell morbidity experienced in needles of <250 μm diameter (Nair *et al.*, 2009).

Thermally driven systems have been used earlier and with more success than electromechanical systems to eject viable cells or clusters of cells. Several groups are using systems that employ the heat of a laser to eject cells embedded in a matrix. Early work examined bacteria (Ringeisen *et al.*, 2002), but was later perfected to include rat Schwann and astroglial cells and pig lens epithelial cells (Hopp *et al.*, 2005). More recently, other groups have improved the resolution of these systems to the single-cell level and reported 97% viability of rat olfactory ensheathing cells (Othon *et al.*, 2008). Thermal inkjet printers have been used successfully to deliver cells into pre-described patterns. Earlier work examined the viability of Chinese hamster ovary (CHO) cells (Xu *et al.*, 2005), but was later extended to neuronal cells (Xu *et al.*, 2006). This work showed that printed ganglial neurons did not show statistically different growth rates from pipetted cells.

Moreover, since presumably the cell membrane may be the most vulnerable part of a cell exposed to shear forces, the electrophysiology of primary hippocampal and cortical neurons was examined after 2 weeks of culturing. No difference between printed and normal cultured cells was found in terms of their resting membrane potential, input resistance and membrane capacitance, as well as a host of other electrophysiological quantities (Xu *et al.*, 2006). The short-term integrity of the membrane of printed CHO cells was investigated more recently. This study found that the printing process caused pores of 100 Å to develop, which close up after about 2 hours in culture as shown in Figure 12.3. This study also found that no overexpression of heat shock proteins occurs due to the printing of those cells compared to normally cultured cells (Cui *et al.*, 2010).

High-concentration cellular inks have also been used successfully. Spheroids of cell aggregates up to 300 μm thick can be prepared by gyratory shaking. These aggregates can be assembled into structures within a sacrificial hydrogel. A variation of this method is the extrusion of pelletted cell suspensions through a capillary tube. Forgacs and coworkers have created vessels by horizontally laying the cell tubes onto agarose rods (Norotte *et al.*, 2009). After removal of the rods and maturation in a bioreactor, a cell tube is obtained. Presumably the morbidity observed with other extrusion techniques is avoided here by using 300–500 μm large extrusion pipettes. However, oxygen diffusion though such thick cellular aggregates may be a problem if necrosis is not desired.

12.4.4 Cell Attachment and Growth

There are no simple answers to identifying a suitable material to support cell growth or to enable growth of embedded cells. The first consideration is whether the material releases cytotoxic compounds, either due to residual contamination or from a degradation process. If the cells are not killed outright, many cell types are attachment-dependent and will not grow on a surface unless they can attach. Cells embedded in a matrix will also be dependent on attachment sites if they are to multiply. Embedded cells also need nutrition and so depend on diffusion through the matrix. The following brief summary is intended to act as a guide to the complexities of the current literature.

Cytotoxicity can readily be tested with cells grown on conventional tissue culture plates in the presence of solutions of extracts from the matrix material to be tested. The most common problem arises from residual solvents. If solvent is used in depositing solid polymer layers, some will be trapped and slowly be released.

Most human cell types will not multiply without attachment to a surface. Those cells that do not attach will eventually undergo apoptosis and die. Cells do not attach

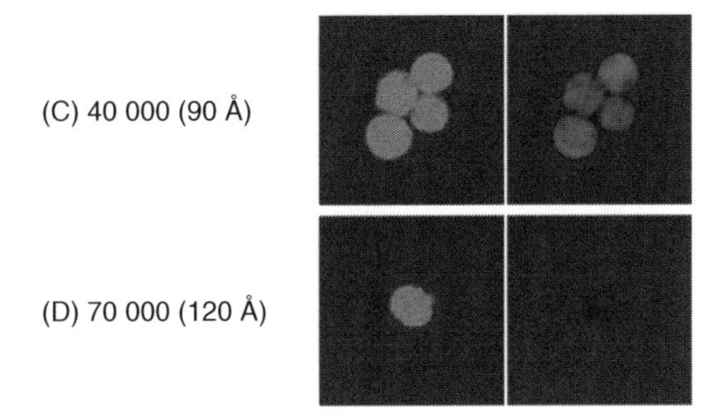

Figure 12.3 *Inkjet-printed cells labeled with high-molecular-weight fluorescent probe, showing that the small (40 kDa) probe penetrates pores in the cell membrane caused by the printing process but the larger probe does not (Reproduced with permission from Cui (2010) Copyright (2010) John Wiley & Sons Inc.) (See plate section for coloured version).*

to most neutral and anionic gel surfaces, but attachment may be induced through chemical modification with specific amino acid recognition sequences, such as RGD (arginine–glycine–aspartic acid) (Rowley, Madlambayan, and Mooney, 1999), or through cationic surface groups (De Rosa *et al.*, 2004; Gatta *et al.*, 2009). Chondrocytes and some types of tumor cells are exceptions and can multiply without attachment. A recent review of cell growth on polyelectrolyte gels formed by layer-by-layer deposition addresses these questions (Detzel, Larkin, and Rajagopalan, 2011).

While neutral and anionic gel surfaces do not promote cell attachment, the equivalent solid surface may, and there is no clear distinction between a gel and a solid except an increase in elastic modulus. As cells are normally cultured from plasma containing a complex mix of proteins, cell attachment is preceded by adsorption of proteins such as fibronectin that then promote cell attachment.

There is not much systematic understanding of the requirements for attachment of cells embedded in matrices. Fibrin gels can provide suitable attachment sites to promote differentiation of embedded mesenchymal stem cells (Ho *et al.*, 2010; Huang and Li, 2011). Cell growth in matrices must also depend on oxygen and nutrient supply. As a rule of thumb, these factors limit growth to a distance of about 100 μm below the matrix surface (Bhatia, Khattak, and Roberts, 2005).

Stem cells are a special case in that they can survive without attachment but do not multiply until they attach and differentiate into a specific cell type. The differentiation is driven by chemical and mechanical signals that are not yet fully identified.

12.4.5 Biocompatibility in the Body

Even if cytocompatibility is established, this does not ensure that the system will be accepted when implanted in the body. In particular, most nonbiodegradable materials set off a foreign body response, although this is reduced for gels (Liu *et al.*, 2008; Zhang *et al.*, 2009). This response leads to formation of a collagen capsule surrounding the implant that will isolate it from the system and limit sensing, drug delivery, or other functions.

12.5 Reactive Inks

Printing a solution of cross-linking calcium ions into alginate solutions to form a hydrogel was explored a few years ago (Varghese *et al.*, 2005). This type of ink reacts with the target substrate, solidifying it. Reactive ink is an attractive concept, because it will allow for building three-dimensional structures while limiting the amount of reactive compounds used. The amounts can thus be targeted to react only locally, producing structures that closely mimic the printed patterns. To date, this concept of reactive ink is only beginning to be explored. Some simple patterns such as hollow spheres and hollow tubes have been described (Boland *et al.*, 2006; Xu *et al.*, 2009a) (Figures 12.4 and 12.5). However, the technique holds promise for building vascular capillary structures, as was demonstrated recently (Cui and Boland, 2009).

In nature, in order to avoid premature precipitation, insoluble biopolymer structures are formed by various strategies involving a soluble precursor polymer followed by

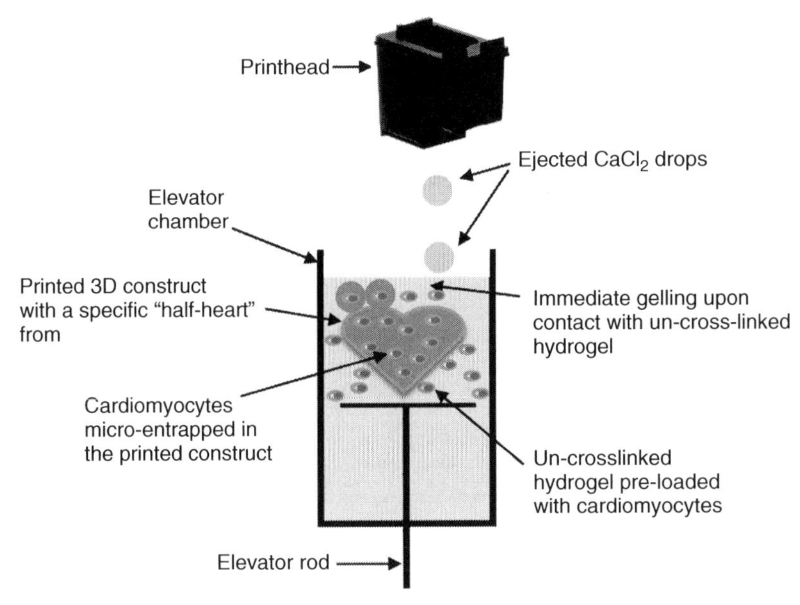

Figure 12.4 *Schematic showing inkjet printing of calcium chloride solution to locally gel alginate in a bath (Reproduced with permission from Boland (2006) Copyright (2006) Wiley-VCH and Xu (2009) Copyright (2009) IOP).*

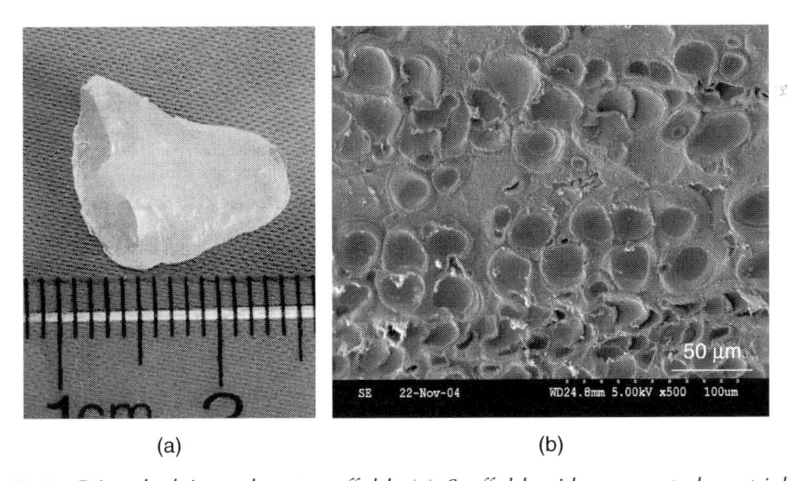

Figure 12.5 *Printed alginate heart scaffold. (a) Scaffold with connected ventricles; and (b) scanning electron microscopy (SEM) image of cross-section of the 3D structure (Reproduced with permission from Boland (2006) Copyright (2006) Wiley-VCH and Xu (2009) Copyright (2009) IOP).*

self-assembly (Silver, Freeman, and Seehra, 2003; Huang, Foo, and Kaplan, 2007). Synthetic polypeptides containing large fractions of cationic and anionic amino acids can be induced to self-assemble by a pH shift from strong acid or strong base where only the amines or acids are ionized to neutral pH, where both are ionized and form an ionic bond (Zhang, 2002; Zhao and Zhang, 2007). A similar strategy has been used to form structures from pairs of synthetic cationic polymers and anionic polymers by careful choice of the polymer ratios (Gratson, Xu, and Lewis, 2004). Capsules made of biopolymer gels can be formed by dropping a solution of cationic polymer, such as chitosan, into an anionic polymer such as alginate (Simsek-Ege, Bond, and Stringer, 2002, 2003). These self-assembled gels can also be formed as thin films using layer-by-layer dipping methods (Decher, Hong, and Schmitt, 1992; Bertrand *et al.*, 2000).

Inkjet printing can be used to form similar self-assembled hydrogels from cationic and anionic polymers (Limem *et al.*, 2009). Alternately printing layers of the two polymers or simultaneous printing from two cartridges results in thin multilayer sandwiches of the two polymers, which can then diffuse and combine to form a gel. Figure 12.6 shows a line of gel formed from polylysine hydrochloride and sodium polyglutamate printed onto glass. Subsequent washing removes the sodium chloride but leaves the insoluble hydrophilic gel. This approach can clearly be extended to a wide range of synthetic and biological polyelectrolytes, as shown by the printed film (Figure 12.7). With droplets in the 10–100 μm range, inkjet printing sits between the very slow layer-by-layer process and the dropping capsule process where the structures are non-uniform and uncontrolled. The effective diffusion distance of 1 μm or less in the flat discs formed by superimposed drying drops allows uniform self-assembly to occur.

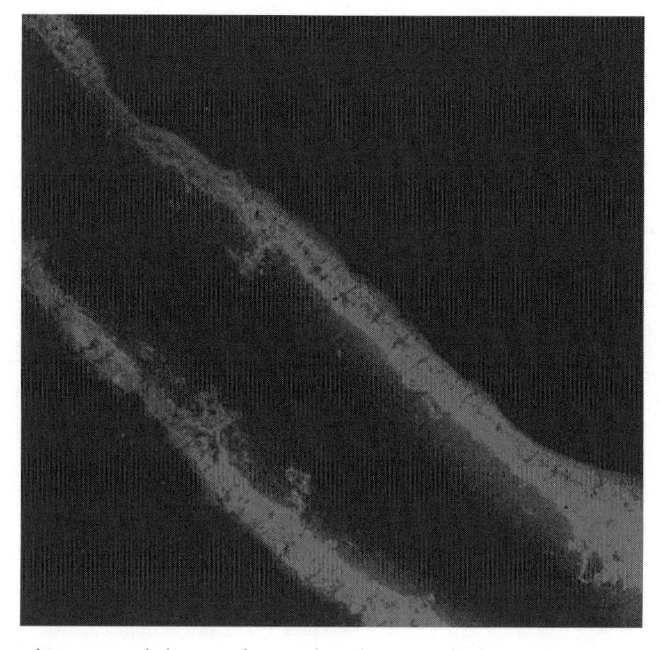

Figure 12.6 *Inkjet-printed lines of a poly-L-lysine/polyglutamate complex, fluorescent labeled. Line width is about 200 μm.*

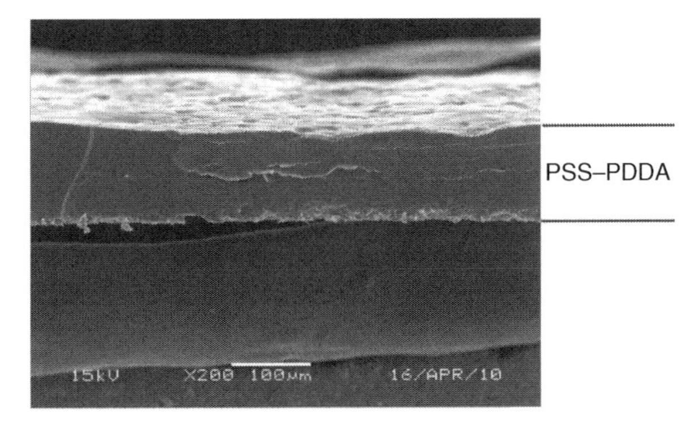

Figure 12.7 *Thick film of polystyrene sulfonate–polydimethyldiallylamine (PSS–PDDA) printed as alternate layers to form an insoluble complex (Reproduced with permission from Limem et al. (2009) Copyright (2009)).*

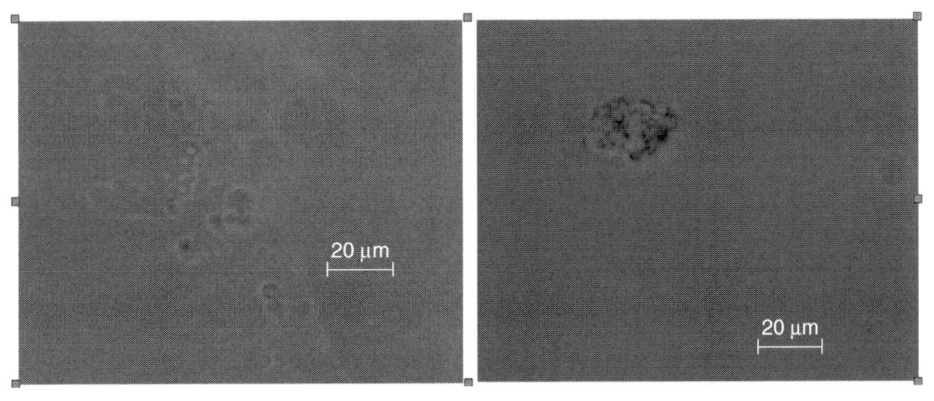

Figure 12.8 *Yeast colonies growing embedded in gels of polyethylene glycol diacrylate (PEGDA). Cured by blue light after co-extrusion of cells and gel (Mishra and Calvert, 2009). After 7 hours, cells started to grow in all three dimensions and formed a colony. Cells are entrapped and grown in a PEGDA thin-film hydrogel sample (20 × 10 × 9 mm). Left after 1 day; right after 7 days (Reproduced with permission from Mishra (2009) Copyright (2009) IEEE) (See plate section for coloured version).*

Inkjet printing is commonly used to print hot waxes that freeze on contact with the substrate, and this could be applied to thermally gelling biopolymers such as agarose or gelatin. Apart from the need to use a heated printed head, there would be severe limitations on molecular weight and concentration of the biopolymer.

UV-curable inkjet inks are widely used, and this approach could be applied to printing hydrogels. UV-cured gels with embedded cells have been formed by stereolithography (Arcaute, Mann, and Wicker, 2006), and Mishra *et al*. have made gels containing cells using blue light to cure polyethylene glycol diacrylates (PEGDAs) after delivery by a 3D extrusion system (Mishra and Calvert, 2009) (Figure 12.8). The main barrier to inkjet-printed gels or cell supports cured with blue or UV light would be oxygen inhibition of the curing reaction. It might therefore prove difficult simultaneously to retain cell viability and monomer curability.

There is a real need for better gel matrix materials to support bioprinting of cells. Calcium cross-linked alginate is widely used, but these gels are very weak and will solubilize in solutions that are low in calcium. The target support should mimic extracellular matrix, for which the best model is Matrigel, a commercial protein mixture secreted by Engelbreth–Swarm–Holm (ESH) mouse sarcoma cells and marketed by BD Biosciences (East Rutherford, NJ, United States). The chief components of Matrigel are structural proteins such as laminin and collagen which present cultured cells with the adhesive peptide sequences that they would encounter in their natural environment. Also present are growth factors that promote differentiation and proliferation of many cell types.

Functionalized polysaccharides can be cross-linked by the Michael-type addition reaction (Vernon *et al*., 2003; Cellesi, Tirelli, and Hubbell, 2004) without damage to entrained cells. Long-wavelength UV (Arcaute, Mann, and Wicker, 2006) or blue light (Mishra and Calvert, 2009; Biase *et al*., 2011) can be used to cross-link cytocompatible PEGDA solutions. It should also be possible to inkjet thermally gelling systems to print cold or hot gelling polymers over cells at room temperature. In addition to processability and mechanical strength, matrix gels will need well-defined transport properties for oxygen, nutrients, and other small and large molecules; stability of size in a range of solution environments; compatibility at implant sites; anchoring surfaces for cell attachment; and probably more. The immediate need is to expand the range of available printable matrices that can be adapted for different systems.

12.6 Substrates for Printing

Applications for bioprinting will include both 2D devices such as sensors and 3D tissues. In the first case the substrate will be a major component of the final device, whereas it will simply act as a temporary support for 3D structures. The interactions of conventional inks with paper have long been studied by the inkjet industry and are a complex mixture of wicking, barrier layer formation, absorption, and drying (Zhmud, 2003). Other studies have modeled ink absorption on textiles (Calvert and Chitnis, 2009). In the long term, one can envisage design of a layered substrate specific to any particular bioprinting application.

Most cell printing studies use a hydrogel target, because it allows the cells to be kept hydrated while providing active epitopes for cell attachment. A second reason may be that the mechanical properties are such that the cells are less damaged upon impact. It may be of interest to deliver hydrogels along with cells onto a target. Several studies have attempted to jet collagen and alginate solutions. However, due to their rheology only diluted concentrations have been printed to date. Thermal printers may have the edge over electro-mechanical printers; however, care must be taken to achieve low viscosities. A more successful approach has been to print cross-linking components into liquid targets as described in this chapter.

Cell survival after printing depends largely on the availability of water and nutrients. In some cases, a wet paper is sufficient to keep cells alive, as shown with *Escherichia coli* bacteria printed onto nylon paper (Xu *et al*., 2004). Mammalian cells are typically printed onto collagen hydrogels, which are subsequently incubated for one or more hours in a saturated humid environment to prevent drying (Boland *et al*., 2006). That typically is sufficient time for attachment of the cells to the hydrogel material. Cells can then be cultured as usual under liquid medium while maintaining the printed patterns. Some researchers have printed cells into dishes filled with media or directly onto tissue culture plastic (Saunders, Gough, and Derby, 2008); however, no patterns were obtained in these studies. As one key advantage of inkjet printing over conventional pipetting is the patterning, suitable hydrogel targets need to be identified.

12.7 Applications

12.7.1 Tissue Engineering

Tissue engineering as normally conceived involves the growth of cellular tissue by seeding cells onto a scaffold for growth in the laboratory and later implantation. Variations on this can be the formation of a scaffold which is seeded in the laboratory, then implanted for the cells to grow either at the final site of the repair or at a convenient low-stress site from which it can be moved when ready. The concept typically involves some type of support scaffold or matrix (Langer and Vacanti, 1993).

Of the many explored and potential applications of tissue engineering, two are really established so far. One is a family of skin prosthetics based on either a biopolymer scaffold or a combination of a scaffold and skin cells. Examples are the dermal regeneration matrix (Integra Life Sciences, Plainsboro, NJ, United States), Dermagraft (Advanced Biohealing, La Jolla, CA, United States), and Apligraf (Organogenesis Inc., Canton, MA, United States). Many other products are at various stages of development or early commercialization. The scaffold of the Integra product is a cross-linked matrix of bovine tendon collagen and shark cartilage glycosaminoglycans covered by a thin layer of silicone; the Apligraft product uses a bovine type I collagen matrix as scaffold (Eaglstein and Falanga, 1998); and the Dermagraft product employs knitted polygalactin 910, a co-polymer of lactic and glycolic acids (Marston, 2004). The other family of successful applications could be viewed as deriving from the use of treated porcine heart valves to replace damaged human valves in that a cell-free animal tissue is used as a scaffold. Acellular porcine small intestinal submucosa (SIS) is one example of a tissue that can be used for surgical repairs, with or without added adult stem cells from the patient or embryonic stem cells (Atala, 2009).

Although there have been many attempts to develop other organs, they have been largely unsuccessful so far. One particular target, which is greatly needed and might appear to be simple, is a small-diameter blood vessel lined with endothelial cells (L'Heureux, McAllister, and de la Fuente, 2007). Experience with skin has shown that cytokines, cell-signaling proteins, are very important for successful implantation but they are not well understood.

Skin is a simple tissue in that it is essentially 2D, so the problems of providing the cells with nutrition and oxygen are minimal. Many of the desired structures are 3D and will need some pore structure on the scale of a few hundred micrometres to allow a vascular structure to develop and supply the cells. This requires the ability to build complex 3D patterns from matrix, different cell types, and pores, which is a natural application for 3D multi-material printing processes. The immediate need is to develop reliable, low-resolution inkjet printing methods for viscous aqueous solutions and a palette of cytocompatible matrix materials.

There are many potential applications for the same approach. These include tissue parts such as patches to replace damaged heart muscle, corneas, cartilage, and many other tissue types in addition to the major aim of replacing whole organs. Separately there are potential applications for engineered tissue in experimental systems such as toxicity tests or studies of intercellular communication.

12.7.2 Bioreactors

One step further away from medicine, the ability to maintain animal cells, yeasts, or bacteria in an artificial structure would allow them to be used as continuous-flow bioreactors for the production of pharmaceuticals. There are also many applications for embedded cells outside the body as biosensors, and there should be many laboratory applications for cells embedded in 3D scaffolds if they could be readily formed.

While cells can be applied to a surface by allowing them to settle from suspension, inkjet printing or nozzle deposition should be preferred as a more controllable method for any production process. By combining the deposition of cells and matrix, there is also much more opportunity to build a structure which protects the cells and allows the tissue to be handled.

The resulting structures can have immediate applications in medicine, but combinations of cells and soft materials also have potential in sensors, in energy conversion such as biofuel cells and artificial photosynthesis, and in muscle-like actuators and robotics. These may be wholly synthetic, may be synthetic with embedded cells, or may be the results of structures built by cells that have been deposited in the right patterns.

12.7.3 Printed Tissues

While the potential for applying printing methods to tissue engineering is clear, there is no established approach or reliable equipment for doing this. Inkjet methods are very versatile and allow structures to be created as different materials mix on the substrate, but they are temperamental in practice. Extrusion methods are limited to materials that set slowly and it is difficult to deliver more than one material at a time, but they are very reliable. If we take the example of rapid prototyping methods, inkjet methods have tended to be the second generation of systems, after simpler methods of materials delivery had established the approach. True multi-material rapid prototyping is still very limited.

In principle, tissue engineering may follow the same path, with extrusion methods preceding printing methods. Several companies have already built extrusion systems for biopolymer printing, but there is not yet an established application that would drive these systems toward standardization of performance. The first need is probably for a photocurable biopolymer gel that can be packaged with cells by the user and used to build 3D scaffolds with embedded cells for laboratory study. Subsequent to this, one could envisage a disposable, sterile, fillable inkjet cartridge system that would allow users to load their own cells. The photocurable approach allows the rheology of the material prior to printing to be divorced from the properties of the gel after printing and so gives much more freedom in ink development.

12.8 Conclusions

Inkjet printing is a natural choice for building biological structures in the laboratory. This method allows complex architectures to be built as a series of patterned layers. The resolution is similar to the size of cells and, as in biology, self-assembly can generate finer structures. Also the whole process can run under wet, benign conditions.

Inkjet printing has only recently been applied to the manufacture of devices of any sort, so the equipment is not well developed for processing materials and the details are not well understood of how multiple layers of different materials will form and behave. These unknowns and uncertainties apply even more to biological systems.

It may be sensible to draw on the analogy with the development of integrated circuit processing and expect an interplay of process development leading device development.

Among the advances needed to open the field may be:

- the development of a printer system that allows simple loading and sterile storage of cells and biopolymers;
- the development of a method for controlling the humidity and water content of biological materials as they are deposited;
- identification of a range of suitable gel matrix materials to support cell growth.

References

A-Alamry, K., Nixon, K., Hindley, R. *et al*. (2010) Flow-induced polymer degradation during ink-jet printing. NIP 26: Digital Fabrication 2010, Technical Program and Proceedings, Society for Imaging Science and Technology, pp. 284–287.

Arcaute, K., Mann, B.K. and Wicker, R.B. (2006) Stereolithography of three-dimensional bioactive poly(ethyleneglycol) constructs with encapsulated cells. *Annals of Biomedical Engineering*, **34**, 1429–1441.

Asai, A., Hara, T. and Endo, I. (1987) One-dimensional model of bubble-growth and liquid flow in bubble jet printers. *Japanese Journal of Applied Physics Part 1*, **26**, 1794–1801.

Atala, A. (2009) Engineering organs. *Current Opinion in Biotechnology*, **20**, 575–592.

Bee, J.S., Stevenson, J.L., Mehta, B. *et al*. (2009) Response of a concentrated monoclonal antibody formulation to high shear. *Biotechnology and Bioengineering*, **103** (5), 936–943.

Bekard, I.B. and Dunstan, D.E. (2009) Shear-induced deformation of bovine insulin in couette flow. *Journal of Physical Chemistry B*, **113**, 8453–8457.

Bertrand, P., Jonas, A., Laschewsky, A. and Legras, R. (2000) Ultrathin polymer coatings by complexation of polyelectrolytes at interfaces: suitable materials, structure and properties. *Macromolecular Rapid Communications*, **21**, 319–348.

Bhatia, S.R., Khattak, S.F. and Roberts, S.C. (2005) Polyelectrolytes for cell encapsulation. *Current Opinion in Colloid and Interface Science*, **10**, 45–51.

Biase, M.D., Saunders, R.E., Tirelli, N. and Derby, B. (2011) Inkjet printing and cell seeding thermoreversible photocurable gel structures. *Soft Matter*, **7**, 2639–2646.

Boland, T., Xu, T., Damon, B. *et al*. (2006) Application of inkjet printing to tissue engineering. *Biotechnology Journal*, **1**, 910–917.

Buchholz, F. and Wilson, L. (1986) High shear rheology and shear degradation of aqueous polymer solutions. *Journal of Applied Polymer Science*, **32**, 5399–5413.

Calvert, P. and Chitnis, P. (2009) Mathematical modeling and numerical simulation of transport of inkjet printed suspensions into textiles. *Research Journal of Textiles and Apparel*, **13**, 46–52.

Calvert, P. and Crockett, R. (1997) Chemical solid free-form fabrication: making shapes without molds. *Chemistry of Materials*, **9**, 650–663.

Calvert, P.D., Lin, T.L. and Martin, H. (1997) Extrusion freeform fabrication of chopped-fibre reinforced composites. *High Performance Polymers*, **9**, 449–456.

Calvert, P. and Liu, Z. (1998) Freeform fabrication of hydrogels. *Acta Materialia*, **46**, 2565–2571.

Cellesi, F., Tirelli, N. and Hubbell, J.A. (2004) Towards a fully-synthetic substitute of alginate: development of a new process using thermal gelation and chemical cross-linking. *Biomaterials*, **25**, 5115–5124.

Chan, P.S-K., Chen, J., Ettelaie, R. *et al*. (2009) Filament stretchability of biopolymer fluids and controlling factors. *Food Hydrocolloids*, **23**, 1602–1609.

Chen, P., Chen, W. and Chang, S-H. (1997) Bubble growth and ink ejection process of a thermal ink jet print-head. *International Journal of Mechanical Sciences*, **39**, 683–695.

Chen, P-H., Chen, W-C., Ding, P-P. and Chang, S.H. (1998) Droplet formation of a thermal sideshooter inkjet print-head. *International Journal of Heat and Fluid Flow*, **19**, 382–390.

Cooper, G., Miller, E., Decesare, G. *et al*. (2010) Inkjet-based biopatterning of bone morphogenetic protein-2 to spatially control calvarial bone formation. *Tissue Engineering Part A*, **16**, 1749–1759.

Cui, X. and Boland, T. (2009) Human microvasculature fabrication using thermal inkjet printing technology. *Biomaterials*, **30**, 6221–6227.

Cui, X., Dean, D., Ruggeri, Z. and Boland, T. (2010) Cell damage evaluation of thermal inkjet printed Chinese hamster ovary cells. *Biotechnology and Bioengineering*, **106**, 963–969.

Decher, G., Hong, J.D. and Schmitt, J. (1992) Buildup of ultrathin multilayer films by a self-assembly process: III. Consecutively alternating adsorption of anionic and cationic polyelectrolytes on charged surfaces. *Thin Solid Films*, 210–211, 831–835.

Delaney, J., Smith, P. and Schubert, U. (2009) Inkjet printing of proteins. *Soft Matter*, **5**, 4866–4877.

Delehanty, J.B. and Ligler, F.S. (2003) Method for printing functional protein microarrays. *BioTechniques*, **34**, 380–380.

Derby, B. (2010) Inkjet printing of functional and structural materials: fluid property requirements, feature stability, and resolution. *Annual Review of Materials Research*, **40**, 395–414.

De Rosa, M., Carteni, M., Petillo, O. *et al*. (2004) Cationic polyelectrolyte hydrogel fosters fibroblast spreading, proliferation, and extracellular matrix production: implications for tissue engineering. *Journal of Cellular Physiology*, **198**, 133–143.

Detzel, C.J., Larkin, A.L. and Rajagopalan, P. (2011) Polyelectrolyte multilayers in tissue engineering. *Tissue Engineering: Part B*, **17**, 101–113.

Dhariwala, B., Hunt, E. and Boland, T. (2004) Rapid prototyping of tissue-engineering constructs using photopolymerizable hydrogels and stereolithography. *Tissue Engineering*, **10**, 1316–1322.

Dimitrov, D., Schreve, K. and De Beer, N. (2006) Advances in three dimensional printing: state of the art and future perspectives. *Rapid Prototyping Journal*, **12**, 136–147.

Eaglstein, W.H. and Falanga, V. (1998) Tissue engineering and the development of Apligraf a human skin equivalent. *Advance Wound Care*, **11** (Suppl. 4), 1–8.

Elbing, B.R., Winkel, E.S., Solomon, M.J. and Ceccio, S.L. (2009) Degradation of homogeneous polymer solutions in high shear turbulent pipe flow. *Experiments in Fluids*, **47** (6), 1033–1044, doi: 10.1007/s00348-009-0693-7

Ferguson, J., Hudson, N. and Warren, B. (1987) Phase changes during elongational flow of polymer solutions. *Nature*, **325**, 234.

Fittschen, U.E. and Havrilla, G.J. (2010) Picoliter droplet deposition using a prototype picoliter pipette: control parameters and application in micro X-ray fluorescence. *Analytical Chemistry*, **82**, 297–306.

Gatta, A.L., Schiraldi, C., Esposito, A. *et al*. (2009) Novel poly(HEMA-co-METAC)/alginate semi-interpenetrating hydrogels for biomedical applications: synthesis and characterization. *Journal of Biomedical Materials Research, Part A*, **90A**, 292–302.

Goodall, S., Chew, N., Chan, K. *et al*. (2002) Aerosolization of protein solutions using thermal inkjet technology. *Journal of Aerosol Medicine*, **15**, 351–357.

Gratson, G.M., Xu, M. and Lewis, J.A. (2004) Direct writing of three dimensional webs. *Nature*, **428**, 386.

Hall, R.P., Ogilvie, C.M., Aarons, E. and Jayasinghe, S.N. (2008) Genetic, genomic and physiological state studies on single-needle bio-electrosprayed human cells. *The Analyst*, **133** (10), 1347–1351.

Hancock, A. and Lin, L. (2004) Challenges of UV curable inkjet printing inks – a formulator's perspective. *Pigment and Resin Technology*, **33**, 280–286.

Harris, M., Doraiswamy, A., Narayan, R. *et al*. (2008) Recent progress in CAD/CAM laser direct-writing of biomaterials. *Materials Science and Engineering C, Biomimetic and Supramolecular Systems*, **28**, 359–365.

Ho, S., Cool, S., Hui, J. and Hutmacher, D. (2010) The influence of fibrin based hydrogels on the chondrogenic differentiation of human bone marrow stromal cells. *Biomaterials*, **31**, 38–47.

Hoath, S.D., Hutchings, I.M., Martin, G.D. *et al*. (2009) Links between ink rheology, drop-on-demand jet formation, and printability. *Journal of Imaging Science and Technology*, **53** (4), 041208.

Hoath, S.D., Martin, G.D. and Hutchings, I.M. (2010a) Effects of fluid viscosity on drop-on-demand ink-jet break-off. NIP 26: Digital Fabrication 2010, Society for Imaging Science and Technology, pp. 10–13.

Hoath, S.D., Martin, G.D. and Hutchings, I.M. (2010b) Improved models for drop-on-demand ink-jet shortening. NIP 26: Digital Fabrication 2010, Society for Imaging Science and Technology, pp. 353–355.

Hopp, B., Smausz, T., Kresz, N. *et al*. (2005) Survival and proliferative ability of various living cell types after laser-induced forward transfer. *Tissue Engineering*, **11**, 1817–1823.

Huang, J., Foo, C. and Kaplan, D. (2007) Biosynthesis and applications of silk-like and collagen-like proteins. *Polymer Reviews*, **47**, 29–62.

Huang, N. and Li, S. (2011) Regulation of the matrix microenvironment for stem cell engineering and regenerative medicine. *Annals of Biomedical Engineering*, **39**, 1201–1214.

Jaspe, J. and Hagen, S.J. (2006) Do protein molecules unfold in a simple shear flow? *Biophysical Journal*, **91**, 3415–3424.

Jorgensen, T.E., Sletmoen, M., Draget, K.I. and Stokke, B.T. (2007) Influence of oligoguluronates on alginate gelation, kinetics, and polymer organization. *Biomacromolecules*, **8**, 2388–2397.

van Krevelen, D.W. (1976) *Properties of Polymers*, Elsevier, Amsterdam.

Langer, R. and Vacanti, J.P. (1993) Tissue engineering. *Science*, **260** (5110), 920–926.

Lausted, C., Dahl, T., Warren, C. *et al*. (2004) POSaM: a fast, flexible, open-source, inkjet oligonucleotide synthesizer and microarrayer. *Genome Biology*, **5** (8), R58.

Lee, S., Yim, Y., Ahn, K. and Lee, S. (2009) Extensional flow-based assessment of red blood cell deformability using hyperbolic converging microchannel. *Biomedical Microdevices*, **11**, 1021–1027.

L'Heureux, N., McAllister, T.N. and de la Fuente, L.M. (2007) Tissue-engineered blood vessel for adult arterial revascularization. *New England Journal of Medicine*, **357**, 1451–1453.

Liao, Y.C., Subramani, H.J., Franses, E.I. and Basaran, O.A. (2004) Effects of soluble surfactants on the deformation and breakup of stretching liquid bridges. *Langmuir*, **20**, 9926–9930.

Limem, S., Calvert, P., Kim, H.J. and Kaplan, D.L. (2006) Differentiation of bone marrow stem cells on inkjet printed silk lines. Digital Fabrication, DF2006, Society for Imaging Science and Technology, pp. 99–102.

Limem, S., Li, D.P., Iyengar, S. and Calvert, P. (2009) Multi-material inkjet printing of self-assembling and reacting coatings. *Journal of Macromolecular Science Part, A: Pure and Applied Chemistry*, **46**, 1205–1212.

Liu, L., Chen, G., Chao, T. *et al*. (2008) Reduced foreign body reaction to implanted biomaterials by surface treatment with oriented osteopontin. *Journal of Biomaterials Science, Polymer Edition*, **19**, 821–835.

Liu, Z.S., Erhan, S.Z., Xu, J. and Calvert, P.D. (2002) Development of soybean oil-based composites by solid freeform fabrication method: epoxidized soybean oil with bis or polyalkyleneamine curing agents system. *Journal of Applied Polymer Science*, **85**, 2100–2107.

Liu, C., Sachlos, E., Wahl, D. *et al*. (2007) On the manufacturability of scaffold mould using a 3D printing technology. *Rapid Prototyping Journal*, **13**, 163–174.

Lozo, B., Stanic, M., Jamnicki, S. *et al*. (2008) Three-dimensional ink jet prints-impact of infiltrants. *Journal of Imaging Science and Technology*, **52** (5), 051004.

Marston, W.A. (2004) Dermagraft, a bioengineered human dermal equivalent for the treatment of chronic nonhealing diabetic foot ulcer. *Expert Review of Medical Devices*, **1** (1), 21–31.

Melendez, P.A., Kane, K.M., Ashvar, C.S. *et al*. (2008) Thermal inkjet application in the preparation of oral dosage forms: dispensing of prednisolone solutions and polymorphic characterization by solid-state spectroscopic techniques. *Journal of Pharmaceutical Sciences*, **97**, 2619–2636.

Mishra, S. and Calvert, P. (2009) Blue light cured 3D living catalysts. IEEE Proceedings 35th Annual Northeast Bioengineering Conference, Massachusetts, 3–5 April, 2009, pp. 1–2.

Mongkoldhumrongkul, N., Flanagan, J.M. and Jayasinghe, S.N. (2009) Direct jetting approaches for handling stem cells. *Biomedical Materials*, **4**, 15018.

Morissette, S.L., Lewis, J.A., Cesarano, J. *et al*. (2000) Solid freeform fabrication of aqueous alumina-poly(vinyl alcohol) gelcasting suspensions. *Journal of the American Ceramic Society*, **83**, 2409–2416.

Nair, K., Gandhi, M., Khalil, S. *et al*. (2009) Characterization of cell viability during bioprinting processes. *Biotechnology Journal*, **4**, 1168–1177.

Nakamura, M., Kobayashi, A., Takagi, F. *et al*. (2005) Biocompatible inkjet printing technique for designed seeding of individual living cells. *Tissue Engineering*, **11**, 1658–1666.

Ng, K., Joly, P., Jayasinghe, S. *et al*. (2011) Bio-electrospraying primary cardiac cells: in vitro tissue creation and functional study. *Biotechnology Journal*, **6**, 86–95.

Nishioka, G.M., Markey, A.A. and Holloway, C.K. (2004) Protein damage in drop-on-demand printers. *Journal of the American Chemical Society*, **126**, 16320–16321.

Nishiyama, Y., Nakamura, M., Henmi, C. *et al*. (2009) Development of a three-dimensional bioprinter: construction of cell supporting structures using hydrogel and state-of-the-art inkjet technology. *Journal of Biomechanical Engineering: Transactions of the ASME*, **131** (3), 035001.

Norotte, C., Marga, F.S., Niklason, L.E. and Forgacs, G. (2009) Scaffold-free vascular tissue engineering using bioprinting. *Biomaterials*, **30**, 5910–5917.

Odell, J., Muller, A. and Keller, A. (1990) Degradation polymer during extensional flow. *Macromolecules*, **23** (12), 3092–3103.

Okamoto, T., Suzuki, T. and Yamamoto, N. (2000) Microarray fabrication with covalent attachment of DNA using Bubble Jet technology. *Nature Biotechnology*, **18**, 438–441.

Othon, C.M., Wu, X., Anders, J.J. and Ringeisen, B.R. (2008) Single-cell printing to form three-dimensional lines of olfactory ensheathing cells. *Biomedical Materials*, **3**, 034101.

Panhuis, M., Heurtematte, A., Small, W.R. *et al*. (2007) Inkjet printed water sensitive transparent films from natural gum-carbon nanotube composites. *Soft Matter*, **3**, 840–843.

Peng, J., Lin, T.L. and Calvert, P. (1999) Orientation effects in freeformed short-fiber composites. *Composites A*, **30**, 133–138.

Ringeisen, B.R., Chrisey, D.B., Pique, A. *et al*. (2002) Generation of mesoscopic patterns of viable *Escherichia coli* by ambient laser transfer. *Biomaterials*, **23**, 161–166.

Roth, E.A., Xu, T., Das, M. *et al*. (2004) Inkjet printing for high-throughput cell patterning. *Biomaterials*, **25**, 3707–3715.

Rowley, J.A., Madlambayan, G. and Mooney, D.J. (1999) Alginate hydrogels as synthetic extracellular matrix materials. *Biomaterials*, **20**, 45–53.

Ruiz, O. (2007) CFD model of the thermal inkjet droplet ejection process. Proceedings of the ASME/JSME Thermal Engineering Summer Heat Transfer Conference, vol. 3, pp. 357–365.

Sanjana, N.E. and Fuller, S.B. (2004) A fast flexible ink-jet printing method for patterning dissociated neurons in culture. *Journal of Neuroscience Methods*, **136**, 151–163.

Saunders, R.E., Gough, J.E. and Derby, B. (2008) Delivery of human fibroblast cells by piezoelectric drop-on-demand inkjet printing. *Biomaterials*, **29**, 193–203.

Sercombe, T.B., Schaffer, G.B. and Calvert, P. (1999) Freeform fabrication of functional aluminium prototypes using powder metallurgy. *Journal of Materials Science*, **34**, 4245–4251.

Shore, H.J. and Harrison, G.M. (2005) The effect of added polymers on the formation of drops ejected from a nozzle. *Physics of Fluids*, **17**, 033104.

Silver, F., Freeman, J. and Seehra, G. (2003) Collagen self-assembly and the development of tendon mechanical properties. *Journal of Biomechanics*, **36**, 1529–1553.

Simsek-Ege, F., Bond, G. and Stringer, J. (2002) Matrix molecular weight cut-off for encapsulation of carbonic anhydrase in polyelectrolyte beads. *Journal of Biomaterials Science, Polymer Edition*, **13**, 1175–1187.

Simsek-Ege, F., Bond, G. and Stringer, J. (2003) Polyelectrolye complex formation between alginate and chitosan as a function of pH. *Journal of Applied Polymer Science*, **88**, 346–351.

Sirringhaus, H., Kawase, T., Friend, R.H. *et al*. (2000) High-resolution inkjet printing of all-polymer transistor circuits. *Science*, **290**, 2123–2126.

Tekin, E., Smith, P.J. and Schubert, U.S. (2008) Inkjet printing as a deposition and patterning tool for polymers and inorganic particles. *Soft Matter*, **4**, 703–713.

Tirella, A., Vozzi, F., De Maria, C. *et al*. (2011) Substrate stiffness influences high resolution printing of living cells with an ink-jet system. *Journal of Bioscience and Bioengineering*, **112**, 79–85.

Tsai, M.H. and Hwang, W.S. (2008) Effects of pulse voltage on the droplet formation of alcohol and ethylene glycol in a piezoelectric inkjet printing process with bipolar pulse. *Materials Transactions*, **49**, 331–338.

Varghese, D., Deshpande, M., Xu, T. *et al*. (2005) Advances in tissue engineering: cell printing. *Journal of Thoracic and Cardiovascular Surgery*, **129**, 470–472.

Vernon, B., Tirelli, N., Bachi, T. *et al*. (2003) Water-borne, in situ crosslinked biomaterials from phase-segregated precursors. *Journal of Biomedical Materials Research, Part A*, **64A**, 447–456.

Xu, T., Baicu, C., Aho, M. *et al*. (2009a) Fabrication and characterization of bioengineered cardiac pseudo tissues. *Biofabrication*, **1**, 1–6.

Xu, T., Rohozinski, J., Zhao, W.X. *et al*. (2009b) Inkjet-mediated gene transfection into living cells combined with targeted delivery. *Tissue Engineering Part A*, **15**, 95–101.

Xu, T., Gregory, C.A., Molnar, P. *et al*. (2006) Viability and electrophysiology of neural cell structures generated by the inkjet printing method. *Biomaterials*, **27**, 3580–3588.

Xu, T., Jin, J., Gregory, C. *et al*. (2005) Inkjet printing of viable mammalian cells. *Biomaterials*, **26**, 93–99.

Xu, T., Kincaid, H., Atala, A. and Yoo, J.J. (2008a) High-throughput production of single-cell microparticles using an inkjet printing technology. *Journal of Manufacturing Science and Engineering – Transactions of the ASME*, **130** (2), 021017.

Xu, T., Olson, J., Zhao, W.X. *et al*. (2008b) Characterization of cell constructs generated with inkjet printing technology using *in vivo* magnetic resonance imaging. *Journal of Manufacturing Science and Engineering – Transactions of the ASME*, **130**, 021013.

Xu, T., Petridou, S., Lee, E.H. *et al*. (2004) Construction of high-density bacterial colony arrays and patterns by the ink-jet method. *Biotechnology and Bioengineering*, **85**, 29–33.

Yan, K., Nair, K. and Sun, W. (2010) Three dimensional multi-scale modelling and analysis of cell damage in encapsulated alginate scaffolds. *Journal of Biomechanics*, **43**, 1031–1038.

Zein, I., Hutmacher, D.W., Tan, K.C. and Teoh, S.H. (2002) Fused deposition modeling of novel scaffold architectures for tissue engineering applications. *Biomaterials*, **23**, 1169–1185.

Zeng, J., Schmidt, C.G., Liu, H. and Jilani, A. (2009) Multi-disciplinary simulation of piezoelectric driven microfluidic inkjet. ASME 2009 International Design Engineering Technical Conference and Computers and Information in Engineering Conference (IDETC), San Diego, August 30, 2009.

Zhang, S. (2002) Emerging biological materials through molecular self-assembly. *Biotechnology Advances*, **20**, 321–339.

Zhang, Z., Chao, T., Liu, L. *et al*. (2009) Zwitterionic hydrogels: an in vivo implantation study. *Journal of Biomaterials Science, Polymer Edition*, **20**, 1845–1859.

Zhao, X.J. and Zhang, S.G. (2007) Designer self-assembling peptide materials. *Macromolecular Bioscience*, **7**, 13–22.

Zhmud, B. (2003) Dynamic aspects of ink-paper interaction in relation to inkjet printing. Pira International Conference: Ink on Paper Brussels.

13

Tissue Engineering: A Case Study

Makoto Nakamura
Graduate School of Science and Engineering for Research, University of Toyama, Japan

13.1 Introduction

13.1.1 Tissue Engineering and Regenerative Medicine

Tissue engineering and regenerative medicine make up a rapidly developing field of science and technology and are expected to have great importance in the 21st century due to improved understanding and control of living cells and biomaterials. This chapter describes an application of inkjet technology in this area of research.

Tissue engineering and regenerative medicine aim to develop functional cell, tissue, and organ substitutes to repair, replace, or enhance biological functions that have been lost through injury, disease, congenital abnormality, or aging. Artificially developed tissues and organs, such as engineered hearts, livers, or kidneys, will solve many of the present problems with organ transplantation and the currently unsatisfactory artificial organ therapies.

Each year, many patients die while waiting for organ transplantation. Organ transplantation is both the final treatment for end-stage organ failure and the most effective method to save such patients. However, this treatment is intrinsically dependent on the availability of healthy organs. Healthy organs must be obtained from donors, for example a brain-dead patient with appropriate advance consent as a donor candidate, or a healthy volunteer relative for certain organs. The number of donor organs is very low compared with the number of organs actually needed, and so organ transplantation from donors is clearly an inadequate solution. Alternatives to donor organs, or some radical solution for the lack of donor organs, have been sought for a long time.

Inkjet Technology for Digital Fabrication, First Edition. Edited by Ian M. Hutchings and Graham D. Martin.

Various types of artificial organs have been developed, such as artificial hearts and livers, dialyzers for artificial kidneys, and oxygenators for artificial lungs. Although these make a valuable contribution to clinical medicine, many are suitable only for temporary "bridging" use or as extracorporeal devices, not as alternatives to transplantation. There are several technological limitations to current artificial organs, such as their durability, implantability, and inadequacy as an alternative to donor organs. In addition, the strategy of bridging does not present a reasonable solution when there is a serious lack of donor organs. The most problematic failing of current artificial organs is their inability to emulate the metabolic responses of real organs, such as *in vivo* energy generation, *in vivo* production of various humoral factors and metabolites, and so on. Artificial organs cannot carry out such biological metabolic functions at all. Moreover, there is still no clear path toward using artificial technologies to perform such biological and physiological functions *in vivo*.

Biotechnology, including molecular biology, decoding of the genome, genomic engineering, and stem cell biology, has recently been developing with remarkable rapidity. This progress inspires the hope that in the near future, donor organs can be developed from living cells by biotechnology.

With these driving forces, research has boomed in tissue engineering and regenerative medicine. To date, engineered skin has been clinically applied instead of donor skin and has successfully saved many severe burn patients. Extension to other tissues and organs is expected, based on current research being carried out worldwide.

13.1.2 The Third Dimension in Tissue Engineering and Regenerative Medicine

In tissue engineering and regenerative medicine, 3D structures form a central feature. Every organ, such as the heart, liver, kidney, or lung, has a 3D morphological structure. However, that structure is also composed of specific 3D tissues, which in turn are composed of specific 3D microstructures, each with several particular types of cells. Each cell plays a specialized role at its particular location in such 3D microstructures, and as a result, highly specialized physiological tissue and organ functions are generated. 3D microstructures and the 3D spatial correlations between the composite cell structures are therefore the key elements in the integration of the physiological functions of organs. For this reason, three-dimensionality is an essential factor in engineering biologically functional tissues and organs.

In addition, three-dimensionality is essential in cell biology, although cell culture is usually performed under 2D conditions. It has been reported that the functions and morphological features of cultured cells are different in 3D cultures from 2D cultures in a culture dish. For example, drug metabolism activities of human hepatocytes were enhanced in 3D spheroidal cultures within alginate scaffolds (Elkayam *et al*., 2006). 3D environments, such as a 3D culture using collagen gels, promote cell adhesion and morphological changes similar to those seen *in vivo* (Elsdale and Bard, 1972; Cukierman *et al*., 2001; Cukierman, Pankov, and Yamada, 2002). This is thought to be because cells are intrinsically situated within a 3D environment *in vivo*, and thus the 3D condition is closer to the physiological case.

13.1.3 The Current Approach for Manufacturing 3D Tissues

To date, the most common approach for manufacturing 3D tissues is by using 3D scaffolds. A scaffold is a temporary substrate which is necessary for cells to adhere and proliferate, as well as for defining and retaining a 3D morphological architecture. A 3D scaffold is usually made of biodegradable material such as poly-glycolic acid or poly-lactic acid, and is destined to degrade gradually and finally to vanish after implantation in the body. Many researchers have used this approach successfully for some simple tissues such as bone and cartilage. In addition, recently, several 3D fabrication techniques have been applied to fabricate 3D scaffolds (Griffith and Naughton, 2002; Landers *et al.*, 2002; Vozzi *et al.*, 2002; Yang *et al.*, 2002; Hutmacher, Sittinger, and Risbud, 2004). This approach is shown in Figure 13.1. However, it is essentially based on a method in which cells are seeded afterwards onto the acellular scaffold. Cells can be seeded only onto the surface of the scaffold, not within it. In addition, cells are generally seeded randomly and homogeneously onto the whole of the scaffold. In the case of tissues composed of multiple cell types, seeding is carried out with the various cell types all at once. For this reason, we cannot control the cell distribution and cell composition inside the scaffold at all. Although this approach can create some simple tissues with a single type of cell, it will not provide the capacity to organize the internal structure of the scaffolds, or a micro-scaled cell structure, especially with multiple different types of cells. In addition, the morphogenesis process in essential tissue architectures is completely dependent on the cells and their *in vivo* patho-physiological responses after *in vivo* implantation, which technology cannot control. The limitations of this approach are summarized in Figure 13.2.

Other, intrinsically different approaches are therefore needed for manufacturing complicated 3D tissues.

13.1.4 A New Approach of Direct 3D Fabrication with Live Cell Printing

Considering the limitations of the scaffold-based approach, we have proposed a new approach in which the tissue structure is arranged, assembled, and cultured prior to implantation, by technological means.

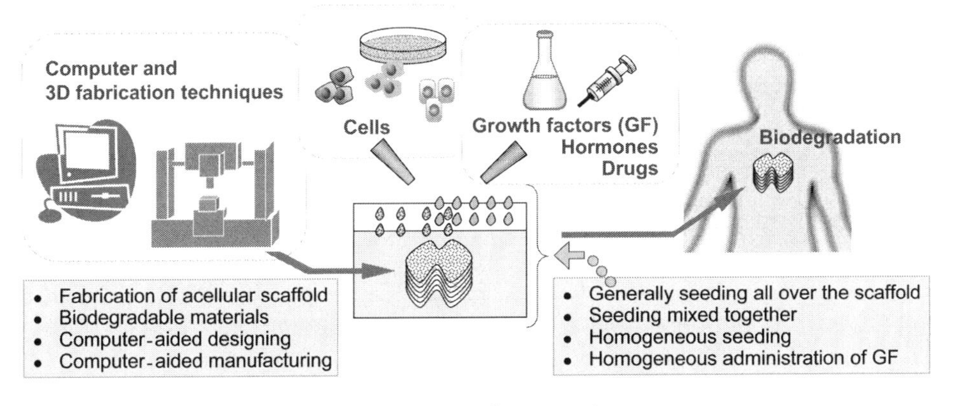

Figure 13.1 Conventional scaffold-based tissue engineering.

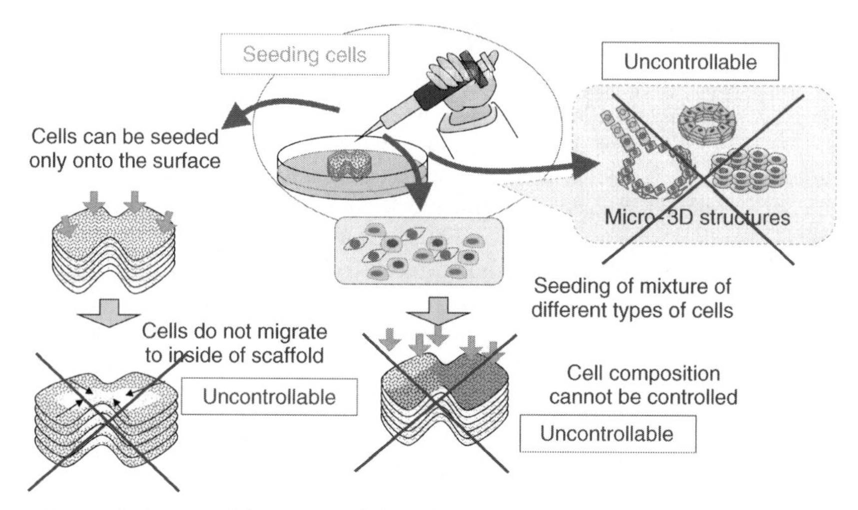

Figure 13.2 *Limitation of the approach based on subsequent cell seeding (See plate section for coloured version).*

Histological observations show each tissue and organ to be composed of multiple types of cells and extracellular matrix proteins such as collagen, forming different specialist 3D structures for each type of tissue. In addition, physiologically functional tissues contain many capillary vessels. Such a perfusion system is the most important structure for tissue manufacturing. Based on these histological considerations, we speculated that an effective approach for manufacturing biological tissues with such characteristics would be one in which composite materials, including living cells and extracellular matrices, are directly positioned and produced in a "bottom-up" process. We considered the following four major technologies to be needed: (i) the technology to arrange those components onto targeted positions with the necessary microscopic-length resolution, (ii) the technology to arrange multiple different materials into their respective positions, (iii) the technology to fabricate 3D structures, and (iv) the technology to induce blood vessels or to provide perfusion systems to deliver oxygen and nutrition to all of the cells in the engineered tissues.

Those speculations prompted us to apply the inkjet printing technique to handling cells and positioning cells directly, because the ability of inkjet printing meets many of these requirements. This was the beginning of our research on inkjet-based direct 3D fabrication with live cell printing. The concept of inkjet 3D biofabrication is illustrated in Figure 13.3.

13.2 A Feasibility Study of Live Cell Printing by Inkjet

As mentioned in Section 13.1, the technology to deposit materials directly onto a particular position is essential and the living cell is the most important material. Therefore, we first planned to examine the feasibility of direct printing of live cells.

We looked for an available inkjet system from the literature. As described in Chapter 2, inkjet printers fall into two major groups, continuous inkjet and drop-on-demand (DOD),

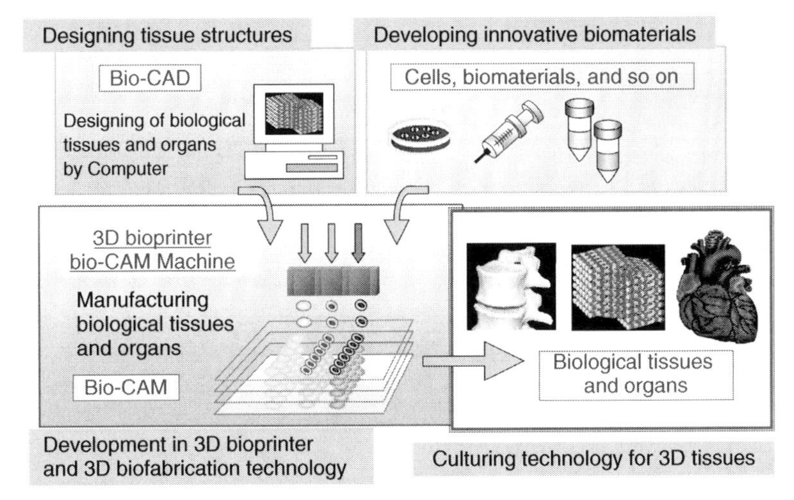

Figure 13.3 *Concept of inkjet-based direct 3D biofabrication. The original data is designed on a computer (bio-CAD), and based on the digital data, the 3D bioprinter fabricates 3D biological structures as a bio-CAM machine. The materials for 3D biofabrication are living cells and various biomaterials including natural and synthetic materials. Fabricated biological structures are cultured in succession to form functional tissues and organs (See plate section for coloured version).*

with the latter being subdivided into piezoelectric and thermal types. Piezo print-heads do not use heat to eject the ink, whereas thermal inkjet generates temperatures of up to 300 °C in order produce the transient bubbles needed to generate the droplets. For the biological materials which would be used, including living cells, we first decided to use a piezo-type inkjet system for considerations of biocompatibility and the ease of color (i.e., multiple-ink) printing. However, other researchers have reported that living cells can also be printed safely with thermal inkjet systems (Xu *et al*., 2005, Xu *et al*., 2006; Boland *et al*., 2006). This is thought to be because the heat is generated for a very short time (within a millisecond) and only at the surface of the heating element.

Initial experiments with a commercial inkjet printer failed after only a few trials because of blockage of the nozzles. Further successful tests used a design of a print-head actuated electrostatically and based on micro-electro-mechanical systems (MEMS) technology (SEAJet™ system, Seiko Epson Corp., Nagano, Japan). Details of the SEAJet system are described elsewhere (Kamisuki *et al*., 1998).

We examined the feasibility of live cell printing with this inkjet system (Nakamura *et al*., 2005). A suspension of bovine endothelial cells was jetted into culture dishes. We observed the ejected ink droplets and confirmed that cells existed in the jetted droplets. Time-lapse monitoring of jetted cells showed that they were alive and adhered to and moved in the culture dish. Observation by scanning electron microscopy (SEM) revealed no significant differences between jetted cells and nonjetted cells. Next, we investigated the viability of jetted cells quantitatively by live-dead assay using fluorescent dyes. We examined several varieties of cells. Human fibrosarcoma cell line HT1080 (JCRB 9113), human hepatoma cell line HepG2 (JCRB 1054), and human cervical carcinoma cell line HeLa (JCRB 9004) were examined as cell-line cells, which are often used in drug safety evaluations, while human umbilical vein endothelial cells (HUVECs), human aortic smooth muscle cells (HASMCs), and neonatal normal human dermal fibroblasts (NHDFs) were also examined as normal cells. The results are shown in Figure 13.4. In total, more than 90% of cells were found to be alive after inkjet deposition.

Their proliferation capacity was also investigated for HeLa cells and HUVECs. Figure 13.5 shows the growth of ejected cells compared with non-ejected cells. No significant difference was found. It was concluded that the inkjet-printing process has no significant harmful effects on living cells and that direct printing of live cells by inkjet is possible.

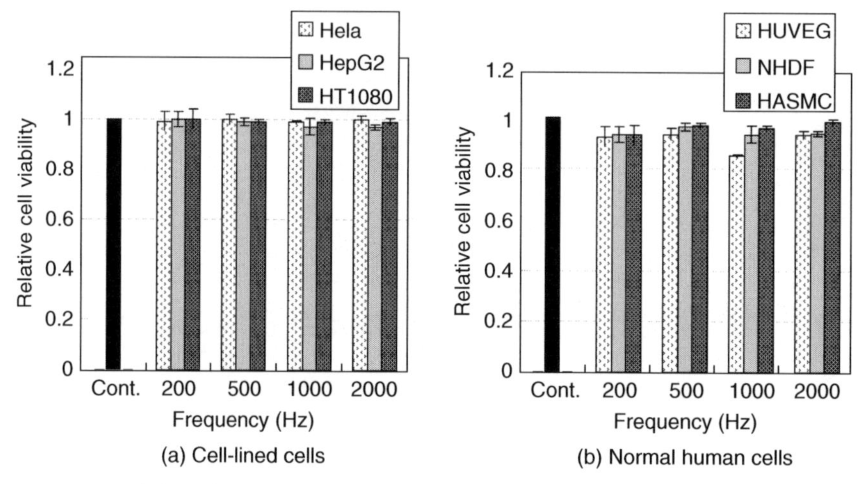

Figure 13.4 Viability of various different cell types after inkjet printing. Viability of (a) jetted cells for cell-line cells; and (b) normal human cells for different jetting frequencies.

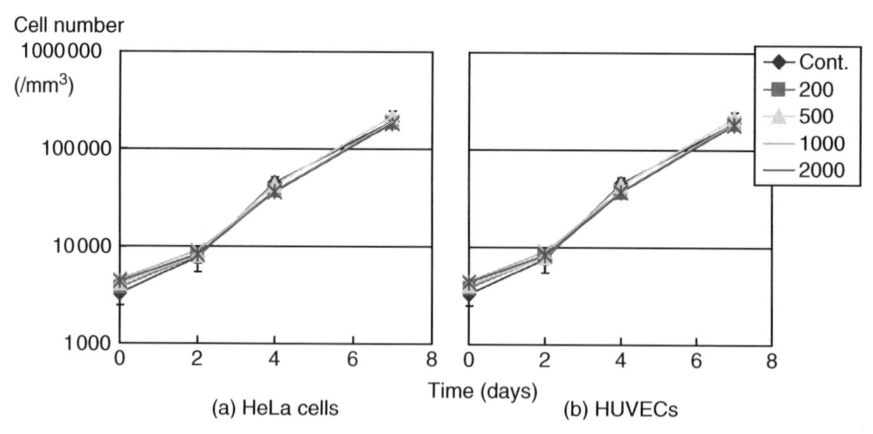

Figure 13.5 Cell proliferation rate after inkjet printing. Proliferation rate after jetting for (a) HeLa cells; and (b) HUVECs.

13.3 3D Biofabrication by Gelation of Inkjet Droplets

Although the work described in Section 13.2 showed inkjet technology to be useful for direct printing of live cells, 3D fabrication was still the unsolved problem. We therefore developed a method of 3D fabrication by gelation of inkjet droplets. We examined alginate hydrogel, which is well known as a biocompatible hydrogel. Alginate hydrogel is formed by mixing two liquid solutions: aqueous sodium alginate and calcium chloride. The former acts as a gel precursor, and the latter as a gel reactor. When we jetted sodium alginate solution into stirred calcium chloride solution by inkjet printing, we found that micro-gel beads were formed by gelation of each inkjet droplet. Using cell-suspended sodium alginate solution, cell-containing gel beads were also obtained (see Figure 13.6c). In addition, we could make micro-gel fibers even in a liquid substrate when lines were printed onto the calcium chloride solution. The hydrogel successfully prevented the lines from merging. In addition, the use of hydrogels is valuable, because they simultaneously provide both the semisolid structures and liquid environments which are required for 3D fabrication and for cell viability, respectively. We found that, when crossing lines were drawn, two lines could cross three-dimensionally without merging and mixing. Using this gelation technique, we recognized the significant feasibility of 2D and 3D fabrication by inkjet. In addition, time-lapse monitoring of cell-containing micro-gel beads also showed that the cells were alive and moved within the gel beads. Cell division was also observed in the micro-gel beads. Thus we confirmed the feasibility of direct 3D biofabrication by live cell printing and by this gelation technique using inkjet droplets.

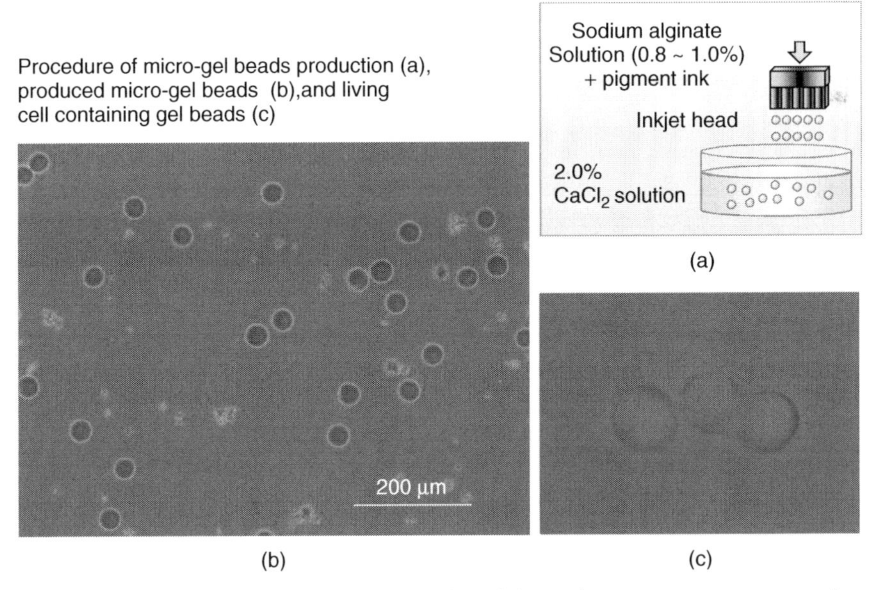

Figure 13.6 *Alginate micro-gel beads produced by inkjet printing. (a) Procedure for micro-gel beads production; (b) micro-gel beads produced; and (c) micro-gel beads containing living cells.*

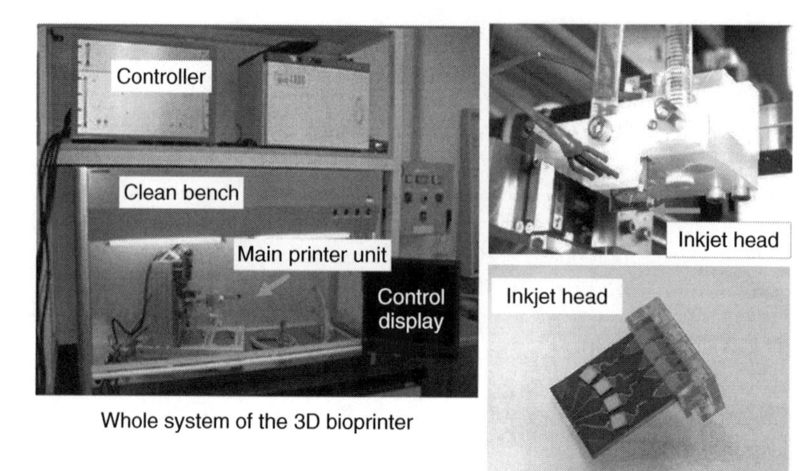

Figure 13.7 Custom-made 3D bioprinter. Whole system of 3D bioprinter and inkjet print-head (piezo-type).

Based on these preliminary results, we developed a custom-made 3D bioprinter to fabricate designed 3D hydrogel structures containing living cells (Nishiyama *et al.*, 2007). The first prototype is shown in Figure 13.7. Both the hardware and software of this bioprinter were developed to meet the special requirements for live cell printing and 3D biofabrication. The overall size of the 3D bioprinter was designed for use in a bio-clean bench, because living cells are handled. The print-head moves in three dimensions in order to print on a wide variety of objects, not only solid sheets but also solid mass, liquid substrate, and biological objects. In our inkjet gelation technique, the substrate is a liquid and the 3D structures are constructed into the liquid.

In this printer, we used two types of inkjet systems. One was the SEAJet system, described in Section 13.2. The other was a piezo-type print-head developed by Fuji Electric Corp. (Tokyo) (as shown in Figure 13.7). Neither type of print-head is thermal, and thus they are particularly suitable for this application. In addition, in the piezo-inkjet system, the ejection performance can readily be controlled by changing the input electric signal, which is advantageous when using a wide variety of biomaterials as inks.

13.4 2D and 3D Biofabrication by a 3D Bioprinter

With the prototype 3D bioprinter, we attempted to fabricate several 2D and 3D structures from hydrogel and living cells (Nakamura *et al.*, 2006, Nakamura *et al.*, 2007; Nishiyama *et al.*, 2007; Henmi *et al.*, 2008).

13.4.1 Micro-Gel Beads

Figure 13.6a,b shows the production of micro-gel beads by inkjet. These beads can form the elementary units for 2D and 3D gel structures. As the micrograph shows, homogeneous micro-gel beads were obtained. Image analysis showed their average diameters to be 38.5 μm (standard deviation 3.6 μm), and more than 94% were from 35 to 45 μm in diameter (Nakamura *et al*., 2007). This suggests that 3D structures can be manufactured by inkjet from such homogeneous micro-gel beads. However, several experiments have also shown that the jetting performance is very delicate and is influenced by several factors, such as the materials used in the ink and the ambient conditions during the jetting experiments. It is therefore very important to maintain stable printing conditions, including both the ink composition and the environment. Figure 13.6c shows micro-gel beads containing living cells. In this case, a suspension of bovine endothelial cells was mixed with sodium alginate solution and jetted into calcium chloride solution. By time-lapse monitoring, we also observed the cells to be living and moving in the gel beads.

13.4.2 Micro-Gel Fiber and Cell Printing

When lines were printed onto a glass slide covered with calcium chloride solution, micro-gel fibers were formed. In this case, adjustment of the relationship between the ejection frequency and head movement speed is important for forming smooth, narrow gel fibers. When using cells suspended in sodium alginate solution, the cells were embedded in the micro-gel fibers (Figure 13.8), which had a thickness of approximately 30 μm. Inkjet gel formation was shown to be effective for printing cell patterns without blurring or merging, even in the calcium chloride solution or under wet conditions.

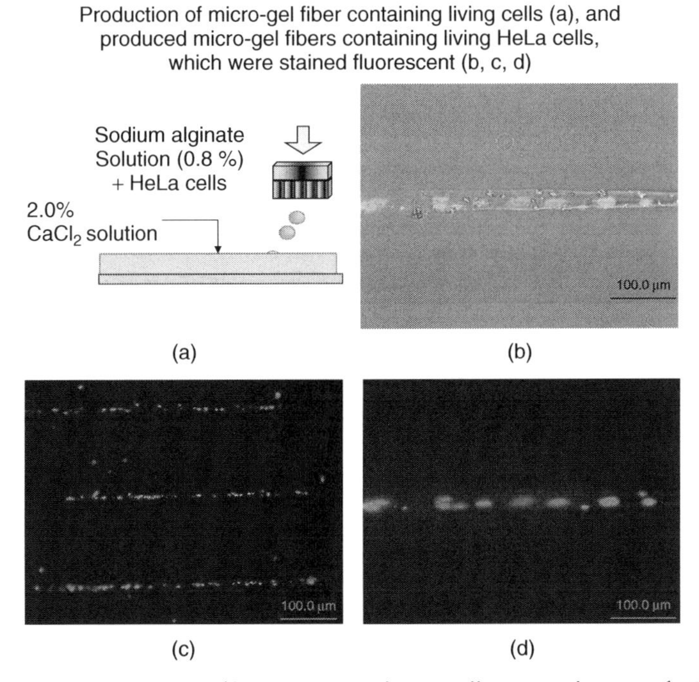

Figure 13.8 *Alginate micro-gel fiber containing living cells. (a) Production of micro-gel fiber containing living tissues; and (b–d) micro-gel fibers containing living HeLa cells (fluorescence stained).*

13.4.3 2D and 3D Fabrication of Gel Sheets and Gel Mesh

Using the inkjet bioprinter, we were able to produce 2D gel sheets. A square area was made in the same manner as discussed using sodium alginate solution, which resulted in a 2D gel sheet (Figure 13.9a–c). The sheets were composed of densely arranged micro-gel fibers, each made of micro-gel beads. We also successfully layered gel sheets by printing onto a previously printed gel sheet. Thus, it was shown that layer-by-layer inkjet printing of gel was possible. In addition, when micro-gel fibers were printed at some intervals, 2D and 3D gel mesh structures were obtained (Figure 13.9d–f). When the intervals were changed, fine-mesh structures and large-mesh structures could be fabricated (Figure 13.9e, f).

13.4.4 Fabrication of 3D Gel Tubes

Tubular structures are often seen in the human body, such as the blood vessels, lymph vessels, trachea and bronchus, bile ducts, renal tubules, various glands, and so on. We attempted to fabricate 3D gel tubes with the 3D bioprinter. The inkjet head was moved in a continuous circular motion, and sodium alginate solution was ejected into the calcium chloride solution (Figure 13.10a). A 3D gel tube structure was successfully formed and grew longer into the calcium chloride solution as the procedure advanced (Figure 13.10b). A representative fabricated gel tube with a diameter of 1 mm and total length of approximately 20 mm that contained living HeLa cells is shown in Figure 13.10c. We have also been able to fabricate 1 mm diameter tubes with a length greater than 50 mm.

13.4.5 Multicolor 3D Biofabrication

To investigate the potential for manufacturing composite tissues with multiple cell types, the hardware and software of the 3D bioprinter system was extended from a monochrome printer to a multicolor system. Using two different-colored gel precursor inks and two

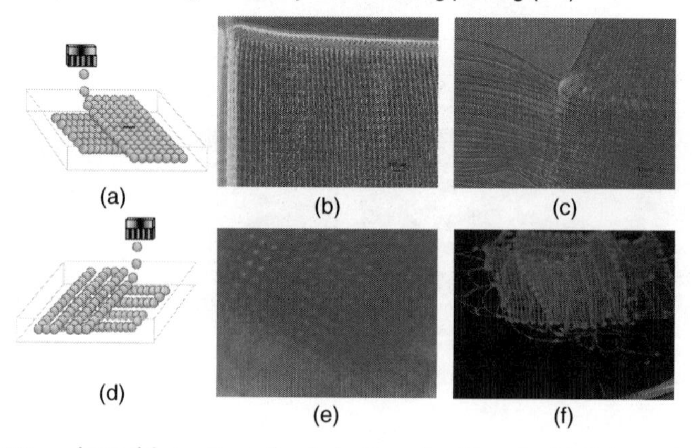

Figure 13.9 2D and 3D fabrication of gel sheet and gel mesh. (a) Schematic showing the production of gel sheet; (b, c) laminated sheets fabricated by 3D bioprinter; (d) schematic showing the production of gel mesh; and (e, f) gel mesh fabricated by 3D laminating printing.

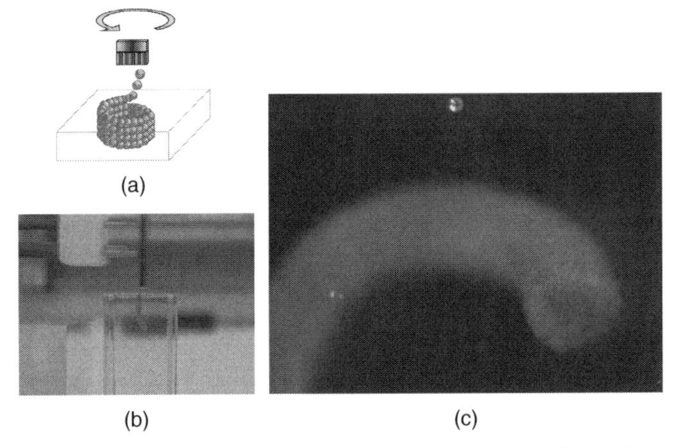

Figure 13.10 Fabrication of 3D gel tube containing HeLa cells. (a,b) Schematic and photograph showing fabrication method; and (c) 3D gel tube containing HeLa cells. See http://www.youtube.com/watch?v=g2ZTWHsO8l0&feature=player_embedded (See plate section for coloured version).

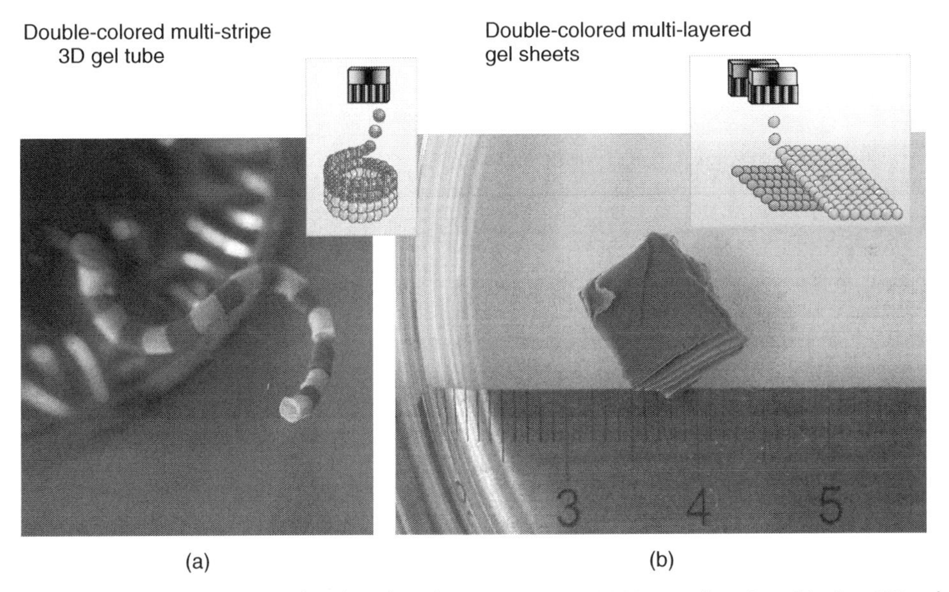

Figure 13.11 Fabrication of multicolored 3D structures. (a) Two-colored multistripe 3D gel tube; and (b) two-colored multilayered gel sheets (See plate section for coloured version).

separate inkjet nozzle systems, 2D and 3D gel structures were fabricated with different colored gels. Figure 13.11a, b shows a two-colored multistriped 3D tube and 3D lamination of different colored gel sheets. A double-walled 3D tube with different colored gels could also be fabricated by alternate printing of inner and outer circles successively with different inkjet nozzles. Fluorescent images are shown in Figure 13.12. The diameters of the inner and outer walls were 1 and 1.3 mm, respectively, and the thickness of each

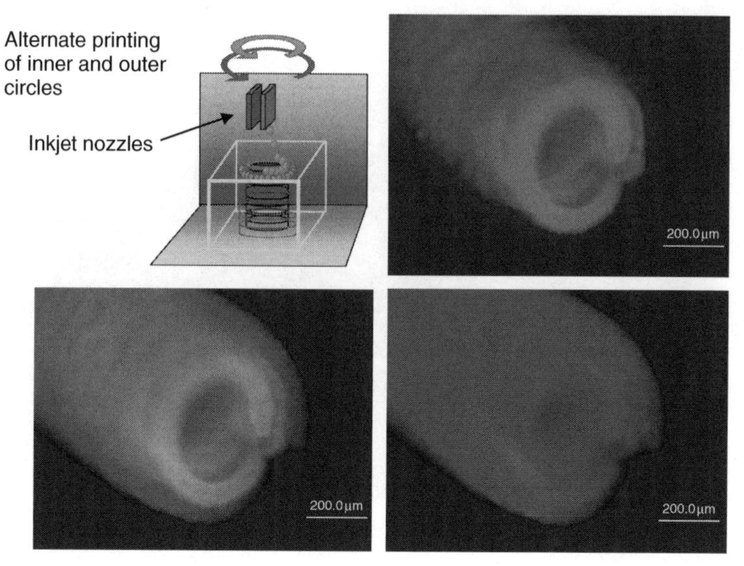

Figure 13.12 *Fabrication of double-walled 3D gel tube with different fluorescent colored inks. (a) Top left: Schematic showing fabrication method. (b–d) Top right, bottom left and bottom right: Double-walled 3D gel tube with different fluorescent colored inks (See plate section for coloured version).*

wall was about 100–150 μm. This double-walled tube was designed and fabricated to mimic the structure of blood vessels, in which endothelial cells are situated at the inner layer and smooth muscle cells form the outer layer, respectively. We then succeeded in fabricating a double-walled tube with living endothelial cells in the inside wall and smooth muscle cells in the outside wall (Calvert, 2007).

We have demonstrated that color inkjet and 3D laminating and printing can successfully realize organized cell structures. In these cases, the alginate hydrogel functions as the cell-supporting medium, not only morphologically as a scaffold but also to maintain the cells under physiological conditions. Although the cells were merely embedded and arranged in the gel structure in these experiments, it is hoped that such organized cell structures can be developed to become significant biological structures by establishing cell–cell connections, and by cell proliferation and growth in or on the hydrogel. The next stage of research should address how to develop such organized cells to form significant biological structures.

13.4.6 Viscosity in Inkjet 3D Biofabrication

The viscosities of both the jetted ink materials and the liquid substrate are important factors in the inkjet 3D biofabrication process. First, as noted in other chapters, jetting performance is influenced by the viscosity of the ink. Many print-heads are not capable of jetting solutions with a viscosity higher than ~10 mPa s. Our work has shown that nozzles are likely to be more easily blocked when using ink with high viscosity. From the point of view of jetting performance, the lower the viscosity of the ink, the better. Therefore, research and development to reduce the viscosity of the ink materials is important, as well as improvement in the ability to jet higher viscosity materials. Meanwhile, in the present procedure, the resulting gel structure is unfixed and floats in the liquid substrate. Control of the viscosity of the liquid substrate can be effective: a substrate with 15% polyvinyl alcohol (PVA) added, with an estimated viscosity of about 400 mPa s, showed good ability to support the gel structures.

13.5 Use of Inkjet Technology for 3D Tissue Manufacturing

Inkjet technology was historically developed for printing text and graphic images. However, the performance of inkjet technology is also applicable to tissue manufacturing. The characteristics and the advantages of inkjet are summarized in Table 13.1. Here we explain the rationale for the use of inkjet for tissue engineering in more detail.

13.5.1 Resolution and DOD Color Printing

Recent inkjet printers have very high resolution and are able to print photo-quality pictures. Typical inkjet-printed dots are 25–30 μm in size, very similar to the size of many biological cells (10–30 μm). The high-resolution and color-printing (i.e., multiple inks) capabilities of a modern DOD printing system will be vital for the fabrication of tissue microstructures: for example, very fine capillaries, such as blood capillaries, are of the order of 10 μm inner diameter. In addition, the cells should be located reasonably close (within 100 μm) to the perfusing capillary systems in order to ensure cell viability. Drop-on-demand color printing should make it possible to arrange different types of cells arbitrarily, effectively cell-by-cell. Because few other technologies can achieve hetero-material fabrication with cell-by-cell resolution, these are unique advantages that are considered essential to the construction of complicated micro-scaled tissue architectures with multiple cell types.

13.5.2 Direct Printing of Live Cells

As mentioned in this chapter, the internal structures and internal compositions of 3D structures cannot be controlled by conventional scaffold-based tissue engineering, in which cells are seeded afterwards onto 3D scaffolds. In contrast, a direct positioning and laminating approach enables 3D structures to be constructed with controlled internal composition. It is therefore a very promising and effective approach. Direct cell positioning is essential because the living cell is the most important component material, but biocompatibility and cell compatibility are also important. As demonstrated in Section 13.2, cell viability and cell proliferation are not significantly affected by the inkjet printing process used in this work. Although very high temperatures are generated locally, other researchers have also reported the high viability of cells when using thermal-type inkjet systems (Xu *et al*., 2005, Xu *et al*., 2006; Boland *et al*., 2006). Inkjet-based direct cell printing is thus a safe procedure for direct cell positioning.

13.5.3 High-Speed Printing

There are some other technologies for handling cells, such as aspiration with micropipettes and laser tweezers. But they can deal with only one or a few cells at a

Table 13.1 *Characteristics of inkjet technology and advantages for tissue engineering.*

Characteristics of inkjet technology	Advantages for tissue engineering
High resolution	For manufacturing of microscopic structures with cellular-sized resolution
Extremely small ink droplets	Micro to macro, multiscaled fabrication
Drop-on-demand printing	Enabling on-demand direct cell printing
Direct printing of ink droplets	For direct arrangement of cells and materials for biofabrication
Color printing	For fabrication of composite products with different cells, materials, and growth factors
High-speed printing more than 10 kHz per one nozzle	Handling massive amount of individual cells Rapid fabrication
Multi-nozzle system can be integrated	Lessens cell damage during fabrication
3D fabrication using hydrogels	Enabling 3D construction by layer-by-layer printing Enabling to print living cells For prevention from drying Enabling 3D fabrication into the liquid
Linkage to digital data sources	For digital printing Easy to apply to computer-aided biofabrication For CAD-, CAM-, and CAE-based biofabrication
Noncontact printing	Usability of reactive materials Preventive effects for friction or contact damages
Printability of several inks; aqueous inks, pigment inks, suspension of several materials, and reactive solution	Printing biological materials; cells proteins, DNAs, biopolymers, humoral factors, drugs, and nanomaterials
Printability onto several subjects; papers, solid mass, disk, dishes, gels, and aqueous solution	Printable onto gels, aqueous solution, cell sheets, directly printing onto the tissues, organs, and wounds during surgical operation

time. On the other hand, several thousand ink droplets can be ejected per second from each nozzle in an inkjet printer. As some printers use print-heads with several hundred nozzles, inkjet printing has the capacity to handle several hundred thousand cells per second. The human body is composed of about 60 trillion cells, with the heart, liver, and kidneys being composed of 300 billion, 6 trillion, and 200 billion cells, respectively. The high printing speed of inkjet is therefore very advantageous. Inkjet technology provides a unique technology which can deliver many cells within a short time.

13.5.4 3D Fabrication Using Hydrogels

As shown in this chapter, we have manufactured several 2D and 3D hydrogel structures by inkjet technology, such as micro-gel beads, gel fibers, gel sheets, and 3D gel structures including 3D laminated sheets and tubes. In these cases, alginate hydrogel provides semisolid properties for 3D fabrication as a scaffold material, and it is also possible to fabricate 3D structures in the liquid. Inkjet gelation is also possible using fibrinogen solution as a gel precursor and thrombin solution as the gel reactant. In this case fibrin hydrogel, which is one of the natural extracellular matrices, is formed, onto which cells can adhere and within which they can live. Vascular endothelial cells can also degrade it. In addition, hydrogel can also function as a supporting medium for other ink components, not only living cells but also various biological materials. As shown, different-colored pigments can be positioned at different positions supported in the hydrogel. If proteins and humoral factors are used instead of the color pigments, the hydrogel can provide microenvironments for the individual cells. This approach has the potential to provide 3D positioning of growth factors, which is one of the challenging issues in tissue engineering.

13.5.5 Linkage to Digital Data Sources

An inkjet printer is a standard computer output device, and thus a digital data connection has been already established. This is advantageous for the application of a biological CAD–CAM–CAE technique (computer-aided design, computer-aided manufacturing, and computer-aided engineering) to tissue engineering. Although we have developed inkjet 3D biofabrication technology as a promising tissue–CAM method (Figure 13.3), this concept has also been proposed by Sun *et al*. as computer-aided tissue engineering (CATE) (Sun and Lal, 2002).

In general terms, both the overall shapes and internal structures are first designed in the computer (bio-CAD). These designs are evaluated and optimized based on several engineering theories by high-performance computation (bio-CAE). CT (computer tomography) scans and MRI (magnetic resonance imaging) scans can measure and obtain 3D spatial data of the body and can generate 3D images. Such 3D spatial data are also available in CATE. However, 3D spatial data is required down to the position of the individual cells for 3D biofabrication using individual cells. Next, the bio-CAM machine constructs designed 3D structures based on the digital data of the final design (bio-CAM). In bio-CAM, 3D structures are fabricated using several biological materials, including living cells, proteins, and growth factors as the materials for fabrication.

The use of computers and programmed CAM machines is very important for the development of tissue engineering. Numerical analysis, for example by finite element methods and computational fluid dynamics, will be useful for designing the optimal design of engineered tissues. Histological studies show that the resolution of the bio-CAM machine needs to be better than 10 μm. Computer control is needed to achieve the necessary resolution, high speed of manufacturing, and high reproducibility. Research on inkjet biofabrication has wide potential for applications not only in tissue engineering but also in cell biology and cell physiology.

13.5.6 Applicability to Various Materials including Humoral Factors and Nanomaterials

Inkjet technology has already been used to print several biological materials instead of ink, such as nucleic acid (Goldmann and Gonzalez, 2000), growth factors (Watanabe, Miyazaki, and Matsuda, 2003; Campbell *et al*., 2005; Miller *et al*., 2006), and biological cells (Nakamura *et al.*, 2005; Xu *et al.*, 2005, Xu *et al.*, 2006; Boland *et al.*, 2006), mainly in the fabrication of bio-chips. Cells, scaffolds, and growth factors are three major factors for tissue engineering. Scaffolds, including extracellular matrices such as collagen, are the substrates for maintaining the shape and structures and providing the locations for cell adhesion and proliferation. Growth factors are biochemical agents with biological functions, such as the promotion of cell proliferation and migration, and sometimes induction of apoptosis or cell death. Both are important to control cell behavior. In *in vivo* conditions, each cell is influenced by growth factors through direct contact or diffusion phenomena from the 3D microenvironment.

For these applications, DOD 3D bioprinting by inkjet is very promising. As mentioned in this chapter, when several proteins and humoral factors are used in a color (multiple-ink) printer, they can be positioned accurately as required at the intended 2D and 3D locations, together with individual cells. Hydrogel can ensure their spatial 2D and 3D interactions, and finally provide the correct microenvironments for the individual cells. In such extracellular matrices, proteins and humoral factors have functions in controlling cellular behavior, and their response over time can be controlled by the spatial positioning of these materials. This approach also offers good potential to provide 3D positioning of growth factors, which is one of the difficult current issues in tissue engineering.

13.5.7 Use of Pluripotent Stem Cells in Bioprinting

In 2007, two research teams (Yamanaka *et al.* at Kyoto University, Japan, and Thomson *et al.* at University of Wisconsin–Madison, United States) reported the major advance of creating human induced pluripotent stem (iPS) cells (Takahashi *et al.*, 2007; Yu *et al.*, 2007). iPS cells have the ability to differentiate to all kinds of body component cells (pluripotency) just like embryonic stem (ES) cells, but this ability is artificially induced in cells from the adult human. As iPS cells and the cells derived from them are the patient's own cells, they have no immuno-discrepancy to the original patient. The technology of iPS cells has much potential, and obtaining differentiable cells for tissue engineering and transplantation is one of its most exciting prospects. To exploit this, we need to develop effective technologies for manufacturing functional tissues and organs from such cells. Inkjet 3D bioprinting holds great promise for manufacturing tissues and organs with such cells for transplantation.

13.6 Summary and Future Prospects

We have developed a custom-made 3D bioprinter and used it to demonstrate several structures, including micro-gel beads, micro-gel fibers, gel sheets, and 3D structures. The many advantages and approaches to the application of this technology in 3D tissue engineering have also been demonstrated. 3D-programmed inkjet-based direct cell printing promises to provide a key to overcoming intrinsic problems in conventional tissue engineering by means of high-resolution positioning of multiple cell types, including the internal composition of 3D structures. We are currently taking just the first steps toward the final goal of manufacturing alternative tissues and organs for transplantation.

As well as research into tissue engineering, multidisciplinary technologies are also needed. These include not only inkjet technologies but also biomaterial technologies such as the synthesis of effective hydrogel and extracellular matrices, and cell culture technologies including the control of growth factors. We hope that many researchers will participate in this research; such collaboration will undoubtedly accelerate its development.

Acknowledgements

The author thanks Dr. Y. Nishiyama, Dr. C. Henmi, Dr. K. Suyama, Dr. S. Iwanaga, and several co-members of the Bioprinting project of the Kanagawa Academy of Science and Technology for their expert technical assistance and helpful support. The author also thanks Seiko-Epson Corp. for their support in regard to the inkjet head system. This work was supported by the Kanagawa Academy of Science and Technology and Grants-in-Aid for Scientific Research (#17300146 and #18880042) from the Japan Society for the Promotion of Science.

References

Boland, T., Xu, T., Damon, B. and Cui, X. (2006) Application of inkjet printing to tissue engineering. *Biotechnology Journal*, **1**, 910–917.

Calvert, P. (2007) Printing cells. *Science*, **318**, 208–209.

Campbell, P.G., Miller, E.D., Fisher, G.W. *et al*. (2005) Engineered spatial patterns of FGF-2 immobilized on fibrin direct cell organization. *Biomaterials*, **26**, 6762–6770.

Cukierman, E., Pankov, R., Stevens, D.R. and Yamada, K.M. (2001) Taking cell-matrix adhesions to the third dimension. *Science*, **294**, 1708–1712.

Cukierman, E., Pankov, R. and Yamada, K.M. (2002) Cell interactions with three-dimensional matrices. *Current Opinion in Cell Biology*, **14**, 633–639.

Elkayam, T., Amitay-Shaprut, S., Dvir-Ginzberg, M. *et al*. (2006) Enhancing the drug metabolism activities of C3A – a human hepatocyte cell line – by tissue engineering within alginate scaffolds. *Tissue Engineering*, **12**, 1357–1368.

Elsdale, T. and Bard, J. (1972) Collagen substrata for studies on cell behavior. *The Journal of Cell Biology*, **54**, 626–637.

Goldmann, T. and Gonzalez, J.S. (2000) DNA-printing: utilization of a standard inkjet printer for the transfer of nucleic acids to solid supports. *Journal of Biochemical and Biophysical Methods*, **42**, 105–110.

Griffith, L.G. and Naughton, G. (2002) Tissue engineering – current challenges and expanding opportunities. *Science*, **295**, 1009–1014.

Henmi, C., Nakamura, M., Nishiyama, Y. *et al*. (2008) Mini-review: new approaches for tissue engineering: three dimensional cell patterning using inkjet technology. *Inflammation and Regeneration*, **28** (1), 36–40.

Hutmacher, D.W., Sittinger, M. and Risbud, M.V. (2004) Scaffold-based tissue engineering: rationale for computer-aided design and solid free-form fabrication systems. *Trends in Biotechnology*, **22** (7), 354–362.

Kamisuki, S., Hagata, T., Tezuka, C. *et al*. (1998) A low power small, electrostatically-driven commercial inkjet head. Proceedings IEEE The Eleventh Annual International Workshop on Micro Electro-Mechanical Systems (MEMS'98), pp. 63–68.

Landers, R., Hubner, H., Schmelzeisen, R. and Mülhaupt, R. (2002) Rapid prototyping of scaffolds derived from thermoreversible hydrogels and tailored for applications in tissue engineering. *Biomaterials*, **23**, 4437–4447.

Miller, E.D., Fisher, G.W., Weiss, L.E. *et al*. (2006) Dose-dependent cell growth in response to concentration modulated patterns of FGF-2 printed on fibrin. *Biomaterials*, **27**, 2213–2221.

Nakamura, M., Kobayashi, A., Takagi, F. *et al*. (2005) Biocompatible inkjet printing technique for designed seeding of individual living cells. *Tissue Engineering*, **11**, 1658–1666.

Nakamura, M., Nishiyama, Y., Henmi, C. *et al*. (2006) Inkjet bioprinting as an effective tool for tissue fabrication. Proceedings of Digital Fabrication, DF2006, Society for Imaging Science and Technology, pp. 89–92.

Nakamura, M., Nishiyama, Y., Henmi, C. *et al*. (2007) Application of inkjet in tissue engineering and regenerative medicine: development of inkjet 3D biofabrication technology. Proceedings of Digital Fabrication, DF2007, Society for Imaging Science and Technology, pp. 936–940.

Nishiyama, Y., Nakamura, M., Henmi, C. *et al*. (2007) Fabrication of 3D cell supporting structures with multi-materials using the bio-printer. Proceedings of MSEC 2007, MSEC2007, p. 31064.

Sun, W. and Lal, P. (2002) Recent development on computer aided tissue engineering – a review. *Computer Methods Programs Biomedical*, **67**, 85–103.

Takahashi, K., Tanabe, K., Ohnuki, M. *et al*. (2007) Induction of pluripotent stem cells from adult human fibroblasts by defined factors. *Cell*, **131** (5), 861–872.

Vozzi, G., Previti, A., De Rossi, D. and Ahluwalia, A. (2002) Microsyringe-based deposition of two-dimensional and three-dimensional polymer scaffolds with a well-defined geometry for application to tissue engineering. *Tissue Engineering*, **8**, 1089–1098.

Watanabe, K., Miyazaki, T. and Matsuda, R. (2003) Growth factor array fabrication using a color inkjet printer. *Zoology Science*, **20** (4), 429–434.

Xu, T., Jin, J., Gregory, C. *et al*. (2005) Inkjet printing of viable mammalian cells. *Biomaterials*, **26** (1), 93–99.

Xu, T., Gregory, C.A., Molnar, P. *et al*. (2006) Viability and electrophysiology of neural cell structures generated by the inkjet printing method. *Biomaterials*, **27** (19), 3580–3588.

Yang, S., Leong, K.F., Du, Z. and Chua, C.K. (2002) The design of scaffolds for use in tissue engineering. Part II. Rapid prototyping techniques. *Tissue Engineering*, **8**, 1–11.

Yu, J., Vodyanik, M.A., Smuga-Otto, K. *et al*. (2007) Induced pluripotent stem cell lines derived from human somatic cells. *Science*, **318** (5858), 1917–1920.

14

Three-Dimensional Digital Fabrication

Bill O'Neill
Department of Engineering, University of Cambridge, United Kingdom

14.1 Introduction

Over the past 10 years, low-cost three-dimensional (3D) printing technologies have accelerated the growth of the rapid prototyping (RP) equipment market. The continuous development of 3D printers, driven by price pressures, has dramatically increased the demand from end users. In addition, the use of additive fabrication technology for practically every new product introduced today has increased the pressure on manufacturers to supply a greater range of technology and services to the end user. There is a particularly high demand for low-cost 3D printers. The increased focus on direct part manufacture, especially for low-volume products, has identified the need for a wider range of functional materials and provided system manufacturers with significant growth opportunities. Rapid manufacturing is expected to be the principal means of manufacturing mass-customized and low-volume goods in the future, with shorter product lifecycles becoming the norm rather than the exception. These new technologies have provided manufacturers with a real opportunity to cut costs and reduce lead times, while increasing their responsiveness to rapidly changing consumer trends.

This chapter explores the current state of the art in 3D printing. The origins and evolution of additive manufacturing are presented along with the development of 3D printing technology. Emphasis is placed on the technology of 3D printing systems, rather than print-head technology which has been reviewed in the inkjet context in Chapter 2. The leading-edge 3D printing system technologies are presented in terms of their production

Inkjet Technology for Digital Fabrication, First Edition. Edited by Ian M. Hutchings and Graham D. Martin.

methodology and capabilities. The chapter concludes with a discussion of future trends and research challenges that must be met in order to deliver 21st-century manufacturing solutions: solutions that offer reduced costs, increased capabilities, the provision of new business models, and even the personal factory so often cited by pundits in this fast-moving industry.

14.2 Background to Digital Fabrication

Today, there exists a plethora of techniques for the production of 3D components. Manufacturing engineers are spoilt for choice, with a vast array of techniques available for the production of metallic, polymeric, ceramic, and composite parts. Manufacturing engineers usual classify production methods in terms of the production methodology. These are:

- **subtractive**: the removal of elements from stock material,
- **formative**: the reshaping of stock material in a mould or die, and
- **additive**: the addition of simple components through welding and joining techniques to form a more complex component.

Manufacturing process selection is a critical decision in all production-based enterprises and will often mean the difference between profit and loss. Productivity is also one of the primary concerns of the manufacturing engineer. The drive toward higher levels of productivity has increased the need for automated production systems and has, in turn, led to the development of revolutionary changes in factory production technologies. The need to reduce lead times, lower production overheads such as tooling costs, and minimize material usage, while seeking rapid product development cycles, has allowed new manufacturing philosophies to gain a firm foothold in the modern factory. This revolutionary change requires all aspects of production and management to be computerized. Computer-integrated manufacturing has come a long way since the 1960s, with every aspect of design, manufacturing, assembly, service, and customer support being reliant on digital information. In terms of actual digital production, the latest additive technologies known as rapid prototyping, solid free-form fabrication (SFF), layered manufacturing (LM), or desktop manufacturing (DM) have offered a step change in production capabilities, and as such offer exciting promise for future manufacturing operations. These terms all refer to additive manufacturing methodologies that are directly driven by digital product information in the form of 3D virtual data sets. For the purposes of this chapter, we will use the term SFF, although it is worth noting that all terms listed are widely used by other authors. They do in fact refer to the same collection of fabrication technologies and involve putting the right amount of material where you want it, when you want it, preferably with lights-out operation and with minimal overhead costs.

The generic process of layer-wise addition of material to develop a 3D object has roots in the 19th century when Blanther (1892) proposed a wax plate lamination method for the production of 3D relief maps. The first use of photo-reactive polymers for the production of laminates, that were subsequently stacked to form a casting mould, was proposed by Matsubara (1974). DiMatteo (1976) used layer-wise sheet metal construction techniques in the production of engineering components such as mould tools and press tools.

This was followed by further developments in laminated construction (Nakagawa *et al.*, 1979; Kunieda and Nakagawa, 1984; Nakagawa, Kunieda, and Liu, 1985). The use of lasers to polymerize liquid monomers was first proposed by Swainson (1977), followed by additional developments from Schwerzel *et al.* (1984). Both techniques employed intersecting laser beams but were not developed to a commercial level.

These advances had no tangible impact on the manufacturing arena, largely due to the lack of computing capabilities that were necessary to translate digital information into controllable motions for guiding energy sources such as laser beams or mechanical tools. By the end of the mid-1980s, the personal computer was becoming prevalent through the provision of relatively low-cost computer-aided design (CAD) tools, some 20 years after the first introduction of interactive graphic systems. CAD systems were essential not only for creating 3D virtual models that described the details of the product to be built, but also for providing the data outputs necessary for precision control of energy sources. By this time, lasers were also reaching high levels of industrial capability and no longer needed laser physicists to operate them. Technologically, the scene was set for the emergence of groundbreaking developments in SFF. One could argue that the starting gun for SFF was fired with the foundation of 3D Systems Inc. in 1986 by Charles W. Hull and Raymond S. Freed. Their first pioneering laser-based photo-curable system went on sale in 1988, the "stereolithography apparatus" or SLA™. This was a machine capable of building polymer parts within a process volume of $190 \times 190 \times 190\,mm^3$ using acrylic resins and a 12 mW HeCd laser. This technology was quickly followed by a team from the University of Texas at Austin with a prototyping system based on the fusion of polymer powder beds by infrared CO_2 lasers (Deckard, Beaman, and Darrah, 1990). The system was capable of constructing components in functional polymers with a build volume of $250 \times 250 \times 250\,mm^3$. DTM Corporation was established to commercialize the selective laser sintering (SLS) technology, and their first machine was released in 1992.[1] Figure 14.1 shows a schematic of the basic SLA and SLS technologies. For SLA, a build platform starts at the surface of the liquid, where a level control, usually a wiper blade, sets the height of liquid monomer to be polymerized by the scanning ultraviolet (UV) laser beam according to the 2D profile of the part. The platform then drops by the layer height, and the liquid height above the part is set again. This process sequence is repeated layer by layer until the part is complete. For SLS, the powder feed

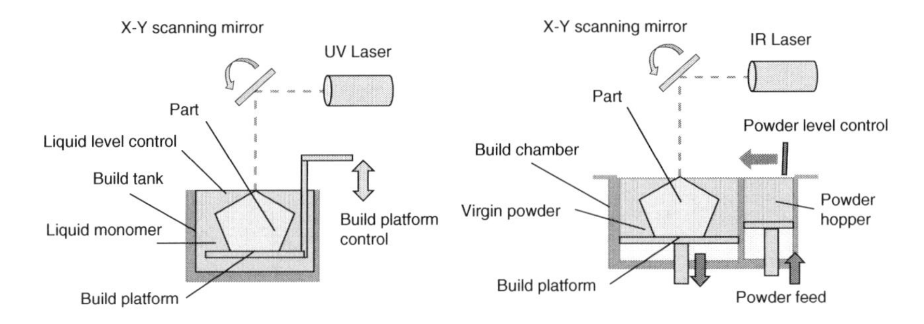

Figure 14.1 *Schematic of the SLA (right) and SLS (left) system technologies.*

[1] DTM was subsequently sold to 3D Systems in 2001 having gained a significant global market share.

piston loads the surface of the chamber with powder, whereupon a powder-level control system sets the height of powder to be fused to the platform using an infrared CO_2 laser according to the 2D profile of the part, the platform drops by the chosen layer height (tens of micrometers), a new powder layer is deposited, and the process is repeated layer-by-layer until the part is complete.

The introduction of these machine concepts stimulated a significant global research and development effort over the past 20 years in order to establish further SFF capabilities. There exist many SFF variants, with liquid-based, powder-based, and film-based devices; many of them rely on laser energy sources or plastic extrusion devices. A selection of current system providers is shown in Table 14.1. A comprehensive listing with company and production information can be viewed at the 'Castle Island's Worldwide Guide to Rapid Prototyping' website (Castle Island, n.d.). These innovations have been driven by the strong demand from industrial users who have benefited greatly from the SFF production systems (Table 14.2). The benefits are immense and can be found across all aspects of manufacturing operations, including applications in ergonomic studies, tool components, casting patterns, visual aids for communicating ideas across departments or to the customer, functional models, jig design, and even direct production of parts. Components can be formed without limits on geometric complexity or intricacy, without the need for costly and complex tooling or final assembly. The technologies offer considerable time compression strategies and can take months off product development lead times compared to conventional routes. There are, of course, some limitations such as material options, resolution (tens of micrometers), and scale (hundreds of millimeters), although the current offerings can deliver useful parts in metals, polymers, and ceramics. A comprehensive review of the industry, technology, and applications can be found in the following sources: Grimm (2004); Hopkinson, Hague, and Dickens (2006); Kamrani and Nasr (2006); and Noorani (2006).

Table 14.1 *SFF system suppliers by classification.*

Liquid-based	Solid-based	Powder-based
3D Systems (US)	Boxford Co (UK)	3D Systems (US)
CMET (Japan)	Cubic Technologies (US)	Arcam (Sweden)
CMET (Japan)	Kira (Japan)	Concept Laser (Germany)
Next Factory (Italy)	Mcor Technologies (Ireland)	EOS (Germany)
Envisiontec (Germany)	Solidica (US)	MTT (UK)
Objet Geometries (Israel)	Toyoda Machine Works (Japan)	Sintermask (Germany)
Huntsman (Switzerland)	Stratasys (US)	3D-Micromac (Germany)

Table 14.2 *Examples of companies exploiting SFF technologies.*

Apple	Ford	Lockheed Martin
AT&T	Gillette	Mercedes Benz
AMP	GE	Nokia
Boeing	Hasbro	Pratt & Whitney
BMW	IBM	Rockwell
Chrysler	Intel	Stryker Howmedica

14.3 Digital Fabrication and Jetted Material Delivery

The early days of SFF relied heavily on laser-based systems. These offered the most accurate means of part production, and delivered solutions that were highly capable although very expensive. Typical systems costs ran to £200 000[2] in the mid-1990s and required costly maintenance contracts due the complexity of their configurations as well as the need to replace laser units after several thousand hours of operation. System developers were seeking alternative technologies that could offer similar performance with reduced cost. In the early days of SFF, product designers were the most common users. Their workplace was the design studio, and all SFF systems operated in workshop or even laboratory conditions. Liquid monomers are inherently toxic, and powders were difficult to handle, requiring sophisticated powder-handling systems and restrictive health and safety practices to protect workers. These new machines were definitely not studio-based production systems.

The use of liquid jets of polymer or wax to deposit material was seen as an obvious step forward, one which would provide real desktop capability and give designers easy access to part-building capabilities, while reducing part cost and capital expenditures. 3D Systems were the first to offer their multi-jet modeling (MJM) technology in 1996, the Actua 2100 (Leyden *et al*., 1997). The print-head was configured for jetting wax-like thermoplastic materials. Print-heads (model no. HDS96i) were supplied by Spectra Corporation (now Fujifilm Dimatix) and employed 96 jets under computer control. Raster scanning of the head over a build plate with height control enabled parts to be created. The system could deliver 12 000–16 000 commands per second to each jet, with around 1.2–1.6 million firing commands to the head each second. Typical materials were paraffin wax having a melting point of 60 °C (20–44% by weight). The units cost around £60 000 at the time but did not sell well largely due to the limited strength of the materials produced. They did, however, set the scene for a new type of SFF system that was suitable for the production of concept models, and well suited for office environments. This technology was followed by improvements in speed with the introduction of the Thermojet™ concept modeler in 1999. The technology did not offer enough advances in material properties for it to increase the market share of jet-based systems. However, it was quite clear that the user base welcomed inkjet-based systems in principle: they were simple, with adequate resolutions of $300 \times 400 \times 600$ dpi, good for most modeling purposes, and had build volumes of $250 \times 190 \times 200\,\text{mm}^3$, close to those of midrange SLA and SLS systems at 10 times the price.

Over the past 10 years a number of advances have been realized, not particularly in the applied jet or print-head technology, but in the material formulations that are delivered through them. The following sections present current state-of-the-art inkjet-based SFF systems. These are grouped in terms of liquid-based systems and powder-based systems. The general technical details are presented along with information on build materials and a summary discussion of their particular strengths and weaknesses. It is worth noting that 3D printing materials, as with printing inks, are proprietary and formulated by the manufacturer or materials supplier: only general descriptions are given by vendors, although they do supply materials data sheets with basic performance specifications.

[2] These prices are similar to those commanded today for equivalent build volumes.

14.4 Liquid-Based Fabrication Techniques

Liquid-based SFF systems have the initial form of the build material in a liquid state, either as a monomer that is polymerized on deposition or as a molten thermoplastic that solidifies on impact.

14.4.1 PolyJet™: Objet Geometries

In 1999, Objet Geometries of Israel submitted a patent for a new form of 3D printing technology (Gothait, 1999). A schematic from the original patent application is shown in Figure 14.2. The system consisted of a print-head (5) with multiple nozzles (6), a materials dispenser (4), a print positioner (3), and a process controller (2) driven by a CAD system (1). The print-head dispensed material (7) in successive layers to produce a part (8) placed on a platform (9). What is particularly novel here is the use of a UV curing lamp that is located in the print-head assembly (10). This system jets photopolymer that is cured on impact under the low-level UV lamp. The depth of each deposited layer can be controlled by selectively controlling the output from each nozzle. Raster scanning the print-head across the build platform delivers the capability to produce very good models that look and behave like those produced by SLA, since they share the same material forms of photo-curable monomers.

As with SLA systems, it is necessary to provide the means of supporting the object being built in order to produce non-re-entrant features or overhangs. In this case two

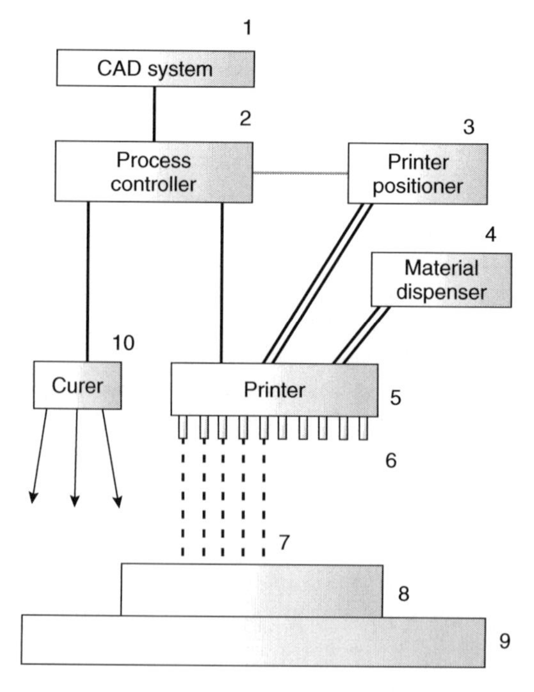

Figure 14.2 Schematic from the patent application for PolyJet™.

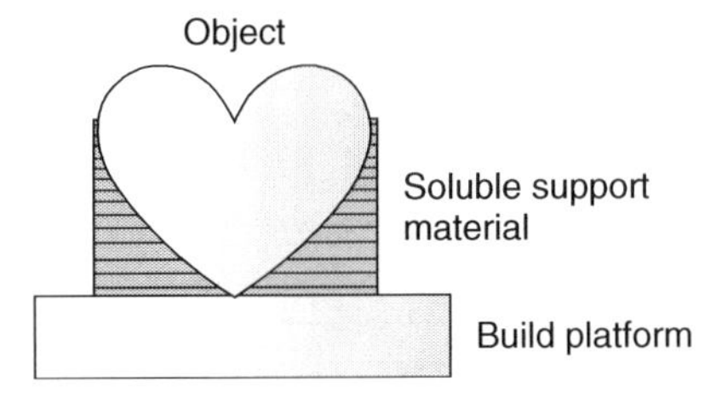

Figure 14.3 *Schematic showing the support and build structures.*

material types are deposited: one that forms a support and the other for the object material itself, as shown in Figure 14.3.

14.4.1.1 Support Materials

The PolyJet™ system uses materials that possess reverse thermal gelation (RTG) properties for the manufacture of supports. These polymers have extreme sensitivity to temperature, showing significant change in properties for a small rise in temperature. They exhibit a low critical solution temperature in aqueous environments and can be easily dispensed from inkjet nozzles. Once they are raised above their gelation temperature, on impact, solidification of the composition is the result, thus offering good characteristics such as toughness and dimensional stability which are ideal for a support material. After the build is complete, the gel is cooled below its gelation temperature and can then be removed with a rinsing step. This innovation is a vast improvement on SLA systems which have to build a scaffold around the object. As the scaffold material is the same photo-curable resin used to make the object, it leaves much work for the post-processing phase in having to remove the scaffold by hand, often compromising the dimension of the part. Figure 14.4 shows a schematic of the PolyJet building process, detailing the essential elements of the technology.

14.4.1.2 System Characteristics

Objet's family of 3D printing systems offers high-resolution part building for detailed models that are ostensibly good for the office or even workshop environment. The technology has two variants: PolyJet and the PolyJet Matrix™. The PolyJet Matrix system offers the ability to jet two building materials in preset combinations for the production of objects that require variation in material properties. This dual-jet process can combine materials in many ways: for instance, two rigid materials, two flexible materials, or one of each type. The addition of transparent or opaque material options allows the designers to test material combinations that are copies of actual production assemblies. Each material is funneled to a liquid handling system connected to the printing assembly. This assembly contains eight print-heads, each containing 96 print nozzles. The printing controls allow the structure of the jetted material to be tailored according to the location of the deposit, allowing specific values to be obtained, such as tensile strength, elongation to break, and even Shore A hardness values. Although there are many models in the current Objet range, Table 14.3 presents the general system characteristics of the PolyJet

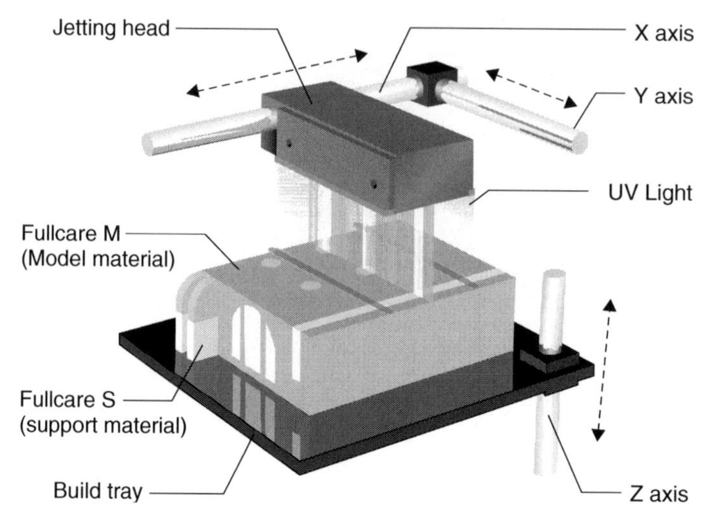

Figure 14.4 *PolyJet™ 3D printing technology (Reproduced from Objet Geometries Copyright (2012) Objet Geometries).*

Table 14.3 *General system characteristics of the PolyJet™, ProJet™, Solidscape, and Z-Corporation 3D printers.*

System	Build envelope (mm)	Printing resolution	Materials	Maximum build rate
PolyJet™ Objet Inc.	(XYZ) 490 × 390 × 200	X-axis: 600 dpi: 42 μm Y-axis: 600 dpi: 42 μm Z-axis: 1600 dpi: 16 μm	Acrylate photopolymers: colored, opaque, flexible, rigid	20 mm/h/strip
ProJet™ 3D Systems Inc.	(XYZ) 298 × 185 × 203	X-axis: 656 dpi Y-axis: 656 dpi Z-axis: 1600 dpi: 16 μm	Acrylate photopolymers: colored, opaque, waxes	Not cited
Solidscape Inc.	(XYZ) 152 × 152 × 100	X-axis: 5000 dpi: 25 μm Y-axis: 5000 dpi: 25 μm Z-axis: 13 μm	Polyester thermoplastic Casting wax	Not cited
ZPrinter Z Corporation Inc.	(XYZ) 254 × 381 × 203	X-axis: 600 dpi Y-axis: 540 dpi Z-axis: 89–102 μm	Composites Foundry sand mixes Elastomerics	Two to four layers per minute

Figure 14.5 A wheel produced using a PolyJet™ Matrix system (Reproduced from Objet Geometries Copyright (2012) Objet Geometries).

technologies. Figure 14.5 shows an example of a part made by a PolyJet Matrix system using a Eden 260 machine. The black rim is produced using an elastomeric material.

14.4.2 ProJet™: 3D Systems

The ProJet™ series of technologies are the latest MJM developments from 3D Systems. This technology uses a larger print-head than the PolyJet systems of Objet, which pass over the full width of the build. These systems have a build speed that is independent of part size, since the PolyJet systems require side-stepping actions to cover the whole cross-sectional area of the build, although build rates are not reported. With the incorporation of UV curing lamps, as in the Objet systems, MJM also uses an increased range of model materials such as hard acrylate photopolymers and waxes for subsequent investment casting operations. The support waxes can be melted away offering hands-free support removal. Figure 14.6 shows an image of a ProJet 3D printer. Figure 14.7 shows a quite complex object printed with a colored photopolymer. Table 14.3 presents the general system characteristics of the ProJet™ technologies.

14.4.3 Solidscape 3D Printers

Solidscape Inc. provides the most accurate 3D printers available today. A schematic of their technology is shown in Figure 14.8. This system employs a single jet for each of the

Figure 14.6 *A ProJet™ series 3D printer (Reproduced from 3-D Systems Copyright (2012) 3-D Systems).*

Figure 14.7 *A complex object, 150 mm wide, produced using ProJet™ series printer in colored photopolymer (Reproduced from 3-D Systems Copyright (2012) 3-D Systems).*

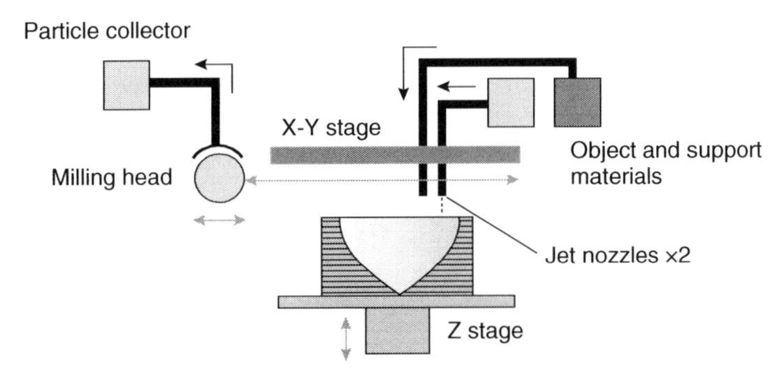

Figure 14.8 *Schematic of the Solidscape 3D printing system (Reproduced from Solidscape Copyright (2012) Solidscape).*

thermoplastic build and wax support materials. The materials are held in heated reservoirs and are then dispensed as micro-droplets as a two-jet head is raster-scanned across the build platform. The materials solidify on impact giving very accurate deposition. To improve accuracy further, once the layer is complete, a milling head and vacuum suction system are traversed across the layer to leave a uniform thickness of 13 μm as the deposit. Particle debris is removed by the vacuum, and the process is repeated until the object is complete.

The Solidscape series of plotters have the ability to produce extremely fine details, giving high resolution and fine surface finish. Precision comes at a penalty in this industry and the units are very slow. The parts are removed from the support structures by dissolving the part covered in support material in a solvent held at around 55 °C. Material selection is limited and the parts are fragile, although the devices do find widespread application in the jewelry industry for the production of lost-wax (investment-casting) patterns, as shown in Figure 14.9. Table 14.3 presents the general system characteristics for Solidscape 3D printers.

14.5 Powder-Based Fabrication Techniques

This powder-based group of inkjet-based SFF systems utilizes inkjets to deliver liquid-phase binders to aid selective consolidation of a power bed. Unlike liquid-based SFF systems, powder beds offer the added simplicity of not requiring support structures, as the surrounding, unbonded powder bed automatically supports the part being built. They do, of course, require post-processing since the powder encasing the final object must be removed. This is almost like engaging in an archeological dig, although system vendors provide enclosed glove-box units to make this operation relatively office friendly.

14.5.1 ZPrinter™: Z Corporation

Z Corporation was established in 1994 by Bornhost, Anderson, and Brett. Their technology platform was invented by Sachs *et al*. (1994) of the Massachusetts Institute of Technology (MIT) and subsequently licensed to Z Corporation. The first system,

Figure 14.9 *Lost-wax pattern produced using the Solidscape 3D printing process (Reproduced from Solidscape Copyright (2012) Solidscape).*

the Z™ 402, was launched in 1997. Figure 14.10 shows the largest machine currently available, the ZPrinter™ 650, with a build chamber on the left and a powder removal chamber on the right.

The process is similar to SLS in its powder lay-up and build strategy, the difference being that an inkjet print-head is used to deposit a binding liquid that consolidates the powder layer-by-layer to produce the 3D part. The technology was conceived for entry-level modeling. In the early systems, the building materials consisted of an inexpensive starch-based powder and a higher definition gypsum-based powder. These have been superseded by polymer-based powders consolidated with proprietary polymer binders. Starch and sand mixes can also be consolidated with polymer binders to produce moulds for casting of metal alloys. In addition, cellulose and fiber mixes are also available. They can be impregnated with elastomers to create rubber-like properties. Low-cost gypsum powders when consolidated with aqueous inks rehydrate and recrystallize the gypsum, producing a selective "setting" of the powder. Careful control of the gypsum powder size (10–50 μm) and droplet volumes produces components of sufficient rigidity allowing them to hold dimensions when handling. A number of print-heads have been associated with the technology including the Tektronix PHASOR 340 from Xerox (Stamford, CT, United States); the PJN 320 print-head from PicoJet, Inc. (Hillsboro, OR, United States); and the Epson 900 print-head from Epson America, Inc. (Long Beach, CA, United States). The system is the fastest and least expensive route for the production of models. One very notable feature of the technology is the use of colored inks that allow full-color models to be built with up to 6 million colors being reported. Table 14.3 presents some general system characteristics for the technology. Figure 14.11 shows a good example of a colored model. The decals and screen images have been built into the model, perfect for marketing and product design communications.

Figure 14.10 *A Z Corporation ZPrinter 650 with build chamber on the left and powder removal chamber on the right (Reproduced from Z Corporation Copyright (2012) Z Corporation).*

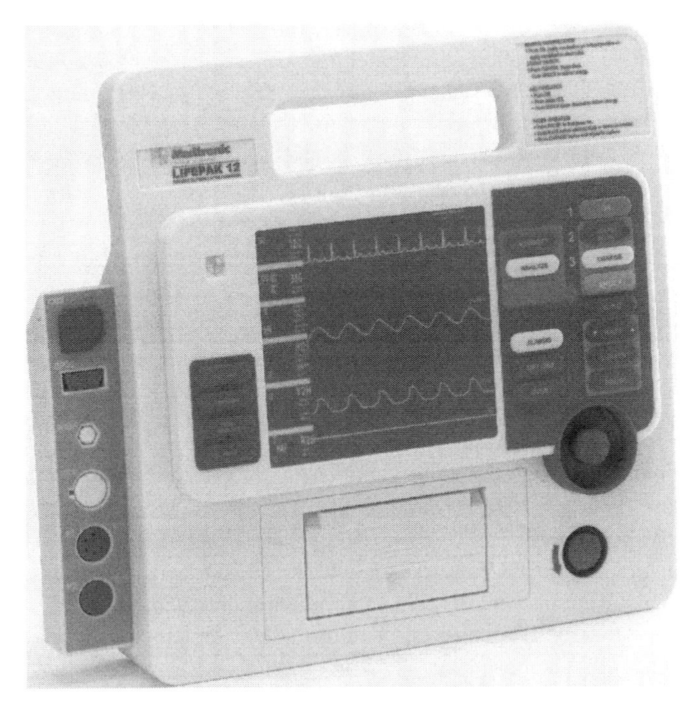

Figure 14.11 *A colored model produced by a Spectrum Z510 (Reproduced from Z Corporation Copyright (2012) Z Corporation) (See plate section for coloured version).*

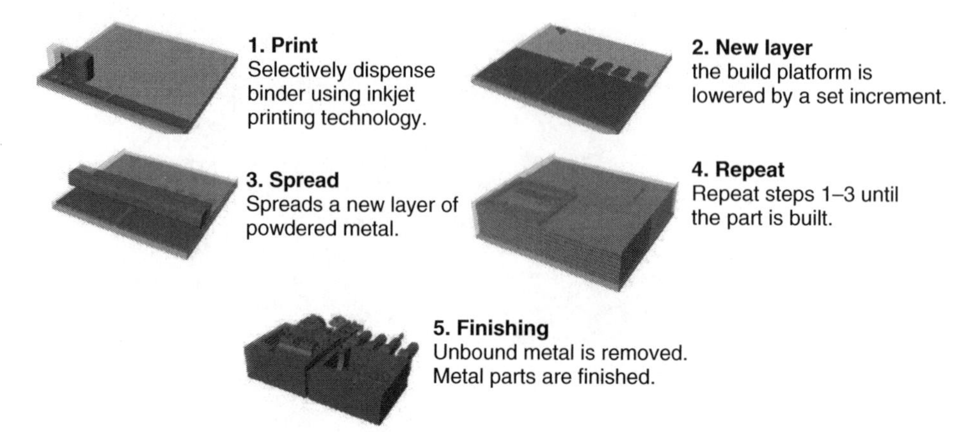

Figure 14.12 *A schematic of the ProMetal™ process (Reproduced from Prometal Inc. Copyright (2012) Prometal Inc.) (See plate section for coloured version).*

14.5.2 Other Powder-Based 3D Printers

A number of other vendors have licensed the MIT 3D printing process. The ProMetal RCT™ (Troy, MI, United States) process uses polymeric binders to consolidate sand cores for casting applications in stainless steel, tool steel, and gold, in addition to the consolidation of bronze-based metal powders for further sintering and infiltration steps to produce complex metal parts for general applications. Figure 14.12 shows a schematic of the process. Voxeljet Technology GmbH (Friedberg, Germany) offers a range of small-series 3D printing systems that produce parts in bonded poly(methyl methacrylate) (PMMA), sand cores for casting, and indirect metal parts. It is perhaps somewhat surprising to see the broad range of materials and applications that low-cost 3D printing systems can offer. Indeed, many of these newer systems provide real part-manufacturing capabilities rather than just concept models. Figure 14.13 shows a complex bronze part produced by the ProMetal RCT process.

14.6 Research Challenges

Market forecasts for 3D printers, consumables, and services are shown in Figure 14.14, with data sourced from I.T. Strategies (Hanover, MA, United States). One can see from the numbers that the overall market size is small compared to the global market in 2009 for 2D inkjet technologies, of the order of US$50 billion (including hardware, media, and chemistry). 3D printers currently have an approximately 12% share of the overall RP and manufacturing market of around US$1 billion. The predicted growth is not startling, largely coming from an increased market share taken from competing technologies and being limited by the lack of foreseen technology innovations. This view is perhaps a little short-sighted as the sector has seen 10–40% growth rates depending on the technology under consideration. The performance of 3D printing systems has improved enormously over their first 10 years of operation. High-end system technologies such as SLA and

Figure 14.13 A steel part produced with the ProMetal™ process (Reproduced from Prometal Inc. Copyright (2012) Prometal Inc.).

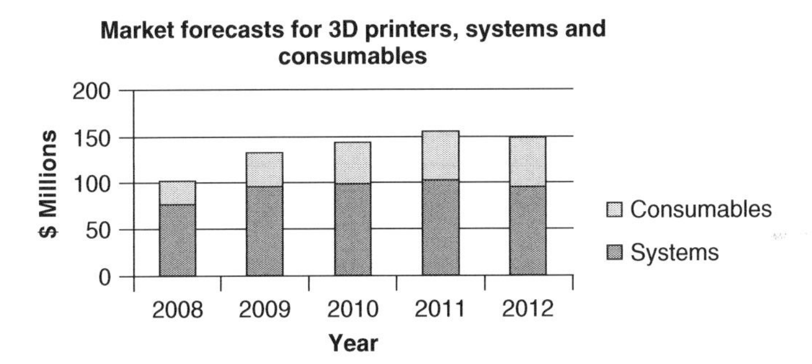

Figure 14.14 Market forecasts for 3D printing systems and consumables (Data sourced from I.T Strategies, Hanover, MA, United States).

SLS now have direct counterparts in the 3D printing world. Major growth in the 3D printing market requires an improvement in the technology and operations. The 3D printing materials on offer today largely only mimic standard engineering materials. This may be acceptable for product modeling, but is not acceptable for direct manufacturing of functional parts. The economic viability of materials used for 3D printing will be driven by volume and capability. This has already been demonstrated in the production of hearing aid shells (Tognola, Parazzini, and Svelto, 2004). For designers and manufacturers to be able to construct parts in the same number of materials available to conventional manufacturing will take some considerable time. This offers the first major research challenge: the development of new materials that offer the performance of traditional materials but can be delivered by 3D printing technologies. They must have longevity and stability equivalent to today's engineering materials.

Part-building times need to be reduced by several orders of magnitude. The current price break-even point, in terms of batch size, for rapid manufactured parts (with direct production using additive manufacturing) versus injection moulded parts ranges from hundreds to several thousands-off, depending on the part size and material (Hopkinson, Hague, and Dickens, 2006, pp. 147–157). The picture is different for some applications. In the case of patient-specific healthcare components such as dental implants, the slow build rates of RP systems on offer today are super-fast compared to handmade dental inserts such as bridges and polymer-based vacuum-formed orthodontic supports (Kruth *et al.*, 2003). Traditional manufacturing techniques are rapidly becoming obsolete for these applications. There are clearly opportunities for 3D printing techniques to manufacture goods more efficiently and economically, although these opportunities are the exception and not the rule. We can now pose the second research challenge: the development of higher speed 3D printing systems that can increase build rates by several orders of magnitude without compromising on material properties and resolution.

One can see from the images presented in this chapter that the build quality of 3D printed components is not comparable to that from conventional routes. Inkjet printing resolutions of 50 μm in the xy plane and 16 μm in the z plane are typical of the industry. In practice, splat geometries, substrate interactions, and machine dynamics lead to even lower resolutions when the cumulative errors are considered over large build volumes. The third research challenge is, then, the development of print resolutions approaching 5 μm, without compromising on material properties and speed.

There are clearly many more research challenges to consider, such as machine format, build strategy, and so on. However, these three main pillars of research must be addressed if the 3D printing industry is to increase its market share of global manufacturing operations.

14.7 Future Trends

Like most technology sectors, the 3D printing industry is highly competitive. Research work being carried in corporate or university laboratories is secretly pursued and often never finds its way to the public domain until the intellectual property (IP) is secure. Patent applications or full patent publications are the best way of determining an organization's technology developments. In this respect, what follows is a basic description of some new 3D printing patents. The search was completed in March 2012, and the details that follow refer to IP that was publicly available from the European Patent Office database search website (n.d.). It must be stated that it does not necessarily follow that the IP holders cited here will bring these developments to market. The information is purely indicative of their IP position at the time of the publication. The information offers a glimpse into the 3D printing technologies that may just be in development. It is possible that two IP positions may change the whole industry.

Bai (2009) described a 3D printing process that utilizes nanoparticle-loaded inks for printing onto a powder bed in the usual way. The objective of the invention is to dramatically improve the properties of the final part by increasing the structural strength and rigidity. Lower sintering temperatures are a direct result of the nanoparticles that bridge the gap between the micrometer-sized particles of the powder. These undergo phase change at much lower temperatures due to their high surface energies (Koga, Ikeshoji, and Sugawara, 2004). Nanoparticles offer key advances in materials development and are already used for printing thin-film metallic conductive layers for polymer electronics, as discussed in Chapter 7. Their application in this industry is just beginning.

Silverbrook (2007) describes a 3D printing system that incorporates multiple high-resolution print-heads that deliver a range of materials from various units in a single pass. The print-head technology is a micro-electro-mechanical system (MEMS) device, probably the Memjet technology given the author and the number of sub-citations in this patent (http://www.memjet.com). The device is designed to print objects that are functional with print-heads some 295 mm wide. Examples cited include personal digital assistants (PDAs), flat-panel displays, calculators, and so on. Many materials are cited including low-melting-point metals for conductive tracks, polymers, waxes, and so on. The suggestion is that with technologies like the Memjet systems, high printing speed ($208\,\mathrm{mm\,s^{-1}}$) and high-resolution ($6\,\mu\mathrm{m}$) 3D printing is just around the corner. It is worth noting that the width of the print-head in this application is limited not by printing technology, but by the ability to store and manage sufficient data sets that describe the object being printed. It is stated that "A typical PDA has dimensions of $115 \times 80 \times 10\,\mathrm{mm}$. Using hexagonal voxels $10\,\mu\mathrm{m}$ with a side length of $6\,\mu\mathrm{m}$, a total of about 98 billion voxels are required to define each product. This requires 98 Gbytes of data, if eight different materials are used in the product." At a scanning speed of $208\,\mathrm{mm\,s^{-1}}$, Silverbrook cites the ability to make 4.3 PDAs per second. Other descriptions are given of alternative devices, for example a 53 cm flat-panel display made in 0.37 seconds, and one production line producing 12 million panels per year. This technology, if it exists, is cloaked in secrecy. Let us hope that we see it before too long.

If one considers the union of advanced nano-scale inks for polymer, semiconductor, metallic, and ceramic materials, delivered through the technology described by Silverbrook (2007), we may have all of our research challenges met in a single platform. The future looks very bright indeed, and the concept of printing certain products (TVs, mobile phones, etc.) in a modern retail outlet actually looks more likely. It has major implications for manufacturing, global supply chains, product delivery, business models, and revenue streams. There is clearly a lot more research and development success required to deliver these concepts, although 3D printing innovations will play a significant part in future manufacturing enterprises.

References

Bai, J.G. (2009) Nanoparticle suspensions for use in three dimensional printing processes. PCT WO 2009/017648.

Blanther, J.E. (1892) Manufacture of contour relief maps. US Patent 473,901.

Castle Island, Castle Island's World Wide Guide to Rapid Prototyping, http://www.additive3d.com (viewed July 2012).

Deckard, C.R., Beaman, J.J. and Darrah, J.F. (1990) Method for selective laser sintering with crosswise laser scanning. US Patent 5,155,324.

DiMatteo, P.L. (1976) Method of generating and constructing three-dimensional bodies. US Patent 3,932,923.

European Patent Office, Database Search Engine (EPO) (N.d.) http://v3.espacenet.com (viewed 20 March 2012).

Gothait, H. (1999) System for three-dimensional printing. International Patent WO 1999/0126023.

Grimm, T. (2004) *Rapid Prototyping*, Society of Manufacturing Engineers, Dearborn, MI.

Hopkinson, N., Hague, R.J.M. and Dickens, P.M. (eds) (2006) *Rapid Manufacturing: An Industrial Revolution for the Digital Age*, John Wiley & Sons, Ltd, Chichester.

Kamrani, A.K. and Nasr, E.A. (eds) (2006) *Rapid Prototyping: Theory & Practice*, Springer, New York.

Koga, K., Ikeshoji, T. and Sugawara, K. (2004) Size and temperature-dependent structural transitions in gold nanoparticles. *Physical Review Letters*, **92** (11), 115507_1–115507_4.

Kruth, J.P., Vaerenbergh, J., Mercelis, P. *et al*. (2003) Selective laser sintering of dental prostheses. Proceedings of EuroRapid 2003 International Users Conference on Rapid Tooling and Rapid Manufacturing. Frankfurt, Germany, December 1–2, 2005.

Kunieda, M. and Nakagawa, T. (1984) Development of laminated drawing dies by laser cutting. *Bulletin of JSPE*, 353–354.

Leyden, R.N., Thayer, J.S., Bedal, B.J.L. *et al*. (1997) Selective deposition modeling method and apparatus for forming three-dimensional objects and supports. Patent WO 1997/011837.

Matsubara, K. (1974) Moulding method of casting using photocurable substance. Japanese Kakai Patent Application, Sho 51[1976]-10813.

Nakagawa, T. *et al*. (1979) Blanking tool by stacked bainite steel plates. *Press Technique*, pp. 93–101.

Nakagawa, T., Kunieda, M., and Liu, S.D. (1985) Laser cut sheet laminated forming dies by diffusion bonding. Proceedings of the 25th MTDR Conference, pp. 505–510.

Noorani, R. (2006) *Rapid Prototyping*, John Wiley & Sons, Inc., Hoboken, NJ.

Sachs, E., Haggerty, J.S., Cima, M.J. and Williams P.A. (1994) Three dimensional printing techniques. US Patent US5340656.

Schwerzel, R.E., Wood, V.E., Mcginniss, V.D. and Verber, C.M. (1984) Three-dimensional photochemical machining with lasers. *Applications of Lasers to Industrial Chemistry, SPIE*, 90–97.

Silverbrook, K. (2007) Three dimensional object printing system. US Patent 2007/0182799.

Swainson, W.K. (1977) Method, medium and apparatus for producing three-dimensional figure product. US Patent 4,041,476.

Tognola, G., Parazzini, M. and Svelto, C. (2004) Design of hearing aid shells by three dimensional laser scanning and mesh reconstruction. *Journal of Biomedical Optics*, **9**, 834.

15

Current Inkjet Technology and Future Directions

Mike Willis
Pivotal Resources Ltd, United Kingdom

15.1 The Inkjet Print-Head as a Delivery Device

The inkjet print-head (Figure 15.1) can be considered basically as a fluid delivery device, capable of generating drops of liquid of a precise size and in a controlled fashion, subject of course to the properties of the liquid itself. As almost all kinds of current inkjet print-heads have been developed for printing text and images, the drop sizes they produce reflect those needed for impressive images at the viewing distances of the application. Therefore standard drop volumes are in the range of 2–20 pl for print-heads designed for document printing, and 5–50 pl for those intended for outdoor and industrial printing applications.

At present the production quantities of print-heads required for inkjet digital fabrication are quite small, so there is little incentive to produce special designs. Therefore, at this stage of the evolution of print-heads the choice of dedicated heads available to developers for fabrication projects is quite small, and the products are adapted from graphics print-heads. However, we expect that as the market for digital fabrication grows, print-head suppliers will develop more custom heads with some of the special features required, such as smaller drop volumes and better control over drop velocity, drop volumes and drive waveform.

Inkjet Technology for Digital Fabrication, First Edition. Edited by Ian M. Hutchings and Graham D. Martin.

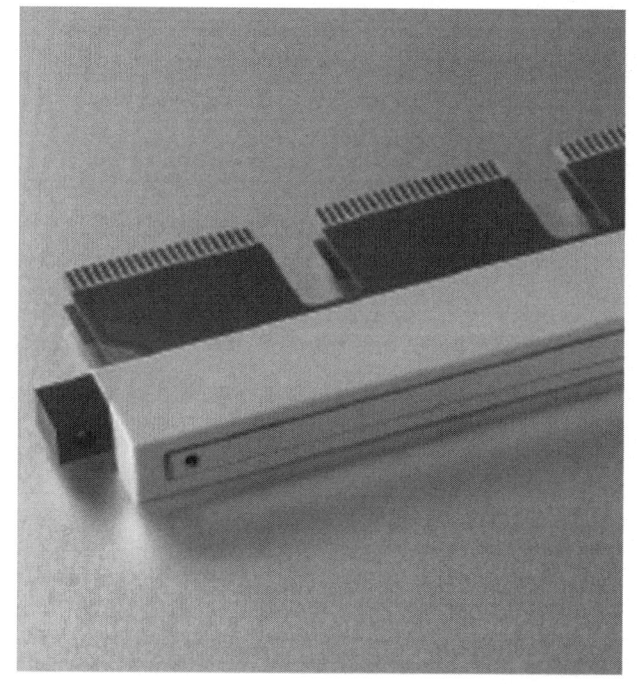

Figure 15.1 FUJIFILM Endura SX-3 printhead
The SX3 is a compact and lightweight jetting assembly designed specifically for deposition applications requiring multiple piezoelectric micropumps packed tightly together. There are 128 jets, and the nozzles are arranged in a single line with a 508 μm spacing between nozzles.

15.2 Limitations of Inkjet Technology

As described in Chapter 1, inkjet technology has some great advantages, including the complete flexibility of a digital process, and as a non-contact process it can allow drop deposition onto non-planar surfaces, wet surfaces and surfaces already printed without the risk of removing the previous deposits. However, there are some drawbacks too. Some of these are fundamental to the technology, while others are limitations for digital fabrication applications due to the print-heads being currently optimised for graphics printing.

15.2.1 Jetting Fluid Constraints

Inkjet technology imposes limitations on the physical properties of the inks that can be used, in general to relatively low viscosities, normally below 20 mPa s. Other parameters such as surface tension need also to be held within certain limits, as noted in Chapter 2. To achieve a low viscosity, the jetting fluids are usually formulated with the functional materials dissolved or dispersed in a carrier liquid, such as water or organic solvent. If the volatility of the solvent is high, this can create operating problems within drop-on-demand (DOD) inkjet print-heads due to the ink drying in the nozzles. The low dilution of active materials can also cause drying problems on the substrate, as a large amount of liquid is needed to deposit a small amount of material.

15.2.2 Control of Drop Volume

There are many applications where it is desirable to vary the amount of material deposited at each pixel. Multiple drops can be put down at each position, but then the incremental amount is one drop. Most of today's technologies are binary, that is, they generate only a single drop volume. Some inkjet technologies allow direct control over the drop volume, but only over a small range. There are a couple of emerging technologies that can put down an infinitely variable range of fluid volumes.

15.2.3 Variations in Drop Volume

With multi-nozzle print-heads, there will be small channel-to-channel variations in drop generation performance introduced during manufacturing. These may be variations in actuator efficiency, alignment tolerances and variations in the physical dimensions of the pressure chambers and nozzles. The result is a variation in drop volume and velocity from channel to channel across an inkjet print-head. End effects can also occur, where the efficiency of channels at the ends of rows is different from those in the centre. To overcome this, some print-heads are manufactured with dummy channels at each end.

To reduce these problems, some print-heads have the capability of adjusting the drive waveform on a channel-by-channel basis so that the variations between channels are minimised.

Variations in drop volume can also arise from the drop formation process according to the ink formulation. If particulates or high-molecular-weight polymers are included in the formulation, for instance, then these can introduce small variations in the break-off of the ligament at the nozzle.

15.2.4 Jet Directionality and Drop Placement Errors

The flight path taken by drops from the print-head to the substrate is dependent on several factors (Figure 15.2), in particular the nozzle exit shape and condition. There will be small variations in the jetting angle from nozzle to nozzle as a result of manufacturing processes. The quality of nozzles varies between different print-head types. In addition, further errors can occur from build-up of contamination inside and around the nozzle rim.

Further variations occur according to the way that drops break off from the nozzles. Several factors affect the size and shape of the ligament that connects the front of the ejected drop to the nozzle during the drop formation process, in particular the ink formulation and drive waveform. If the position at which the tail of the drop breaks from the nozzle varies, then variations in trajectory and landing position of the drop are also likely.

Once the trajectory of the drops has been determined by the print-head and drop break-off has occurred, then aerodynamic effects come into play.

For some digital fabrication applications where drop landing accuracy is paramount, for instance the printing of flat panel displays, a technique of pre-patterning the substrate has been used (Figure 15.3), as described in Chapter 10. The patterning process locally changes the wettability of the substrate. The surface tension of the jetted fluid then causes the fluid to align and fill the spaces defined by the patterning.

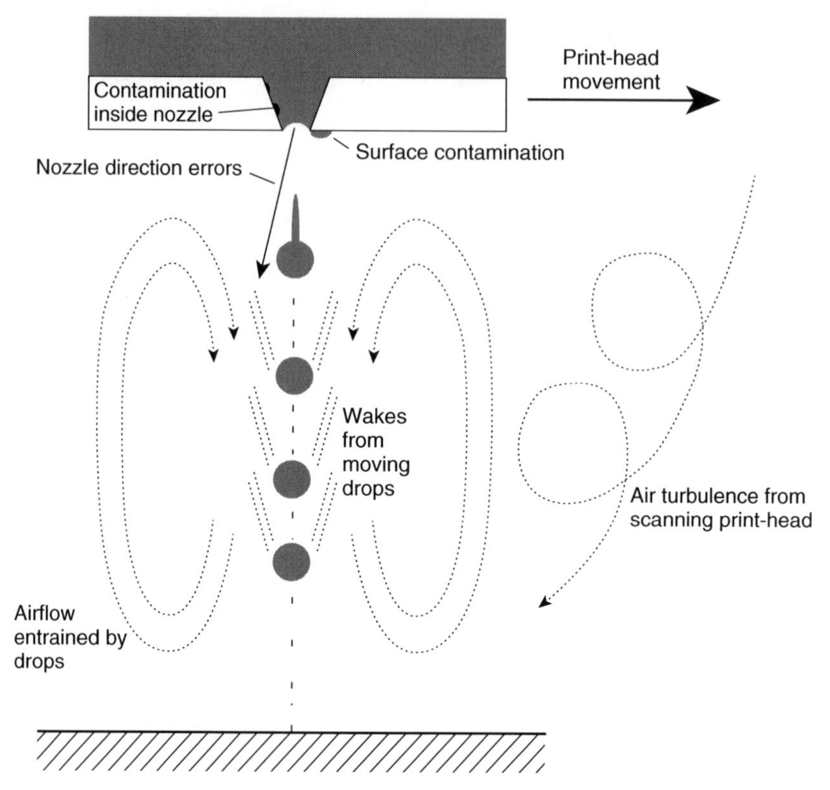

Figure 15.2 Some of the factors that can affect the trajectory of drops in flight.
Shown here are just some of the factors, due to manufacturing tolerances and finish of nozzles, contamination inside and around the outer rim of the nozzles, airflows caused by the movement of the printhead and by the drops themselves. In addition there are end effects at the ends of rows of nozzles.

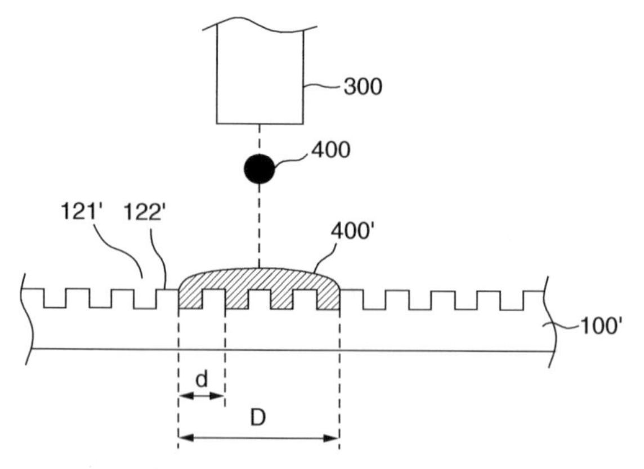

Figure 15.3 Substrate patterning to improve drop placement
A laser interferometric lithographic method is used to pre-pattern a substrate coated with a photoresist. Conductive ink printed onto the substrate preferentially fills the valleys rather than the ridges, so the line width is constrained to a width D (from US 2010/0055396 Method for high resolution ink-jet print using pre-patterned substrate and conductive substrate manufactured using the same, LG Chemical Ltd.).

15.2.5 Aerodynamic Effects

As the drops pass through air on their way to the substrate, they encounter aerodynamic drag. At small drop volumes, perhaps 5 pl or less, the air drag is sufficient to decelerate the individual drops strongly, and depending on the spacing between the print-head and the substrate, they may not even reach the substrate. The situation is confusing, as when a series of drops is jetted from the same nozzle, the first drop is slowed by drag but the following drops slip-stream behind it and longer working distances are possible. In addition, if many drops are jetted simultaneously, air is entrained towards the substrate, which also helps to carry the drops towards it.

However, the greater the influence of the air, the worse the placement error of the drops is likely to be. Because of the differences in geometry at the ends of the print-heads, the airflows will be different there from those in the centre of a nozzle row. There have been some proposals to enclose inkjet systems within vacuum chambers in order to eliminate the influence of aerodynamic effects entirely (Figure 15.4).

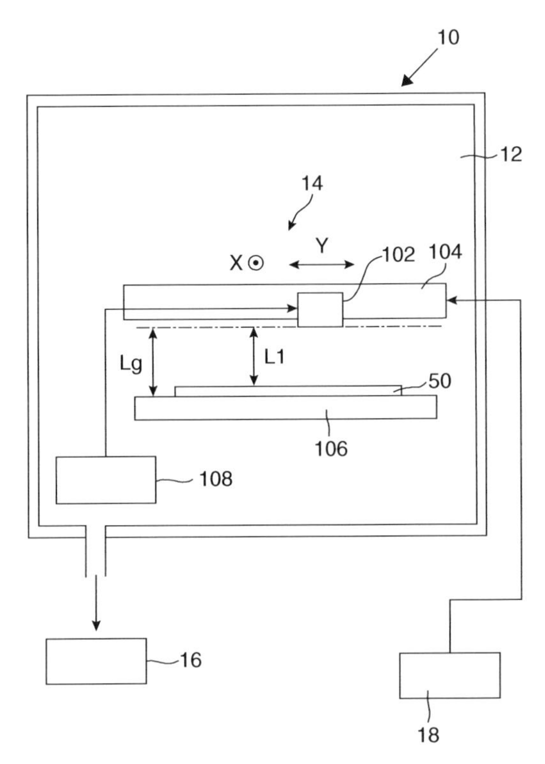

Figure 15.4 Enclosing an ink jet system within a vacuum to reduce or eliminate aerodynamic effects of drops in flight
A pump is used to reduce the pressure between the medium and the nozzles to below atmospheric pressure (from US 2009/0256880 Printing system, inkjet printer and method for printing, Mimaki Engineering Co., Ltd).

15.2.6 Impact and Surface Wetting Effects

As the relatively low-viscosity fluid impacts the substrate, it spreads and wets the surface as described in Chapter 5. The amount of spreading varies according to the ink formulation, viscosity and surface tension, drop impact velocity, substrate hardness, surface micro-profile and surface energy. If drops are jetted onto previously deposited wet drops, then splashing may occur. If printing the same or other materials has already patterned the surface, then the wetting behaviours may vary between the virgin substrate and the patterned regions.

If, during the jetting process, long ligaments are formed, then the ligament may not have been drawn into the main drop by the time of impact. If the print-head and substrate have relative motion, then the ligament may impact outside the main drop, misshaping the printed dot.

If satellite drops are formed in the jetting process, then these too may land outside the periphery of the main drop. Although not important for most graphics applications, drop debris outside the intended impact area may lead to unacceptable image quality for some digital fabrication applications.

15.3 Today's Dominant Technologies and Limitations

Most inkjet devices in use today are of the DOD type, where each nozzle has its own actuator and drops are generated only as required, either as the print-head moves over the substrate or as the substrate moves under the print-head. For DOD print-heads, the two main actuator types are thermal or bubble-jet, and piezoelectric, as described in Chapter 2.

15.3.1 Thermal DOD Inkjet

We recall that in the thermal or bubble-jet process, a heater inside the actuator cavity is in contact with the liquid to be ejected. When an electrical pulse is applied to the heater, the surface temperature rises rapidly to around 350–400 °C. This initially causes the nucleation of bubbles, followed by coalescence into one large bubble. The rapid bubble growth has been likened to a micro-explosion, and it causes a very rapid pressure increase within the actuator chamber. Most of this pressure is released by the flow of ink from the nozzle, although inevitably some flow occurs back into the ink manifold of the print-head. In addition, the print-head structure itself will flex slightly. As the electrical pulse ends, the heater cools rapidly (as it has minimal thermal mass), the bubble collapses and the pressure falls quickly within the actuator chamber. This stops the flow of ink through the nozzle, causing the ink already ejected to break away and fly towards the substrate.

Today's devices are designed and optimised for desktop graphics printing, which requires a high nozzle density and low manufacturing costs. To achieve a high nozzle density, thermal management within the print-head is key. Much of the heat generated during the actuation cycle is wasted, and the management of this waste heat is critical to continued operation of the print-head. By keeping the ink viscosity low, the required drop ejection energy is also kept low, reducing the thermal management problems. Therefore, thermal or bubble-jet print-heads are designed to operate with low-viscosity inks. For graphics printing, ink containing a small percentage by weight of a colorant is sufficient to form an acceptable image. Pigment-based inks also include some resins, but still the formulations of these inks require tight viscosity specifications.

Using thermal or bubble inkjet technology for digital fabrication applications would ideally require:

- redesign to allow higher viscosity fluids to be jetted;
- improvements to robustness and life;
- better stability and control: thermal inkjet (TIJ) tends towards the formation of long tails and satellite drops.

Both Canon and Hewlett-Packard began using new manufacturing processes around 2003–2004 where the nozzle plate is directly formed on the silicon substrate rather than added as a separate item. This allows all of the print-head to be manufactured in a serial production line, and allows for much greater freedom in changing the design for different applications. However, at the time of writing little evidence of customisation has emerged, although Hewlett-Packard has a series of patents on a small replaceable dispensing-type print-head and drive module referred to as TIPS (Thermal Inkjet Pico-Fluidic System) (Figure 15.5). A variety of fluids are claimed capable of being jetted, such as poly(3,4-ethylenedioxythiophene) (PEDOT), conductive polymers and organic solvents. The drop volume is around 1 pl.

It is possible to redesign thermal or bubble-jet print-heads to jet higher viscosity fluids, and there have been reports of this.

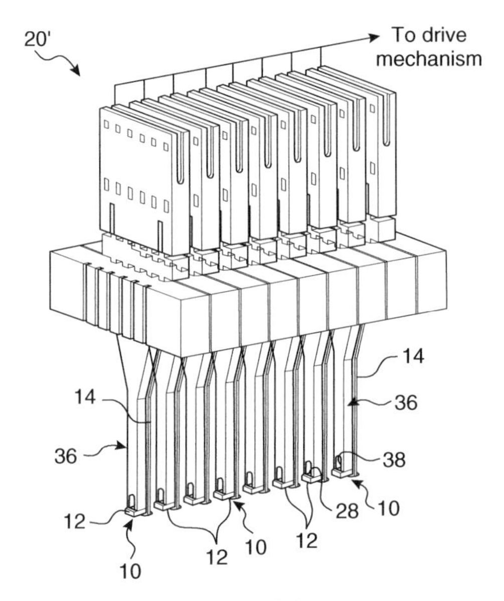

Figure 15.5 Dispensing fluids using thermal ink jet
The nozzles 10 are first immersed in the fluids, and a small quantity wicks into the firing chambers and manifolds by capillary forces. Drops of the fluid can then be dispensed by firing drops (from US 2009/0047440 Fluid delivery system, Hewlett-Packard).

Of course, materials within any jetted fluid need to be compatible with temperatures reached within the print-head. Only a very small amount of the ink is vaporised during the bubble-jet process, but over time deposits may build up.

Although there are some non-graphics applications of the current technologies where low-viscosity fluids can be used, we think it unlikely that special print-heads will be developed unless some high-volume applications emerge. The strength of thermal or bubble inkjet is that it can be manufactured in high volume at low cost. If home or personal dispensing uses could be found, then this could be the application to make it viable.

15.3.2 Piezoelectric DOD Inkjet

In a piezoelectric DOD print-head, drops are generated, as in bubble-jet, by a rapid pressure increase of the ink inside the actuator chamber. The difference is that a small piezoelectric actuator, one for each chamber, generates the pressure. When a voltage is applied across the actuator the piezoelectric material distorts, and this change in shape is used to displace the chamber wall or roof into the chamber to generate the required pressure for ink ejection. Because of limitations of the activity of piezo materials, practical drive voltages and other considerations, the area of piezo material required to generate drops is much higher than the area of a bubble-jet heater. Therefore the nozzle-to-nozzle spacing and the area of piezo print-heads tend to be greater, leading to a lower native print resolution.

Although heat is not used to generate drops, in operation the piezo actuators generate waste heat and removing this is a consideration for high-duty cycle operation.

A challenge when designing piezo print-heads is how to manufacture the relatively complex ink paths from a central manifold to the actuator chambers and hence the nozzles. Traditionally, stacked layers of stainless steel or ceramic have been separately patterned, then bonded together to achieve this. Recent advances in silicon fabrication techniques, particularly deep reactive ion etching (DRIE), have led to a trend in the use of silicon micro-electro-mechanical systems (MEMS) techniques to fabricate these complex 'channel' plates. The use of silicon enables stiffer structures, higher channel resolution and higher drop frequencies.

In order to decrease the channel spacing, the actuator dimensions have to be reduced. To maintain usable drop volumes, this means that the displacement of the actuator into the chamber needs to be increased. This necessitates a reduction in the thickness of the actuator roof and also the piezo material itself.

In the past, print-heads have used pieces of piezoelectric material such as ceramic lead zirconate titanate (PZT) sliced into wafers and then diced to form individual actuators. As the need for thinner piezo layers has grown, and also to aid cost savings for mass manufacturing, alternative means of forming piezo layers have been developed. Thick-film deposition of piezo layers has been used commercially since the mid-1990s, and more recently thin-film deposition techniques have been used. Thin-film piezo layers can be moisture sensitive when electric fields are applied, so the piezo material has to be protected in hermetically sealed covers.

Generally it is possible to apply more energy to the ink in a piezo actuator than in a bubble-jet device, and hence to jet higher viscosity materials. In addition, there is no need to include a volatile material within the ink for bubble generation, and the ink ingredients do not have to withstand high-temperature excursions.

Most of today's piezo print-heads use what is known as roof-mode architecture. One of the industrial print-head suppliers, Xaar, uses a moving wall architecture where the channel walls are made from piezo material and flex to produce pressure in the ink.

15.4 Other Current Technologies

Apart from bubble and piezo DOD inkjet, continuous inkjet (CIJ) has been employed for many years for coding, marking, addressing and form printing. Development work has also been ongoing in the search for a way of generating drops from nozzles using electrostatic forces. We shall discuss these technologies here.

15.4.1 Continuous Inkjet

In CIJ technology, as described in Chapter 2, drops are generated continuously, charged and deflected to separate printing drops from non-printing drops which are then collected and recirculated. This technology does not lend itself to high channel densities, due to the high voltages required for charging the drops. The system is additionally complex and relatively expensive. An added drawback for some applications using expensive materials is the large volume of ink required within the feed and recirculating system.

However, Eastman Kodak has developed a new type of CIJ (called Stream), which we will discuss in Section 15.5.1 as an emerging technology. This has the potential to offer higher nozzle densities, smaller drops and faster print speeds than traditional CIJ, although it still retains a large-volume ink supply.

15.4.2 Electrostatic DOD

Methods of generating drops using electrostatic forces have been known for some time. An external electrostatic field is applied to a raised meniscus in a nozzle. Above a certain field strength, the meniscus forms a cone shape as the electrostatic forces reduce the surface tension forces of the liquid. Further increasing the field results in the emission of a thin thread of fluid from the tip of the cone. The drop size is controlled by the length of the field pulse, and it is infinitely variable. This electrospray process has also been described in Chapter 2.

There are several advantages of this technology. The drop diameter is only a fraction of the nozzle diameter, and it is not affected by nozzle quality or contamination as in conventional DOD. It is also of great current interest because of its potential to create very small drop sizes, for instance in the femtolitre range.

However, the process has some drawbacks. The high electric fields involved, together with the need for a bulging meniscus in the nozzle, means that nozzle densities are quite limited. The high voltages are more expensive to generate and switch than for piezo or bubble-jet. Finally, the conductivity of the fluid needs to be within a certain range, and so it is difficult to jet either highly conducting or insulating fluids.

Several teams are working with this technology, for instance AIST in Japan (Murata *et al*., 2005) and Queen Mary College, London, United Kingdom (Paine *et al*., 2007), aiming to develop it for direct writing electronic applications, and for the dispensing of very small quantities of fluids (Figure 15.6).

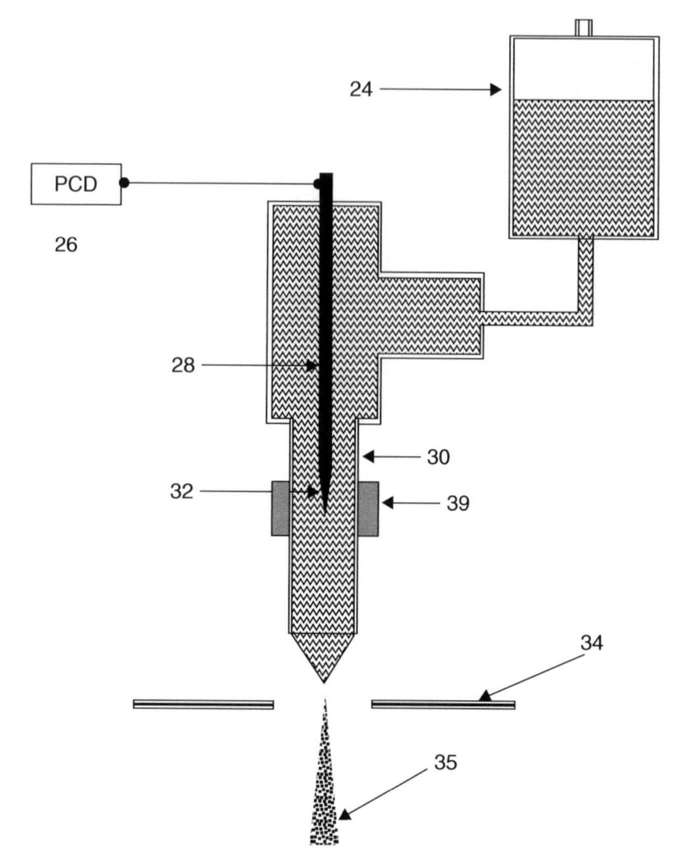

Figure 15.6 Electrostatic spray jetting
The liquid to be jetted, in this case silicone oil, fills the emitter tube 30. A piezoelectric charging device is connected to electrode tip 32, and the resultant charged fluid forms a cone in the emitter nozzle where a highly asymmetric field is generated by aperture electrode 34. A stable cone of fluid is sprayed from the tip of the fluid cone (from WO 2008/142393 An electrostatic spraying device and a method of electrostatic spraying, Queen Mary and Westfield College).

15.4.3 Acoustic Drop Ejection

There are systems that produce drops directly from high-frequency acoustic generators. Xerox developed a nozzle-less inkjet technology where the acoustic generators are submerged in the ink, with the acoustic energy focused on the ink surface. There are no nozzles or indeed a nozzle plate. Fine drops of ink are emitted in a stream at the points of focus.

Similar principles have been used by Picoliter Inc., now called Labcyte, to cause the non-contact transfer of fluids in volumes from 2 nl to 10 μl. Acoustic droplet ejection (ADE) technology (Figure 15.7) is claimed to be a very gentle process. Labcyte has demonstrated that it can be used to transfer proteins, high-molecular-weight DNA and live cells without damage or loss of viability.

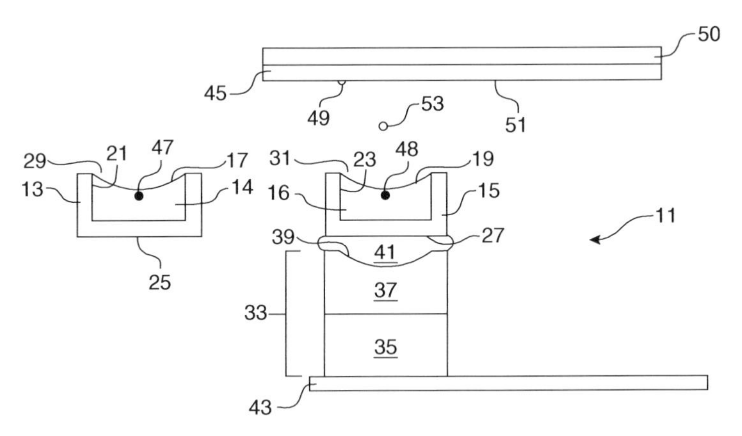

Figure 15.7 Acoustic drop generation
Acoustic ejector 33 comprises an acoustic generator 35 and a focusing component 37. This is acoustically coupled to a liquid reservoir 15 by fluid 41. The acoustic energy is focussed to a point 48 just below the liquid surface, and drops 53 are ejected from the free surface of the liquid (from EP 1 337 325 Acoustic ejection of fluids from a plurality of reservoirs, Picoliter, Inc.).

Like electrospray printing, it is capable of generating very small drops and the drive pulse width can directly control the amount of fluid deposited.

15.5 Emerging Technologies

For around a decade there has been very little development of new inkjet technologies, the focus being on the evolution of piezo and bubble jet DOD technology for office and graphics applications. But more recently, driven by the need for higher print speeds in the graphics area, and fine drops for digital fabrication applications, we are now seeing the introduction of new technologies. In most cases, these overcome some of the restrictions of today's inkjet devices.

15.5.1 Stream

Eastman Kodak has been developing a new CIJ technology for a number of years. At the time of writing, it has been publicly demonstrated printing at substrate speeds of 200 m per min ($3.3\,m\,s^{-1}$) at 600 dpi with variable drop sizes in full colour, using pigmented inks. The print-head technology is considerably less complex than that of traditional binary CIJ, consisting of annular ring heaters surrounding nozzles formed in a silicon nozzle plate.

As with other CIJ technologies, the ink is pressurised and flows continuously from the nozzles during operation. To stimulate the liquid flowing from each nozzle to break into regular-sized drops, the heater surrounding the nozzle exit is pulsed. The temperature of the ink surface is raised by just a few degrees, which changes the surface tension of the ink stream enough to provide discontinuities at which the stream can break up. In this way a of drops is produced, and by turning off the heater pulses for three to four cycles, larger drops can be formed. In the current implementation the larger drops pass to the

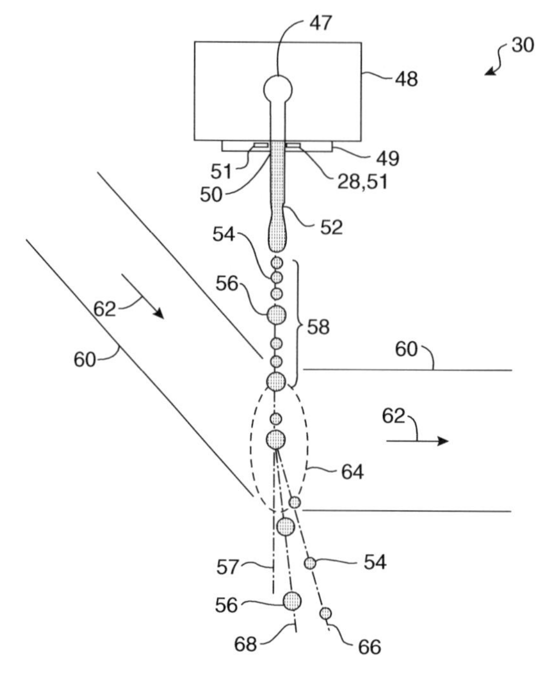

Figure 15.8 Kodak Stream technology
Ink under pressure flows through nozzle 50 that is surrounded by heater 51. Pulses of heat stimulate the jet 52 to form small drops 54. If the pulses cease for a few cycles, larger drops 56 are formed. A laminar transverse airflow 62 then deflects the smaller drops to a gutter, while the larger drops continue to the substrate (from US 2011/0242169 Continuous printer with actuator activation waveform, Eastman Kodak Company).

substrate to print and the smaller drops are deflected by a transverse air flow, collected and recirculated (Figure 15.8).

The result is a print-head technology capable of high nozzle densities, and the nozzle heater drivers can be fabricated on the nozzle front face along with the heaters. The transverse air flow system can be built up as a structure directly onto the nozzle plate.

For potential deposition applications, a key advantage of this technology may be its freedom to use a wide range of inks. It is believed that continuous jets of most inks can be broken up with thermal pulses without any special formulation being required, and the pump pressure can be adjusted to accommodate different ink viscosities. The break-up length of the ink stream can be adjusted by varying the amplitude of the heater pulse. Therefore, beyond the current graphics printing uses of Stream, it may have a future application in high-speed printing for manufacturing.

15.5.2 MEMS

MEMS technology has been commercialised for many applications during the past decade, and Chapter 6 discusses how inkjet technology can be used as a process tool in the manufacture of MEMS. MEMS devices, usually made by modified semiconductor processes such as wet etching, dry etching and electro-discharge machining, also have application for the generation of liquid jets and drops.

On 15 July 1997, Silverbrook Research filed 372 patent applications on a single day at the US Patent and Trademark Office. By the end of July 2010, they had filed just over 5000 US patent applications. The majority of these applications relate to MEMS inkjet devices, with around one quarter relating to the actuator design.

Initially Silverbrook Research was pursuing a wide range of possible MEMS actuator designs, but the first one chosen for commercialisation is based on a suspended heater TIJ design. By suspending the heater within the chamber, the efficiency of the device is improved by around one order of magnitude compared to conventional TIJ devices with surface-mounted heaters. This high efficiency allows a very high density of actuators without thermal management problems, as the ink carries all of the excess heat away. However, we understand that this design as commercialised may be limited to relatively simple dye-based inks.

Another design that Silverbrook has demonstrated is a moving nozzle operated by a thermal bend actuator (Figure 15.9). The nozzle is driven downwards into the ink, displacing an equal volume from the nozzle. Compared to a conventional actuator, the pressure exerted on the ink is considerably less. Because there is no heating within the chamber, this method is probably suited for more complex inks. Until the technology is commercialised, it is too early to tell what role it may have for digital fabrication.

Other companies have also explored new print-head designs using MEMS manufacturing techniques. For instance, during 2008 Xerox published a number of US patent applications covering a roof-mode actuator where the diaphragm is moved electrostatically (Figure 15.10).

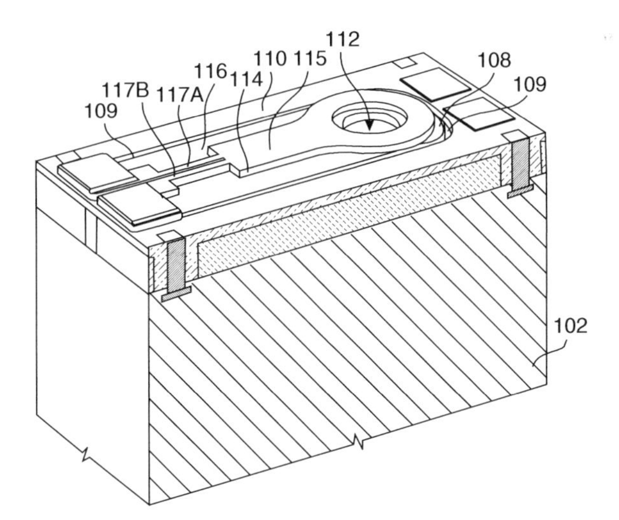

Figure 15.9 MEMS printhead technology

Heaters 117 cause a deflection of nozzle plate 115 downwards into the fluid chamber below (not shown). This causes some of the fluid to be pushed up through nozzle 112 that forms a drop. The top surface is planarised with a compliant layer of PDMS (not shown) that also provides a non-wetting surface (from US 2009/0278876 Short pulsewidth actuation of thermal bend actuator, Silverbrook Research Pty Ltd).

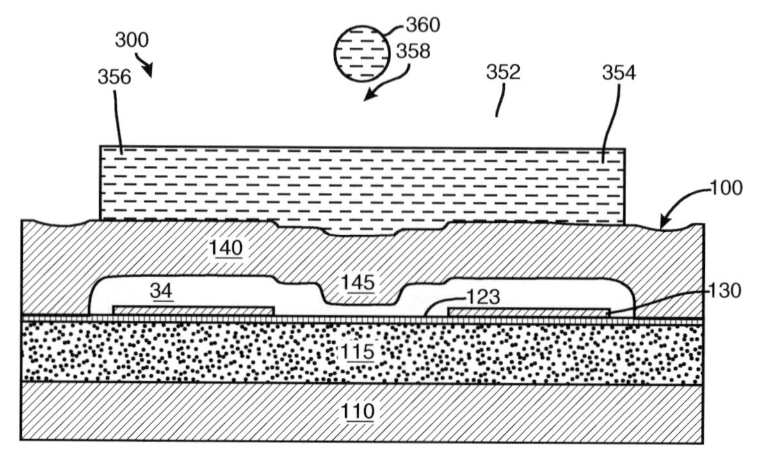

Figure 15.10 Electrostatic diaphragm roof mode actuator
Diaphragm 140 is attracted to ring electrode 130 by the electrostatic field between them. The protrusion 145 restricts the diaphragm movement and prevents it shorting against the electrode. When the field is removed the diaphragm springs back up and drop 360 is ejected from the nozzle (from US 7,942,508 Decreased actuation voltage in MEMS devices by constraining membrane displacement without using conductive "landing pad", Xerox Corporation).

15.5.3 Flextensional

Flextensional devices consist of a piezo actuator coupled to a flexible mechanical device. Some of the MEMS devices proposed by Silverbrook are flextensional, and there have been other designs in the past from Hewlett-Packard.

The Technology Partnership also has a series of patent applications covering a flextensional design. The piezo ceramic is bonded to an electroformed nozzle plate. There are slots on each side of the piezo material allowing flexure of the nozzle layer (Figure 15.11). When a pulse is applied, a downwards deflection of the nozzle occurs and ink is displaced out of the nozzle. A second pulse is applied to damp the return movement. As the drop ejection takes place under relatively low pressures, then it should be possible to jet more difficult ink types, for instance those containing heavy pigments.

15.5.4 Tonejet

Tonejet ink is similar to liquid toner, consisting of charged pigments and polymers in solid form suspended in an insulating carrier liquid. The ink wicks up electrodes that also form ejection points (Figure 15.12). When a voltage is applied to these electrodes, the particles agglomerate at the electrode tips. Once the accumulated charge reaches a high enough level, the particles are ejected towards the substrate, with only a small amount of the carrier liquid being carried over too.

Although for graphics applications the particles consist of pigmented polymer resins, it is also possible to fabricate toners as functional materials and incorporate higher density pigments.

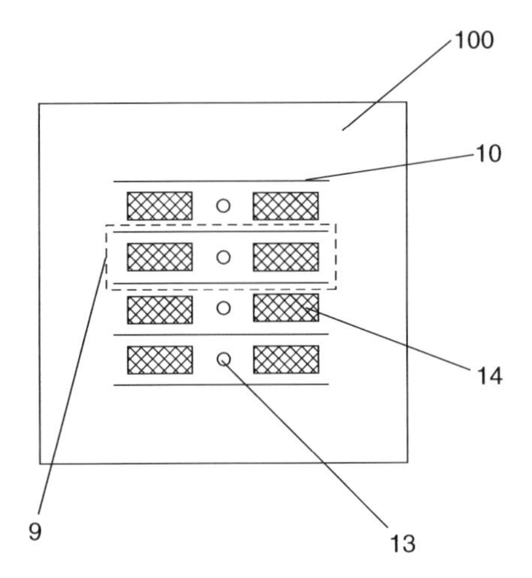

Figure 15.11 Moving nozzle printhead
Piezo actuators (shaded) are fixed to a nozzle plate, which are able to move due to slits 10. The actuators cause the nozzle plate to move downwards, forcing ink up through nozzles 13 (from WO 2008/044069 Liquid projection apparatus, The Technology Partnership plc).

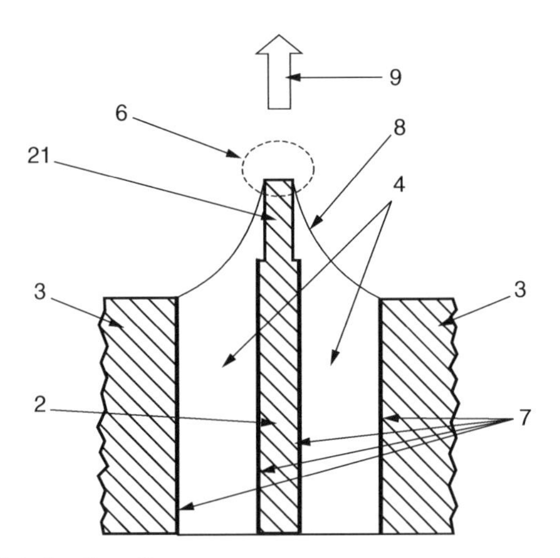

Figure 15.12 Tonejet ejection site
Charged pigment particles 6 agglomerate at electrode tip 21 in response to an electric field, and are then ejected 9 to the substrate (from WO 2011/154334 Image and printhead control, Tonejet Limited).

15.6 Future Trends for Print-Head Manufacturing

In the past, print-head manufacturers designed and manufactured their own print-heads. Print-head manufacturing requires a huge investment to provide capital equipment and establish production. With inkjet increasingly sharing the manufacturing technologies of the semiconductor industry, however, it is becoming possible to outsource at least part of the manufacturing process. As an example, it is believed that Hewlett-Packard now outsources over 50% of its TIJ print-head dies from ST Microelectronics.

This is particularly the case now that MEMS manufacturing processes can be incorporated into silicon processing as well as the fabrication of integrated circuitry. Therefore, there is the possibility of the emergence of fab-less print-head companies, analogous to fab-less chip companies. The Silverbrook MemJet print-head dies are made under a partnership with Taiwan Semiconductor Manufacturing Company (TSMC): Silverbrook Research designs and models the print-head designs, which TSMC then makes.

At the moment there are few print-head designs specifically for non-graphics applications. Print-head manufacturers are reluctant to invest in the design, development, tooling and manufacture of devices specifically for digital fabrication applications as the customer base is fragmented and the market at present unknown. Examples of current products include the Fujifilm Dimatix D-128 DPN, the Konica Minolta 128SNG-MB and internal-use only designs at Seiko Epson and Sharp.

Print-head manufacturers are addressing other issues for graphics applications. For piezoelectric print-heads, there is regulatory and environmental pressure to move towards using lead-free materials (instead of the current standard, PZT). At the moment it is difficult to achieve the current amount of piezoelectric activity from lead-free materials, but within a few years we expect practical lead-free devices to become available.

Another trend is the development of thin-film piezo material deposition techniques enabling higher channel densities. Several of the existing print-head manufacturers are using thin-film piezo, for instance Seiko Epson, Panasonic and Fujifilm, with others likely to follow.

15.7 Future Requirements and Directions

Graphics applications for inkjet use inks suitable for printing on paper, both coated and uncoated, and non-porous substrates such as boards, ceramic tiles, plastics and metals. In general, the printed spot size needs to be correct for viewing at an appropriate working distance. The printed spot size is a function of the ink interaction with the substrate among other factors.

For digital fabrication the fluids used and the printed spot sizes required vary widely, over a much wider range than for graphics applications. In some cases the spot size must be very well controlled, for instance when printing fine conductive tracks. For other applications the drop volume is important, for instance when printing micro-arrays for biomedical applications.

15.7.1 Customisation of Print-Heads for Digital Fabrication

At present, little customisation of print-head technology is practicable. Print-head manufacturers apply great development effort into yield improvement and manufacturing consistency, and are reluctant to make even small changes. What is needed are print-heads designed for relatively easy customisation in nozzle count and layout, drop size, fluid handling capability and so on.

15.7.2 Reduce Sensitivity of Jetting to Ink Characteristics

For virtually every print-head available at the moment, the fluid must be formulated to work within the chosen print-head. Substituting the print-head for a different model may well require changes to the fluid formulation to maintain acceptable jetting. This constrains the development of new applications.

When scientists focus on developing a new application, they wish to optimise the fluid for its function, not for the print-head. They are side-tracked into having to learn the special requirements of the inkjet process. Print-head manufacturers may supply a basic specification for fluids, for instance the viscosity and surface tension limits, but this is often just a starting point on the way to satisfying the requirements for consistent jetting without satellite drops.

What is needed are print-heads that are much more tolerant of fluid formulation, both by the design of the print-head and by the control of its operating parameters. At the moment there are some heads where the temperature of the print-head can be controlled to adjust the ink viscosity and bring it within the print-head operating range. In others there may be control over the drive voltage or the drive waveform shape. At present, experience and knowledge are required to find the best operating parameters for jetting fluids, but the hope is that at some stage in the future this can be fully automated. In fact, research is ongoing in an attempt to predict the operating parameters directly from the fluid composition.

15.7.3 Higher Viscosities

A typical room-temperature viscosity that can be jetted with a piezo-based DOD inkjet print-head is 5–15 mPa s. For some graphics applications, for instance with UV-curable inks, it is not possible to formulate a suitable ink at such low viscosity. To overcome this limitation some print-heads can be heated, for instance to 50–100 °C, significantly reducing the fluid viscosity so that it can be successfully jetted.

However, this approach may not be suitable for some digital fabrication applications, where some of the materials in the formulation are temperature-sensitive. In addition, the fluid may be held within the heated print-head and fluid supply system for a considerable time, leading to accelerated aging or deterioration of the ingredients.

To eject higher viscosities from nozzles requires higher pumping pressures. It is possible to design print-heads to do this, but it adds to the cost per actuator and increases the nozzle-to-nozzle spacing. It also becomes more difficult to maintain a streamlined fluid flow through the nozzles, which is essential for consistent jetting.

Thermal or bubble inkjet is normally associated with low jetting viscosities, but some years ago Canon developed a system incorporating a valve at the back of the channel to increase the jetting efficiency. Claims were made that fluid with a viscosity of 100 mPa s was jettable. The technology was not commercialised due to life issues – the valve was prone to breakage – but it did demonstrate that new designs could allow much higher viscosities.

There is perhaps more potential for CIJ systems to jet higher viscosities. At present, they are limited to 1–3 mPa s. The patents for the Kodak Stream technology discussed in Section 15.5.1 suggest that 100 mPa s may be possible, by using a suitable nozzle design and applied pressure to achieve the flow, and increasing the 'pinch' amplitude.

But generally, today's thinking is that the viscosity of jettable fluids will always need to be below about 100 mPa s to satisfy the laws of physics for consistent jet break-up.

15.7.4 Higher Stability and Reliability

For graphics applications it often does not matter if there are a few satellite drops, or if the jet break-up has some instability, causing small amounts of fluid to scatter around the main drop on the substrate.

Graphics images are often printed in such a way to deliberately mask jetting defects like these, yet they still produce acceptable image quality. However, this does not apply to most digital fabrication and dispensing applications where fluid outside the 'image' boundaries may be unacceptable and lead to low production yields.

With today's devices, various tricks are used to enhance jetting stability and performance, but these create significant reductions in throughput. For instance, print-heads may be operated at a low drop frequency, at which the jetting will be more stable. To keep all of the nozzles working, drops may be fired from every nozzle and the print-head wiped after every scan of the substrate. Although this may succeed in maintaining jetting stability and the integrity of the image, excessive spitting of ink drops will be wasteful of fluid and increased wiping of the nozzle plate may lead to excessive wear and a shortened print-head life.

15.7.5 Drop Volume Requirements

The requirements for drop volume vary considerably for current and future applications of inkjet for digital fabrication. Today's devices can jet drops from as small as 1 up to 100 pl or more. But this is for different print-heads. At present the range of drop sizes that can be jetted from the same nozzles is much smaller, and the number of drop-size increments limited to five (Epson) or eight (Xaar).

Sub-picolitre drops are needed for some applications. It is likely that current piezo print-head technology can generate drops down to 0.1 pl. More interesting is the electrostatic (electrospray) technology discussed in Section 15.4.2 that allows the formation of drop volumes down to 0.01 pl (10 fl); by adjusting the pulse width, it is also possible to continuously vary the amount of fluid deposited.

Perhaps more important for digital fabrication work is not the precise drop volume but the stability of the drop production process and the variations from nozzle to nozzle. The drop volume from a single nozzle can also vary from drop to drop. The variation is likely to increase with the complexity of the ink formulation. Unfortunately, most methods used to measure drop volume, such as jetting many thousands of drops and then weighing them, measure only the average drop volume and not the variation.

The drop volume is also likely to vary from nozzle to nozzle due to small manufacturing variations that affect the efficiency of each drop generator and channel dimensions. In addition, cross-talk effects between channels can and will affect drop volumes according to the pattern of drops printed, and at the ends of the print-heads there may be end effects again causing variations in drop volume.

15.7.6 Lower Costs

In most cases within a manufacturing environment, digital fabrication applications will be part of the overall factory processes, and the print-heads will be used in a similar fashion to machine tools. In this respect the cost of the print-head will be amortised over its period of use.

However, there may be applications where the inkjet system is designed to be relatively low-cost and installed and maintained by the user, in which case the cost of the print-head and its replacement becomes much more significant.

At present the lowest print-head cost, in terms of cost per nozzle, is thermal or bubble inkjet. This has been taken to extremes by the Memjet technology from Silverbrook Research, where the very low cost per nozzle has been achieved by making the cell size (the area occupied by a single actuator, driver and nozzle) as small as possible. Cell densities as high as 48 000 per cm^2 are achieved. Assuming the rest of the print-head package is also low cost, then this could be very valuable if high throughput using a relatively large number of nozzles is required.

There have also been proposals to use less expensive materials than silicon to construct print-heads. Eastman Kodak has proposed making a CIJ Stream head from sheets of polymers rather than silicon (Vaeth *et al.*, 2007).

For the dominant technology for digital fabrication – piezo DOD – a major way to reduce costs is to miniaturise the whole design. This is likely to require etched silicon channel plates and thin-film piezo. For relatively low-volume industrial print-heads, it is likely that the unit cost will remain the same or perhaps increase; the 'cost saving' will be represented by an increase in the number of nozzles for the price.

15.8 Summary of Status of Inkjet Technology for Digital Fabrication

As we have discussed in this chapter, inkjet technology has some key advantages for digital fabrication, although there are some constraints. Some of the constraints are fundamental to inkjet, while others can be solved providing there is sufficient incentive to fund the development and manufacture of new printing devices. Current trends in inkjet graphics printing are focused on achieving high image quality and high reliability, both welcome attributes for digital fabrication.

We shall summarise here the improvements that might follow from our discussion of emerging technology.

Firstly, let us consider the costs of inkjet technology. If the costs of the print-head and peripheral technology can be reduced, then almost certainly their use would become more widespread. Some would argue that the costs of inkjet are small compared to those of an overall production system, with the benefit of flexible manufacturing and reduced waste. However, if digital fabrication is going to move from factories to localised production and become mainstream, it has to become more accessible. At present the cost of even an off-the-shelf development system with a print-head, electronics and ink system can approach US$100 000.

When inkjet technology is capable of handling a wider range of fluids without extended fluid development, we will again see a greater market uptake for digital fabrication. Such flexibility has not been required for graphics printing, and only relatively recently has it drawn the attention of research and development initiatives.

For some digital fabrication applications, a further breakthrough will occur when smaller drop volumes are available. Smaller drops are required to form narrower conductive tracks and finer feature sizes, and to deposit smaller amounts of functional materials. However, the ability to form smaller drops is only one step to a solution. To achieve the throughput possible with larger drops, the number of nozzles must be increased. It is also much harder to achieve controlled drop trajectories to the substrate with smaller drops.

Therefore we shall hopefully see the day in the near future when scientists and engineers wishing to develop a new digital fabrication process do not have to become inkjet experts first. Print-heads, drive electronics and software and ink supply systems can already be sourced for graphics applications. What the digital fabrication industry now needs are the tools for their world, enabling them to focus on the application rather than on inkjet development.

References

Murata, K., Matsumoto, J., Tezuka, A. *et al*. (2005) Super-fine ink-jet printing: toward the minimal manufacturing system. *Microsystem Technologies Micro and Nanosystems Information Storage and Processing Systems*, **12** (1–2), 2–7.

Paine, M.D., Alexander, M.S., Smith, K.L. *et al*. (2007) Controlled electrospray pulsation for deposition of femtoliter fluid droplets onto surfaces. *Journal of Aerosol Science*, **38**, 315–324.

Vaeth, K., DeMejo, D., Dokyi, E. *et al*. (2007) MEMS-Based inkjet drop generators fabricated from plastic substrates. NIP23: International Conference on Digital Printing Technologies and Digital Fabrication 2007, DF2007, Society for Imaging Science and Technology, pp. 297–301.

Index

References to figures are given in *italic* type. References to table are given in **bold** type.

accelerometers 141
acid etch **199**
acoustic drop ejection 33, 352
acrylate UV inks 101–3, *102*
active electronics *see* electronic components, active
air bubbles 38
alginates 286, 296
aliasing 35
alkaline etch **199**
aluminium oxide 6, 10
ammonia sensors 217
antennae (RFID) 260–3, *261*, 264–5
anti-aliasing 35
aqueous inks 96–100, 210–11, 277
- dyes 100
- polymeric additives 100

asynchronous communication 267
azo dyes 97, *98*

barium strontium titanate 6
batteries 214–15
bead formation
- stable 128–30, *129*
- unstable 130–2

Bernoulli underpressure refilling 68–9
binary printers 20, 35
biocides 172
biomaterials 3, 151–2, 223–4, 275
- alginates 287
- applications
  - bioreactors 298
  - tissue engineering 297–8
    - *see also* tissue engineering
  - tissue printing 298
- cells *see* cells
- corrosion 285
- drying behaviour 285
- humoral factors 322
- ink formulation 282
  - reactive inks 292–6
- polymers 295
- print heads 280–1
- proteins 287–9
- rapid prototyping 281–2
- substrates 296–7
- thermal inkjet for 279

bioreactors 298
biosensors 146–7
2,4-biphenylyl-5,4-tertbutylphenyl-1,3,4-oxadiazole (PBD) 240
Boltzmann constant 125
Bond number 114
Boston Microsystems 145, *146*
bubble-jet *see* thermal inkjet
Burroughes, Jeremy 238

cadmium selenide 108, 212
calcium acetate 99
Canon 349
capacitors 176, 214–15
capillary blood vessels 310

Inkjet Technology for Digital Fabrication, First Edition. Edited by Ian M. Hutchings and Graham D. Martin.

capillary forces 73
  coalescence 126
  contact line motion 124–5
  refilling 67–8
  spreading 122–4
capital costs 162
carbon nanoparticles **165**
carbon nanotubes (CNT) 106, 165–6, *166*, 213–14, 239
Carnegie Mellon chemical sensor 144, *144*, *146*
Cassie-Baxter relation 123
Castle Island's Worldwide Guide to Rapid Prototyping 328
Cavendish Laboratory 240
cells (biological) 275–6
  attachment and growth 291–2
  biocompatibility 292
  cell-directing patterns 289
  inks containing 290
  matrix 276
  multiple types 316–17
  print heads 276–7, 290–1, 311–12
  see also tissue engineering
cells (electrical) 214–15
cellulose ester 91
ceramics 31–2, 281
  applications 10
  deposition methods 6–7, 10
  as substrate **173**
  see also lead zirconate titanate
cerium oxide 6
chalcogenides 216–17
chemical sensors 144–7
  optically based *148*
chemoresistive materials 144
Chinese hamster ovary (CHO) cells 291
circuit breakers 176–7
coalescence 128, 248, 284
  bridge formation and broadening 126–7
  droplet relaxation 127–8
coffee stain effect 11, 209, 221, 243–4
colourants 97–9
  substrate interactions 99
complementary metal-oxide semiconductor (CMOS) process 220–1
complexity 42
computational fluid dynamics (CFD) 208
computer-aided design (CAD) 188–9, 321, 327
Conductive Inkjet Technology, Inc. 204
conductive inks 164–72
  see also conductive polymers; metals
conductive polymers 105–6, 172, 221, 228
  see also poly(3,4-ethylenedioxythiophene)
conductive tracks 5, 160–1, 163, 222
  inks
    dispersants 168–70
    nanoparticle properties 164–8
  materials 165–6
  protective layers 174–5, 175
  resistors 175–6
  sintering 174–5
  substrates **173**
  three-dimensional 163
  see also electronic components; printed circuit boards
conductivity 162–3
contact angle 242–3
contact angle hysteresis 125–6
contact line motion 124–5, 130–1
contact line retraction 132–3
contacts 163–4
continuous inkjet (CIJ) 351
  applications 28
  drop generation 27–8
  fundamentals 21–2
  ink formulation 88
  mask printing 12
  metals 4
  multiple jet 28
  single-jet 27–8
copper 165, 203–4, 260
copper indium gallium selenide (CIGS) 216–17
copper zinc tin sulfide (CZTS) 216–17
corrosion, biological materials 285

costs 340, 361
cytokines 276
cytotoxicity 291

data formats and sources 321–2
Deerac system 280–1
degassing 38
DemaMatrix 289–90
denaturation 288
deposition maps 114–16
diamond 7
Dimatix printer 278
diodes 265, 271
direct deposition 9, *9*
  cells 319
  ceramics 10
  metals 4, 9
  polymers 10–11
dispersants 168–9
displays 212–13, 237–8, 345
  geometry 241
  ink formulation 243–5
  process reliability 246–9
dodecylbenzene sulfonate acid (DBSA) 217
drive-per-nozzle system 247
drop-on-demand (DOD) 1–2, 28–9
  binary 35
  droplet formation 89–95
  electrostatic 33
  fundamentals 21–2
  piezoelectric 30–1
  thermal 29–30
  see also piezoelectric-actuated printing; thermal inkjet
droplet properties and behaviour
  coalescence 126–8
  deceleration due to elongational effects 70–1
  deposition
    deposition maps 114–16
    unstable 120–1
  formation 21–2, 89–96, 248–9
    aerodynamic drag 90
    fluid behaviour regimes *92*
    piezoelectric print head 60–6
  impact 90–1, 114, 116–20, 348
    gravitational potential energy 117
    inkjet-sized drops 119–20
    millimetre-sized drops 116–19
    spreading 117
  placement 16, 345
    aerodynamic effects 347
  satellite formation 38–9, 90, 277, 348
  size 22, **23**, 41, 201
    grey-scale printing 35–6
    volume 247, 248–9, 283, 345, 361
  spacing 40
    see also resolution
  speed/velocity 22, 37
  spreading
    heterogeneous substrates 123
  surface tension 23–4
  trajectory 346, *346*
  see also ink formulation; substrate interactions; viscosity
drug delivery systems 152
drying behaviour 243
  biological materials 285
dust 126
dyes 97–9

Eastman Kodak 353–4
Eisler, Paul 184
electrohydrodyamic jet printing 210
electronic components
  active 208
    biological materials 224–5
    ink formulations 210–11
    memory storage 221
    organic LEDs 211–14
    photovoltaics *see* photovoltaics
    sensors 217–19
    substrate requirements 209
    transistors 219
  passive 159, 221–2
    applications 159–60
    capacitors 176
    circuit breakers 176–7
    electrostatic discharge 177–8
    fuses 176
    inductors and transformers 177

electronic components (*continued*)
passive filters 177
resistors 175–6
switches 177
thermal management 178
transistors 269
see also conductive tracks
electronic interconnections 9
electronics industry 207–8
electrostatic discharge (ESD) 177–8
electrostatic spray printing 33–4, *34*, 351, *352*
elongation effects 70–1
Escherichia coli 223, 296
etch resist 197–202
etching *9*, 13–14, *14*
evaporation 132–4
extrusion fabrication 281

face shooter 30
failure modes 37–8
fifth pen approach 100
flat-panel displays *see* displays
flexible substrates 214
flexstensional devices 356
fluid properties
surface tension 23–4
viscosity *see* viscosity
see also droplet properties and behaviour; ink formulation
fullerenes 215–16
fuses 176–7

gearwheels *11*
gelation 132
gels *see* hydrogels
gene chips 289
glass **173**
globular proteins 287–8
gold 5, 108, 144, 165, **165**, 221
graphene 221
grey-scale printing 35–6, 201
growth factors 322

hard disk drives 153, 154
HeLa cells 312
Helmholtz resonator 47–8, *48*
multi-cavity 71–7
Hewlett-Packard 349
high-definition TVs (HDTV) 241
high-frequency RFID 257, 262, 264–6
human umbilical vein endothelial cells (HUVECs) 312
humectants 170–1
hydrogels 12, 286–7, 294, 296, 320–1

in situ synthesis 223
indium tin oxide (ITO) 213–14, 239
inductors 177
ink formulation 208
added polymer 91, *93*, 100
aqueous inks 96–100
biomaterials 282–9, 318–20
conductive inks 164–72
liquid medium 170–2
continuous inkjet 88
design considerations 95
dispersants 168–9
effect of polymer addition on rheological properties 91–2
etch resists **199**
instability 37
jetting sensitivity 359
oil-based inks 101
organic light-emitting diodes (OLED) 243–5
phase change inks 101
solvent-based inks 103–4
surfactants 171
UV cure processes 101–3
see also droplet behaviour and properties; surface tension; viscosity
inkjet printing
advantages 2–3
limitations 344–8
principles 1–2
insulin 289

integrated circuits (IC) 153
  see also conductive tracks; electronic components; printed circuit boards
interconnects 125, 128, 221, 242
International Energy Agency (IEA) 215
inverse printing *9*, 14–15
iron oxide 221

jettability 243, 280

kinetic energy 61, 62
Kodak Stream *354*
kogation 30

Labcyte 280, 352
laser direct imaging (LDI) 102–3, 188
lead zirconate titanate (PZT) 6, 31, 221, 350
legend printers 91
lenses 148–9
light-emitting diodes *see* organic light-emitting diodes
light-emitting polymers (LEP) 150–1
linear array print heads 46, *47*
Litrex printer 248
low-frequency RFID 256–7

machine integrators 192
manufacturing 8, 223
  history of 3D printing 325–8
  limitations 16–17
  process requirements 40
  process selection 326–7
  see also solid freeform fabrication
MapleDW 280–1
Marangoni flows 135
mask printing *9*, 12–13, 101
material deposition 3
matrix print head 47, *47*
mechanical components *11*
melt printing *4*
Memjet 341, 358
meniscus formation 73, 95
metals 3–6
  conductive tracks 165
  direct deposition 4, *4*
  fluid properties 24
  melt printing 3, 9
  nanoparticle suspensions 5, 9, 107–8, 221
  precursor printing 5
  ProMetal process 338
micro-electro-mechanical systems (MEMS) 141–2, 340–1, 350–1
  advantages of inkjet 143
  assembly and packing 152–5
  bioactive 151–2
  chemical sensors 144–7
  limitations 142–3
  optoelectronic 142, 147–9
  resonant 145
  see also electronic components
micro-gel beads 314–15
micro-gel fibre 315
micro-satellites 38
micro-sieves 14, *15*
Microcraft JetPrint **191**
Microdrop Technologies 278
MicroFab 278
microwave RFID 257–8
Moore's law 259
multiwall carbon nanotubes (MWCNT) 213

nanoparticles
  biomaterials 285
  dispersants 169
  metallic 5, 107–8, 164–6, 221, 268–9
  particle shape 166
nanotubes 166
nickel 165, **165**
nonplanar surfaces 143
nozzle-plate flooding 37

Objet Geometries systems 330–2, *332*
  ProJet 333

Ohnesorge number 25–6, 61, 89–90, 114
in deposition maps 114–16, *115*
open-ended print head 48, *49*
optical fibre sensors 146–7, *147*
optoelectronics 142, 147–9
see also organic light-emitting diodes; photovoltaics
Optomec 280
Orbotech Sprint printers 191
organ transplantation 307
organic light-emitting diodes (OLED) 106–7, 211, 238–9
display elements, containment and solid content 241–3
for displays 212–13
ink formulation 243–5
for lighting 211–12
print defects 246–9
quantum dot 212, *214*
substrates 239
see also displays
oscillatory motion 51–2
overprinting 41

paired emitter detector diode (PEDD) 217
papers 99, 214
passive electronics *see* electronic components, passive
pattern formation 128
patterning 105
PEDOT *see* poly(3,4-ethylenedioxythiophene)
phase change inks 101, 130, 132–3, **199**, 295
phenyl-C_{61}-butyric acid methyl ester (PCBM) 104, 215
photodetectors 149, 217
photolithography 160–1, 190
phototool imaging 187
photovoltaics 106, 151, 215–17
phthalocyanine dyes 97, *98*
Picofilter Inc. 352
piezoelectric actuated printing 30–1, 46–7
active electronics fabrication 211
actuation mechanisms 31–3
for cells and biopolymers 277–8, 279–80, 282–3, 290–1, 311
drop formation 60–1
characteristic frequency 61–2
elongational effects 70–1
negative pulse 64–6
positive pulse 62–4
dynamics
damping 66–7
large print head 53–4, 56–8, *57*
long ducts 77–82
multiple pumps 71–7
pressure change in pump chamber 50–1
pulse formation 52–3
refilling 67–70
small print head 53–5, *54*, *55*
volume change in print head 50–1
ink formulation 89
for laboratory use 278
limitations 350
long duct theory
equations of motion 81
pulse response 81–2
meniscus formation 73
multi-cavity resonators 71–7
eigenmodes 76
resonance frequencies **76**
negative pulse 54
proteins 288
pulse length 53
pump *50*
pump chamber pressure 55, *56*
valveless pump *78*
pigments 98–9
pixels 241
pluripotent stem cells *see* stem cells
Poiseuille law 68
polyacrylamide 91
polyaniline (PANI) **173**, 217
poly(ethylene imine) 99–100
poly(ethylene oxide) 91
poly(ethylene terephthalate) (PET) 108, 187

poly(3,4-ethylenedioxythiophene) (PEDOT) **173**, 244, *245*, 349
  doped with polystyrenesulphonate (PEDOT:PSS) 7, 105, 106, 214–15, 215, 239
polyhedral oligomeric silsesquioxane (POSS) 240
PolyJet devices 330–2, *332*
polymers
  as aqueous ink additives 92–4, 100
  biological 285–6, 294
    coiled 286–7
    globular proteins 287–9
  deposition methods 7–8
  dyes 98–9
  effect on ink droplet formation 92–4
  etching 14–15, *14*
  molecular weight degradation 94
  organic conducting 105–6
  for organic LEDs, performance assessment 244–5
  in solvent-based inks 103–5
poly[2-methoxy-5-(2′-ethylhexyloxy)-1,4-phenylenevinylene] (MEH-PPV) 91, 107, 208
poly(methyl methacrylate) (PMMA) 338
poly(*para*-phenylene vinylene) (PPV) 240
polypeptides 294
poly(phenylene ethylene)-poly(phenylene vinylene) (PPE-PPV) 211
polypyrrole **173**
polysaccharides 286, 296
polystyrene 91
polythiophene **173**
polyvinylidene fluoride trifluoroethylene (PVDF-TrFE) 217
poly(4-vinylphenol) (PVP) 219
powder bed printing *8*, 15, *15*, 335–8
pre-patterned substrates 209
prednisolone 289
pressure waves 72
print heads 343
  arrangements 41–2
  biological ink 290–1
  biomaterials 280–1
    cell printing 276–7, 290–1, 311–12
  cells (biological) 276–7, 290–1, 311–12
  continuous inkjet 88
  failure modes 37–8
  future trends 357
  linear array 46, *47*
  matrix array 47, *47*
  motion 39–42
  nozzle size 41
  open-ended 48, *49*
  piezoelectric 47–8, *48*, *49*, 277–8, 350
    dynamics 49–60
    small 53–5, *54*, *55*
  thermal inkjet 88–9, 279
print quality 34–5
  see also resolution
print speed 319–20
Printar LGP/GreenJet **191**, 194–5
printed circuit boards (PCB) 102, *103*, 183–5, *184*
  complexity 201
  conventional manufacturing 185, *186*
    imaging 185–7
    laser direct imaging 188
    phototool imaging 187–8
  data formats 188–9
  design formats 188–9
  etch resist 196–202
    commercial implementation 202
    requirements **198–9**
  flexible substrates 204
  future possibilities 202–5
  inkjet applications 189–90
    disadvantages 196–7
    legend printing 190–3
    soldermask 194–6
  multi-layer *203*
  print resolution 197–201
  RFID tags 260
  substrate preparation **200**
  vari-layer 205
  see also conductive tracks

ProMetal RCT 338, *339*
protective layers 222
proteins 287–9
pulse generation *90*
PZT *see* lead zirconate titanate

quantum dots (QD) 212, *214*, 240

radiofrequency identification (RFID) tags 155, 252
  conventional 256, 258–60
  printed 263
    antenna stage 264–5
    antenna structures 260–3
    applications 260
    demonstrators 272
    digital section 265–7
    drawbacks of inkjet 261–2
    implementation issues 270–2
    materials 268
    rectifier, power supply, clamp 265
    transistors 269
  standards and classification 256–8
    13.56 MHz 257, 265
    135 kHz 256–7
    ultra-high frequency 257–8
ramp functions 58, *59*, 74
rapid prototyping 326, 328
  biomaterials 281–2
raster image processor 189
raw materials 160–1
Rayleigh-Plateau instability 24, 38–9, 130
reactive inks 292–6
rectification 265, 266
refilling 67–70
registration **198**, 220
released layers 143
reliability 36–8, **198**
  organic LED manufacture 246–9
repair **198**
resistors 5, 175–6
resolution 34–5, 39, 40–1, **40**
  biomaterials printing 283, 319
  printed circuit board etch resist 197–201
  solid freeform fabrication (SFF) 340
  see also accuracy
resonant MEMS structures 145
reverse thermal gelation (RTG) 331–2
Reynolds number 25, *26*, 61, 89, 114

satellite drops 38–9, 90, 277, 348
screen printing 190–1
selective laser sintering (SLS) 327–8
semiconductors 212–13, 219–20
  as ink 106–7
  stability 270
  as substrate 149, 154–5
sensors 217–19
shear strain 24, 25, 91, 287, 291
silicon chips 258–60, 263
silicon nitride 11
silver 108, 165, 221
  resin coated *210*
Silverbrook Research 355
sintering 25–6, 185–6, 194
sol-gels 6–7
solar cells 106, 151, 215–17
solder bumps 9, *10*
Solder Jet 153
soldermask 194–6
solid freeform fabrication (SFF) 160, 326–8
  challenges 338–40
  future trends 340–1
  introduction of inkjet technology 329–30
  laser-based 329
  liquid-based
    Objet Geometries PolyJet 330–2
    Solidscape Inc. 333–5
  market size 338–40
  powder-based *8*, 15, 335–8
  resolution 340
  for tissue engineering 313–14, 315–16
    direct printing 319
    high-speed 319–20
    resolution 319
  see also tissue engineering
solidification 132

Solidscape Inc. 333–5, *335*, *336*
solute segregation 134
solvent-based inks 103–4
  nanoparticle suspensions 170
solvents 219–20
  inkjet etching 13–14
splashing 121
stem cells 292, 322
stent 152
stereolithography apparatus (SLA) 327
Stokes' law 107
strain hardening 91
stream printing 354
substrate
  biological applications 296–7
  conductive tracks 172, 173
  motion 39–42
  surface energy 209
substrate interactions 113–14
  bead formation
    stable 128–30
    unstable 130–1
  capillarity-driven contact-line motion 124–5
  capillary-driven spreading 122–4
  coalescence *see* coalescence
  contact angle hysteresis 125–6
  droplet coalescence 126–8
  evaporation 132–4
  ink adhesion **199**
  solidification 132
  see also droplet properties and behaviour, impact
superconducting quantum interference device (SQUID) 221
support materials 330–1
surface energy 117–18, 209
surface tension 23–4, 284
surfactants 171–2, 210
surfactants 63
switches 177

Taiwan Semiconductor Manufacturing Company 358
Taylor cone 33
Technology Partnership 356
text printing 28
textiles **173**
thermal inkjet 29–30, 277
  for biological materials 279, 279–80
    effect on cells 290–1
    proteins 287–9
  ink formulation 88–9
  limitations 348–50
  thermal management 348
Thermal Inkjet Picofluidic System (TIPS) 279
thermo-cleavable solvents 104–5, *105*
thin film deposition 17, 350
3D printing *see* solid freeform fabrication
3D Systems Inc 327
tissue engineering 297–8, 298, 307–8
  3D printers 313–14
    micro-gel beads 314–15
    viscosity of materials 318–19
  applications 319–22
  current methods 309
  data source linkage 321
  inkjet printing feasibility study 310–12
  multiple cell types 316–18
  stem cells 322
  see also cells
titanium oxide *11*
Tonejet 356
transformers 177
transistors 219–20, 265, 269–70
trichlorobenzene 107
Trident heads 279
Trouton ratio 25
tumor cells 275

ultra-high frequency (UHF) RFID 257–8, 262–3
ultraviolet (UV) cure inks 101–3, **199**, 285–6, 295, 330
  ink formulation 101–3
Underwriters Laboratories 192

valveless pump 79, 79–81

van der Waals forces 168–9
vari-layer PCBs 205
vertical cavity surface-emitting laser (VCSEL) 148, *149*, 153
viscosity 24–5, 55–6, 89, 91, 348
  biological materials 283–4, 318–19
  droplet coalescence 126
  effect of polymer addition 92–4, *93*
  energy dissipation 75
  high 359
  in piezo-driven print head 50
  thermal inkjet 349
volatile organic compounds (VOC) 144
VoxelJet Technology GmbH 338

water 20, 23
  viscosity 24–5
waveguides 150–1, *151*
wax plate lamination 326–7
waxes *see* phase change inks
Weber number 25, 61, 114
  in deposition maps 114–16, *115*
  droplet impact 119–20
  solidification 132
Weissenberg number 93
Wenzel relation 123
wetting behaviour 123–4, 242–3
white light interferometry 249

Xaar 351
Xerox 352

yeast 283
Young equation 121, 127

Z Corporation printers 335–8, *337*
Z-Corp process 16
Zimm non-free draining time 93
zinc sulfide 212

Printed and bound by CPI Group (UK) Ltd, Croydon, CR0 4YY